OPTICAL PROPERTIES OF MIXED CRYSTALS

MODERN PROBLEMS IN CONDENSED MATTER SCIENCES

Volume 23

Series editors

V.M. AGRANOVICH

Moscow, USSR

A.A. MARADUDIN

Irvine, California, USA

Advisory editorial board

F. Abelès, Paris, France
F. Bassani, Pisa, Italy
N. Bloembergen, Cambridge, MA, USA
E. Burstein, Philadelphia, PA, USA
I.L. Fabelinskii, Moscow, USSR
P. Fulde, Stuttgart, FRG
M.D. Galanin, Moscow, USSR
V.L. Ginzburg, Moscow, USSR
H. Haken, Stuttgart, FRG
R.M. Hochstrasser, Philadelphia, PA, USA
I.P. Ipatova, Leningrad, USSR
A.A. Kaplyanskii, Leningrad, USSR
L.V. Keldysh, Moscow, USSR
R. Kubo, Tokyo, Japan
R. Loudon, Colchester, UK
Yu.A. Ossipyan, Moscow, USSR
L.P. Pitaevskii, Moscow, USSR
A.M. Prokhorov, Moscow, USSR
K.K. Rebane, Tallinn, USSR
J.M. Rowell, Red Bank, NJ, USA

NORTH-HOLLAND
AMSTERDAM · OXFORD · NEW YORK · TOKYO

OPTICAL PROPERTIES OF MIXED CRYSTALS

Volume editors

R.J. ELLIOTT

Department of Theoretical Physics
University of Oxford, UK

I.P. IPATOVA

A.F. Ioffe Physical Technical Institute
Academy of Sciences of the USSR
Leningrad 194021, USSR

N·H
P·C

1988

NORTH-HOLLAND

AMSTERDAM · OXFORD · NEW YORK · TOKYO

ISBN: 0 444 87069 5

Published by:

North-Holland Physics Publishing
a division of
Elsevier Science Publishers B.V.
P.O. Box 103
1000 AC Amsterdam
The Netherlands

Sole distributors for the USA and Canada:
Elsevier Science publishing Company, Inc.
52 Vanderbilt Avenue
New York, NY 10017
USA

Library of Congress Cataloging-in-Publication Data

Optical properties of mixed crystals / volume editors, R.J. Elliott,
I.P. Ipatova.

p. cm. — (Modern problems in condensed matter sciences; v. 23)
Bibliography: p.
Includes indexes.
ISBN 0-444-87069-5
1. Semiconductors—Optical properties. 2. Insulating materials—Optical properties.
3. Solids—Optical properties. 4. Crystal optics. 5. Phonons.
I. Elliott, R.J. (Roger J.), 1928. II. Ipatova, I.P. (Ija P.), 1929. III. Series.
QC611.6.060663 1988
548.9—dc 19 87-32923
CIP

Printed in The Netherlands

MODERN PROBLEMS IN CONDENSED MATTER SCIENCES

Vol. 1. SURFACE POLARITONS
V.M. Agranovich and D.L. Mills, *editors*

Vol. 2. EXCITONS
E.I. Rashba and M.D. Sturge, *editors*

Vol. 3. ELECTRONIC EXCITATION ENERGY TRANSFER IN CONDENSED MATTER
V.M. Agranovich and M.D. Galanin

Vol. 4. SPECTROSCOPY AND EXCITATION DYNAMICS OF CONDENSED MOLECULAR SYSTEMS
V.M. Agranovich and R.M. Hochstrasser, *editors*

Vol. 5. LIGHT SCATTERING NEAR PHASE TRANSITIONS
H.Z. Cummins and A.P. Levanyuk, *editors*

Vol. 6. ELECTRON–HOLE DROPLETS IN SEMICONDUCTORS
C.D. Jeffries and L.V. Keldysh, *editors*

Vol. 7. THE DYNAMICAL JAHN–TELLER EFFECT IN LOCALIZED SYSTEMS
Yu.E. Perlin and M. Wagner, *editors*

Vol. 8. OPTICAL ORIENTATION
F. Meier and B.P. Zakharchenya, *editors*

Vol. 9. SURFACE EXCITATIONS
V.M. Agranovich and R. Loudon, *editors*

Vol. 10. ELECTRON–ELECTRON INTERACTIONS IN DISORDERED SYSTEMS
A.L. Efros and M. Pollak, *editors*

Vol. 11. MEDIUM-ENERGY ION REFLECTION FROM SOLIDS
E.S. Mashkova and V.A. Molchanov

Vol. 12. NONEQUILIBRIUM SUPERCONDUCTIVITY
D.N. Langenberg and A.I. Larkin, *editors*

Vol. 13. PHYSICS OF RADIATION EFFECTS IN CRYSTALS
R.A. Johnson and A.N. Orlov, *editors*

Vol. 14. INCOMMENSURATE PHASES IN DIELECTRICS
(Two volumes)
R. Blinc and A.P. Levanyuk, *editors*

Vol. 15. UNITARY TRANSFORMATIONS IN SOLID STATE PHYSICS
M. Wagner

Vol. 16. NONEQUILIBRIUM PHONONS IN NONMETALLIC CRYSTALS
W. Eisenmenger and A.A. Kaplyanskii, *editors*

Vol. 17. SOLITONS
S.E. Trullinger, V.L. Pokrovskii and V.E. Zakharov, *editors*

Vol. 18. TRANSPORT IN PHONON SYSTEMS
V.L. Gurevich

Vol. 19. CARRIER SCATTERING IN METALS AND SEMICONDUCTORS
V.F. Gantmakher and I.B. Levinson

Vol. 20. SEMIMETALS – 1. GRAPHITE AND ITS COMPOUNDS
N.B. Brandt, S.M. Chudinov and Ya.G. Ponomarev

Vol. 21. SPECTROSCOPY OF SOLIDS CONTAINING RARE EARTH IONS
A.A. Kaplyanskii and R.M. Macfarlane, *editors*

Vol. 22. SPIN WAVES AND MAGNETIC EXCITATIONS
(Two volumes)
A.S. Borovik-Romanov and S.K. Sinha, *editors*

Vol. 23. OPTICAL PROPERTIES OF MIXED CRYSTALS
R.J. Elliott and I.P. Ipatova, *editors*

In preparation

Vol. 24. LANDAU LEVEL SPECTROSCOPY
G. Landwehr and E.I. Rashba, *editors*

Oh, how many of them there
are in the fields!
But each flowers in its
own way —
In this is the highest achievement
of a flower!

Matsuo Bashō
1644–1694

PREFACE TO THE SERIES

Our understanding of condensed matter is developing rapidly at the present time, and the numerous new insights gained in this field define to a significant degree the face of contemporary science. Furthermore, discoveries made in this area are shaping present and future technology. This being so, it is clear that the most important results and directions for future developments can only be covered by an international group of authors working in cooperation.

"Modern Problems in Condensed Matter Sciences" is a series of contributed volumes and monographs on condensed matter science that is published by North-Holland Physics Publishing, a division of Elsevier Science Publishers. With the support of a distinguished Advisory Editorial Board, areas of current interest that have reached a maturity to be reviewed, are selected for the series. Both Soviet and Western scholars are contributing to the series, and each contributed volume has, accordingly, two editors. Monographs, written by either Western or Soviet authors, are also included. The complete series will provide the most comprehensive coverage available of condensed matter science.

Another important outcome of the foundation of this series is the emergence of a rather interesting and fruitful form of collaboration among scholars from different countries. We are deeply convinced that such international collaboration in the spheres of science and art, as well as other socially useful spheres of human activity, will assist in the establishment of a climate of confidence and peace.

The publishing house "Nauka" publishes the volumes in the Russian language. This way the broadest possible readership is ensured.

The General Editors of the Series,

V.M. Agranovich A.A. Maradudin

CONTENTS

Preface to the series vii

Contents ix

Introduction xi

1. Universal parameters in mixed crystals 1
   I.P. Ipatova
2. Phonon response theory and the infrared and Raman experiments . . 35
   D.W. Taylor
3. Effect of composition disorder on the electronic properties of semiconducting mixed crystals 133
   A.L. Efros and M.E. Raikh
4. Infrared and Raman studies of disordered magnetic insulators 177
   W. Hayes and M.C.K. Wiltshire
5. Spectroscopy of excitons in disordered molecular crystals 215
   E.I. Rashba
6. Phonon multimode spectra: biphonons and triphonons in crystals with defects 297
   V.M. Agranovich and O.A. Dubovsky

Author index 399

Subject index 417

Materials index 423

Cumulative index 427

INTRODUCTION

The optical properties of solids are dominated by processes in which the light quanta interact with the elementary excitations of the material. These excitations are related to the motions of the constituent particles, the electrons and the nuclei. Under usual conditions the nuclear motions, attended by their adjacent charge clouds, can be separated, using the adiabatic approximation, from the more rapid electronic motions. They are well described by simple harmonic motions whose quantum excitations are phonons, and except in special circumstances anharmonicity can be taken as a small perturbation. The dominant optical processes create or destroy phonons in small numbers.

The allowed electronic states in a solid form bands of energies which may be separated by forbidden bands of energies. Since optical processes only transfer energies of a few electron volts to the electrons, we are only concerned with the outer electrons of the atoms with states within a few eV of the Fermi energy. In semiconductors and insulators there are an integral number of full bands, separated from the empty bands by an energy gap. The simplest optical process is the promotion of an electron from the full to the empty band leaving behind a hole. In solids in general, because of the high densities of electrons, the electron interactions are important but in semiconductors and insulators they are readily described in terms of the effective overall crystal potential. However, the residual interaction between the electron and the hole can give an important final state interaction. Indeed it can give rise to bound hole–electron pairs called excitons. If this residual interaction is strong the exciton can be regarded as a single uncharged unit which itself has a band of allowed energies associated with its kinetic motion. If the residual binding is weaker, as it is in semiconductors because of the screening of the other charges, both weakly bound excitons and unbound hole–electron pairs play an important role in determining the optical properties. There are then broad bands of allowed energies for the pairs comparable to those of the fundamental electron bands.

Electrons have other motions associated with their intrinsic spins. In solids where the spins are ordered, i.e. in magnetic materials, there are well-defined collective excitations of spin deviations from the ordered state, called spin waves, and these will also be quantised into magnons. Here also there are bands of allowed excitation energies and the magnons can be created from light quanta to contribute to the optical properties of these materials.

In metals, or indeed in heavily doped semiconductors, where there are high densities of electrons in unfilled bands, the residual interaction between the electrons plays an important role. Although the dominant consequence is the production of collective density fluctuations of the plasma (plasmons) the remaining screened Coulomb interaction is significant. Thus optical processes in metals which involve the promotion of electrons from within the Fermi sea into unoccupied states, leaving behind a hole, are strongly affected by this interaction.

This volume is concerned with the description of optical processes in semiconductors and insulators which can be basically described through their elementary excitations. We shall be concerned with the creation of phonons, electron–hole pairs (excitons) and magnons in these materials. A proper description of the properties of metals requires a somewhat different approach which cannot be dealt with within the limited scope of a single volume.

Excitations

The theory of elementary excitations in perfect ordered solids is enormously simplified by the high symmetry. This requires that every state, and in particular every excitation, is wave-like since it must have equal amplitude on each equivalent atom. It can therefore be specified by the wave vector $\boldsymbol{k}$ whose allowed values are confined to the Brillouin zone. This restricts the energies of the excitations $\epsilon(\boldsymbol{k})$ to well-defined bands. Since the light in optical processors also has a well-defined wave vector $\boldsymbol{q}$ there are definite selection rules. For example in the simplest process where a single photon is destroyed and a single excitation created this requires $\boldsymbol{k}=\boldsymbol{q}$. Since the wavelength of light used in optical experiments is always very much longer than the characteristic length (the lattice spacing) in the crystal only $\boldsymbol{k}$ values very near the centre of the zone will be involved.

Any disorder in a crystal breaks the translation symmetry and hence destroys the description of the elementary excitations in terms of a wave vector $\boldsymbol{k}$. Elementary excitations still exist with normal modes and eigen states which reflect the actual distribution of atoms in the systems. These excitations are still confined to bands since the spectrum of allowed energies is still controlled by the rate at which the excitations can pass from one atom to another. Optical processes are still looking for the response of the system to a stimulus with wave vector $\boldsymbol{q}$. However, the simple selection rule has broken down and in principle transitions involving any of the excitations will be allowed. Exploration of the optical properties of mixed crystals therefore allows us in principle to investigate the whole excitation spectra and gives some detailed information about the nature of the excitations themselves.

In the first chapter Ipatova describes some of the universal properties of excitations in mixed crystals. But it may be useful to summarise these in a more general form in this introduction. If we consider first a single impurity atom we know that it will modify the nature of the crystal excitations only in its immediate vicinity. Excitations will still exist throughout the allowed bands and these wave states will be scattered by the impurity and will therefore have a modified amplitude close to the impurity. This modification may be large for some regions on the spectrum where the wave excitation energy matches the characteristic excitation energy of the defect. Such effects are sometimes called resonances. In addition strongly perturbing defects can allow excitations outside the normally allowed bands, which are localised in the vicinity of the defect. Such localised modes or bound states occur at single isolated energies.

If we increase the concentration of such defects the wave states are further scattered and their energies are modified to an extent which depends on the concentration of defects. In addition any localised states also now appear in numbers dependent on the number of defects, they interact and spread into a band. Thus in mixed crystals with finite defect concentrations we revert to a band structure, which is modified from that in the perfect lattice. In recent years there have been many attempts to produce working theories of this type of excitation spectra, when the overall translational symmetry is absent. The most successful are effective medium theories which reimpose apparent translational symmetry by assuming a complex energy-dependent potential which mimics the scattering effects to as good a degree as possible. The most widely used is the coherent potential approximation and a summary of these methods can be found in the review article of Elliott, Krumhansl and Leath (1974).

Such theories deal, however, with the average properties of the medium and do not treat adequately the important effects which arise from fluctuations in the composition of the crystal. Thus for example, a region of a mixed AB crystal which is particularly rich in B atoms might be expected to have states very close to those of a B crystal. Such extreme fluctuations are rare so that the number of states in the extreme edges of the bands will be small. However it is sometimes possible in experiments to pick up such states with extra weight, for example in fluorescence from an excited band. In the first chapter Ipatova discusses the general properties of the band tails which arise from the fluctuations.

The structure of the actual states and the way in which these are correlated with the particular distribution of defects which exist in the real system is a complex problem. In particular a detailed description is necessary in order to understand the phenomenon of localisation first introduced by Anderson. He pointed out that the energy states of disordered systems would be divided into those which covered the whole crystal and those which were confined to a small region of the crystal and decayed exponentially away from it. We have already noted that such a division exists for a single defect but it is not obvious that it does so when there is a finite concentration of such defects. For example the localised states referred to above are initially in resonance with each other and are expected to broaden into an impurity band. Anderson showed that such localised states are most likely to occur in the tails of the bands where the properties of the states are dominated by fluctuations in the composition of the crystal.

Observation of excitations

We pointed out above that the simplest optical process involved a single photon and a single excitation; however many more complicated processes are allowed. They can be usefully classified, first of all, into two types which depend on the nature of the optical event. Single-photon processes involve either absorption or emission of the light. Two-photon processes involving the inelastic scattering of light give rise to the Raman effect. Two-photon absorption processes are also possible using high-powered lasers and they have been used to determine some useful properties of solids. They are, however, still relatively rare.

The dominant processes of either type are usually those that create or destroy a single excitation within the material. In the case of lattice vibrations we have single infrared absorption arising from one-phonon processes and also Raman scattering arising from similar processes. These effects are described in detail in the paper by Taylor. Similar processes involving magnons are described in the paper by Hayes and Wiltshire. Here the single magnon process is relatively weak because it is a magnetic excitation. In both of these cases excitations involving two phonons and two magnons can also be observed. These can arise in direct processes where the light interacts with two excitations or indirectly through anharmonic interactions. These sorts of effects are considered further in the paper by Agranovich and Dubovsky. They are particularly concerned with processes which involve several phonons and they show interesting effects involving the interplay of the anharmonic coupling between the phonons and the interaction with the defects in the disordered system.

In electronic transitions the fundamental states are those of the electrons but the optical process creates hole–electron pairs. These are discussed in the paper by Efros and Reich. The situation which arises when the electron–hole binding dominates so that the excitations created are excitons is discussed in the paper by Rashba. Both of these papers also include interesting information about other aspects of the properties of these crystals.

The fundamental ideas involved in the description of all the optical properties of mixed crystals are the same. The fundamental structure of the excitation spectrum including band formation, band tailing and localisation is common to every type of excitation. The fundamental selection rules are also the same. This volume shows how these ideas can be widely applied to bring coherence and understanding to a diverse area of solid state physics.

R.J. Elliott

Reference

Elliott, R.J., J.A. Krumhansl and P.L. Leath, 1974, Rev. Mod. Phys. **46**, 465.

CHAPTER 1

Universal Parameters in Mixed Crystals

I.P. IPATOVA

A.F. Ioffe Physical Technical Institute
Academy of Sciences of the USSR
Leningrad, 194021, USSR

Optical Properties of Mixed Crystals
Edited by
R.J. Elliott and I.P. Ipatova

Contents

1. Introduction 3
2. Effects of the mixed crystal composition disorder on the continuous spectrum of the electron energy 5
3. Electron localization by composition fluctuations 10
4. Qualitative approach to Anderson localization 13
5. Effects of composition disorder on vibrational spectra of mixed crystals 17
6. Uniaxial and biaxial mixed crystals 26
References 32

1. *Introduction*

Mixed crystals are disordered systems with constituent atoms of several types randomly distributed over the sites of the crystal lattice. Some examples of mixed crystals are solid solutions of III–V compounds ($Al_xGa_{1-x}As$), binary alloys (Ge_xSi_{1-x}, Au_xAg_{1-x}), molecular crystals with isotope substitution, $HgBr_{2x}Cl_{2(1-x)}$ and so on. There are also quaternary compounds such as $Al_xGa_{1-x}P_yAs_{1-y}$.

Mixed crystals are of interest for many reasons. First of all, they are convenient model systems for studying general properties of disordered systems. The fluctuating crystal potential caused by the isoelectronic substitution of components has a short-range nature. Therefore the effects of disorder are not complicated by long-range effects. Secondly, mixed crystals are of great importance for modern electronics. Their band structure may be varied over a wide range of energies by a change in composition of the mixed crystal. There are many technical applications of semiconductor mixed crystals such as light emitting diodes and semiconductor lasers.

Much experimental evidence exists at present of the random atomic placements over the lattice sites and of the short-range correlations which may be observed in light absorption and light scattering spectra of mixed crystals.

In order to find physical parameters which govern the effects of disorder in mixed crystals we consider for simplicity the binary mixed crystal A_xB_{1-x} where A- and B-atoms are randomly distributed over the lattice sites. The composition of the mixed crystal is specified in this case by the fractional concentration of A-atoms $x = n_A/n$, where n_A is the concentration of A-atoms and n is the concentration of lattice sites. The fractional concentration of B-atoms is $(1-x) = n_B/n$, where n_B is the concentration of B-atoms. In this chapter we call A-atoms the impurity atom in an ideal crystal consisting of B-atoms. In fact B-atoms could be equally taken as impurities in an ideal crystal of A-atoms.

The energy gap in semiconductors is typically less than the energy band width, and when the second A-component of the mixed crystal is introduced in the B-crystal the single impurity A-atom is not able to create the bound state of the electron. At the same time the clustering of A-atoms, that is the fluctuation of the mixed crystal composition, does create bound states of an electron to composition fluctuations. It was shown by Alferov et al. (1968) that since the binding energy in the case of a state bound by fluctuation is less than for a state bound to the single impurity atom, the fluctuation bound states form a tailing of the electron density of states in the band gap.

The experimental data show (Onton and Chicotka 1971, Alferov et al. 1972, Tzarenkov et al. 1972, Li et al. 1976, Pikhtin 1977, Nelson 1982, Evtichiev et al. 1983) that the band-tailing is very small in III–V mixed crystals. Nevertheless it is quite clear now that the bound exciton spectra in III–V and II–VI mixed crystals are affected by the disorder (Nelson and Holonyak 1976, Suslina et al. 1978, 1979, Goede et al. 1978, Shui Lai and Klein 1980, Cohen and Sturge 1982, Mach et al. 1982a, b, Permogorov et al. 1983a, b, Wo Hoang Thai and Miloslavskii 1983).

The relation between the energy band width and the gap in bimetallic alloys varies considerably (see, e.g., Ehrenreich and Schwartz 1976). In particular there are alloys with the energy gap larger than the energy band width, and a single impurity atom is able to bind an electron in this case. The localized states contribute to the density of states.

The composition of the mixed crystal is well described by the deviation Δx of the fractional concentration of A-atoms and of average value x. To describe the random distribution of the mixed crystal components we choose a certain volume measuring R^3, R being the linear size of the volume. The average number $\overline{N}_A$ of A-atoms in this volume is

$$\overline{N}_A = n_A R^3. \tag{1.1}$$

The total number N of lattice sites in the volume is

$$N = nR^3. \tag{1.2}$$

The local deviation Δx of concentration of A-atoms from the mean value (the fluctuation of composition) is given by

$$\Delta x = \frac{N_A - \overline{N}_A}{N} = \frac{\Delta N_A}{nR^3} \approx \frac{\sqrt{xnR^3}}{nR^3}. \tag{1.3}$$

It follows from eq. (1.3) that the fluctuation Δx depends on the excess number ΔN_A of A-atoms and on the linear size R of the fluctuation volume. If the A-atoms in the lattice are located at random a distribution of fluctuations Δx with various ΔN_A and R occurs. Fluctuations in the crystal composition create fluctuations of the lattic periodic potential ΔU which distort the electronic states in the energy band and produce new electronic states in the energy gap.

2. *Effects of the mixed crystal composition disorder on the continuous spectrum of the electron energy*

One can find physical parameters which govern the composition dependence of the electron density of states by considering the low concentration of substitutional isoelectronic A-impurities in a B-crystal. Electrons of a B-crystal are scattered by A-impurities. The scattering in the effective mass approximation is described by the Schrödinger equation with the short-range impurity potential $U(\boldsymbol{r})$ with a radius of the order of the lattice constant

$$-\frac{\hbar^2}{2m}\Delta\Psi(\boldsymbol{r}) + \int \mathrm{d}^3r' \, n_{\mathrm{imp}}(\boldsymbol{r}-\boldsymbol{r}')U(\boldsymbol{r}')\Psi(\boldsymbol{r}') = E'\Psi(\boldsymbol{r}'). \tag{2.1}$$

Here $\Psi(\boldsymbol{r})$ is the electron wave function, E' is the electron energy, m is the electron mass. The concentration of A-impurities n_{imp} has the form

$$n_{\mathrm{imp}}(\boldsymbol{r}) = \sum_l \delta(\boldsymbol{r}-\boldsymbol{r}_l), \tag{2.2}$$

where the sum runs over all lattice sites occupied by impurities.

The scattering of an electron from the impurity atom can be conveniently presented in terms of the scattering amplitude $f(\theta)$. This represents the distortion of the Bloch plane wave by the impurity centre. The explicit form of $f(\theta)$ is defined from asymptotic behaviour of the wave function at large distances from the impurity centre (see, e.g., Landau and Lifshitz 1974, §123)

$$\Psi(\boldsymbol{r}) = \mathrm{e}^{ikz} + \frac{f(\theta)}{r}\mathrm{e}^{ikr}. \tag{2.3}$$

Here the electron wave vector $\boldsymbol{k}$ is equal to

$$k = \frac{1}{\hbar}\sqrt{2mE}, \tag{2.4}$$

where E differs from E' in eq. (2.1) by the shift of the continuous spectrum edge due to the presence of impurities. The effect of the scattering centres depends on the electron energy and on the possibility of the creation of an electron state bound to the impurity atom.

If the potential energy of the electron in the field of the impurity centre is larger than the kinetic energy of the electron, an electron state bound to the centre appears. The condition for electron localization has the form

$$U_0 \gtrsim \hbar^2/mR_0^2, \tag{2.5}$$

where U_0 is the electron potential energy near the impurity centre, R_0 is the radius of the potential which in the case of isoelectronic substitution is of the order of the lattice constant a. When condition (2.5) holds, the scattering amplitude f has a singularity at the electron bound state energy (Landau and Lifshitz 1974, §132)

$$f = \frac{1}{g + \mathrm{i}k}, \tag{2.6}$$

where g is the real function of the energy.

If inequality (2.5) does not hold there is no electron bound to an individual impurity centre. To estimate the effect of the impurity centre one can use the perturbation theory

$$f(q) = \frac{m}{2\pi\hbar^2} \int \mathrm{d}^3 r \; U(\boldsymbol{r}) \; \mathrm{e}^{\mathrm{i}qr}, \tag{2.7}$$

where $q = \frac{1}{2}k \cos\theta$. The scattering amplitude f for the low-energy electrons does not depend on the energy (Landau and Lifshitz 1974, §123). $f(0)$ is called the scattering length and has a physical meaning relating to the effective linear size of the scatterer. It follows from condition (2.5) and eq. (2.7) that to an order of magnitude the scattering length equals

$$f(0) \equiv \tilde{f} = \frac{m}{2\pi\hbar^2} \int \mathrm{d}^3 r \; U(\boldsymbol{r})$$

$$\approx \left[\frac{mR_0^2}{\hbar^2} U_0 \right] R_0 < R_0 . \tag{2.8}$$

Since $R_0 \sim a$ the scattering length $\tilde{f}$ is less than the lattice constant

$$\tilde{f} < a. \tag{2.9}$$

On the other hand the average separation between impurities $\overline{R}$ is always larger than the lattice constant

$$\overline{R} \gg a. \tag{2.10}$$

It is seen from conditions (2.9) and (2.10) that

$$\overline{R} \gg f. \tag{2.11}$$

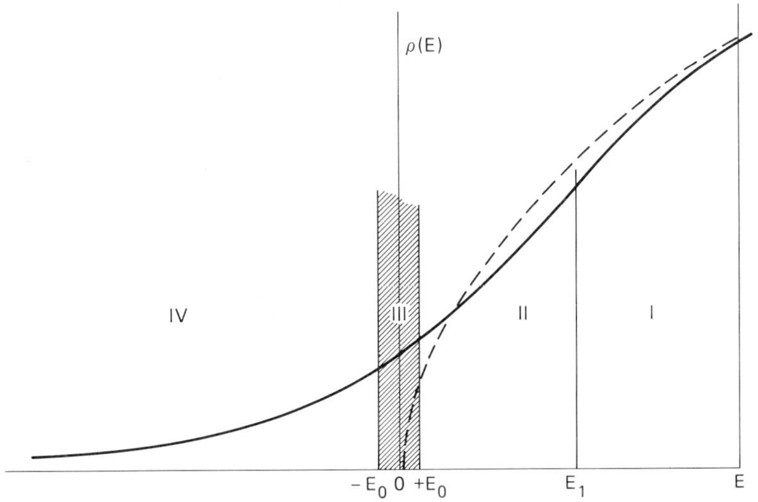

Fig. 1. Qualitative energy dependence of the electron density of states in the case when there is no localized electron state by the individual impurity centre: I – region of electron scattering from isolated impurity centres; II – region of electron scattering from composition disorder in mixed crystals; III – diffusion region; IV – region of localized electron states by macroscopic fluctuations of composition.

When the electron wavelength

$$\lambda = \hbar/\sqrt{2mE'} \tag{2.12}$$

is less than the average separation of impurities

$$\lambda/\bar{R} \ll 1, \tag{2.13}$$

scattering of the electron from the individual impurity centres takes place without interference. Since the perturbation due to the isolated centre is small with respect to the electron energy E', the electron density of states is very close to that for the ideal crystal. Region I where relation (2.13) holds is schematically shown in fig. 1.

The electron with small positive energy from region II in fig. 1 has a wavelength that obeys the condition

$$\lambda/\bar{R} > 1. \tag{2.14a}$$

Substitution of eq. (2.12) in condition (2.14a) leads to the equivalent parameter

$$E' < \frac{\hbar^2}{m\bar{R}^2} \equiv E_1. \tag{2.14b}$$

When conditions (2.14a, b) hold, the electrons are scattered from the macroscopic clusters of the impurity atoms. The impurity concentration n_{imp} in eq. (2.1) should be replaced in that case by the average value taken over the volume that is larger than R^3 and less than λ^3. After averaging, one obtains from eq. (2.1)

$$-\frac{\hbar^2}{2m}\Delta\Psi(\boldsymbol{r}) + n\left[\int \mathrm{d}^3r'\; U(\boldsymbol{r}')\right]\Delta x\Psi(\boldsymbol{r})$$

$$= (E' - E_c)\Psi(\boldsymbol{r}), \tag{2.15}$$

where the average concentration shift of the electron energy is equal to

$$E_c = n_A\int \mathrm{d}^3r\; U(\boldsymbol{r}) = 2\pi\frac{\hbar^2}{m}n_A\tilde{f}. \tag{2.16}$$

Everywhere below we use the electron energy E which is referred to E_c, i.e. $E = E' - E_c$.

When impurities are randomly distributed over the lattice sites fluctuations $\Delta x(\boldsymbol{r})$ and $\Delta x(\boldsymbol{r}')$ are statistically independent. This means that the pair correlation function is approximated by

$$\langle\Delta x(\boldsymbol{r})\Delta x(\boldsymbol{r}')\rangle = \nu\delta(\boldsymbol{r} - \boldsymbol{r}'), \tag{2.17}$$

where $\langle\cdots\rangle$ denotes an averaging over the random distribution of impurity atoms. It will be shown in section 3 that $\nu = (1/n)x(1-x)$.

It was assumed here that in addition to condition (2.14a) the wavelength of an electron is less than the mean free path of the electron l:

$$\lambda/l \ll 1. \tag{2.18}$$

Condition (2.18) enables us to consider the electron scattering from fluctuations of the mixed crystal composition with the help of the perturbation theory. This calculation has been done by Petukhov et al. (1967). Relation (2.18) can be rewritten in the form

$$(E/E_0)^{1/2} \gg 1, \tag{2.19}$$

where the characteristic energy E_0 is given by

$$E_0 = \frac{1}{\hbar^6}\left[\int \mathrm{d}^3r\; U(\boldsymbol{r})\right]^4 n^2x^2(1-x)^2m^3. \tag{2.20}$$

Comparison of condition (2.14b) and eq. (2.20) shows that

$$\frac{E_0}{E_1} \approx \left(\frac{\tilde{f}}{R}\right)^4 \ll 1. \tag{2.21}$$

Inequality (2.21) shows that there is an electron energy interval

$$E_0 \ll E \ll E_1, \tag{2.22}$$

where the effect of scattering of electrons from fluctuations of the mixed crystal composition is small. The corresponding wave function is a weakly damped plane wave. The electron density of states is given by

$$\rho(E) = \rho_0(E)\left[1 - \frac{1}{\pi}\sqrt{\frac{E_0}{E}}\right], \tag{2.23}$$

showing a small deviation from (see e.g., Anselm 1978)

$$\rho_0(E) = \frac{\sqrt{2E}\, m^{3/2}}{\hbar^3\, 2\pi^2}. \tag{2.24}$$

Near the edge of the density of states ($E \to 0$) the electron wavelength λ becomes comparable with the mean free path l

$$\lambda \gtrsim l. \tag{2.25}$$

The corresponding electron energy satisfies the condition

$$|E| \lesssim E_0. \tag{2.26}$$

The interval (2.26) is labelled in fig. 1 by III and is the hatched area. This is the diffusion region where the electron motion is controlled by collisions. Notice that at any low concentration of impurities there is always an energy interval (2.26) where the collision controlled regime occurs. There are no small parameters in the region (2.26). An analytic approach to the problem is not possible. On the other hand this interval of energies is of considerable interest. According to modern understanding developed originally by Anderson (1958) the localization threshold which separates localized and delocalized states occurs in the region (2.26). The qualitative discussion of Anderson localization is given in section 4.

3. Electron localization by composition fluctuations

Localized electron states in mixed crystals occur not only due to the individual impurity centres but also due to macroscopic fluctuations of the mixed crystal composition. These states occur in the energy gap and they create the exponentially decaying tails of the electron density of states in the energy gap (see region IV in fig. 1).

This tailing can be estimated using the optimum fluctuation method (Halperin and Lax 1966, 1967, Zittartz and Langer 1966, Lifshitz 1967). The probability that the fractional concentration $(x+\Delta x)$ will occur is defined by the entropy of the system $S(x+\Delta x)$. The probability density W of finding an excess concentration Δx equals (see, e.g., Landau and Lifshitz 1976, §110)

$$W(x+\Delta x) = \mathrm{e}^{S(x+\Delta x)}. \tag{3.1}$$

The maximum value of W corresponds to the maximum of S and therefore

$$\left.\frac{\partial S}{\partial x}\right|_{\Delta x \to 0} = 0; \qquad \left.\frac{\partial^2 S}{\partial x^2}\right|_{\Delta x \to 0} < 0. \tag{3.2}$$

The simplest way to find W is by considering the small fluctuation limit $\Delta x \ll x$. The small argument expansion of S is given by

$$S(x+\Delta x) = S(x) - (\beta/2)(\Delta x)^2, \tag{3.3}$$

where $\beta > 0$. Substituting eq. (3.3) into eq. (3.2) gives

$$W \sim \mathrm{e}^{-(\beta/2)(\Delta x)^2}. \tag{3.4}$$

The omitted prefactor of W in eq. (3.4) is usually found from the probability normalization condition. The expansion coefficient β is connected in a standard way to the dispersion $\langle(\Delta x)^2\rangle = 1/\beta$. If Δx is the excess concentration of A-atoms in the volume R^3 the dispersion $\langle(\Delta x)^2\rangle$ depends on R. Using the same arguments that have been used to estimate Δx from eq. (1.3) one obtains

$$\langle(\Delta x)^2\rangle = \frac{1}{(nR^3)^2}\left[\langle N_\mathrm{A}^2\rangle - \langle N_\mathrm{A}\rangle^2\right] = \frac{x(1-x)}{nR^3}. \tag{3.5}$$

Substituting eq. (3.5) into eq. (3.4) one obtains

$$W \sim \exp\left\{-\frac{(\Delta x)^2}{2x(1-x)(1/nR^3)}\right\}. \tag{3.6}$$

The fluctuations of composition Δx create the short-range fluctuating crystal potential ΔU. The characteristic radius of ΔU is less than the radius of the wave function of the electron bound to fluctuation. Therefore ΔU is the local potential of the form [see eq. (2.15)]

$$\Delta U = \alpha \Delta x = \alpha \Delta N_A / nR^3. \tag{3.7}$$

The potential ΔU effects the electron states in the energy band and creates new localized states in the gap. Here α is taken to be constant.

We now find the probability of splitting the electron level with energy $E < 0$ in the band gap. The level occurs when the kinetic energy of the electron $\hbar^2/mR^2$ is less than the potential energy ΔU from eq. (3.7). The binding energy is approximately equal to the difference of kinetic and potential energies

$$|E| = \alpha \frac{\Delta N_A}{nR^3} - \frac{\hbar^2}{mR^2}. \tag{3.8}$$

The first localized level occurs when $E \approx 0$. The corresponding ΔN_A is given by

$$\Delta N_A = \frac{nR^3}{\alpha}\left[|E| + \frac{\hbar^2}{mR^2}\right]. \tag{3.9}$$

Introducing eq. (3.9) into eq. (1.3), and then into expression (3.6), we obtain the probability density (3.6) in the form

$$W \sim \exp\left\{-\frac{nR^3(|E| + \hbar^2/mR^2)^2}{\alpha^2\, 2x(1-x)}\right\}. \tag{3.10}$$

If R is small enough the binding energy satisfies the condition

$$|E| \ll \hbar^2/mR^2.$$

Neglecting $|E|$ in expression (3.10) we can see that W is proportional to $\exp\{-1/R\}$ and increases with increasing R. If R is large enough the binding energy satisfies the condition

$$|E| \gg \hbar^2/mR^2.$$

The probability W is proportional to $\exp\{-R^2\}$ and W increases with decreasing R. Therefore there exists an optimum dimension of the fluctuation

R_{opt} which corresponds to the maximum of W and which minimizes the argument of the exponential in (3.10):

$$R_{opt} \approx \hbar/\sqrt{m|E|}. \tag{3.11}$$

Substitution of eq. (3.11) into eq. (3.10) leads to

$$W_{max} \sim \exp\left\{-\sqrt{|E|/E_0}\right\}, \tag{3.12}$$

where

$$E_0 = \frac{\alpha^4 m^3 x^2 (1-x)^2}{\hbar^6 n^2} \tag{3.13}$$

is the same characteristic energy that appeared in (2.21) when considering a continuous spectrum with positive electron energies in section 3. Equation (3.13) holds when

$$(|E|/E_0)^{1/2} \gg 1. \tag{3.14}$$

Our qualitative approach allows us to find the exponential behaviour of W. To calculate the prefactor one should incorporate in the theory fluctuations close to the optimum ones. An example of the prefactor calculation is given by Efros and Raikh in ch. 3 of this volume. Nevertheless eq. (3.12) shows the qualitative dependence of W on E and R.

The exponential in probability (3.12) is the same one that enters the expression for the electron density of states

$$\rho(E) \sim W_{max} \sim \exp\left\{-\sqrt{|E|/E_0}\right\}. \tag{3.15}$$

The exponential tail of the density of states in the forbidden band is built from the states of electrons bound to different optimum fluctuations. Gaussian fluctuations (3.10) create shallow potential wells for the electron and contribute to $\rho(E)$ near the bottom of the conduction band. The density of states $\rho(E)$ from (3.15) is shown in fig. 1 in region IV.

Electron states with energies lying deeper in the band gap were considered by Lifshitz et al. (1982). It was shown that they originate from a Poisson distribution of fluctuations. The low-energy ending of the density of states is defined by the low probable fluctuations with $\Delta x \sim x \sim 1$ (Lifshitz et al. 1976). The fluctuation distribution was shown to be a step function of Fermi–Dirac type in this case. They are called Fermi fluctuations.

Equations (2.21), (2.22) and (3.14) show that in the rather wide region of positive and negative electron energies near the edge the density of states is the function of the ratio E/E_0 only. This behaviour follows from the possibility of introducing the dimensionless variables (Shklovskii and Efros 1979)

$$r = r'\hbar/\sqrt{mE_0}\,; \qquad \Delta x(\boldsymbol{r}) = \Delta x(\boldsymbol{r}')E_0/n\int \mathrm{d}^3r\; U(\boldsymbol{r}) \tag{3.16}$$

into the Schrödinger equation (2.15) and the correlation function (2.17). Substituting eq. (3.16) in eqs. (2.15) and (2.16) results in

$$-\tfrac{1}{2}\Delta\Psi + \Delta x\Psi = \frac{E}{E_0}\Psi \tag{3.17}$$

and

$$\langle \Delta x(\boldsymbol{r})\Delta x(\boldsymbol{r}')\rangle = \delta(\boldsymbol{r}-\boldsymbol{r}'). \tag{3.18}$$

Since eqs. (3.17) and (3.18) are functions of parameter E/E_0 alone the density of states is also the universal function of E/E_0 in the energy region where the electrons are scattered from macroscopic fluctuations of impurities.

The numerical solution of eq. (3.17) (Baranovskii and Efros 1978) shows that there is an extra numerical factor 1/178 in the dimensionless parameter (3.14). It means that the characteristic energy $E_0/178$ is smaller than the qualitative value (3.13). This factor reduces the band tailing of mixed crystals to 10^{-2} meV in III–V compounds and up to 3 meV in II–VI materials. As a result the exciton optical spectra are well resolved in mixed crystals.

If the electron interaction with the fluctuation potential satisfies inequality (2.5) localized electron states occur due to the isolated impurity atom. The scattering length approximately equals the radius of the electron localized state. Equation (2.10) holds in this case only at low concentrations of impurities $x \ll 1$ or $(1-x) \ll 1$. The extra feature of the electron density of states in this case is a peak near the binding energy of the electron. The peak is broadened due to the small overlap of electron wave functions of neighbouring impurities. This overlapping increases with concentration growth. When $\tilde{R} \approx \tilde{f}$ an impurity band appears (Shklovskii and Efros 1979, Lifshitz et al. 1982).

4. *Qualitative approach to Anderson localization*

According to the modern picture there exists a localization threshold that separates localized and delocalized electron states. It occurs at $\overline{R} \lesssim \tilde{f}$ in the diffusion region (2.25) (see, e.g., Efros 1978). Up to this point we have

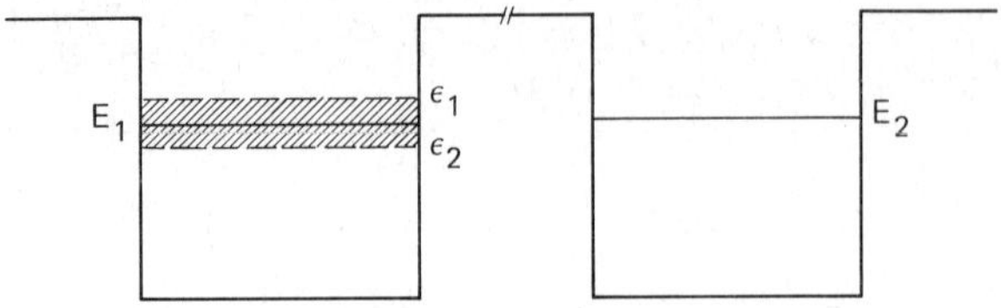

Fig. 2. Two resonant potential wells.

discussed the electron localization by the single impurity or by the potential wells arising from individual fluctuations. There is an infinite number of equivalent wells in a homogeneous macroscopic medium. The resonant interaction between equivalent wells leads to the extension of the electron wave function over the whole medium so that, at first glance, it might seem that there is an extended wave function corresponding to a delocalized state. However, the real situation is more complicated.

To elucidate the situation we start with two equivalent fluctuation potential wells separated by spacing R which is larger than the radius of the electron state in a single potential well (see fig. 2). We take E_1, Ψ_1 and E_2, Ψ_2 to be the electron eigenstates and wave functions in the first and second potential wells, respectively. The small overlap of Ψ_1 and Ψ_2 results in the splitting of the initial energy level $E_1 = E_2 = E$ into two levels

$$\epsilon_1 = E + \tfrac{1}{2}J_{12},$$

$$\epsilon_2 = E - \tfrac{1}{2}J_{12}. \tag{4.1}$$

Here the overlap integral J_{12} equals

$$J_{12} = \mathscr{E}_0 \int \mathrm{d}^3 r \; \overset{*}{\Psi}_1(\boldsymbol{r}) \Psi_2(\boldsymbol{r}), \tag{4.2}$$

where $\mathscr{E}_0$ is an energy constant. The corresponding extended wave functions are symmetric and antisymmetric linear combinations of Ψ_1 and Ψ_2

$$\Psi_S = \frac{1}{\sqrt{2}}(\Psi_1 + \Psi_2); \qquad \Psi_A = \frac{1}{\sqrt{2}}(\Psi_1 - \Psi_2). \tag{4.3}$$

We turn now to the case of three equivalent potential wells separated by spacings R_{12}, R_{23} and R_{13}. Suppose $R_{12} < R_{23}$, R_{13}. It means that the mutual interaction of states Ψ_1 and Ψ_2 is stronger than the effect of Ψ_3. Therefore in the first approximation the overlapping of Ψ_1 and Ψ_2 should be taken into account. The corresponding splitting of E_1 and E_2 is equal to J_{12} from eq.

(4.2). As a result the level of the third potential well E_3 becomes nonresonant. The next order correction to Ψ_3 due to the influence of Ψ_1 and Ψ_2 follows from nondegenerate perturbation theory

$$\Delta\Psi_3 \approx \frac{J_{13}}{\epsilon_1 - E_3}\Psi_1 \approx \frac{J_{13}}{\frac{1}{2}J_{12}}\Psi_1. \tag{4.4}$$

The overlap integral J_{ik}, i, k being 1, 2, 3, decreases with R_{ik} as e^{-R/R_0}, where R_0 is the radius of the electron wave function. If $R_{13} > R_{12} > R_0$ the ratio J_{13}/J_{12} is small:

$$\frac{J_{13}}{J_{12}} \approx \exp\left\{-\frac{R_{13} - R_{12}}{R_0}\right\} \ll 1. \tag{4.5}$$

Therefore $\Delta\Psi$ contains only an exponentially small admixture of states Ψ_1 and Ψ_2. This means that the resulting wave function is localized either by the individual potential well (singlet state Ψ_3) or by the pair of potential wells (doublet state (4.1)). In the case of three potential wells the extended state is possible only in a special case $R_{12} = R_{13}$.

One can generalize the picture for any number of randomly distributed potential wells. All the states are either singlet or doublet type. The probability of an ordered configuration with the delocalized electron state extended over many equivalent wells is very small. This follows from the fact that the separation between potential wells is taken to be large and the overlap integrals are small.

If the separation between potential wells R is of the order of the electron state radius (high concentration limit) the mutual interaction of the fluctuation wells is large. Neither the tight binding approximation nor the perturbation theory are valid in this case. However, the localization threshold occurs just in

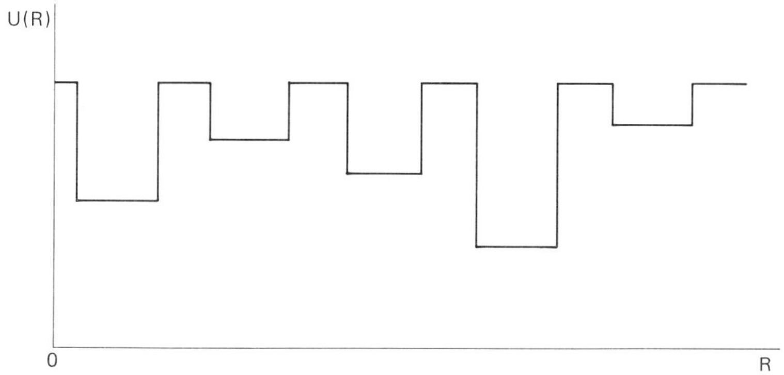

Fig. 3. Anderson model of a disordered crystal.

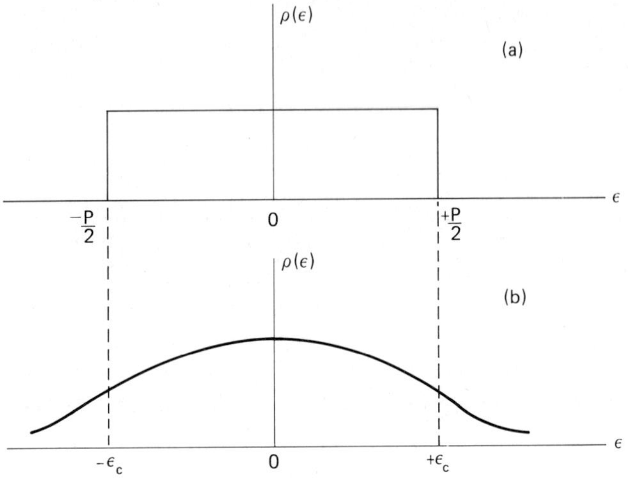

Fig. 4. Density of states for the Anderson model: (a) noninteracting potential wells; (b) interacting wells at $J \gtrsim P$.

this very region, as was shown by Anderson (1958) for a special model of disordered systems. In Anderson's model there is a set of regularly distributed short-range potential wells with a depth which fluctuates within the interval P (see fig. 3) with equal probability $1/P$. The short-range overlap J-integrals for the neighbouring wells are taken to be constant. The model bears a strong resemblance to the short-range fluctuating potential in mixed crystals. The corresponding density of states for noninteracting wells is shown in fig. 4a.

If the electron levels for two isolated potential wells have a separation in energy as large as

$$(\epsilon_1 - \epsilon_2) \gg J,$$

energy levels ϵ_1 and ϵ_2 are nonresonant. The first-order corrections to Ψ_1 or Ψ_2 are small according to eq. (4.4). There are no extended states (see fig. 5a).

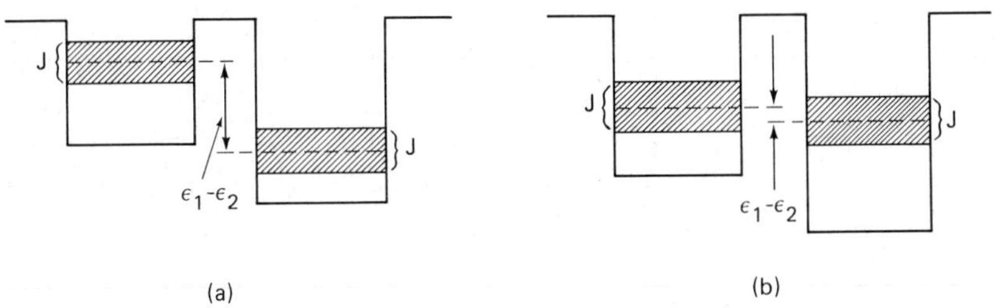

Fig. 5. Extended electron states for two potential wells: (a) $\epsilon_1 - \epsilon_2 > J$; there is no extension, (b) $\epsilon_1 - \epsilon_2 < J$; there is an extension.

When

$$(\epsilon_1 - \epsilon_2) \ll J$$

the energy levels ϵ_1 and ϵ_2 are resonant (see fig. 5b). Extended states with wave functions Ψ_S and Ψ_A are obtained from eq. (4.3).

In the case of three potential wells the extended electron states occur when the maximum separation between three levels is less than J. Notice that two potential wells must be neighbours of the third one.

In a real crystal where there is an infinite number of potential wells, extended states occur for those wells which have electron energy levels within the band J. These wells are resonant. Two electron states are of resonant type if corresponding potential wells are either neighbours or are connected by the chain of resonant wells. The whole complex of resonant wells is named a cluster. The electron states in such a cluster are extended states.

The initial distribution of levels within the interval P is taken to be the uniform one. Therefore the fraction of resonant wells with electron energy within the interval J equals J/P. When resonant potential wells are distributed randomly and when $J/P \ll 1$ the resonant wells are arranged one by one with large spacing. The probability of two resonant wells, and a fortiori of several resonant wells, is very low. As the ratio J/P increases there first appear pairs, then triplets, and finally an infinite cluster of overlapping resonant wells. This cluster covers only a fraction of the lattice sites. The extended wave functions of the cluster do not fall off with the distance and they correspond to delocalized states. The threshold value $J_c/P = k_c$ corresponds to an infinite cluster; the k_c value is called the Anderson transition point.

The electron density of states in the Anderson model is shown in fig. 4b. The number of resonant wells is maximum in the central part of the distribution. The first delocalized state emerges in the central part of the density of states. If $J/P \gg 1$ then the localized states correspond to the tails of the density of states. The existence of the localization threshold is connected with the problem of a metal–dielectric transition. It also affects the optical properties of mixed crystals. This aspect is discussed in several chapters in this volume.

5. *Effects of composition disorder on vibrational spectra of mixed crystals*

The most remarkable fact about lattice optical properties of the mixed crystals is that the optical spectra consist of relatively narrow spectral lines with a line width of the same order as that in the pure crystal while the line shift of the vibration frequency can be large. This means that the broadening of the phonon distribution function caused by composition fluctuations is small.

This picture has allowed the development of useful semiempirical descriptions of the optical spectra of mixed crystals with the help of the set of oscillators with frequencies which, in cases of low ($x \ll 1$ or $(1-x) \ll 1$) concentrations, transform into the long-wavelength optical frequencies of pure crystals. In the single-site approximation the oscillator approach was suggested by Chang and Mitra (1968). In the double-site approximation a similar oscillator approach was suggested by Zinger et al. (1976) for III–V mixed crystals, by Zinger et al. (1977a) for GeSi alloys, and by Zinger et al. (1984) for quaternary compounds. A similar approach has been used even earlier by Broude and Rashba (1961) for a very successful interpretation of molecular exciton optical spectra.

More elaborate calculations of mixed crystals vibrational spectra were carried out with the help of several techniques: the virtual crystal approximation (Nordheim 1931), the cluster model (Verleur and Barker 1966), the coherent potential approximation (Taylor 1967) and the average T-matrix approach (Kamitohara and Taylor 1974). One can find the details in review papers by Elliott, Krumhansl and Leath (1974), Barker and Sievers (1975) and Belousov (1982).

In this section we consider again the qualitative physical parameters that define the vibrational spectra behaviour with the change of the fractional concentration in mixed crystals. As a simplification we assume only variation of the masses (as if we were concerned with isotopic substitution).

The crystal lattice Hamiltonian with substitutional isotope impurities has the form

$$H = \tfrac{1}{2}\sum_{\alpha l s} M_{ls}\left[\dot{u}^{\alpha}_{ls}\right]^2 + \tfrac{1}{2}\sum_{\substack{\alpha l s\\ \beta l' s'}} \Phi^{\alpha\beta}_{ls;l's'} u^{\alpha}_{ls} u^{\beta}_{l's'}, \tag{5.1}$$

where $\boldsymbol{u}_{ls}$ is the ls atom displacement, $\Phi^{\alpha\beta}_{ls;l's'}$ is the force constant. The atomic mass in the ls site is

$$M_{ls} = M^0_s(1 - \epsilon c_{ls}), \tag{5.2}$$

where

$$\epsilon = \frac{M_s - M_{s'}}{M_s}. \tag{5.3}$$

The quantity c_{ls} represents the random distribution of isotopes

$$c_{ls} = \begin{cases} 1 & \text{if the } ls \text{ site is occupied} \\ & \text{by isotope impurity A} \\ 0 & \text{if the } ls \text{ site is occupied} \\ & \text{by the host atom B.} \end{cases} \tag{5.4}$$

The corresponding equations of motion are the following

$$\sum_{\beta l' s'} \left\{ M_{ls} \delta_{\alpha\beta} \delta_{ll'} \delta_{ss'} \omega^2 - \Phi^{\alpha\beta}_{ls;l's'} \right\} u^{\beta}_{l's'} = 0. \tag{5.5}$$

Substituting (5.2) into (5.5) one obtains

$$\sum_{\beta l' s'} \left\{ M_s \omega^2 \delta_{\alpha\beta} \delta_{ll'} \delta_{ss'} - \Phi^{\alpha\beta}_{ls;l's'} - V_{ls;l's'} \delta_{\alpha\beta} \right\} u^{\beta}_{l's'} = 0. \tag{5.6}$$

The matrix

$$V_{ls;l's'} = M_s \epsilon \omega^2 c_{ls} \delta_{ss'} \delta_{ll'} \tag{5.7}$$

is equivalent to the local impurity potential (3.7) and eq. (5.6) is an analogue of Schrödinger equation (2.1). We shall show that the composition dependence of the frequency distribution function is governed by the same universal parameter as that which defines the behaviour of the electron density of states.

We start with the case of low concentrations of the isotope component. When the isotope impurities are introduced in the crystal the phonons are scattered by the impurity potential (5.7). The T-matrix of the scattering has the following diagrammatic representation

$$T(\boldsymbol{q}, \omega) = \cdot + \cdot\!\!-\!\!\!-\!\!\cdot + \cdot\!\!-\!\!\!-\!\!\cdot\!\!-\!\!\!-\!\!\cdot + \cdots. \tag{5.8}$$

Here the solid line corresponds to the Green function of the $\boldsymbol{qj}$ phonon of the ideal crystal of B-type

$$D^0_{ls;l's'}(\boldsymbol{q}) = \frac{1}{Nr\sqrt{M_s M_{s'}}} \sum_{\boldsymbol{q}j} \frac{\boldsymbol{e}_s \cdot \boldsymbol{e}_{s'}}{\left(\omega^2 - \omega_j^2(\boldsymbol{q}) \right)} \mathrm{e}^{\mathrm{i}\boldsymbol{q}(\boldsymbol{l}-\boldsymbol{l}')}, \tag{5.9}$$

where $\omega_j^2(\boldsymbol{q})$ is the host lattice phonon frequency, $(\boldsymbol{q}j)$ are the phonon wave number and phonon branch, and $\boldsymbol{e}_s$ and $\boldsymbol{e}_{s'}$ are unit vectors of corresponding vibrations. The dot on the diagram (5.8) represents potential (5.7), and r is the number of atoms in the the unit cell.

Physical quantities such as the absorption coefficient or the light scattering cross section are directly connected with the T-matrix (Elliott et al. 1974).

Therefore expansion (5.8) can be averaged over the random distribution of impurities (Leath 1970). The pair correlation function has the form

$$\langle c_{ls} c_{l's'} \rangle = x\delta_{ll'}\delta_{ss'} + x^2(1-\delta_{ll'}\delta_{ss'}). \tag{5.10}$$

The isotope impurities may or may not give rise to the localized vibration near the impurity atoms in the lattice. The summation of the diagrams corresponding to the multiple scattering by a single impurity in expansion (5.8) leads to the scattering amplitude in the form

$$f(\boldsymbol{q}, \omega) = \frac{\epsilon\omega^2}{3Nr} \sum_{\boldsymbol{q}j} \frac{1}{\omega^2 - \omega_j^2(\boldsymbol{q})} \times \left[1 - \frac{\epsilon\omega^2}{3Nr} \sum_{\boldsymbol{q}j} \frac{1}{\omega^2 - \omega_j^2(\boldsymbol{q})}\right]^{-1}. \tag{5.11}$$

The scattering amplitude has a singularity when

$$1 - \frac{\epsilon\omega^2}{3Nr} \sum_{\boldsymbol{q}j} \frac{1}{\omega^2 - \omega_j^2(\boldsymbol{q})} = 0 \tag{5.12}$$

which corresponds to the frequency of the local vibration of the impurity atom.

We consider the effect of the impurities on the frequency distribution function of the ideal lattice near the edge of the phonon spectrum which is generated by optical phonons. For simplicity's sake we take the upper optical branch in the form

$$\omega^2(\boldsymbol{q}) = \omega_0^2 - \eta(aq)^2, \tag{5.13}$$

where η characterises the dispersion of the optical branch (see fig. 6).

Substituting eq. (5.13) in eq. (5.12) leads to an equation for defining the local frequency (Lifshitz 1942, 1943)

$$1 - \epsilon\omega^2 \int \mathrm{d}\omega'^2 \frac{g_0(\omega'^2)}{\omega^2 - \omega'^2} = 0. \tag{5.14}$$

Here $g_0(\omega^2) \approx \text{const}\sqrt{\omega_0^2 - \omega^2}$ is the distribution function of the ideal crystal near the phonon spectrum edge ω_0^2. It follows from eq. (5.14) that the local frequency at $\omega^2 > \omega_0^2$ occurs only in cases of a light isotope when ϵ from (5.3)

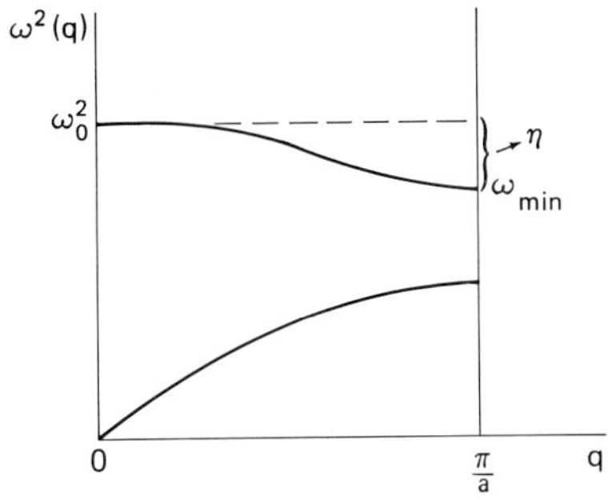

Fig. 6. Model of the optical branch used for estimations.

is positive. What is more, ϵ for a three-dimensional crystal should be larger than the critical value ϵ_{cr}

$$\epsilon > \epsilon_{cr} = \left\{ \int d^3q \, a^3 \frac{\omega_0^2}{\omega_0^2 - \omega_j^2(\boldsymbol{q})} \right\}^{-1} \approx \frac{\eta}{\omega_0^2}. \tag{5.15}$$

Condition (5.15) is similar to the electron localization criterion (2.5).

Near the edge ($\omega^2 \to \omega_0^2$) the scattering amplitude (5.11) does not depend on the phonon frequency and transforms into the scattering length R_ℓ

$$R_\ell = \frac{\epsilon/\epsilon_{cr}}{1 - \epsilon/\epsilon_{cr}} a. \tag{5.16}$$

If there is no local mode of the individual impurity atom ($\epsilon < \epsilon_{cr}$) it follows from (5.16) that $R_\ell < a$. The average separation is always larger than a. Therefore

$$\bar{R} \gg R_\ell. \tag{5.17}$$

Condition (5.17) is similar to condition (2.11) as far as the electron density of states is concerned.

Phonons are scattered, besides, by fluctuations of the mixed crystal composition. This scattering leads, first, to the shift of the edge of the continuous spectrum caused by a change in the average atomic mass with the increase in concentration of impurities. The frequency shift is proportional to ϵ:

$$\Delta\omega_j^2(\boldsymbol{q}) = x\epsilon\omega_0^2. \tag{5.18}$$

Secondly, the phonon scattering leads to small damping Γ of plane waves corresponding to phonons. The scattering is elastic and the probability of scattering is proportional to the distribution function of the final state

$$g_0(\omega^2) \sim \sqrt{\omega^2 - \omega_0^2 - x\epsilon\omega_0^2}$$

(see, e.g., Krivoglas 1967)

$$2\omega(\boldsymbol{q})\Gamma(\boldsymbol{q}) = x(1-x)\epsilon^2 a^3 \omega_0^4 g_0(\omega^2). \tag{5.19}$$

The frequency spectrum $g_0(\omega^2)$ decreases near the edge in the continuous spectrum. Therefore the damping (5.19) is small and the perturbation theory is applicable for calculating the phonon distribution function near the edge of the continuous spectrum. The analysis of the perturbation expansion made by Zinger et al. (1977b) allows the identification of a small dimensionless parameter for the expansion. The extraction of the parameter requires the special "renormalization" procedure in which the shift of the spectrum edge was eliminated in each order of perturbation theory. The dimensionless parameter appears to be

$$\frac{\left(\omega_0^2 - \tilde{\omega}^2\right)^{1/2}}{x(1-x)(\epsilon/\epsilon_{\rm cr})^{3/2}\left(\epsilon\omega_0^2\right)^{1/2}} \gg 1. \tag{5.20}$$

Here $\tilde{\omega}^2$ is the phonon frequency with reference to the true edge of the phonon continuous spectrum. Condition (5.20) is similar to condition (2.19) in the discussion of the electron properties in section 2. Condition (5.20) holds over a wide region of concentrations x and $(1-x)$ due to the small ratio $(\epsilon/\epsilon_{\rm cr}) < 1$.

The concentration dependence of the phonon frequency (5.18) represents the gradual transformation of $(\omega_{\rm LO}^2, \omega_{\rm TO}^2)$ frequencies of one of the end-members of the mixed crystal into $(\omega_{\rm LO}^2, \omega_{\rm TO}^2)$ frequencies of the second end-member. A dependence of this type is called one-mode behaviour. Many examples of one-mode crystals are discussed in the review paper by Chang and Mitra (1968). The schematic composition dependence of $\omega_{\rm LO}$ and $\omega_{\rm TO}$ frequencies for a one-mode mixed crystal is shown in fig. 7.

When $\tilde{\omega}^2 \to \omega_0^2$ condition (5.20) is violated. The phonon wavelength λ becomes comparable with the phonon mean free path l. The damping $\Gamma(\boldsymbol{q})$ in this limit is of the order of magnitude of the phonon frequency $(\omega_0 - \tilde{\omega}(\boldsymbol{q}))$. This is the collision controlled regime for phonons. Condition (5.20) means that the diffusion region is less than the separation from the edge $(\tilde{\omega}^2 - \omega_0^2)$.

When $\epsilon < \epsilon_{\rm cr}$ there is no phonon localization by the single impurity atom. At the same time the macroscopic fluctuations of the composition contribute local vibrations to the phonon distribution function. The optimum fluctuation

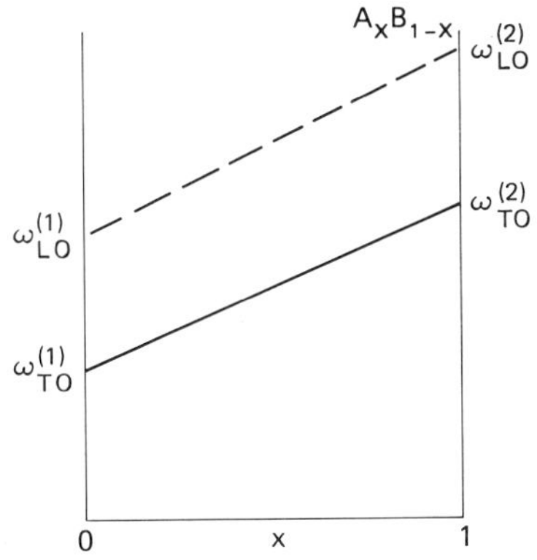

Fig. 7. Qualitative composition dependence of ω_{LO} and ω_{TO} frequencies of phonons for one-mode crystal.

method discussed in section 3 allows an estimate to be made of the corresponding contribution to the distribution function. The tail of the distribution function appears in the gap caused by localization of the phonon by fluctuation of the composition. The total broadening in the direction of the gap is defined by the maximum frequency shift corresponding to the most probable fluctuation.

To estimate the broadening we consider the fluctuation with linear dimension R. The maximum frequency shift in the case of the optical phonon given in fig. 6 equals

$$\tilde{\omega}^2 - \omega_0^2 \equiv \Delta\omega^2 = \Delta x \epsilon \omega_0^2 - \eta a^2/R^2. \tag{5.21}$$

Here Δx is the fluctuation of the composition from eq. (1.3). The first term in (5.21) takes into account an average mass variation in the volume of fluctuation R^3. The second term shows that the maximum frequency shift corresponds to the phonon wave number which is of the order of $1/R$. In other words the wavelength of the local mode cannot exceed the reciprocal linear size of the fluctuation R.

Estimating $(\Delta x)^2$ from (5.21) and substituting into (3.6) one obtains the probability density W in the form

$$W \sim \exp\left\{ -\left(\frac{\Delta\omega^2 + \eta a^2/R^2}{\epsilon\omega_0^2} \right)^2 \frac{1}{2x(1-x)(1/nR^3)} \right\}. \tag{5.22}$$

The radius of the optimum fluctuation R_{opt} is obtained from the condition $W = W_{max}$

$$R_{opt} \approx \sqrt{\eta/\Delta\omega^2}\, a, \tag{5.23}$$

The substitution of (5.23) into (5.22) leads to

$$W \sim \exp\left\{-\left(\frac{\Delta\omega^2}{x^2(1-x)^2(\epsilon\omega_0^2)^4/n^2a^6\eta^3}\right)^{1/2}\right\}. \tag{5.24}$$

Estimating ϵ_{cr} from (5.15) and taking into account that $na^3 = 1$ one can find the probability (5.24) in the form

$$W \sim \exp\left\{-\left(\frac{\Delta\omega^2}{x^2(1-x)^2(\epsilon/\epsilon_{cr})^3\epsilon\omega_0^2}\right)^{1/2}\right\}. \tag{5.25}$$

The tail of the frequency distribution function is defined by the probability (5.25). The characteristic frequency of the tail is the same as in (5.20). It is the universal parameter (5.20) which defines the behaviour of the frequency spectrum in both regions $\tilde{\omega}^2 < \omega_0^2$ and $\tilde{\omega}^2 > \omega_0^2$. The existence of the universal parameter follows from the possibility of introducing the dimensionless variables in the lattice dynamics equations (5.6) and the correlation function (5.10). The procedure is similar to (3.16) in discussing the electron density of states in section 3. The phonon damping (5.19) and the exponential tail (5.25) enable us to conclude that the line shape of mixed crystal optical spectra is asymmetric.

It follows from eq. (5.16) that at $\epsilon > \epsilon_{cr}$ the multiple phonon scattering by an isolated impurity centre leads to the creation of the local vibration of the single impurity centre. Parameter (5.20) is replaced by another parameter

$$\left(\frac{\tilde{\omega}^2 - \omega_0^2}{x^2(1-x)^2\epsilon_{cr}\omega_0^2}\right)^{1/2} \gg 1, \tag{5.26}$$

which holds in the low-concentration limits ($x \to 0$ or $(1-x) \to 0$). In the case of intermediate concentration $x \sim 0.5$ the mean free path of the phonon is comparable with the phonon wavelength. There is no analytic solution of the problem in this case.

Nevertheless there are reasons to believe that parameters (5.20) and (5.26) in fact contain a small numerical factor such as the number $1/178$ in eq. (3.14). This numerical factor could reduce the distribution function tails (5.25) and the phonon broadening (5.19). It should enhance resolved features in optical spectra.

The increase of concentration x leads to the overlap of local modes. As a result the new frequency band appears. The composition dependence of optical spectra is of the many-mode type. The qualitative composition dependence of three-mode and two-mode optical phonon frequencies are shown in figs. 8 and 9, respectively.

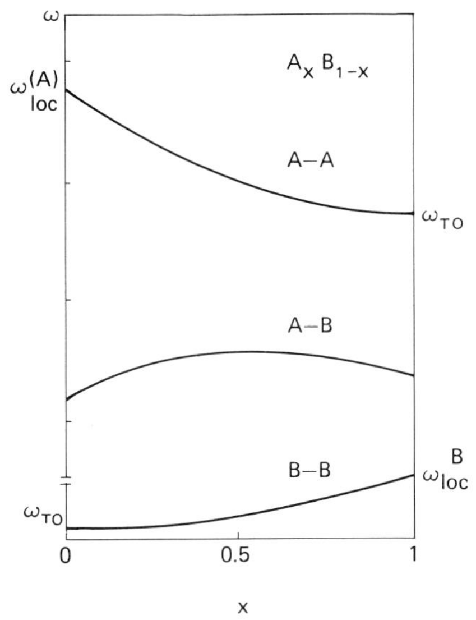

Fig. 8. Qualitative composition dependence of the phonon spectrum for a three-mode crystal.

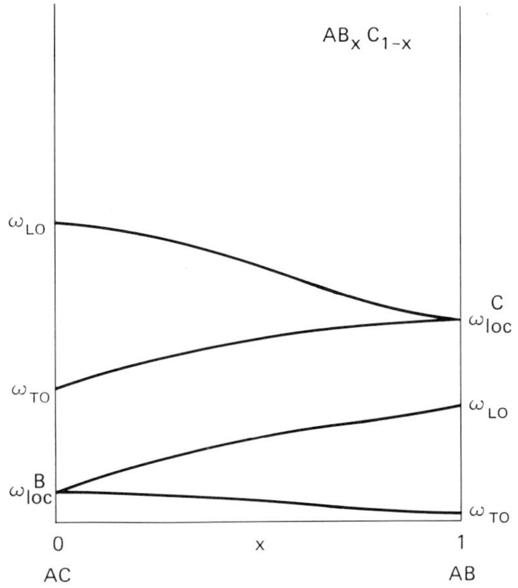

Fig. 9. Qualitative composition dependence of the phonon spectrum for a two-mode crystal.

6. *Uniaxial and biaxial mixed crystals*

So far we have considered the density of electron states and the phonon distribution function for crystals with optical isotropy and with short-range forces. But many mixed crystals with low symmetry have also been studied; e.g. ZnS_xSe_{1-x} in the hexagonal phase, $Hg_2Br_{2x}Cl_{2(1-x)}$, mixed crystals of sheelites such as $PbMo_{1-x}W_xO_4$, and so on. We shall show now that the phonon distribution function and the electron density of states for low-symmetric crystals behave differently.

It is well known that the refractive index of low-symmetry crystals depends on the direction of electromagnetic wave propagation. There are, for example, ordinary and extraordinary electromagnetic waves in uniaxial crystals. Similar directional dependence has been shown to exist (Pekar 1958) for any dipole-active excitation in low-symmetry crystals (polar optical phonon, electronic exciton, plasmon). These excitations are accompanied by a long-range electrical field. In cubic and tetrahedral crystals this field leads to the splitting of the excitation energy into two components corresponding to longitudinal and transverse excitations (e.g., LO–TO splitting of the optical phonon branch). In uniaxial crystals this splitting depends on the direction of propagation of the excitation.

It was shown by Pekar (1958) that near the upper edge of the phonon spectrum in a uniaxial crystal the optical phonon frequency $\omega^2(\boldsymbol{q})$ depends on the angle θ between the wave vector $\boldsymbol{q}$ and the optical axis of the crystal

$$\omega^2(\boldsymbol{q}) = \omega_{\mathrm{LO}}''^2 - \eta(aq)^2 - \gamma \sin^2\theta. \tag{6.1}$$

The dependence of the phonon frequency on θ near the lowest edge of the phonon spectrum has the form

$$\omega^2(\boldsymbol{q}) = \omega_{\mathrm{TO}}''^2 + \eta(aq)^2 + \gamma \cos^2\theta. \tag{6.2}$$

Here ω_{LO}'' and ω_{TO}'' are phonon frequencies of vibrations polarized along the optical axis of the crystal. The second term in eqs. (6.1) and (6.2) represents the usual dispersion over the Brillouin zone (see eq. (5.13)). The third term contains the directional dependence on θ.

The directional dispersion of the optical phonon frequency affects the frequency distribution function $g(\omega^2)$ in an essential way. We shall estimate $g(\omega^2)$ for phonon branch (6.1) near the edge $\omega_{\mathrm{LO}}''^2$ where phonons with $\theta \approx 0, \pi$ contribute mainly to $g(\omega^2)$. If $\theta = 0$ the dependence (6.1) transforms into

$$\omega^2(\boldsymbol{q}) \approx \omega_{\mathrm{LO}}''^2 - \eta(aq)^2 - \gamma\theta^2. \tag{6.3}$$

The definition of the distribution function for a uniaxial crystal is

$$g(\omega^2) = \frac{1}{(2\pi)^3} \int \delta(\omega^2 - \omega^2(\boldsymbol{q})) 2\pi q^2 \, \mathrm{d}q \sin\theta \, \mathrm{d}\theta. \tag{6.4}$$

Since near the edge $\sin\theta \approx \theta$ the differential $2\pi q^2 \, \mathrm{d}q \sin\theta \, \mathrm{d}\theta$ transforms into

$$2\pi q^2 \, \mathrm{d}q \, \theta \, \mathrm{d}\theta, \tag{6.5}$$

which has quasi-five-dimensional form. Substituting eq. (6.1) in eq. (6.4) and taking the integral one obtains (Rashba 1962)

$$g(\omega^2) \approx \frac{1}{12\pi^2 \eta^{3/2} \gamma a^3} \left(\omega_{\mathrm{LO}}''^2 - \omega^2 \right)^{3/2}. \tag{6.6}$$

The frequency dependence of $g(\omega^2)$ from eq. (6.6) corresponds to the distribution function in the space with "effective" dimensionality $d = 5$. One can show in a similar way that near the edge $\omega_{\mathrm{TO}}''^2$ from (6.2) the frequency dependence of $g(\omega^2) \sim (\omega - \omega_{\mathrm{TO}}''^2)^1$ corresponds to the space with $d = 4$.

The concept of increasing dimensionality was advanced by Larkin and Khmel'nitskii (1969) when applied to the theory of phase transitions in uniaxial ferroelectrics. The order parameter fluctuations near the transition point were shown to be smaller than those in the three-dimensional case due to the "effective" dimensionality $d = 4$. For example, the heat capacity has logarithmic singularity in this case instead of a power dependency for $d = 3$. A similar situation for "effective" dimensionality $d = 5$ was considered by Levanyuk and Sobyanin (1970).

The suppression of fluctuations due to the directional dispersion changes the nature of localization by different defects: point substitutional defects (Shchukin 1985), planar stacking faults (Ipatova et al. 1985a, b).

The augmentation of the space dimensionality in low-symmetry crystals also manifests itself in the localization of excitations by fluctuations of composition in mixed crystals. The phonon localization in mixed crystals with directional dispersion was considered by Kusmartsev and Shchukin (1986a, b). They have shown that localized vibrational states which split from the edge of the optical branch (6.1) are formed from the continuous spectrum of phonons with $\tilde{\omega}^2 \approx \omega_{\mathrm{LO}}''^2$, predominantly from those phonons with $\sin^2\theta = \pi q_\rho^2/(q_\rho^2 + q_z^2) \ll 1$. Here q_z is the wave vector component along the optical axis and q_ρ is the wave vector component perpendicular to the optical axis. If $\sin\theta \approx \theta \approx 0$ one finds that $q_\rho \ll q_z$, and therefore vibrations are localized by composition fluctuations of pancake-shape with

$$\rho_0 \gg z_0 . \tag{6.7}$$

Here ρ_0 and z_0 are linear dimensions of the fluctuation. The maximum frequency shift for the phonon (6.1) equals

$$\Delta\omega^2 = \tilde{\omega}^2 - \omega''^2_{\mathrm{LO}} = \Delta x \epsilon \omega''^2_{\mathrm{LO}} - \eta a^2 \frac{1}{z_0^2} - \gamma \frac{z_0^2}{\rho_0^2} \gtrsim 0 . \tag{6.8}$$

The first term in (6.8) is similar to (5.21). It corresponds to the average mass variation in the volume of fluctuation V_0,

$$V_0 = \rho_0^2 z_0 . \tag{6.9}$$

The second and third terms in (6.8) describe the usual dispersion over the Brillouin zone and the directional dispersion term, respectively. The directional dispersion plays an essential role if, in spite of parameter (6.7), the second and third terms are of the same order of magnitude, that is

$$\eta a^2 / z_0^2 \approx \gamma z_0^2 / \rho_0^2 . \tag{6.10}$$

In the case of anisotropic fluctuations (6.7) the probability of Gaussian fluctuations (3.6) has the form

$$W \sim \exp\left\{ - \frac{(\Delta x)^2}{x(1-x)} \frac{V_0}{a^3} \right\} , \tag{6.11}$$

where the width of the Gaussian distribution is equal to

$$\sqrt{\langle (\Delta x)^2 \rangle} = \left(\frac{x(1-x) a^3}{V_0} \right)^{1/2} . \tag{6.12}$$

Estimating ρ_0^2 from (6.10) and substituting in (6.9) one obtains

$$V_0 = \frac{\gamma}{\eta a^2} z_0^5 . \tag{6.13}$$

Note that the volume of the fluctuation V_0 is proportional to the fifth order of linear dimension z_0, that is the volume is five dimensional.

Introducing eq. (6.13) into eq. (6.12) leads to

$$\sqrt{\langle(\Delta x)^2\rangle} = \left(\frac{x(1-x)\eta a^5}{\gamma z_0^5}\right)^{1/2} \approx \Delta x. \tag{6.14}$$

The substitution of eqs. (6.10) and (6.14) in eq. (6.8) gives the frequency shift in the form

$$\Delta\omega^2 = \epsilon\omega_{\mathrm{LO}}''^2\left(\frac{x(1-x)\eta a^5}{\gamma}\right)^{1/2}\frac{1}{z_0^{5/2}} - \eta\frac{a^2}{z_0^2}. \tag{6.15}$$

The first term in (6.15) is the "potential energy" of the phonon in the field of fluctuation. It is proportional to $1/z_0^{5/2}$. The second term is the "kinetic energy" of the phonon and is proportional to $1/z_0^2$. Figure 10a demonstrates qualitatively that there is no crossing of the curves corresponding to potential and kinetic energies and hence there is no localization of the phonon with directional dispersion by Gaussian fluctuations.

It was shown by Kusmartsev and Shchukin (1986b) that localization in low-symmetric crystals with directional dispersion occurs for Fermi fluctuations, i.e. macroscopic fluctuations with low probability, where $\Delta x \sim x \sim 1$, and the cluster is built entirely of atoms of one type in the mixed crystal. The probability of a Fermi fluctuation is described by a binomial distribution

$$W \sim \exp\left\{-\left[(x+\Delta x)\ln\left(\frac{x+\Delta x}{x}\right) + (1-x-\Delta x)\ln\left(\frac{1-x-\Delta x}{1-x}\right)\right]\frac{V_0}{a^3}\right\}. \tag{6.16}$$

The peculiarity of eq. (6.16) in the case of crystals with directional dispersion arises from the five-dimensional nature of the volume from eq. (6.13). The first localized level splits off when $\Delta\omega^2$ from eq. (6.8) is zero. To optimize the exponent in the binomial distribution (6.16) with respect to Δx one should express V_0 from eq. (6.13) in terms of Δx from eq. (6.8) and substitute V_0 into eq. (6.16). The optimization of W with respect to Δx leads to the conclusion that for optimum fluctuation $(\Delta x)_{\mathrm{opt}} \sim x \sim 1$ holds. The "kinetic energy" is then proportional to $1/z_0^2$ and the "potential energy" of fluctuation is a constant. Therefore there is always the intersection of the curves representing the potential and kinetic energies (see fig. 10b). Substituting $(\Delta x)_{\mathrm{opt}} \sim 1$ in

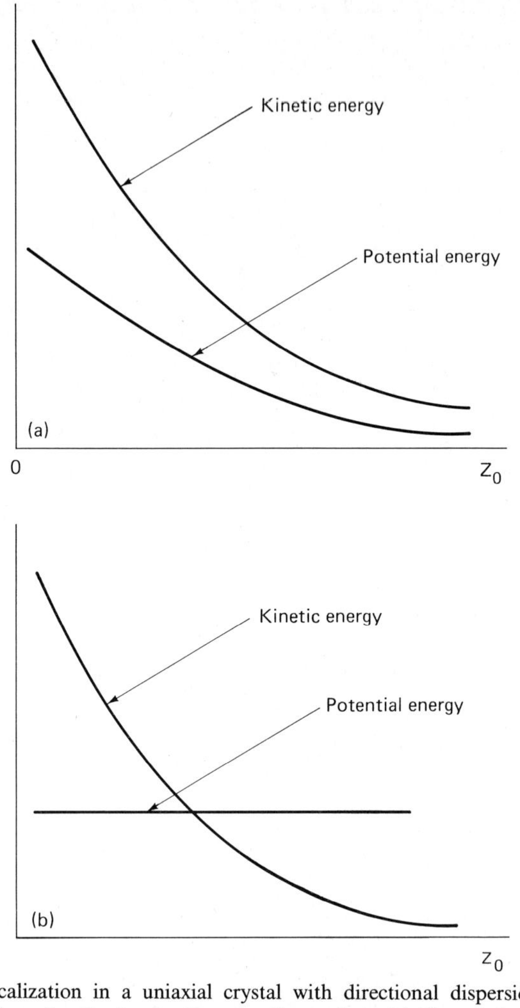

Fig. 10. Phonon localization in a uniaxial crystal with directional dispersion of the phonon spectrum. (a) There is no localization by Gaussian fluctuations. (b) There is localization by Fermi fluctuations.

binomial distribution (6.16) one obtains $W_{\max}$ and the corresponding phonon distribution function

$$g(\omega^2) \sim W_{\max} \sim \exp\left\{ -\frac{\eta^{3/2}\gamma}{\left(\epsilon\omega_{\rm LO}''^2\right)^{5/2}} \left[f_1(x) + f_2(x) \frac{\Delta\omega^2}{\epsilon\omega_{\rm LO}''^2} \right] \right\}. \tag{6.17}$$

Here $f_1(x)$ and $f_2(x)$ are functions of composition x. They are obtained from a numerical solution of equations with the exact distribution of mixed crystal components within the optimum fluctuation.

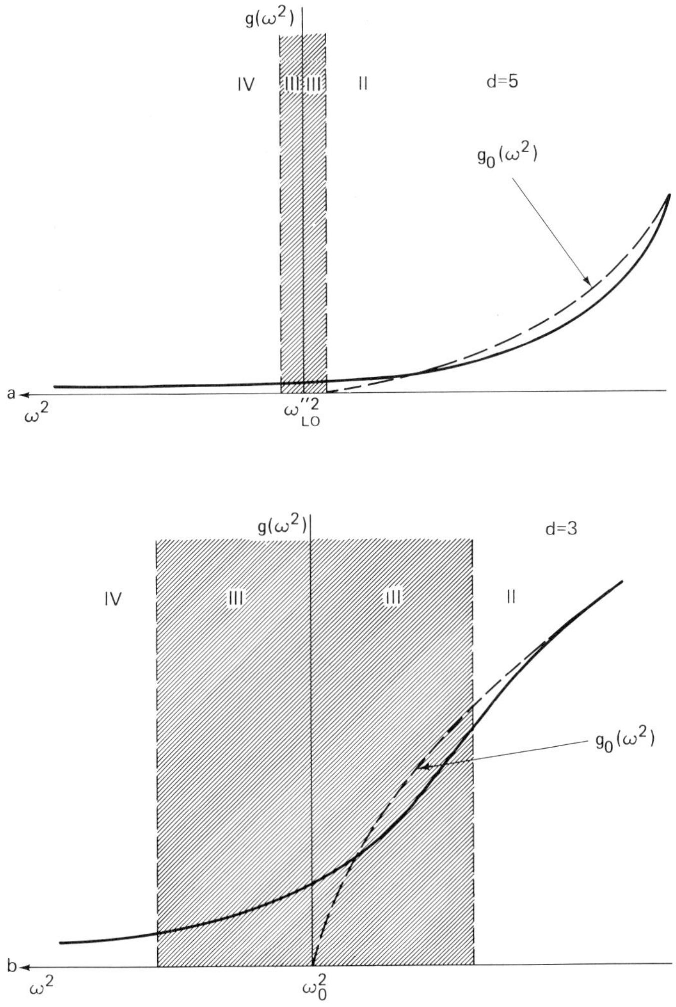

Fig. 11. Schematic comparison of phonon distribution functions for isotropic and uniaxial crystals. (a) The diffusion region is exponentially narrow for localization by Fermi fluctuations. (b) The diffusion region is relatively wide for localization by Gaussian fluctuations.

The most important peculiarity of eq. (6.17) consists of the presence of the frequency independent factor

$$\exp(-S_0) = \exp\left[-\frac{\eta^{3/2}\gamma}{\left(\epsilon\omega''^2_{\mathrm{LO}}\right)^{5/2}} f_1(x)\right]. \tag{6.18}$$

Since $\epsilon < 1$ one has $S_0 > 1$, which means that the localization is strongly suppressed with respect to the usual three-dimensional phonon distribution function corresponding to eq. (5.25). The phonon distribution function for mixed crystals with directional dispersion is therefore exponentially small near the edge of the continuous spectrum. The difference between the phonon distribution functions for fluctuations in 3d and 5d is shown in fig. 11.

Considering a continuous spectrum of phonon states similar to that in section 2 one should find the phonon damping in the second order perturbation theory from eq. (5,19) and put in it the quasi-five-dimensional $g(\omega^2)$ from eq. (6.6). Since in this case $g(\omega^2)$ is proportional to $(\omega''^2_{\mathrm{LO}} - \tilde{\omega}^2)^{3/2}$ the phonon damping $2\omega''_{\mathrm{LO}}\Gamma$ near the edge always remains less then the frequency shift $\Delta\omega^2 = |\omega''^2_{\mathrm{LO}} - \omega^2|$. Analysis of the next order terms of the perturbation theory results in the damping Γ which is weak at all the frequencies up to the edge ω''^2_{LO}. There is no convergence parameter like (5.20).

To find the effect of the impurities on the distribution function of the continuous spectrum we take into account quasilocal vibrations created by Fermi fluctuations in the region of the continuous spectrum. In this case an extra possibility of the phonon resonant scattering appears. An order of magnitude estimate shows that the phonon damping is proportional to the concentration of these fluctuations, i.e., to $\mathrm{e}^{-S_0(x)}$. The corresponding phonon distribution function at $\omega^2 \to \omega''^2_{\mathrm{LO}}$ approaches the edge according to $\mathrm{e}^{-S_0(x)}$ (see region II in fig. 11b). The characteristic separation from the spectrum edge where the phonon mean free path l is of the order of the phonon wavelength λ is also proportional to $\mathrm{e}^{-S_0(x)}$. Therefore the diffusion region in this case appears to be exponentially narrow and its width is proportional to $\mathrm{e}^{-S_0(x)}$. As a result the phonon damping and the width of the spectral lines depend exponentially on the composition of the mixed crystal. A sharp dependence of the phonon damping on composition x, obtained from IR spectra, was reported by Nagiev et al. (1984) in $Pb(MoO_4)_{1-x}(WO_4)_x$.

References

Alferov, Zh.I., E.L. Portnoy and A.A. Rogachev, 1968, Fiz. Tekhn. Poluprovodn. **2**, 1194 [Sov. Phys.-Semicond. **2**, 1001].

Alferov, Zh.I., D.Z. Garbuzov, Yu.V. Zhilyaev and P.S. Kop'ev, 1972, Fiz. Tekhn. Poluprovodn. **6**, 682 [Sov. Phys.-Semicond. **6**, 589].

Anderson, P.W., 1958, Phys. Rev. **109**, 1492.

Anselm, A.I., 1978, Introduction in the Theory of Semiconductors (Nauka, Moscow).

Baranovskii, S.D., and A.L. Efros, 1978, Fiz. Tekhn. Poluprovodn. **12**, 1233 [Sov. Phys.-Semicond. **12**, 1328].

Barker, A.S., and A.J. Sievers, 1975, Rev. Mod. Phys. **47**, S1.

Belousov, M.V., 1982, Vibrational Frenkel Excitons, in: Excitons, eds E.I. Rashba and M.D. Sturge (North-Holland, Amsterdam) ch. 18, pp. 771–807.

Broude, V.L., and E.I. Rashba, 1961, Fiz. Tverd. Tela **3**, 1941 [1962, Sov. Phys.-Solid State **3**, 1415].

Chang, I.F., and S.S. Mitra, 1968, Phys. Rev. **172**, 924.

Cohen, E., and M.D. Sturge, 1982, Phys. Rev. B **25**, 3828.

Efros, A.L., 1978, Usp. Fiz. Nauk **126**, 383 [Sov. Phys.-Usp. **21**].

Ehrenreich, H., and D. Schwartz, 1976, The Electronic Structure of Alloys, in: Solid State Physics, Vol. 31, eds H. Ehrenreich, F. Seitz and D. Turnbull (Academic Press, New York).

Elliott, R.J., J.A. Krumhansl and P.L. Leath, 1974, Rev. Mod. Phys. **46**, 465.

Evtichiev, V.G., D.Z. Garbuzov and A.G. Gorelenok, 1983, Fiz. Tekhn. Poluprovodn. **17**, 1402 [Sov. Phys.-Semicond. **17**, 891].

Goede, O., L. John and D. Henning, 1978, Phys. Status Solidi b **89**, K183.

Halperin, B.I., and M. Lax, 1966, Phys. Rev. **148**, 722.

Halperin, B.I., and M. Lax, 1967, Phys. Rev. **153**, 802.

Ipatova, I.P., A.V. Subashiev and V.A. Shchukin, 1985a, Zh. Eksp. & Teor. Fiz. **88**, 1263 [Sov. Phys.-JETP **61**, 746].

Ipatova, I.P., A.V. Subashiev and V.A. Shchukin, 1985b, Fiz. Tverd. Tela **27**, 1017 [Sov. Phys.-Solid State **27**, 618].

Kamitohara, W.A., and D.W. Taylor, 1974, Phys. Rev. B **10**, 1190.

Krivoglas, M.A., 1967, Theory of X-Rays and Heat Neutrons Scattering in Real Crystals (Nauka, Moscow).

Kusmartsev, F.V., and V.A. Shchukin, 1986a, Pis'ma v Zh. Eksp. & Teor. Fiz. **43**, 126 [Sov. Phys.-JETP Lett. **43**, 161].

Kusmartsev, F.V., and V.A. Shchukin, 1986b, Fiz. Tverd. Tela **28**, 1552 [Sov. Phys.-Solid State **28**, 879].

Landau, L.D., and E.M. Lifshitz, 1974, Quantum Mechanics (Nauka, Moscow).

Landau, L.D., and E.M. Lifshitz, 1976, Statistical Physics (Nauka, Moscow) p. 363.

Larkin, A.I., and D.E. Khmel'nitskii, 1969, Zh. Eksp. & Teor. Fiz. **56**, 2087 [Sov. Phys.-JETP **29**, 1123].

Leath, P.L., 1970, Phys. Rev. **132**, 3078.

Levanyuk, A.P., and A.A. Sobyanin, 1970, Pis'ma v Zh. Eksp. & Teor. Fiz. **11**, 540 [Sov. Phys.-JETP Lett. **11**, 371].

Li, S.S., D.W. Shoenfeld and R.T. Ouen, 1976, Phys. Status Solidi a **34**, 255.

Lifshitz, I.M., 1942, Zh. Eksp. & Teor. Fiz. **12**, 117.

Lifshitz, I.M., 1943, J. Phys. (USSR) **7**, 215.

Lifshitz, I.M., 1967, Zh. Eksp. & Teor. Fiz. **53**, 743 [1968, Sov. Phys.-JETP **26**, 462].

Lifshitz, I.M., S.A. Gredeskul and L.A. Pastur, 1976, Fiz. Nizk. Temp. **9**, 1093 [Sov. J. Low Temp. Phys. **2**, 533].

Lifshitz, I.M., A.S. Gredeskul and L.A. Pastur, 1982, Introduction to the Theory of Disordered Systems (Nauka, Moscow).

Mach, R., P. Flögl, L.G. Suslina, A.G. Areshkin, J. Maegle and G. Voight, 1982a, Phys. Status Solidi b **109**, 607.

Mach, R., L.G. Suslina and A.G. Areshkin, 1982b, Fiz. Tekhn. Poluprovodn. **16**, 649 [Sov. Phys.-Semicond. **16**, 418].

Nagiev, V.M., Sh.M. Efendiev and V.M. Burlakov, 1984, Phys. Status Solidi b **125**, 467.

Nelson, R.J., 1982, Excitons in Semiconductor Alloys, in: Excitons, eds E.I. Rashba and M.D. Sturge (North-Holland, Amsterdam) ch. 8, pp. 319–348.
Nelson, R.J., and N. Holonyak, 1976, J. Phys. & Chem. Solids **37**, 629.
Nordheim, L., 1931, Ann. Phys. (Leipzig) **9**, 607.
Onton, A., and P.J. Chikotka, 1971, Phys. Rev. B **4**, 1847.
Pekar, S.I., 1958, Zh. Eksp. & Teor. Fiz. **35**, 522 [1959, Sov. Phys.-JETP **8**, 360].
Permogorov, S.A., A.N. Resnitskii, V. Travnikov, S.Yu. Verbin and V.G. Lysenko, 1983a, Solid State Commun. **47**, 5.
Permogorov, S.A., A.N. Resnitskii, S.Yu. Verbin and V.G. Lysenko, 1983b, Pis'ma v Zh. Eksp. & Teor. Fiz. **37**, 390 [Sov. Phys.-JETP Lett. **37**, 462].
Petukhov, V.B., V.L. Pokrovskii and A.V. Çhaplik, 1967, Zh. Eksp. & Teor. Fiz. **53**, 1150 [1968, Sov. Phys.-JETP **26**, 678].
Pikhtin, A.N., 1977, Fiz. Tekhn. Poluprovodn. **11**, 425 [Sov. Phys.-Semicond. **11**, 245].
Rashba, E.I., 1962, Fiz. Tverd. Tela **4**, 3301 [1963, Sov. Phys.-Solid State **4**, 2417].
Shchukin, V.A., 1985, Fiz. Tverd. Tela **27**, 3406 [Sov. Phys.-Solid State **27**, 2050].
Shklovskii, B.I., and A.L. Efros, 1979, Electronic Properties of Doped Semiconductors (Nauka, Moscow).
Shui Lai, and M.V. Klein, 1980, Phys. Rev. Lett. **44**, 1087.
Suslina, L.D., A.G. Plyukhin, D.L. Fedorov and A.G. Areshkin, 1978, Fiz. Tekhn. Poluprovodn. **12**, 2238 [Sov. Phys.-Semicond. **12**, 1331].
Suslina, L.G., A.G. Plyukhin, O. Goede and D. Henning, 1979, Phys. Status Solidi b **4**, K185.
Taylor, D.W., 1967, Phys. Rev. **156**, 1017.
Tzarenkov, B.V., Ya.G. Akperov, A.H. Imenkov and Yu.G. Yakovlev, 1972, Fiz. Tekhn. Poluprovodn. **6**, 677 [Sov. Phys.-Semicond. **6**, 682].
Verleur, H.W., and A.S. Barker, 1966, Phys. Rev. **149**, 715.
Wo Hoang Thay, and V.K. Miloslavskii, 1983, Fiz. Tverd. Tela **25**, 3234 [Sov. Phys.-Solid State **25**, 1864].
Zinger, G.M., I.P. Ipatova and A.V. Subashiev, 1976, Fiz. Tekhn. Poluprovodn. **10**, 479 [Sov. Phys.-Semicond. **10**, 286].
Zinger, G.M., I.P. Ipatova and A.V. Subashiev, 1977a, Fiz. Tekhn. Poluprovodn. **11**, 656 [Sov. Phys.-Semicond. **11**, 383].
Zinger, G.M., I.P. Ipatova and A.V. Subashiev, 1977b, Fiz. Tverd. Tela **19**, 1258 [Sov. Phys.-Solid State **19**, 1322].
Zinger, G.M., I.P. Ipatova and A.I. Riskin, 1984, Fiz. Tekhn. Poluprovodn. **18**, 24 [Sov. Phys.-Semicond. **18**, 13].
Zittartz, J., and J.S. Langer, 1966, Phys. Rev. **148**, 741.

CHAPTER 2

Phonon Response Theory and the Infrared and Raman experiments

D.W. TAYLOR *

Physics Department
McMaster University
Hamilton, Ontario, Canada

* Supported by the Natural Sciences and Engineering Research Council of Canada.

Optical Properties of Mixed Crystals
Edited by
R.J. Elliott and I.P. Ipatova

Contents

1. Introduction 38
2 General theory 42
2.1. Pure crystal theory 42
2.2. Infrared absorption 45
2.3. Raman scattering 49
2.4. Green functions and correlation functions 52
3. Disorder theory 54
3.1. Multiple scattering theory 54
3.2. Low concentration Green function approximations 55
3.3. Large concentration Green function approximations 59
3.4. Random element isodisplacement models 65
3.5. Cluster isodisplacement models 71
3.6. Miscellaneous methods 75
4. One-/two-mode criteria 76
5. Mixed crystals systems 83
5.1. Mixed monatomic crystals 83
5.1.1. $Si_{1-c}Ge_c$ 83
5.1.2. Se/Te 86
5.1.3. $Bi_{1-c}Sb_c$ 86
5.2. Mixed I/VII crystals 86
5.2.1. Li systems 86
5.2.2. $NH_4Cl_{1-c}Br_c$ 87
5.2.3. Mixed alkali halides 88
5.2.3.1 Infrared reflectivity 88
5.2.3.2 Raman scattering 89
5.2.4. Cu halides 90
5.2.5. $AgCl_{1-c}Br_c$ 91
5.3. Mixed II/VI crystals 91
5.3.1. $Mg_{1-c}Zn_cS$ 91
5.3.2. $Mg_{1-c}Zn_cTe$ 92
5.3.3. $Mg_{1-c}Cd_cTe$ 92
5.3.4. $Zn_{1-c}Cd_cS$ 92
5.3.5. $Zn_{1-c}Cd_cSe$ 93

5.3.6. $Zn_{1-c}Cd_cTe$ 93
5.3.7. $Zn_{1-c}Hg_cTe$ 93
5.3.8. $ZnS_{1-c}Se_c$ 93
5.3.9. $ZnSe_{1-c}Te_c$ 94
5.3.10. $CdS_{1-c}Se_c$ 94
5.3.11. $CdS_{1-c}Te_c$ 95
5.3.12. $CdSe_{1-c}Te_c$ 95
5.3.13. $Cd_{1-c}Hg_cTe$ 95
5.3.14. $Mn_{1-c}Zn_cTe$ 97
5.3.15. $Mn_{1-c}Cd_cTe$ 97
5.3.16. $Mn_{1-c}Hg_cTe$ 98
5.4. Mixed III/V crystals 98
5.4.1. $Al_{1-c}Ga_cP$ 98
5.4.2. $Al_{1-c}Ga_cAs$ 99
5.4.3. $Al_{1-c}Ga_cSb$ 100
5.4.4. $Ga_{1-c}In_cP$ 101
5.4.5. $Ga_{1-c}In_cAs$ 104
5.4.6. $Ga_{1-c}In_cSb$ 104
5.4.7. $GaP_{1-c}As_c$ 105
5.4.8. $GaAs_{1-c}Sb_c$ 105
5.4.9. $InP_{1-c}As_c$ 106
5.4.10. $InAs_{1-c}Sb_c$ 106
5.4.11. $Al_{1-x}Ga_xP_yAs_{1-y}$ 106
5.4.12. $Ga_{1-x}In_xP_{1-y}As_y$ 106
5.5. Miscellaneous mixed crystals 107
5.5.1. Co/NiO 107
5.5.2. $M(H_c/F_{1-c})_2$, M = Ca, Sr, Ba 108
5.5.3. Ca/SrF_2, Sr/BaF_2 108
5.5.4. Ca/PbF_2 109
5.5.5. Sr/CdF_2 109
5.5.6. KMg/NiF_2 109
5.5.7. $Ge_{1-c}(GaSb)_c$ 109
5.5.8. PbSe/Te 109
5.5.9. Zr, $Hf(S/Se)_3$ 110
5.6. Layer crystals 111
5.6.1. GaS/Se 111
5.6.2. $Sn(S/Se)_2$ 113
References 113
Notes added in proof 121
References added in proof 129

1. *Introduction*

The study of the lattice dynamics of crystals by measuring their infrared reflectivity (IR) and Raman scattering (RS) spectra has long been a popular and productive exercise. Even though such experiments measure just the small wavevector (relative to the Brillouin zone size) response of the crystal, much useful information can be obtained. First-order processes allow just the study of the zone centre phonons but two-phonon processes couple the photons to any combination of phonons that has zero wavevector. The result then resembles a convolution of the phonon density of states, with the resulting peaks often corresponding to combinations of phonons from symmetry points.

For mixed crystals, in which the arrangement of the ions on the lattice sites is disordered, the situation, at least in principle, changes drastically. No longer are the phonon eigenstates labelled by a wavevector so each eigenstate may now have a nonzero projection onto zero wavevector. Hence, subject to restrictions imposed by the nature of the coupling of the phonons to the photons, a response can be expected over a wide range of frequencies even in a first-order process. In fact, as experiment clearly shows, the vibrational states are often reasonably well characterised by a wavevector. In the simplest of cases, as the concentration of components changes, the optical frequencies just move smoothly, often linearly, between those of the end member crystals. This is called one-mode behaviour which we will label as type I. An example is shown in fig. 1, being the behaviour of the reststrahlen band of Zn/CdS.

Two-mode behaviour, or type II, is a more interesting situation. It arises from the localised impurity modes in a crystal. If the resulting perturbation due to the substitution of an impurity atom is sufficiently strong (mass and/or force constant changes sufficiently large) vibrational states can split off the host crystal phonon bands to produce nonpropagating, and hence local, modes. (In the special case where the states split off into a gap they are called gap modes.) These effects have been studied very extensively both experimentally and theoretically. As the basic theory is straightforward, even though there may be problems with the details of the perturbations, much has been learned from such studies about both the impurities and the host crystals. General reviews of this area have been given by Barker and Sievers (1975),

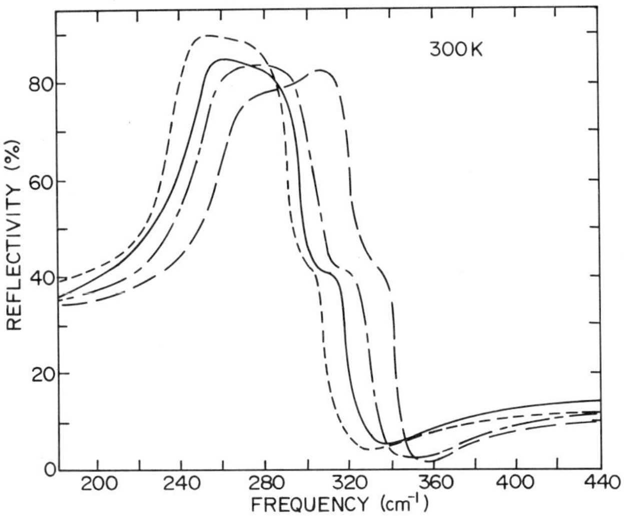

Fig. 1. Type I example: infrared reflectivity of $Zn_{1-c}Cd_cS$; (— — —) $c = 0.12$, (- - - -) $c = 0.45$, (———) $c = 0.64$, (- - - - - -) $c = 0.81$ (Lucovsky et al. 1967).

Taylor (1975) and Bilz et al. (1984). Examples of more specialised reviews are those of Newman (1973) and Farge and Fontana (1979).

However, our present interest is in what happens when the concentration of the impurities is increased all the way across the concentration range. As a convenient reference we will consider a mixture of crystals AB and AC (i.e., AB/C or $AB_{1-c}C_c$) although the discussion is by no means restricted to mixed binary materials. As described above, at very small c a local mode of frequency ω_I may form and in a cubic crystal it will be degenerate. However, as more C atoms are added the electric fields associated with these modes will cause a splitting between the longitudinal and transverse components (Maradudin and Oitmaa 1969). What is usually observed experimentally, as c increases, is that this splitting increases and the spectral peaks progress smoothly with increasing strength to become those seen in pure AC. The peaks are well defined, albeit with some disorder width, indicating that there are modes in the crystal that are dominantly of zero wavevector. It almost seems as though the long wave nature of the probe performs a kind of averaging. The consequence for the IR is that there are two reststrahlen bands approximately corresponding to those of AB and AC. The AB band decreases and narrows as c increases with the AC band doing the opposite. ZnS/Se is an example of this case and is shown in figs. 2 and 3.

There is a variant of type II called partly two-mode (II′) in which case the strength of the low-frequency impurity mode becomes very small well before the dilute limit. This appears to occur when the impurity mode in this limit is

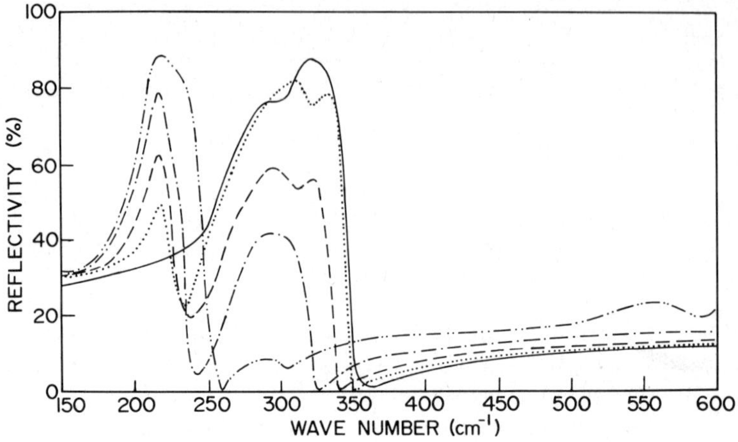

Fig. 2. Type II example: infrared reflectivity of $ZnS_{1-c}Se_c$; (———) $c = 0$, (······) $c = 0.18$, (– – –) $c = 0.40$, (·–·–·) $c = 0.67$, (–···–) $c = 0.985$ (after Brafman et al. 1967).

a resonant mode, i.e., it is within a phonon band of AB. A less obvious situation predicted by the Random Element Isodisplacement (REI) model (section 3.4) occurs when one set of modes behaves in a one-mode manner (i.e., one branch is always dominant) and the other set in a two-mode manner. This is usually referred to as mixed mode (M). These different types of

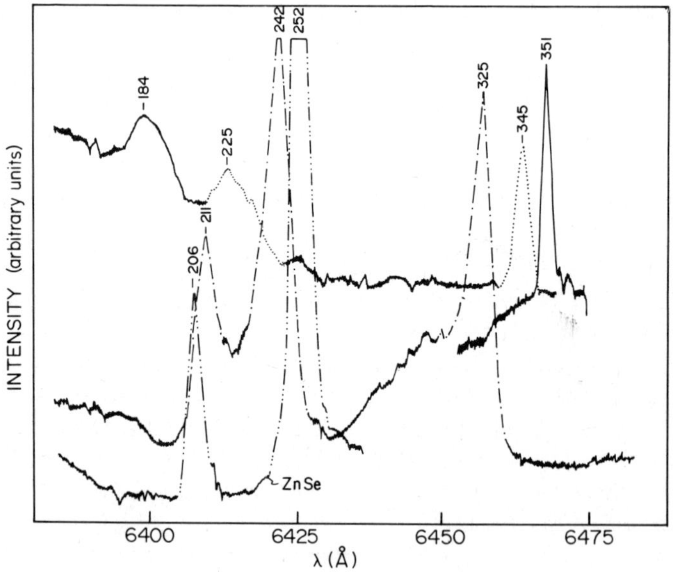

Fig. 3. Type II example: Raman scattering spectrum of $ZnS_{1-c}Se_c$; (–····–) $c = 1.0$, other values of c as in fig. 2 (after Brafman et al. 1967).

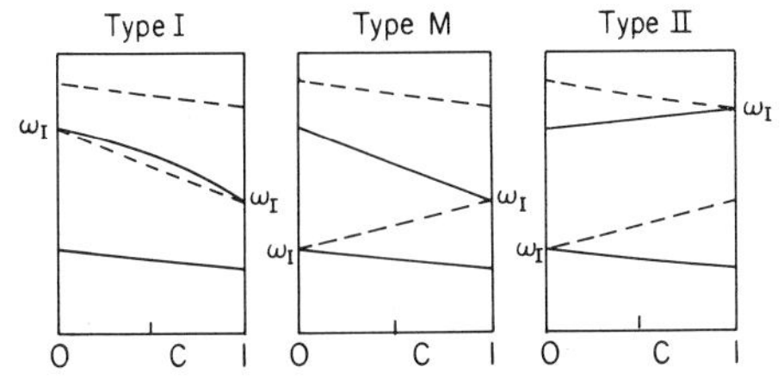

Fig. 4. Sketch of mode behaviours; (———) transverse, (- - - - - -) longitudinal.

behaviour are summarised in fig. 4. In some cases extra impurity modes may appear at finite concentrations and may be associated with clustering effects. They have been seen as weak contributions to the spectra of type I systems. When such impurity modes are particularly strong the behaviour is sometimes referred to as three-mode. The mixed monatomic system described in section 5.1 shows all such modes.

The above discussion has glossed over the effects of disorder on the coupling between the photon and the phonon. This is particularly significant in Raman scattering where, for instance, in alkali halides first-order scattering is forbidden by symmetry. However, in a disordered crystal a typical ion will not see a symmetric ordering of ion types, even as nearest neighbours, and so such a selection rule no longer holds. First-order RS is seen in impure alkali halides. A similar effect occurs in infrared absorption in Si/Ge alloys. Symmetry requires that there be no first-order dipole moment in Si, yet first-order absorption occurs in Si/Ge. This disorder effect is also significant in changing the coupling even when RS or IR is allowed. However, just as with the optic mode response the basic symmetry of the crystal appears to be retained. In pure crystals different symmetry modes can be selected in RS by using different polarisations and scattering geometries. Even at large concentrations these same geometries have strong selective properties in mixed crystals.

As suggested above there are many other modes with zero wavevector projections and they can be seen in favourable cases. Usually this requires that they are not obscured by two-phonon effects. Reflecting a response of all modes other than those of a strong optic nature, it is not surprising that their contributions resemble the host crystal phonon density of states (for relatively small c). Symmetry selection rules still seem to hold in that those peaks in the host crystal density of states which correspond to a given symmetry only appear when that symmetry is selected. These peaks are then attributed to the defect or disorder activation (DA) of these symmetry modes, which are often at the zone boundary.

Although two-phonon contributions can obscure first-order effects, they often contain useful information. When the origin of such peaks can be identified in the pure crystals and then followed as the concentration increases, the concentration dependence of the contributing phonons can be determined. This is often the only source of information about zone boundary phonons in mixed crystals.

In the above we have summarised the general themes connecting most optical studies of mixed crystals. In the remaining sections we will expand on the theories involved and then discuss in varying detail the experimental and theoretical results for the many systems investigated. However, there are a number of very complex mixed crystals for which it is hard to summarise their properties briefly other than to note their mode type. These are collected together, along with systems for which there is little experimental work, in tables 5 and 6 at the end of section 5. Omitted from this review are studies of mixed crystals where phase transitions (structural or ferroelectric) occur. Nor have we referenced resonant RS studies of mixed crystals unless they particularly refer to mode type or DA modes. It is intended that the references included are complete to mid 1984.

In closing this section it should be noted that earlier reviews of this topic have been given by Chang and Mitra (1971) and Barker and Sievers (1975). It is also discussed by Bilz et al. (1984).

2. General theory

2.1. Pure crystal theory

As the description of the properties of mixed crystals is couched in pure crystal terminology we first give some basic results.

The Hamiltonian for lattice vibrations has the form

$$H = \tfrac{1}{2}\sum_{l\kappa} \boldsymbol{p}(l\kappa)^2/M(\kappa) + \tfrac{1}{2}\sum_{\substack{l\kappa,\\ l'\kappa'}} \boldsymbol{u}(l\kappa)\cdot\boldsymbol{\Phi}(l\kappa, l'\kappa')\cdot\boldsymbol{u}(l'\kappa') \tag{2.1}$$

in the harmonic approximation. Here l, κ refer to the unit cell and site within the cell, respectively, and $\boldsymbol{u}$ and $\boldsymbol{p}$ are the displacements from equilibrium and the corresponding momenta. $\boldsymbol{\Phi}$ is the force constant matrix which in general includes both short-range overlap and long-range Coulomb forces. It must satisfy

$$\sum_{\substack{l\kappa,\\ l'\kappa'}} \boldsymbol{\Phi}(l\kappa, l'\kappa') = 0. \tag{2.2}$$

Due to lattice translational invariance the eigensolutions of this Hamiltonian are characterised by a wavevector $\boldsymbol{q}$ which is confined to the first

Brillouin zone. For each such $\boldsymbol{q}$, there are $3s$ different solutions, or modes $j = 1, 3s$, where s is the number of sites per unit cell ($\kappa = 1, s$). The eigenvalues are written as $\omega_j(\boldsymbol{q})$ and the eigenvectors as $e_\alpha(\kappa|\boldsymbol{q}j)$, the standard notation of Maradudin et al. (1971) being used.

Typical maximum eigenfrequencies are of the order of 10 THz but photons of such frequencies have wavevectors of magnitude 3×10^{-5} nm^{-1}, very much smaller than zone boundary wavevectors that have a magnitude of 10 nm^{-1}. Hence when the infrared absorption is due to a one-phonon process only those phonons of essentially zero $\boldsymbol{q}$ can be involved. Further, such absorption is due to just those phonon modes that can interact with photons. These are the transverse optic modes, although this term is often used to refer to all transverse modes that are not acoustic in nature. In polar crystals, the direction of $\boldsymbol{q}$ is retained as the electric field associated with the longitudinal modes splits otherwise degenerate eigenfrequencies into longitudinal optic (ω_{LO}) and transverse optic (ω_{TO}) frequencies. Similarly considerations apply to the inelastic scattering of light by a crystal when the energy change on scattering is due to the creation or destruction of an optic phonon (Raman scattering). As the wavevector of the incident light (from a laser) is typically 10^{-2} nm^{-1} corresponding to a frequency of 5×10^2 THz the wavevector of the scattered light will also have a similar magnitude. Consequently, once more, the phonon involved will have a value of $\boldsymbol{q}$ that is essentially zero.

In a diatomic cubic crystal, AB, the $\boldsymbol{q} = 0$ equation of motion for the relative displacement of the ions, $\boldsymbol{w}$, can be written as

$$\omega^2 \boldsymbol{w} + \mathrm{i}\omega\gamma \boldsymbol{w} - \omega_0^2 \boldsymbol{w} = -e^* \boldsymbol{E}^{\mathrm{L}}/M_{\mathrm{AB}}, \tag{2.3}$$

where e^* is the Szigeti effective charge, γ is a phenomenological damping frequency, $\boldsymbol{E}^{\mathrm{L}}$ is the local field felt by the ions and M_{AB} is the reduced mass given by

$$M_{\mathrm{AB}}^{-1} = M_{\mathrm{A}}^{-1} + M_{\mathrm{B}}^{-1}. \tag{2.4}$$

The short-range force constants give rise to ω_0^2 via

$$M_{\mathrm{AB}}\omega_0^2 = -2\sum_{l'} \Phi_{\alpha\alpha}^{\mathrm{s}}(l1, l'2).$$

Equation (2.3) can be rewritten as

$$\boldsymbol{w} = -g^{\mathrm{s}}(\omega) e^* \boldsymbol{E}^{\mathrm{L}}/M_{\mathrm{AB}},$$

where the Green function $g^{\mathrm{s}}(\omega)$ is given by

$$g^{\mathrm{s}}(\omega) = (\omega^2 - \omega_0^2 + \mathrm{i}\omega\gamma)^{-1}. \tag{2.5}$$

In these terms the polarisation is

$$\boldsymbol{P}=\left[\alpha-e^{*2}g^{\mathrm{s}}(\omega)/M_{\mathrm{AB}}\right]\boldsymbol{E}^{\mathrm{L}}/v,$$

α being the electronic polarisability of the ions in a unit cell of volume v. The local field can be replaced by the macroscopic field using the Lorentz relation to give

$$\boldsymbol{P}=\left[(\epsilon^{\infty}-1)/4\pi-(e^{\mathrm{T}})^{2}g^{\mathrm{T}}(\omega)/vM_{\mathrm{AB}}\right]\boldsymbol{E}, \tag{2.6}$$

where a new Green function has appeared

$$g^{\mathrm{T}}(\omega)=\left(\omega^{2}-\omega_{\mathrm{TO}}^{2}+\mathrm{i}\omega\gamma\right)^{-1}, \tag{2.7}$$

the local field having shifted ω_0 to ω_{TO},

$$\omega_{\mathrm{TO}}^{2}=\omega_{0}^{2}-4\pi e^{*}e^{\mathrm{T}}/3M_{\mathrm{AB}}v.$$

In the above α has been written in terms of ϵ^{∞} using the Clausius–Mosotti relation and e^{T} is the transverse charge

$$e^{\mathrm{T}}=(\epsilon^{\infty}+2)e^{*}/3.$$

In the absence of polariton effects, $\boldsymbol{E}$ is zero for transverse modes and is $-4\pi\boldsymbol{P}$ for longitudinal modes which gives for the longitudinal/transverse splitting (LT)

$$\omega_{\mathrm{LO}}^{2}-\omega_{\mathrm{TO}}^{2}=4\pi(e^{\mathrm{T}})^{2}/M_{\mathrm{AB}}v\epsilon^{\infty}. \tag{2.8}$$

These results can be conveniently summarised in terms of the dielectric constant which is readily obtained from (2.6):

$$\epsilon(\omega)=\omega_{\infty}-fg^{\mathrm{T}}(\omega), \tag{2.9}$$

where the oscillator strength f is given by

$$f=4\pi(e^{\mathrm{T}})^{2}/M_{\mathrm{AB}}v. \tag{2.10}$$

Then ω_{TO} is given by the maximum in $\omega\mathrm{Im}\epsilon(\omega)$ and ω_{LO} by the maximum in ω $\mathrm{Im}(1/\epsilon(\omega))$. For small damping these frequencies are given to a good approximation by the maximum and minimum of $|\epsilon(\omega)|$.

The above results are, of course, long established (Born and Huang 1954) but have been given here in a different form for later convenience.

2.2. *Infrared absorption*

The absorption of infrared light is determined by measuring the reflectivity, and sometimes the transmission coefficient, of a crystal. In either case the fundamental quantity is the complex, frequency-dependent dielectric constant $\epsilon(\omega)$ already introduced in eq. (2.9). This can be extracted from the experimental reflectivity either via a Kramers–Kronig analysis or by fitting the reflectivity using a dispersion formula for the dielectric constant.

In the first method, following the results of the previous section, the positions of the maxima in $\omega \mathrm{Im}\epsilon(\omega)$ can be taken to give the ω_{TO}'s, and the maxima in $\omega \mathrm{Im}(1/\epsilon(\omega))$ the ω_{LO}'s, for the sample under investigation. In disordered crystals there may be several such maxima, even in cubic systems, and much effort has been made to track their behaviour as a function of the concentration of the crystal constituents.

In the second method the dielectric constant is written in terms of several damped harmonic oscillators

$$\epsilon(\omega) = \epsilon^{\infty} - \sum_{j=1}^{n} f_j / \left(\omega^2 - \omega_j^2 + \mathrm{i}\gamma_j\omega\right)^{-1}, \tag{2.11}$$

where ω_j, γ_j, f_j are the oscillator frequency, damping and strength, respectively. These parameters are then used to fit the experimental data. It is quite common for experimental results to be summarised by the concentration dependence of these parameters, with the ω_{LO} determined as above. A review of these methods has been given by Balkanski (1980) who also describes how the transmission coefficient may be used in the analysis. Chang et al. (1968) have investigated a multimode version of the theory in section 2.1, that includes a coupling between the various longitudinal modes, all of which contribute to the polarisation and hence the electric field. They first analysed experimental results for two sets of disordered crystal systems (ZnS/Se, CdS/Se) finding good agreement between the above two methods using the maxima and minima of $|\epsilon(\omega)|$. They further used the fitted optical parameters in their multimode theory, solving the resulting secular equation for ω_{LO}, obtaining good agreement between these values of ω_{LO} and those obtained directly from $|\epsilon(\omega)|$.

In order to give a theory for the dielectric constant in mixed crystals the discussion in section 2.1 must be generalised. However, it will be seen that a full generalisation is difficult and has not yet been carried through.

As we are interested in substitutionally disorder crystals, not amorphous materials, a basic lattice structure is retained albeit with local relaxation from a strictly translationally invariant lattice. The disorder leads to the l dependence of the mass and the force constant matrix $\boldsymbol{\Phi}$ no longer has lattice translational invariance due to being dependent on what atoms happen to be at the $l\kappa$ and $l'\kappa'$ sites. However Φ must still satisfy relation (2.2).

As we are only interested in the $\boldsymbol{q}=0$ solution it is convenient to follow the route given for a pure crystal and introduce the Coulomb field via the local field. The explicit force constant matrix then refers to just short-range forces. The relevant interaction term due to the presence of the Coulomb field has the form

$$-\boldsymbol{\mu}\cdot\boldsymbol{E}^{\mathrm{L}}$$

in terms of a dipole moment operator $\boldsymbol{\mu}$ and the local field $\boldsymbol{E}^{\mathrm{L}}$.

In general the dipole moment $\boldsymbol{\mu}$ can be expanded in terms of the atomic displacements

$$\mu_\alpha = \mu^0{}_\alpha + \sum_{l\kappa\beta} \mu_{\alpha\beta}(l\kappa)\, u_\beta(l\kappa) + \ldots \tag{2.12}$$

to give an interaction term in the phonon Hamiltonian

$$H_{\mathrm{I}} = -\sum_{\substack{\alpha\beta\\ l\kappa}} u_{\alpha\beta}(l\kappa)\mu_\beta(l\kappa)E^{\mathrm{L}}_\alpha(l\kappa). \tag{2.13}$$

$\mu_{\alpha\beta}(l\kappa)$ can be considered to be an effective charge tensor that arises from both the ionic charge (in a polar crystal) and from atomic distortions due to wave function overlap. In general it will depend upon both the kind of atom at $l\kappa$ and on those at neighbouring sites. This can be seen in detail, for instance, either by noting that the effective charge in the shell model (see for instance Venkataram et al. 1975) depends upon the forces between neighbouring atoms, or from the origin of the effective charge in the bond orbital model (Harrison 1980). Usually, μ is replaced by an isotropic effective charge. In some of the simple cluster calculations it is possible to include at least its variation with atom type without much difficulty (see section 3.4). A consequence of retaining just the linear term in (2.12) is that we are ignoring two-phonon processes.

The complication in giving a full formal generalisation of the dielectric constant derivation lies in the inclusion of the variation of μ with atom type in the calculation of the local field. In the derivation below, this difficulty is ignored.

Following Bonneville (1980) the local field can be written in a Lorentz form as

$$\boldsymbol{E}^{\mathrm{L}} = \boldsymbol{E} + (4\pi/3)(\mathsf{\boldsymbol{L}}^{\mathrm{I}}\boldsymbol{P}^{\mathrm{I}} + \mathsf{\boldsymbol{L}}^{\mathrm{e}}\boldsymbol{P}^{\mathrm{e}}) \tag{2.14}$$

where $\boldsymbol{P}^{\mathrm{I}}$ and $\boldsymbol{P}^{\mathrm{e}}$ are the ionic and electronic polarisations, $\mathsf{\boldsymbol{L}}^{\mathrm{I}}$ is a geometric factor being the identity matrix in cubic crystals, and $\mathsf{\boldsymbol{L}}^{\mathrm{e}}$ includes any electronic delocalisation (it is equal to $\mathbf{L}^{\mathrm{I}}$ if the charge is fully localised on an ion). This is really an equation for the averaged quantities which should be a good approximation in that it is the $\boldsymbol{q} \to 0$ quantities that are eventually used. However, a full formalism should allow for the possibility that the local field could depend on the local environment of an ion. In the derivation of (2.14) in a cubic crystal the near field contribution is zero by symmetry, yet this would not be the case at specific sites in disordered crystals.

Noting that

$$\boldsymbol{P}^{\mathrm{e}} = \chi \boldsymbol{E}^{\mathrm{L}}/v, \tag{2.15}$$

where χ is the average high-frequency susceptibility per unit cell, the local field is given, after a little algebra, by

$$\boldsymbol{E}^{\mathrm{L}} = (\boldsymbol{I} - (4\pi/3v)\mathbf{L}^{\mathrm{e}}\chi)^{-1}(\boldsymbol{E} + (4\pi/3)\mathbf{L}^{\mathrm{I}}\boldsymbol{P}^{\mathrm{I}}). \tag{2.16}$$

As

$$\boldsymbol{P}^{\mathrm{I}} = (1/V)\sum_{l\kappa} \mu(\kappa)\boldsymbol{u}(l\kappa)$$

it is seen that the $\boldsymbol{P}^{\mathrm{I}}$ contribution to $\boldsymbol{E}^{\mathrm{L}}$ leads to a harmonic force constant term in H_{I} (2.13),

$$(4\pi/6V)\sum_{\substack{l\kappa,\\ l'\kappa'}} \boldsymbol{u}(l\kappa)\tilde{\mu}^{\mathrm{T}}(\kappa)\mathbf{L}^{\mathrm{I}}\mu(\kappa')\boldsymbol{u}(l'\kappa'). \tag{2.17}$$

A factor of $1/2$ has been introduced to avoid overcounting and $V = Nv$ is the sample volume. This is just the term that appears in the $\boldsymbol{q} \to 0$ Coulomb dynamical matrix for motion transverse to $\boldsymbol{q}$. In the above $\tilde{\mu}^{\mathrm{T}}$ is the transposed transverse effective charge matrix given by

$$\tilde{\mu}^{\mathrm{T}}(\kappa) = \tilde{\mu}(\kappa)(\boldsymbol{I} - (4\pi/3v)\mathbf{L}^{\mathrm{e}}\chi)^{-1}. \tag{2.18}$$

The interaction with the macroscopic field is

$$-\sum_{l\kappa} \boldsymbol{u}(l\kappa)\tilde{\mu}^{\mathrm{T}}(\kappa)\boldsymbol{E}$$

and applying linear response theory (Fetter and Waleka 1971)

$$\boldsymbol{P}^{\mathrm{I}} = -(1/V)\sum_{\substack{l\kappa,\\ l'\kappa'}} \mu(\kappa)\boldsymbol{G}^{\mathrm{T}}(l\kappa, l'\kappa'; \omega)\tilde{\mu}^{\mathrm{T}}(\kappa)\boldsymbol{E}. \tag{2.19}$$

$\boldsymbol{G}^{\mathrm{T}}(\omega)$ is a displacement/displacement Green function evaluated with (2.17) added to the short-range force constant matrix (see section 2.4 for the definition and calculation of such Green functions).

The dielectric constant can now be calculated using relations (2.15), (2.16) and (2.19) to write $\boldsymbol{P}$ in terms of $\boldsymbol{E}$. The final result is

$$\epsilon(\omega) = \epsilon^{\infty} - (4\pi/v)\sum_{\kappa\kappa'} \mu^{\mathrm{T}'}(\kappa)\boldsymbol{G}^{\mathrm{T}}(\kappa, \kappa', \omega)\tilde{\mu}^{\mathrm{T}}(\kappa'), \tag{2.20}$$

where

$$\boldsymbol{G}^{\mathrm{T}}(\kappa\kappa'; \omega) = \frac{1}{N}\sum_{ll'} \boldsymbol{G}^{\mathrm{T}}(l\kappa, l'\kappa'; \omega),$$

being the $\boldsymbol{q} = 0$ transform. $\mu^{\mathrm{T}'}(\kappa)$ is similar to $\tilde{\mu}^{\mathrm{T}}$ but is given by

$$\mu^{\mathrm{T}'}(\kappa) = (\boldsymbol{I} + (\epsilon^{\infty} - \boldsymbol{I})\boldsymbol{L}^{\mathrm{I}}/3)\mu(\kappa). \tag{2.21}$$

Equation (2.20) is the required generalisation of eq. (2.9).

In the cubic case these results are equivalent to those of Bonneville (1981) with $\mu^{\mathrm{T}'}$ reducing to the scalar equation,

$$\mu^{\mathrm{T}'} = (2 + \epsilon^{\infty})e^{*}/3$$

and

$$\mu^{\mathrm{T}} = e^{\mathrm{T}} = (3 + \boldsymbol{L}^{\mathrm{e}}(\epsilon^{\infty} - 1))e^{*}/3.$$

It is often convenient to transform to the basis spanned by the eigenstates of the pure crystal using $\langle \boldsymbol{q}j \mid l\kappa\alpha\rangle$ given by

$$\langle \boldsymbol{q}j \mid l\kappa\alpha\rangle = (1/\sqrt{N})e_{\alpha}(\kappa \mid \boldsymbol{q}j)^{*}\, \mathrm{e}^{-\mathrm{i}\boldsymbol{q}\cdot\boldsymbol{R}(l)}. \tag{2.22}$$

The dielectric constant is now given by

$$\epsilon_{\alpha\beta}(\omega) = \epsilon^{\infty}{}_{\alpha\beta} - (4\pi/v)\sum_{jj'} \mu_{\alpha}^{\mathrm{T}'}(j)G_{jj'}(\omega)\mu_{\beta}^{\mathrm{T}}(j'), \tag{2.23}$$

where

$$\mu_{\alpha}(j) = \sum_{\beta\kappa} \mu_{\alpha\beta}(\kappa)e_{\beta}(\kappa \mid \boldsymbol{q}j)/\left(\sqrt{M(\kappa)}\right)$$

and j refers to modes that have components transverse to $\boldsymbol{q}$, with $\boldsymbol{q} \to 0$.

The Green function in this representation is given by

$$G_{jj'}(\omega) = G_{jj'}(\boldsymbol{q} \to 0, \omega),$$

$$G_{jj'}(\boldsymbol{q}, \omega) = \sum_{\substack{l\kappa\alpha \\ l'\kappa'\beta}} \sqrt{M(\kappa)}\,\langle j\boldsymbol{q} \,|\, l\kappa\alpha\rangle G_{\alpha\beta}(l\kappa, l'\kappa'; \omega)\langle l'\kappa'\beta \,|\, j'\boldsymbol{q}\rangle\sqrt{M(\kappa')} \tag{2.24}$$

being a generalisation of the transformation to be used for the pure crystal Green function in section 2.4.

The absence of an l dependence of the effective charge tensor should be noted. It is physically appealing to restore this dependence in eq. (2.20) in spite of the fact that the above derivation ignores any such variation in its consideration of the local and macroscopic electric fields. This l dependence is essential in materials for which, by symmetry, the effective charge is zero in the pure crystal. The reduction in site symmetry due to nearby impurities leads to the possibility of nonzero dipole moments and hence optical absorption (see, for example, Si/Ge). In such cases there will again be long-range Coulomb fields with the resultant splitting between longitudinal and transverse type modes. At low concentrations and/or for weak effective charges, presumably this is not important and does not seem to have been discussed in the literature.

2.3. Raman scattering

Raman scattering involves the inelastic scattering of light from a crystal and we are interested in the case where the energy loss or gain is due to phonon creation or destruction. The general theory and experimental aspects have been described recently by Hayes and Loudon (1978), see also Pinczuk and Burstein (1975) and Poulet and Mathieu (1976).

The basic method uses second-order time-dependent scattering theory to obtain the inelastic scattering cross section for light of wavevector $\boldsymbol{k}$ and frequency ω_k scattering to wavevector $\boldsymbol{k}' = \boldsymbol{k} - \boldsymbol{q}$ and frequency $\omega_{k'} = \omega_k - \omega$. This cross section is given by

$$\mathrm{d}^2\sigma/\mathrm{d}\Omega'\,\mathrm{d}\omega_{k'} = (V^2/2\pi)(\omega_{k'}/c)^4(\omega_{k'}/\omega_k)\int \mathrm{d}t\; \mathrm{e}^{\mathrm{i}\omega t}\langle \delta\chi(t)\delta\chi\rangle.$$

Within the Born Oppenheimer and dipole approximations the matrix element of $\delta\chi$ between the initial and final states is given by

$$\langle f \,|\, \delta\chi \,|\, i\rangle = -(1/V)(\omega_k/\omega_{k'})\sum_{\alpha\beta}\langle \nu_f \,|\, a_{\alpha\beta}(\omega_k) \,|\, \nu_i\rangle \sigma'_\alpha \sigma_{\beta'}.$$

Here $\boldsymbol{\sigma}'$, $\boldsymbol{\sigma}$ are the polarisation vectors of the reflected and incident light and the polarisability tensor $\boldsymbol{a}(\omega)$ is evaluated between the initial and final phonon eigenstates. $\boldsymbol{a}(\omega)$ itself is given in terms of electronic states and operators by

$$\boldsymbol{a}(\omega) = (e^2/\hbar)\sum_n \left\{ \frac{\langle f|\boldsymbol{r}|n\rangle\langle n|\boldsymbol{r}|i\rangle}{\omega_{ni}-\omega} + \frac{\langle n|\boldsymbol{r}|i\rangle\langle f|\boldsymbol{r}|n\rangle}{\omega_{fn}+\omega} \right\}.$$

In the quasistatic approximation, $\omega_k \ll \omega_{ni}$, ω_{fn}, i.e., less than any electronic excitation energies, and $\boldsymbol{a}(\omega)$ is symmetric and independent of the frequency of the incident light. This is the situation for most of the experiments to be discussed in this chapter. Otherwise the frequency dependence of $\boldsymbol{a}(\omega)$ is important and the scattering process approaches resonant Raman scattering. Although such scattering has been performed on disordered crystals, the interest has been mainly on the electronic states involved and we will not discuss this theory further [see for instance, Ganguly and Birman (1968), Kleinert and Bechstedt (1978)]. However, in a few cases the resonant enhancement has been used to emphasise features such as DATA peaks.

The scattering due to phonons can be introduced by expanding $\boldsymbol{a}$ in terms of the phonon displacements $\boldsymbol{u}$

$$a_{\alpha\beta} = a^0_{\alpha\beta} + \sum_{l\kappa\gamma} a_{\alpha\beta,\gamma}(l\kappa)u_\gamma(l\kappa) + \ldots. \tag{2.25}$$

The final expression for the one-phonon scattering is then

$$\frac{\mathrm{d}^2\sigma}{\mathrm{d}\Omega\,\mathrm{d}\omega_{k'}} = (e^4/\hbar^2)(\omega_{k'}^3\omega_k/c^4)N\bar{S}(\omega), \tag{2.26}$$

where

$$\bar{S}(\omega) = (1/N)\sum_{\substack{l\kappa\alpha\\ l'\kappa'\beta}} \bar{a}_\alpha(l\kappa)S_{\alpha\beta}(l\kappa, l'\kappa'; \omega)\bar{a}_\beta(l'\kappa'),$$

$$\bar{a}_\alpha(l\kappa) = \sum_{\beta\gamma} \sigma'_\beta\sigma_\gamma a_{\beta\gamma,\alpha}(l\kappa).$$

The phonon contribution is contained in the correlation function

$$S_{\alpha\beta}(l\kappa, l'\kappa'; \omega) = (1/2\pi)\int \mathrm{d}t\ \mathrm{e}^{\mathrm{i}\omega t}\langle u_\alpha(l\kappa, t)u_\beta(l'\kappa')\rangle. \tag{2.27}$$

The relation of $\boldsymbol{S}$ to Green functions and the methods of calculating them are discussed in section 2.4. It will be noted that the velocity of light in free space occurs in eq. (2.26). Modifications to the cross section due to the fact that the scattering occurs inside a crystal but is measured outside, are discussed in, for instance, Hayes and Loudon (1978) and Pinczuk and Burstein (1975).

Just as with the dipole moments, μ, appearing in the infrared absorption theory, the third-order polarisation tensor $a_{\alpha\beta,\lambda}(l\kappa)$ depends on the atom at site $l\kappa$ and on the neighbouring atoms. In fact it is the reduced site symmetry due to the local environment that permits one-phonon Raman scattering in impure alkali halides. However, this variation in these tensors is difficult to include except in approximations that are strictly valid only for very dilute impure crystals. However, such approximations have been used at finite concentrations and will be discussed in section 3.2.

When these variations are ignored the result (2.26) can be transformed to the $(\boldsymbol{q}, j)$ representation, just as was done for the dielectric constant. The result is

$$\bar{S}(\omega) = \sum_{jj'} a_j S_{jj'}(\omega) a_{j'}$$

with

$$a_j = \sum_{\kappa\alpha} a_\alpha(\kappa) e_\alpha(\kappa|\boldsymbol{q}j)/\left(\sqrt{M(\kappa)}\right) \tag{2.28}$$

and $S_{jj'}(\omega)$ is defined in exactly the same way as $G_{jj'}(\omega)$. In this case all the optic branches are included in the sum but a distinction should be made between the longitudinal and transverse modes in polar crystals. This is because the macroscopic field associated with polar modes that have longitudinal components also affects the polarizability tensor $\boldsymbol{a}$ and relation (2.25) becomes

$$a_{\alpha\beta} = a^0{}_{\alpha\beta} + \sum_{l\kappa\gamma} a_{\alpha\beta,\gamma}(l\kappa) u_\gamma(l\kappa) + \sum_\gamma b_{\alpha\beta,\gamma} E_\gamma.$$

Following the standard route described, for instance, by Hayes and Loudon (1978), this leads to $a_\alpha(\kappa)$ being replaced by

$$\bar{a}_\alpha(\kappa) = \sum_{\beta\gamma} \sigma'_\beta \sigma_\gamma \left[a_{\beta\gamma,\alpha}(\kappa) - \sum_\lambda 4\pi b_{\beta\gamma,\lambda} \tau_\lambda \tau_\alpha e^*(\kappa)/\bar{\epsilon} \right],$$

where

$$\bar{\epsilon} = \sum_{\alpha\beta} \tau_\alpha \epsilon^\infty{}_{\alpha\beta} \tau_\beta, \qquad \tau_\alpha = q_\alpha/q.$$

For simplicity the effective charge tensor μ has been written as $\mu_{\alpha\beta}(\kappa)=\delta_{\alpha\beta}e^*(\kappa)$. The effect of the vectors τ is to select the longitudinal parts of S. This effect can be introduced via the local field (Ovander and Tyu 1979) as has been done by Martin (1975) within the REI model (section 3.4).

A consequence of retaining just the linear term in eq. (2.25) is that we are ignoring the effect of two-phonon processes. These are important in RS as their contributions often overlap the one-phonon region. However, they have not been treated by disorder theory.

2.4. Green functions and correlation functions

In the previous sections it has been shown that the phonon contributions are given either by a retarded Green function (infrared absorption) or a correlation function (Raman scattering).

This phonon Green function is defined by

$$G_{\alpha\beta}(l\kappa,\, l'\kappa';\, \omega)=\frac{1}{\hbar}\int_{\infty}\langle\langle u_\alpha(l\kappa,\, t);\, u_\beta(l'k')\rangle\rangle\, \mathrm{e}^{\mathrm{i}\omega t}\,\mathrm{d}t, \tag{2.29}$$

where

$$\langle\langle A(t);\, B\rangle\rangle=\mp\mathrm{i}\theta(\pm t)\langle[A(t),\, B]\rangle$$

is a retarded or advanced (upper or lower signs) Green function. It is related to the correlation function $S(\omega)$, (2.27), via the spectral representation

$$G_{\alpha\beta}(l\kappa,\, l'\kappa';\, z)=\frac{1}{\hbar}\int_{-\infty}^{\infty}\frac{(1-\mathrm{e}^{-\beta\omega})}{(z-\omega)}S_{\alpha\beta}(l\kappa,\, l'\kappa';\, \omega)\,\mathrm{d}\omega,$$

where $\beta=\hbar/k_{\mathrm{B}}T$. Continuation into the upper and lower half planes gives, respectively, the retarded and advanced Green functions. A more convenient relation is obtained from the discontinuity of the Green function across the real axis,

$$\mathbf{G}(\omega+\mathrm{i}0^+)-\mathbf{G}(\omega-\mathrm{i}0^+)=(2\pi\mathrm{i}/\hbar)(1-\mathrm{e}^{-\beta\omega})\mathbf{S}(\omega).$$

For clarity a super matrix notation is being used.

The equation of motion for the Green function (2.29) is obtained by using the Heisenberg equation of motion for $A(t)$. To obtain a closed form for the displacement Green function, this procedure has to be applied twice as $\mathrm{d}\boldsymbol{u}/\mathrm{d}t$

yields $\boldsymbol{p}$ [for details on much of the material in this section see Taylor (1975)]. The result is

$$-\boldsymbol{M}\omega^2\boldsymbol{G}(\omega) + \boldsymbol{\Phi}\boldsymbol{G}(\omega) + \boldsymbol{I} = 0 \tag{2.30}$$

or

$$\boldsymbol{G}(\omega)^{-1} = \boldsymbol{M}\omega^2 - \boldsymbol{\Phi}.$$

This is the familiar equation of motion for $\mathrm{d}^2\boldsymbol{u}/\mathrm{d}t^2$ dressed up in Green function language ($\boldsymbol{M}$ is a diagonal matrix)

$$\boldsymbol{M}\omega^2\boldsymbol{u} - \boldsymbol{\Phi}\boldsymbol{u} = 0. \tag{2.31}$$

The isodisplacement approximations to be discussed in sections 3.4 and 3.5 are, in fact, more easily derived using (2.31) rather than (2.30).

The pure crystal Green function $\boldsymbol{G}^0(\omega)$ satisfies the same form of equation as $\boldsymbol{G}(\omega)$ and the two are related by

$$\boldsymbol{G}(\omega) = \boldsymbol{G}^0(\omega) + \boldsymbol{G}^0(\omega)\boldsymbol{C}(\omega)\boldsymbol{G}(\omega), \tag{2.32}$$

where $\boldsymbol{C}(\omega)$ contains the changes due to disorder

$$\boldsymbol{C}(\omega) = (\boldsymbol{M}^0 - \boldsymbol{M})\omega^2 + \boldsymbol{\Phi} - \boldsymbol{\Phi}^0. \tag{2.33}$$

$\boldsymbol{G}^0(\omega)$ is readily calculated from the eigenvalues and eigenvectors (2.22) of the pure crystal dynamical matrix,

$$G^0_{\alpha\beta}(l\kappa, l'\kappa'; \omega) = (M(\kappa)M(\kappa'))^{-1/2} \sum_{\boldsymbol{q}j} \langle l\kappa\alpha|\boldsymbol{q}j\rangle^* G^0_j(\boldsymbol{q}, \omega) \times \langle \boldsymbol{q}j\,|\,l'\kappa'\beta\rangle \tag{2.34}$$

with

$$G^0_j(\boldsymbol{q}, \omega) = 1/\left(\omega^2 - \omega_j(\boldsymbol{q})^2\right).$$

The calculation usually proceeds by calculating the imaginary part of $\boldsymbol{G}^0$, which has the form of a density of states, summing over a large number of $\boldsymbol{q}$ points within the first Brillouin zone. A more recent technique uses special directions in the Brillouin zone (Prasad and Bansil 1980). The real part can then be calculated via a Hilbert transform

$$\mathrm{Re}\,\boldsymbol{G}^0(\omega) = -(2/\pi)\int_0^\infty \omega' \frac{\mathrm{Im}\,\boldsymbol{G}^0(\omega')\,\mathrm{d}\omega}{(\omega^2 - \omega'^2)}.$$

3. Disorder theory

3.1. Multiple scattering theory

The standard approach to solving equations such as (2.32) is to use multiple scattering techniques. These require that the perturbation matrix $\mathbf{C}(\omega)$, rather than spanning the whole crystal space, be written in terms of perturbation matrices for each impurity, $\mathbf{C}^i(\omega)$, that span only a small space around the impurity,

$$\mathbf{C}(\omega) = \sum_i \mathbf{C}^i(\omega). \tag{3.1}$$

In this case (2.32) can be written as a set of coupled equations

$$\mathbf{G}(\omega) = \mathbf{G}^0(\omega) + \mathbf{G}^0(\omega)\sum_i \mathbf{t}^i(\omega)\mathbf{G}^i(\omega),$$

$$\mathbf{G}^i(\omega) = \mathbf{G}^0(\omega) + \mathbf{G}^0(\omega)\sum_{j\neq i} \mathbf{t}^j(\omega)\mathbf{G}^j(\omega), \tag{3.2}$$

where $\mathbf{t}^i(\omega)$ is the t-matrix for the ith impurity

$$\mathbf{t}^i(\omega) = \mathbf{C}^i(\omega)\left[\mathbf{I} - \mathbf{G}^0(\omega)\mathbf{C}^i(\omega)\right]^{-1}. \tag{3.3}$$

An average over all the possible configurations of a mixed crystal of a given concentration of constituents can be written as

$$\langle\mathbf{G}(\omega)\rangle = \mathbf{G}^0(\omega) + \mathbf{G}^0(\omega)\sum_l \left\langle \mathbf{t}^l(\omega)\left\langle \mathbf{G}^l(\omega)\right\rangle_l\right\rangle,$$

$$\langle\mathbf{G}^l(\omega)\rangle_l = \mathbf{G}^0(\omega) + \mathbf{G}^0(\omega)\sum_{l'\neq l} \langle\mathbf{t}^{l'}(\omega)\langle\mathbf{G}^{l'}(\omega)\rangle_{ll'}. \tag{3.4}$$

The sum is now over all unit cells. The symbols $\langle\ldots\rangle_l$, $\langle\ldots\rangle_{ll'}$ indicate configuration averages conditional on what atoms are in the cell l or the cells l and l'. $\langle\ldots\rangle$ indicates either a full average or, in the summation, an average over the explicit conditional cells. The single site approximation replaces $\langle\ldots\rangle_{ll'}$ by $\langle\ldots\rangle_{l'}$, closing what would otherwise be an infinite sequence of equations with progressively more restricted averages. The result is

$$\langle\mathbf{G}(\omega)\rangle = \mathbf{G}^0(\omega) + \mathbf{G}^0(\omega)\Sigma(\omega)\langle\mathbf{G}(\omega)\rangle, \tag{3.5}$$

where the self-energy $\Sigma(\omega)$ is given by

$$\Sigma(\omega) = \sum_{l} \langle \boldsymbol{t}^{l}(\omega)\rangle [\boldsymbol{I} + \boldsymbol{G}^{0}(\omega)\langle \boldsymbol{t}^{l}(\omega)\rangle]^{-1}. \tag{3.6}$$

General reviews of the applications of the Green function methods to the vibrational properties of mixed crystals have been given by Elliott et al. (1974) and Elliott and Leath (1975). In sections 3.2 and 3.3 these methods are outlined within the present context of calculating the $\boldsymbol{q} = 0$ response of a mixed crystal.

3.2. Low concentration Green function approximations

The simplest application of (3.4) is at very low concentrations, replacing $\langle \boldsymbol{G}^{l}(\omega)\rangle_{l}$ by $\boldsymbol{G}^{0}(\omega)$ to give

$$\langle \boldsymbol{G}(\omega)\rangle = \boldsymbol{G}^{0}(\omega) + \boldsymbol{G}^{0}(\omega)\sum_{l} \langle \boldsymbol{t}^{l}(\omega)\rangle \boldsymbol{G}^{0}(\omega) \tag{3.7}$$

in which the correction to $\boldsymbol{G}^{0}(\omega)$ is linear in the concentration. The most dramatic effect due to impurities is readily seen in this equation and comes from poles in the t-matrix. In frequency regions where there are no host crystal phonons and hence Im $\boldsymbol{G}^{0}(\omega) = 0$, for some impurity force constants and masses the t-matrix can have a pole, leading to a nonzero contribution to Im$\langle \boldsymbol{G}(\omega)\rangle$. Hence infrared absorption or Raman scattering may occur in these frequency regions. Due to the nonpropagating nature of these vibrational excitations they are known as local or gap modes, depending on whether they appear above the maximum phonon frequency of the host crystal or in a region between two phonon bands. Further Im $\boldsymbol{t}(\omega)$ will be nonzero throughout the host phonon bands leading to infrared absorption and/or Raman scattering at other than the discrete frequencies characteristic of a pure crystal. This can reflect the density of states through $\boldsymbol{G}^{0}(\omega)$ or impurity resonant modes due to resonant peaks in the t-matrix. This is an explicit demonstration of the effects described in section 1.

There has been an enormous effort expended by many groups on studying such modes, both experimentally and theoretically, in a wide variety of crystals. Most of the studies deal essentially with one impurity in the crystal and it is not the point of this review to discuss this area. Recent surveys are noted in section 1. However, the presence, or not, of local modes is an important consideration in the one-/two-mode question. That will be discussed in section 4.

Further, eq. (3.7) has been used both for one impurity, $l = 0$ say, and for finite concentrations. When force constant changes are included $\boldsymbol{C}^{l}$ and hence

$\boldsymbol{t}^l$ can be rather large matrices. However, for isolated impurities the point symmetry of the crystal is retained and the use of group theory leads to considerable simplifications. For instance, in the zincblende structure, the point symmetry is T_d and for force constant changes to just nearest neighbours, $\boldsymbol{C}^l$ is a 15×15 matrix. This representation decomposes into the irreducible representations.

$$A_1 + E + F_1 + 3F_2$$

so that the largest matrix to be inverted is of size 3×3. This model was first discussed by Grim et al. (1972) using a simple one-parameter scaling for the force constant changes.

In the $(\boldsymbol{q}, j)$ representation the Green function (3.7) becomes ($\boldsymbol{q} \to 0$ to be understood, same convention as 2.24)

$$G_{jj'}(\omega) = G^0{}_j(\omega)\delta_{jj'} + cG^0{}_j(\omega)t_{jj'}(\omega)G^0{}_{j'}(\omega), \tag{3.8}$$

with

$$t_{jj'}(\omega) = \sum_{\substack{\Gamma\mu \\ ab}} (\boldsymbol{q}j \,|\, \Gamma a\mu) t_{ab}(\Gamma, \omega)(\Gamma b\mu \,|\boldsymbol{q}j').$$

In the above Γ labels the representation, μ the row, and $a = 1, \ldots, c_\Gamma$ where c_Γ is the number of times the representation Γ appears. The transformation matrix is given by

$$(\boldsymbol{q}j \,|\, \Gamma_{a\mu}) = \sum_{s\kappa\alpha} \langle \boldsymbol{q}j \,|\, s\kappa\alpha \rangle \psi^{\Gamma a\mu}{}_\alpha(s\kappa) / \left(\sqrt{M(\kappa)}\right),$$

where (s, κ) are confined to the space of one impurity. The transformation ψ is discussed at length by Maradudin et al. (1971) and a number of them are tabulated by Maradudin (1965) (see also Agrawal 1969). For zincblende the $\boldsymbol{q} = 0$ modes have just F_2 symmetry but are infrared and Raman active. Hence, only the F_2 representation contributes to $t_{jj'}(\omega)$. However, experimental Raman scattering indicates other symmetries must be present and these can only come from changes in the polarisation tensors $\boldsymbol{a}$.

In this case a configuration average has to be taken of the function

$$\sum_{\substack{l\kappa\alpha \\ l'\kappa'\beta}} \left[\bar{a}_\alpha(\kappa) + \Delta\bar{a}_\alpha(l\kappa)\right] G_{\alpha\beta}(l\kappa, l'\kappa'; \omega)\left[\bar{a}_\beta(\kappa') + \Delta\bar{a}_\beta(l'\kappa')\right].$$

Once again retaining just those terms linear in the concentration, this becomes

$$\sum_{\substack{l\kappa\alpha\\ l'\kappa'\beta}} \bar{a}_\alpha(\kappa)\big\langle G_{\alpha\beta}(l\kappa,\, l'\kappa;\, \omega)\big\rangle \bar{a}_\beta(\kappa')$$

$$+\sum_{l}\sum_{\substack{s\kappa\alpha\\ s'\kappa'\beta}} \Delta\bar{a}_\alpha(s\kappa)\big\langle G_{\alpha\beta}(s\kappa,\, s'\kappa';\, \omega)\big\rangle_l \Delta\bar{a}_\beta(s'\kappa') \tag{3.9}$$

plus a cross term between $\bar{a}$ and $\Delta\bar{a}$ (Pershan and Lacina 1969). Here $(s\kappa)$, $(s'\kappa')$ are confined to the defect space of the impurity in cell l. The first term is just the case already considered above with $\langle \boldsymbol{G}\rangle$ replaced by (3.7). In the second term the conditionally averaged Green function to first order in the concentration is given by

$$\langle \boldsymbol{G}(\omega)\rangle_l = \boldsymbol{G}^{(1)}(\omega) = c\boldsymbol{G}^0(\omega)\left[\boldsymbol{I} - \boldsymbol{C}^l(\omega)\boldsymbol{G}^0(\omega)\right]^{-1}$$

(see for instance Taylor 1975) evaluated in the defect space of an impurity in the lth cell. Transforming to the irreducible representations gives for the second term of (3.9)

$$Nc\sum_{\substack{\Gamma\lambda\\ ab}} \Delta a(\Gamma a\lambda)G^{(1)}{}_{ab}(\Gamma;\, \omega)\Delta a(\Gamma b\lambda), \tag{3.10}$$

where

$$\Delta a(\Gamma a\lambda) = \sum_{s\kappa a} \Delta\bar{a}_\alpha(s\kappa)\psi^{\Gamma a\lambda}{}_\alpha(s\kappa).$$

For the zincblende example Δa is nonzero for the A_1, E and F_2 representations.

The original Raman calculations are those of Xinh et al. (1965) and Benedek and Nardelli (1967a) for impurities in alkali halides. In this case symmetry precludes Raman scattering in the pure crystal and it is only the term in Δa that contributes. There have been several calculations for impurities in zincblende crystals, all based on Talwar et al. (1980) with the exception of Krol et al. (1978) who only calculated the A_1 contribution. In the form of the F_2 representation chosen by Talwar et al. the third row ($\lambda = 3$) corresponds to a rigid translation of the impurity plus nearest neighbour cage. They claim this should be Raman inactive, but this does not seem valid when it is realised that this motion is embedded in a crystal. Indeed on evaluating (3.10) the present author found that all three rows are Raman active. Further, all calculations in the literature ignore the contribution of the first term in (3.9).

This is correct for the rock salt calculation of Xinh et al. (1965) on which Talwar et al. (1980) based their work, but not so for the zincblende structure.

A more satisfactory low concentration approximation is obtained by replacing $\langle \boldsymbol{G}^{l'}(\omega)\rangle_{ll'}$ by $\langle \boldsymbol{G}^{l'}(\omega)\rangle_{l'}$. The result is

$$\langle \boldsymbol{G}(\omega)\rangle = \boldsymbol{G}^0(\omega) + \boldsymbol{G}^0(\omega)\sum_l \Sigma^l(\omega)\langle \boldsymbol{G}(\omega)\rangle,$$

with

$$\Sigma^l(\omega) = \boldsymbol{C}^l(\omega)[\boldsymbol{I} - (1-c)\boldsymbol{G}^0(\omega)\boldsymbol{C}^l(\omega)]^{-1} \tag{3.11}$$

being a slightly modified form of the t-matrix.

Again a transformation to $(\boldsymbol{q}, j)$ space is appropriate giving

$$G_{jj'}(\omega) = G^0_j(\omega)\delta_{jj'} + \sum_{j''} G^0_j(\omega)\Sigma_{jj''}(\omega)G_{j''j'}(\omega), \tag{3.12}$$

with $\Sigma_{jj'}(\omega)$ being defined in the same way as $t_{jj'}(\omega)$ in (3.8). This result immediately suggests the possibility of mode mixing due to the disorder. In fact even in the simplest impurity model, i.e., just a mass change (mass defect approximation), there is a mixing between acoustic and optic branches of the same symmetry (Vinogradov 1970). Using this approximation in a diatomic cubic structure eq. (3.12) can be solved by hand and the result is given by Taylor (1975).

There are two applications of this approximation to optical properties. Benedek and Nardelli (1967b) give in considerable detail the evaluation of eq. (3.12) using just changes in the mass and the nearest neighbour central force constant for impurities in the rock salt structure. The above mode mixing is also explicit in their work but they later dropped it as it finally leads to a correction of order c^2. Numerical results were presented for the infrared absorption but only for $c < 1\%$ as at that time there was little in the way of finite concentration experiments.

The other application was to Ca/SrF_2 by Lacina and Pershan (1970). In this system the $\boldsymbol{q} = 0$ modes for the pure crystal transform as F_{2g} (Raman) and F_{1u} (infrared). Assuming $\Delta\boldsymbol{a} = 0$ they calculated the appropriate Green functions, neglecting the mode mixing, using force constant changes between the metal ion and the nearest neighbour F^- ions and between these F^- ions. On fitting their parameters to experiment they found that $\Sigma(F_{2g}) \sim c\boldsymbol{C}(F_{2g})$, i.e. a mean crystal behaviour, in agreement with their experimental results. This was not the case for $\Sigma(F_{1u})$ which enters in the infrared absorption. Their results for a Sr concentration of $c = 0.25$ are shown in fig. 5 along with the results of Verleur and Barker (1967b). This low c calculation appears to give better

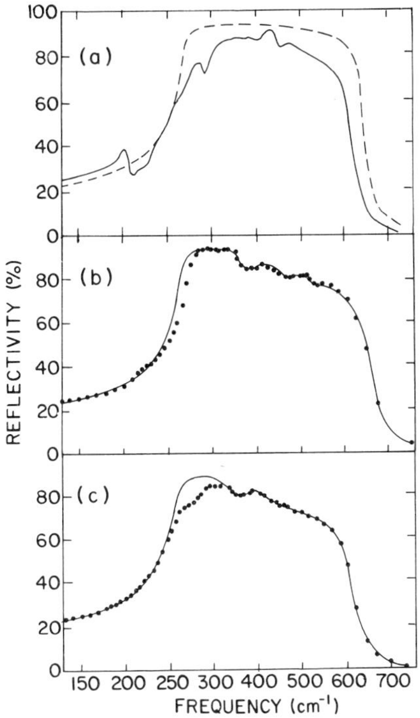

Fig. 5. Reflectivity of $Ca_{1-c}Sr_cF_2$; (a) Calculation (- - - - - -) $c = 0$, (———) $c = 0.25$; (b) experiment for $c = 0$; (c) experiment for $c = 0.25$, lines are oscillator fits (Verleur and Barker (1967b). Taken from Lacina and Pershan (1970).

agreement with experiment than the isodisplacement model used by Verleur and Barker (see section 3.5) although it does predict unobserved structure at low frequencies.

3.3. *Large concentration Green function approximations*

Perturbation theories usually work better, the closer the chosen unperturbed or reference system is to the system to be solved. With this in mind, and in order to obtain approximations valid at large concentrations a Green function for a reference system, $\bar{\boldsymbol{G}}(\omega)$, is introduced. This function is related by a self-energy $\boldsymbol{E}(\omega)$ to the original unperturbed Green's function $\boldsymbol{G}^0(\omega)$, i.e.

$$\bar{\boldsymbol{G}}(\omega) = \boldsymbol{G}^0(\omega) + \boldsymbol{G}^0(\omega)\boldsymbol{E}(\omega)\bar{\boldsymbol{G}}(\omega). \tag{3.13}$$

To fit in with the general formulation $\boldsymbol{E}(\omega)$ must be written as a sum over identical self-energies for each unit cell, maintaining the translational symmetry of the unperturbed crystal,

$$\boldsymbol{E}(\omega) = \sum_l \boldsymbol{E}^l(\omega).$$

Now all cells depart from the unperturbed system with impurity matrices equal to

$$\boldsymbol{C}^l(\omega) - \boldsymbol{E}^l(\omega) \qquad \text{original impurity cell}$$

or

$$-\boldsymbol{E}^l(\omega) \qquad \text{original host cell.}$$

The Average T-matrix Approximation (ATA) uses the virtual or mean crystal as the reference system. In this case

$$\boldsymbol{E}^l(\omega) = \langle \boldsymbol{C}^l(\omega) \rangle. \tag{3.14}$$

For the case of only one kind of original impurity per cell, of concentration c, the self-energy becomes

$$\Sigma^l(\omega) = c\boldsymbol{C}^l(\omega)(\boldsymbol{I} + c\boldsymbol{G}(\omega)\boldsymbol{C}^l(\omega))[\boldsymbol{I} - (1-2c)\boldsymbol{G}(\omega)\boldsymbol{C}^l(\omega)]^{-1},$$

with

$$\Sigma(\omega) = \sum_l \Sigma^l(\omega).$$

The extension of the ATA to quaternary mixed crystals has been performed by Sen and Lucovsky (1975) for the particular case of Ga/Al/As/P.

The better, but harder to calculate, Coherent Potential Approximation (CPA) requires that $\boldsymbol{E}(\omega)$ be determined self-consistently by requiring

$$\langle \boldsymbol{G}(\omega) \rangle = \overline{\boldsymbol{G}}(\omega),$$

i.e.,

$$\Sigma^l(\omega) = \boldsymbol{E}^l(\omega).$$

From (3.6) and (3.13) this is equivalent to

$$\langle \boldsymbol{t}^l(\omega) \rangle = 0.$$

For one type of impurity per unit cell this gives

$$\boldsymbol{E}^l(\omega) = c\boldsymbol{C}^l(\omega)[\boldsymbol{I} - \overline{\boldsymbol{G}}(\omega)(\boldsymbol{C}^l(\omega) - \boldsymbol{E}^l(\omega))]^{-1}. \tag{3.15}$$

The Green function required for optical properties is the $\boldsymbol{q}=0$ transform and is given by

$$\langle \mathbf{G}(\boldsymbol{q},\omega)\rangle = \bar{\mathbf{G}}^0(\boldsymbol{q},\omega) + \bar{\mathbf{G}}^0(\boldsymbol{q},\omega)\Sigma(\boldsymbol{q},\omega)\langle \mathbf{G}(\boldsymbol{q},\omega)\rangle. \tag{3.16}$$

Expanding in terms of pure crystal eigenvectors leads to

$$\langle G_{jj'}(\boldsymbol{q},\omega)\rangle = \bar{G}^0{}_j(\boldsymbol{q},\omega)\delta_{jj'} + \bar{G}^0{}_j(\boldsymbol{q},\omega)\sum_{j''}\Sigma_{jj''}(\boldsymbol{q},\omega)\langle G_{j''j'}(\boldsymbol{q},\omega)\rangle. \tag{3.17}$$

Again mode mixing occurs due to the off-diagonal elements of Σ.

At large concentrations it is important to realise that the need to write $\mathbf{C}(\omega)$ in terms of the $\mathbf{C}^i(\omega)$ (3.1) is very restrictive. It applies only to diagonal disorder which includes mass differences and the shell-model polarisation differences described by Grunewald (1982) and Bonneville (1984). These atomic polarisations are due to differences in the intra-atomic shell/core forces and can be incorporated into a frequency-dependent mass.

The inclusion of force constant changes is much more difficult. The force constant matrix must satisfy relation (2.2). Consequently, $\boldsymbol{\Phi}(l\kappa, l\kappa)$ depends not only on what is at the site $l\kappa$ but also on what are at all the sites to which it is coupled. Along with the fact that a significant fraction of the impurities will be coupled by force constants this means that a rigorous decomposition of the form (3.1) is impossible. In crystals where the structure keeps impurities separated one solution is to ignore this overlap of defect spaces, or local environment effect. This is correct if the forces between the different types of atoms are determined by arithmetic averaging. Kaplan and Mostoller (1974) have performed such a calculation for the neutron scattering from NH_4/KCl. There are methods that attempt to include off-diagonal disorder but none appear to have been applied in the present context with the exception of the 1D Homorphic CPA calculation of Kleinert (1982).

The need to solve both the CPA equation (3.15) and the equation relating $\bar{\mathbf{G}}(\omega)$ to the pure crystal lattice dynamics (3.16), makes the CPA method numerically cumbersome. In general the Green function in (3.15) has to be calculated by a $\boldsymbol{q}$ space sum with the matrix relation between $\mathbf{G}(\boldsymbol{q},\omega)$ and $\mathbf{G}^0(\boldsymbol{q},\omega)$, (3.16), being inverted at each $\boldsymbol{q}$ point. Consequently there have been several calculations using approximate models that simplify the procedure.

The first natural simplification is to work in 1D with a nearest neighbour model. This was done by Sen and Hartmann (1974) who used their results to discuss the 1/2 mode question (section 4). The omission of Coulomb forces in this model means that there is an absence of the LT splitting. Bonneville (1981) has given an approximate method for the inclusion of this effect. The

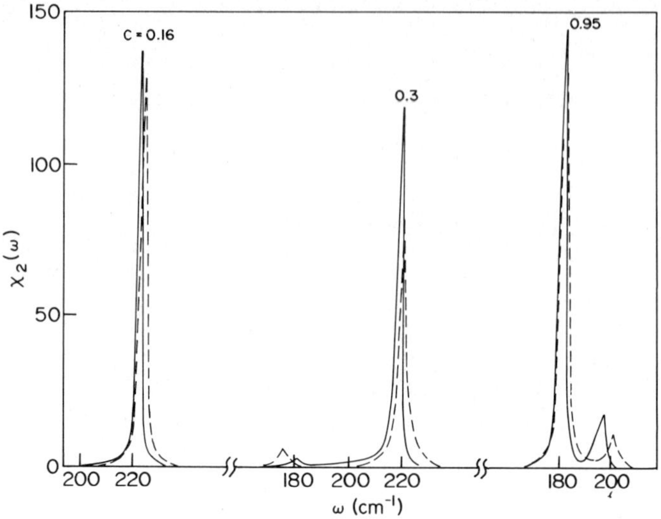

Fig. 6. Infrared response for Ga/InSb; (—) CPA Bethe lattice, (- - - - - -) experiment (Brodsky et al. 1970). From Kleinert (1982).

nearest neighbour model is used within CPA to calculate the equivalent of $g^{s}(\omega)$ with ω_0^2 being shifted by the CPA self-energy $\Sigma_{jj}(\omega)$. Then the local field effects are included in the same manner as was done in going from $g^{s}(\omega)$ to $g^{T}(\omega)$ (eq. (2.6)), with the dielectric constant being given by the equivalent of eq. (2.9). Then the mixed crystal ω_{TO}'s and ω_{LO}'s can be extracted from $\epsilon(\omega)$ in the usual way. A simple extension of the 1D calculation is to work with a Bethe lattice. Results for four nearest neighbours have been given by Kleinert (1982, 1983), and the results are surprisingly good. Figure 6 shows the infrared response for Ga/InSb illustrating the II′ behaviour of this system.

Bonneville (1981, 1984) has extended the above approach to three dimensions. He realised that in the nearest neighbour model it is not necessary to numerically solve the coupled equations (3.15) and (3.16) as was done by Taylor (1976). Rather, the relation between $\bar{\boldsymbol{G}}(\omega)$ and $\boldsymbol{G}^{0}(\omega)$ can be solved by hand and the CPA equation written in terms of a frequency integral over a density of states. This is a great improvement compared with the general procedure mentioned earlier. In his second paper Bonneville (1984) applied his procedure to the zincblende structure also including disorder in the interatomic core/shell coupling which enters through a frequency-dependent mass term. The result for the concentration dependence of ω_{TO} and ω_{LO} in Al/GaAs is shown in fig. 7 and it compares well with the experimental results. Longitudinal and transverse responses were found in the acoustic band regions but were at least two orders of magnitude smaller than the optic branch responses.

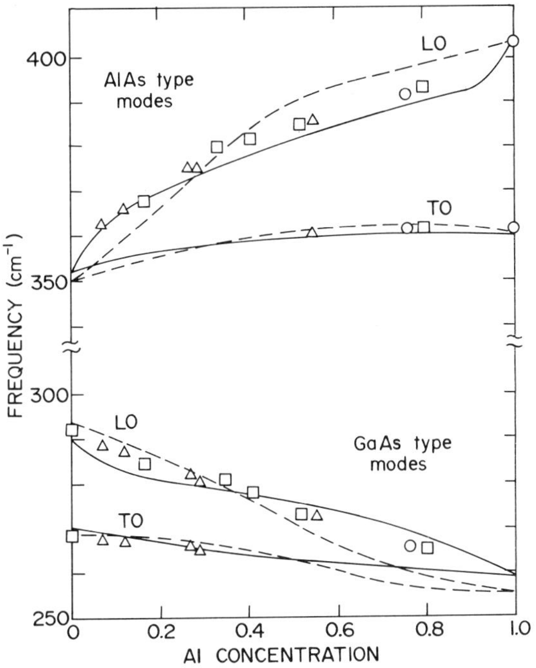

Fig. 7. Comparison of approximate CPA optic mode frequencies (———) with experiment of Jusserand and Sapriel (1981) (□△○) and Tsu et al. (1972) (- - - - - -). From Bonneville (1984).

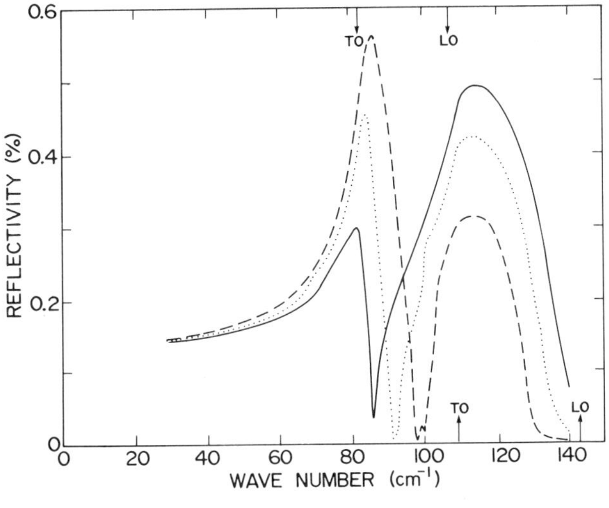

Fig. 8. CPA calculation of reflectivity of $K_{1-c}Rb_cI$, (———) $c = 0.25$, (······) $c = 0.50$, (- - - - - -) $c = 0.75$ (Taylor 1973).

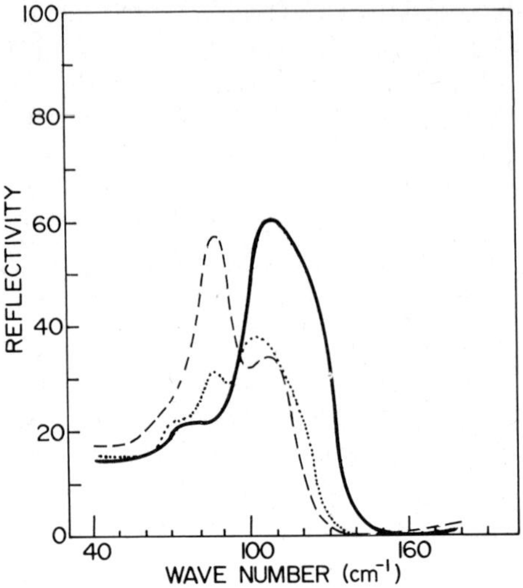

Fig. 9. Experimental reflectivity of $K_{1-c}Rb_cI$; values of c as in fig. 8 (after Fertel and Perry 1969).

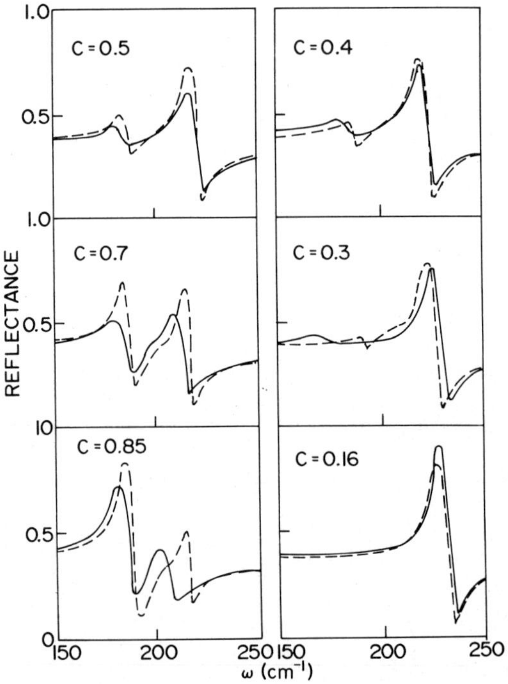

Fig. 10. Reflectivity of $Ga_{1-c}In_cSb$; (------) calculated with CPA, (———) experiment (Brodsky et al. 1970). From Kleinert (1984).

Grunewald (1982) earlier considered a disordered shell model including both breathing and dipole type distortions using a locator rather than a Green function formalism. The full model requires repeated $\boldsymbol{q}$ sums but the inclusion of just dipole distortions gives a result entirely equivalent to that of Bonneville (1984). However, Grunewald did not include the effect of long-range forces. The 1/2 mode question was investigated with the finding that the shell disorder can be significant.

Full CPA mass defect calculations have been carried out by Taylor (1973) for KCl/Br and K/RbI and by Kleinert (1984) for Ga/InP, Ga/InSb and GaAs/Sb. In both cases $\boldsymbol{G}^0(\omega)$ was calculated using a model fitted to the host crystal phonon dispersion curves. The general behaviour of the transverse and longitudinal responses as a function of concentration was in good agreement with experiment. However, the agreement between the calculation of Taylor for KRb/I (fig. 8) and the experiments of Fertel and Perry (1969) (fig. 9), is at best qualitative. Kleinert obtained better agreement with experiment in his calculation for Ga/InSb (fig. 10).

3.4. Random element isodisplacement models

A much simpler approach than the Green function method is that of the various random element isodisplacement (REI) models. The original model was introduced by Chen et al. (1966) and, as it did not include any explicit Coulomb forces, dealt with just the transverse modes. Coulomb forces were added by Chang and Mitra (1968, 1971) who introduced the macroscopic electric field in order to generate the LT splitting. This was called the modified REI (MREI) model. The version to be given below employs the local field and is basically that of Ilegems and Pearson (1970), Gasanly et al. (1971) and Harada and Narita (1971). The authors in the first two papers also included a division between the local and nonlocal charge associated with the ions. The more elaborate cluster isodisplacement models (Verleur and Barker 1966, Zinger et al. 1976, 1977) will be described in section 3.5.

The basic observation is that in a pure crystal all like atoms have the same displacement in the $\boldsymbol{q}=0$ optic modes. This is applied directly to an AB/C mixed crystal by considering an averaged cell. Using an obvious notation the equations of motion are

$$\begin{aligned}
M_A\omega^2 u_A &= (1-c)f_{AB}(u_A-u_B)+cf_{AC}(u_A-u_C)-e_A E_L,\\
M_B\omega^2 u_B &= f_{AB}(u_B-u_A)+cf_{BC}(u_B-u_C)-e_B E_L,\\
M_C\omega^2 u_C &= f_{AC}(u_C-u_A)+(1-c)f_{BC}(u_C-u_B)-e_C E_L,
\end{aligned}\tag{3.18}$$

with $c=c_C$ the concentration of C, E_L is the local field, and e_A etc. are

Szigeti charges. The f_{AB} etc., are effective force constants and due to the isodisplacement approximation are given by

$$f_{\kappa\kappa'} = -\sum_{l'} \Phi_{\alpha\alpha}(l\kappa, l'\kappa'),$$

use having been made of eq. (2.2). Charge neutrality requires that

$$e_A + (1-c)e_B + ce_C = 0$$

and it is then readily seen that the centre-of-mass coordinate

$$M_A u_A + (1-c)M_B u_B + cM_C u_C$$

is a zero frequency solution. Introducing the relative displacement

$$W_X = u_X - u_A, \qquad X = B, C$$

and eliminating the local field in the usual way using the Lorentz relation, the equation of motion for $\boldsymbol{W} = (W_X, W_Y)$ is

$$\omega^2 \boldsymbol{W} = \boldsymbol{K}\boldsymbol{W} - \boldsymbol{Z}^T E. \tag{3.19}$$

Here

$$K_{XX} = F_{AX} c_X / M_A + (F_{AX} + c_Y F_{BC})/M_X,$$

$$K_{XY} = c_X (F_{AY}/M_A - F_{BC}/M_X)$$

with

$$F_{AB} = f_{AB} + 4\pi e_A^T e_B / 3v, \text{ etc.},$$

$$Z_X^T = e_X^T / M_X - e_A^T / M_A,$$

$X, Y = B$ or C and $c_B = 1 - c$. The difference between F and f is just the transverse Coulomb force already met in (2.17). The transverse charge is defined in the usual way, but with an average polarisation,

$$e^T = e/(1 - 4\pi\bar{\alpha}/3v) = e(\epsilon^\infty + 2)/3,$$

$$\bar{\alpha} = \alpha_A + (1-c)\alpha_B + c\alpha_C.$$

The two transverse frequencies ω_{T1}, ω_{T2} are easily extracted by setting $E=0$. For the longitudinal frequencies, $E=-4\pi P$, which is equivalent to setting $E=0$ in (3.19) and replacing F_{XY} by

$$F_{XY}^{L}=F_{XY}+4\pi e_X^{T}e_Y^{T}/\epsilon^{\infty}v. \tag{3.20}$$

Equation (3.19) can be formally rewritten as

$$\boldsymbol{W}=-\boldsymbol{g}^{T}\boldsymbol{Z}^{T}E$$

and consequently the dielectric constant obtained

$$\epsilon(\omega)=\epsilon^{\infty}-4\pi\boldsymbol{e}_M^{T}\boldsymbol{g}^{T}\boldsymbol{Z}^{T}/v \tag{3.21}$$

with

$$\boldsymbol{e}_M^{T}=\left((1-c)e_B^{T},\ ce_C^{T}\right).$$

On writing

$$\boldsymbol{A}^{T}=(\omega^2\boldsymbol{I}-\boldsymbol{K}) \quad \text{and} \quad \boldsymbol{A}^{L}=(\omega^2\boldsymbol{I}-\boldsymbol{K}^{L})$$

where by $\boldsymbol{K}^{L}$ it is understood that the above longitudinal contribution has been added to F_{AB} etc., the dielectric constant becomes

$$\epsilon(\omega)=\epsilon^{\infty}\det\boldsymbol{A}^{L}/\det\boldsymbol{A}^{T}. \tag{3.22}$$

This is useful for extracting oscillator strengths for which it is easy to show that the f sum rule is satisfied,

$$f_1+f_2=4\pi\left[(\epsilon^{\infty}+2)^2/9v\epsilon^{\infty}\right]\left(e_A^2/M_A+(1-c)e_B^2/M_B+ce_C^2/M_C\right). \tag{3.23}$$

It also shows that the generalised Lyddane–Sachs–Teller relation holds

$$\epsilon(0)/\epsilon^{\infty}=\left(\omega_{L1}^2\omega_{L2}^2\right)/\left(\omega_{T1}^2\omega_{T2}^2\right).$$

It is understood in the above that v is the volume at the given concentration c. The model is completed by adding a parameter θ that describes the concentration (volume ?) dependence of the force constants

$$f_{XY}(c)=(1-\theta c)f_{XY}(0). \tag{3.24}$$

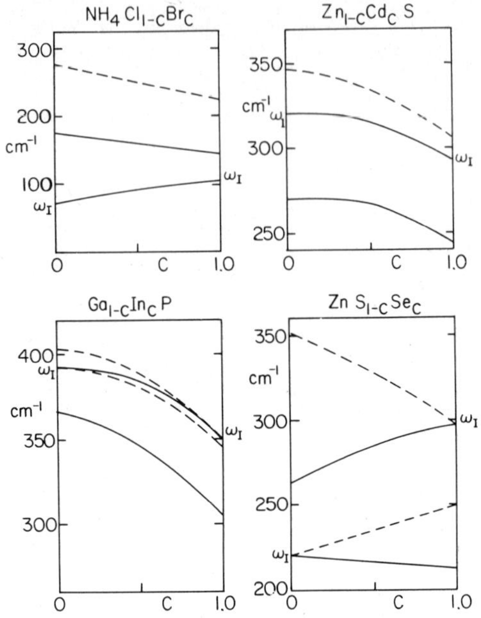

Fig. 11. REI optic mode frequencies: (———) transverse, (- - - - - -) longitudinal. For NH_4Cl/Br and Zn/CdS the LT splitting of the weak mode is too small to show.

A more elaborate scaling would lead to too many unknowns. As it is, the six parameters in the above model (f_{AB}, f_{AB}, f_{BC}, e_B, e_C, θ) require the knowledge of $\omega_L(AB)$, $\omega_L(AC)$, $\omega_T(AB)$ and $\omega_T(AC)$, being the longitudinal and transverse optic frequencies for the end members AB and AC, and, as well, the impurity mode frequencies $\omega_{AB,C}$ and $\omega_{AC,B}$. These are the frequencies of an isolated C in AB and an isolated B in AC. They may be due to gap or local modes such as occur in two-mode systems or in band, presumably resonant, modes occurring within the host phonon bands in one-mode systems. Such modes appear as secondary peaks in Im $\epsilon(\omega)$ and Im$(1/\epsilon(\omega))$ or in the Raman cross sections. For one-mode systems, resonant modes may not be present or identifiable making the model in the present form difficult to apply, but they can be seen in some systems such as GaIn/P (see, for instance, Jahne et al. 1979). In the absence of impurity modes the representation of Jahne (1976a, b), to be introduced below, is the preferable way to describe one-mode systems.

This model can include 4 types of behaviour that can be classified as two-mode, mixed-mode and one-mode (two types). These are illustrated in fig. 11 for the frequencies and fig. 12 for the oscillator strengths. The input data and fitted parameters are given in tables 1a and 1b, respectively.

The one-mode examples are based on NH_4Cl/Br (weak mode below reststrahl) and Zn/CdS (weak mode within reststrahl). They show the ex-

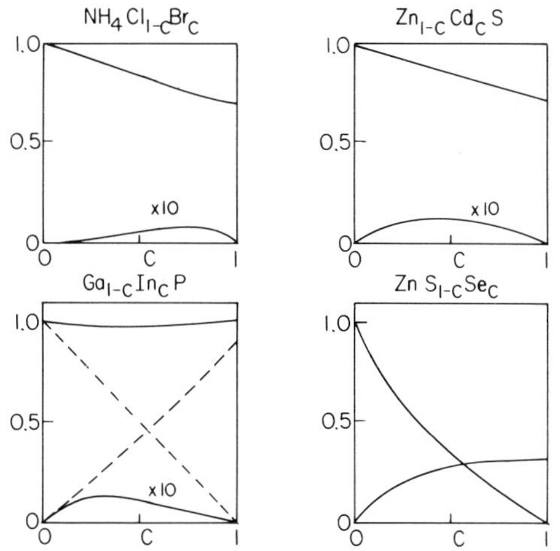

Fig. 12. REI oscillator strengths, referred to unity at $c = 0$. The dashed curves are the longitudinal oscillator strengths.

Table 1a
REI input data.

	M (a.u.)		ϵ^{∞} (Å)	a	ω_{TO}	ω_{LO} (cm^{-1})	ω_I
NH_4Cl	18	35	2.7	3.9	174	275	75
NH_4Br	18	80	2.9	4.1	144	224	105
ZnS	65	32	5.5	5.4	270	347	320
CdS	112	32	5.5	5.9	242	305	291
GaP	70	31	9.1	5.5	367	403	393
InP	115	31	9.6	5.9	304	345	350
ZnS	65	32	5.6	5.4	264	352	220
ZnSe	65	78	5.6	5.7	213	250	297

Table 1b
REI fitted parameters.

	F_{AB}	F_{AC} (dyne cm^{-1})	F_{BC}	θ	e_B/e	e_C/e
NH_4Cl/Br	3.95	3.91	−1.26	0.15	0.42	0.41
Zn/CdS	13.6	23.3	45.3	0.45	0.85	0.88
Ga/InP	20.0	30.7	74.1	0.46	0.57	0.65
ZnS/Se	13.9	15.3	7.0	0.20	0.90	0.70

pected almost straight line interpolation between the values of ω_{TO}, ω_{LO} and f for $c=0$, 1. As an example of the mixed-mode behaviour we have chosen Ga/InP as this is the recent reinterpretation of the experimental spectra. As it is the longitudinal modes that have two-mode behaviour we have given the longitudinal oscillator strength (the f value in a dispersion formula for $\epsilon(\omega)^{-1}$) which shows the crossover of strengths between the longitudinal modes as c increases. However, when compared with experiment (fig. 30) the general shape is not very good. The final example is ZnS/Se which has the clear two-mode behaviour. We have not shown a II′ behaviour as it really needs an examination of the calculated reflectivity to decide between II and II′ (see, for instance, Gasanly et al. 1971). It should be noted that the order of the modes is always TO, LO, TO, LO (frequency increasing).

In the fitting procedure f_{BC} is closely related to the impurity mode frequencies and is somewhat artificial as it has to compensate for a poor approximation for these frequencies. This is seen from the somewhat extreme values of f_{BC} in table 1b. The correct calculation of local mode frequencies looks for poles in the t-matrix (3.3) and needs the full lattice dynamics, not just averaged force constants, in order to calculate the required Green functions. Further, the atomic displacements around a local mode depart strongly from the isodisplacement approximation. With these ideas in mind, Jahne (1976b, c) has investigated an alternative approach. He observed that if the concentration scaling was ignored the solution of (3.19) gives

$$F_{BC}=M_X\left(\omega^2_{A X,Y}-\omega^2_T(A X)\right)$$

for X, $Y=$ B, C or C, B. As a consequence he suggested a concentration average of these two formulae. However, this produced an unsatisfactory nonmonotonic behaviour of the ω_{Li} and ω_{Ti} which was not found experimentally and he was led to introduce a force constant scaling of F_{AB} and F_{AC}.

Jahne (1976a, b) also introduced a different basis in which to examine the equation of motion. This basis is defined by

$$W_1=u_A-\frac{(1-c)M_Bu_B+cM_Cu_C}{(1-c)M_B+cM_C}, \tag{3.25}$$

$$W_2=u_B-u_C$$

replacing our eq. (3.17). W_1 refers to virtual crystal behaviour (VC) with the A sublattice moving against an averaged version of the other sublattice and W_2 refers to motion in just the disordered sublattice (DSL). He found that in one-mode systems there is very little mixing between these modes and the VC modes were dominant. This means that this is the preferable representation to describe one-mode systems as f_{BC} can be ignored as it appears only in the

DSL diagonal term. Consequently no knowledge of impurity modes frequencies is then required. For two-mode systems there is a strong mixing and the frequencies of the VC and DSL modes (defined as the diagonal terms of the equation of motion) cross in going from $c = 0$ to $c = 1$.

In order to produce a general survey of mode behaviours within a simple model Genzel et al. (1974) used just ω_{TO} and ω_{LO} for AB and AC and so avoided the problem of locating the extra modes. They were thus forced to drop f_{BC} and the scaling parameter θ. The consequence is that $\omega_{AB,C}$ is given by

$$\omega_{AB,C} = (f_{AB}/M_C)^{1/2}, \tag{3.26}$$

and similarly for $\omega_{AC,B}$. It is not surprising that these values are not always in good agreement with experiment. However, the general behaviour of the ω_{Li} and ω_{Ti} as a function of c predicted for a number of mixed crystals (alkali halides, II–VI and III–V) is almost always in good qualitative agreement with experiment.

It should be noted that Martin (1975) has used this model to calculate the intensity of Raman scattering from CdS/Se with some success. In this calculation account was taken of the variation of the susceptibility $\boldsymbol{a}$ with respect to both displacements and the local field, the model being used both for the dynamics and the local field.

Kutty (1974) and Bottger (1976) have shown how the REI result can be derived using Green function methods. Working with conditionally averaged Green functions similar to those introduced in section 3.1, Bottger was able to compare the REI, ATA and CPA approximations.

The application of the REI method to nonzero wave vectors is called the pseudoatom approximation. Recently, Massa et al. (1982), using a shell-model generalisation of this approximation, have calculated one- and two-phonon properties of several mixed alkali halides and obtained good agreement with experiment.

3.5. Cluster isodisplacement models

The REI type models described above can produce only two transverse modes, yet sometimes in the experimental data there is clear evidence of more such modes. Such an observation regarding the reflectivity of GaP/As led Verleur and Barker (1966) to introduce a much more elaborate isodisplacement model which, in fact, appeared slightly earlier than the REI model of Chen et al. (1966).

Rather than deal with an averaged cell they introduced five possible units characterised by the number of C atoms that are n.n. (nearest neighbours) of an A atom, separate coordinates ($u_A(i)$, $i = 1, 5$; $u_B(i)$, $i = 1, 4$; $u_C(i) = 2, 5$)

being introduced for each atom in each unit. The equation of motion for $u_A(i)$ has the form

$$\omega^2 M_A u_A(i) = F_1 + F_2 + F_3 - \left[x(i)e_B + y(i)e_C\right]E_L. \qquad (3.27)$$

Here F_1 describes the effect of relative motion within the ith unit,

$$F_1 = x(i)k_1(i)\left[u_A(i) - u_B(i)\right] + y(i)k_2(i)\left[u_A(i) - u_C(i)\right].$$

As the n.n. B atoms of an A atom are also members of the units associated with neighbouring A atoms, a weight of $1/4$ must be introduced. When multiplied by the number of n.n. B atoms in the ith unit this gives the factor $x(i)$. $y(i) = 1 - x(i)$ is the equivalent factor for the C atoms. k_1 and k_2 are force constants.

The interaction between n.n.n. A atoms is contained in F_2,

$$F_2 = k_6 \sum_{j=1}^{5} f_j\left[u_A(i) - u_A(j)\right],$$

where f_j is the probability of the occurrence of unit j.

The remaining interaction with n.n. B and C atoms, considered as members of adjacent units, is contained in F_3,

$$F_3 = \sum_{j=1}^{5} f_j\left[k_7 x(j)(u_A(i) - u_B(j)) + k_8 y(j)(u_A(i) - u_C(j))\right].$$

The equations of motion for $u_B(i)$ and $u_C(i)$ have a similar form and, in total, there are 14 force constants and two effective charges. Verleur and Barker also included a short-range order parameter in the calculation of the f_j. An example of their fit to their experimental data for GaP/As is shown in fig. 13. This model has also been applied to CdS/Se (Verleur and Barker 1967a) as well as to Ca/SrF_2 and Sr/BaF_2 (Verleur and Barker 1967b). It should be noted that if eq. (3.27) is averaged over units and then the i label disregarded on the displacements, the REI equation (3.17) is regained. In this process the f force constants of eq. (3.17) are given in terms of concentration averages of the k's in eq. (3.24). This leads to a concentration dependence of the f's and Jahne et al. (1979) have suggested that this may be a source of the concentration dependence used in the REI models.

A simplified version of this model has been applied by Jahne et al. (1979) to their experimental results for Ga/InP. They reduced the number of coordinates by ignoring the i dependence of u_{Ga} and u_{In}. This was justified by noting that the P atoms ($\equiv$ A) being lighter than Ga and In should dominate

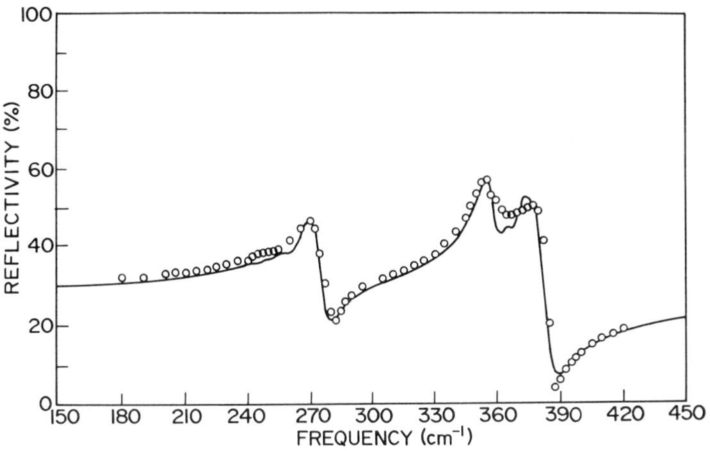

Fig. 13. Theoretical fit of CI model to reflectivity of $GaP_{0.56}As_{0.44}$ (Verleur and Barker 1966).

the modes. A number of approximations were made to reduce the number of independent force constants and the resulting frequencies and oscillator strengths agree well with experiment. However, the model predicts a weak low-frequency mode not seen experimentally.

Zinger et al. (1976) have described an even simpler model taking as basic units an AB unit cell and an AC unit cell. The equation of motion for an A atom in an AB unit cell is

$$\begin{aligned} M_A\omega^2 u_A(l) &= \Phi(lA, lA)u_A(l) + \Phi(lA, lB)u_B(l) \\ &\quad + (1-c)\sum_{l'}[\Phi(lA, l'A)u_A(l') + \Phi(lA, l'B)u_B(l')] \\ &\quad + c\sum_{l'}[\Phi(lA, l'A)u_A(l') + \Phi(lA, l'C)u_C(l')] - e_A E_L \end{aligned} \tag{3.28}$$

with similar equations for the B atom in this unit cell and for the A and C atoms in an AC cell. The first line refers to intra-cell interactions and the second and third lines to inter-cell interactions with AB and AC cells, respectively. A transformation is then made to centre of mass and relative displacement coordinates. For the AB cell these are

$$Z_1(l) = [M_A u(lA) + M_B u(lB)]/(M_A + M_B),$$

$$Z_3(l) = u(lA) - u(lB).$$

Similarly, $Z_2(l)$ and $Z_4(l)$ are introduced for the AC cell. On an assumption of weak inter-cell forces the isodisplacement approximation is then applied to the Z coordinates giving an equation of motion of the form

$$\mu_{AB}\omega^2 Z_3 = \Phi_{31}Z_1 + \Phi_{33}Z_3 + (1-c)\left[\Phi_{31}Z_1 + \tilde{\Phi}_{31}Z_1 + \tilde{\Phi}_{33}Z_3\right]$$
$$+ c\left[\tilde{\Phi}_{32}Z_2 + \tilde{\Phi}_{34}Z_4\right] - e_B E_L,$$

where μ_{AB} is the reduced mass of an AB cell and the Φ's and $\tilde{\Phi}$'s combinations of intra-cell and inter-cell force constants, respectively. There are similar equations for the other three coordinates. Only the relative displacement coordinates, Z_3 and Z_4, couple to photons and again using the assumption of weak intercell forces the Z_1 and Z_2 coordinates are dropped. On applying their results to ZnS/Se and GaP/As they obtained good agreement with experiment. However, their results are no better than those of the REI model described earlier.

These authors have also applied their model to Si/Ge mixed crystals. In this case there are 3 possible unit cells – SiSi, SiGe and GeGe. Consequently

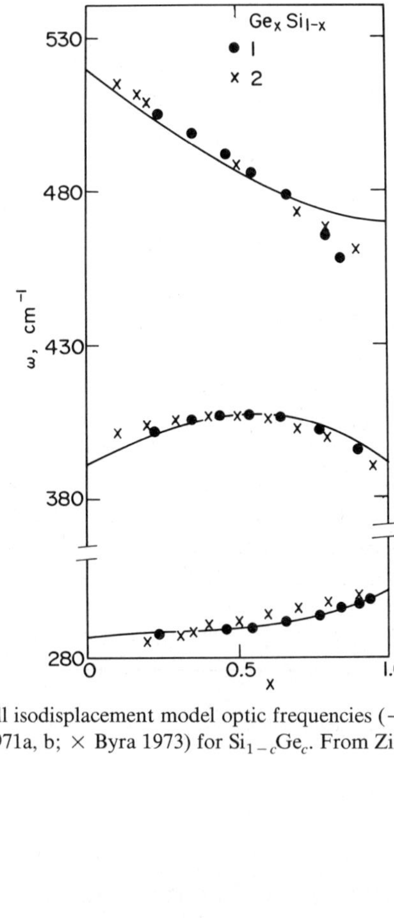

Fig. 14. Comparison of cell isodisplacement model optic frequencies (———) with experiment (● Renucci et al. 1971a, b; × Byra 1973) for $Si_{1-c}Ge_c$. From Zinger et al. (1977).

there are 6 coordinates and, again ignoring inter-cell couplings, they predict three dominant modes. Their fitted result compares well with the Raman experiments of Feldman et al. (1966) and Byra (1973) as can be seen in fig. 14.

3.6. Miscellaneous methods *

Large cluster calculations that do not use the isodisplacement approximation would seem an appropriate approach to elucidate the behaviour of the optical properties of mixed crystals. It should even be possible to include Coulomb forces by using periodic boundary conditions and Ewald sums. This kind of calculation, which would involve considerable computational time, has not yet been done, although there have been a number of cluster-type calculations for the vibrational properties of mixed crystals and amorphous materials.

The only true cluster calculations are in one dimension starting with Matossi (1951) (who used an ordered array of AB and AC cells) and Rosenstock and coworkers (Rosenstock and McGill 1968, Haas et al. 1969). Recently Wanser and Wallis (1979) have used a random chain of 72 atoms with periodic boundary conditions to calculate the optical absorption for NaCl/Br. They included just n.n. and n.n.n. force constants and then used the eigenfrequencies in a dispersion formula, with the eigenvectors determining the oscillator strengths. Their main point was that these eigenvectors showed a large departure from the isodisplacement conditions and further they predicted up to five infrared active modes in contrast with the two obtained from REI calculations. It should be noted, though, that the localisation of eigenstates is much stronger in 1D than in 3D. Unfortunately there are no experiments on mixed crystals of NaCl/Br.

A much larger chain was used by O'Hara et al. (1981) to construct just the density of states for quaternary alloys. As do all exact 1D calculations, their results have much fine detail. In particular, it is interesting to note that the GaP/As example demonstrates a type II behaviour, in agreement with experiment, both at the $\boldsymbol{q} = 0$ and the zone boundary.

A related calculation is that of Kleinert (1979a) who considered the $\boldsymbol{q} = 0$ susceptibility of a 1D chain but only within a cluster approximation. This method did produce considerable structure. An alternative form of 1D cluster approximation is that used by Myles (1983). In this case a cluster of 8 atoms was embedded within a CPA medium. The results agree well with the exact 50 000 atom chain calculations showing something like two-mode behaviour for Ga/InSb and one-mode for Zn/CsS (small separation of the inevitable local mode peaks for the host bands) at low concentrations.

A more 3D type of calculation is that described by Kleinert (1979b) who used a cluster Bethe lattice approximation. Clusters of 5 atoms and 17 atoms

* See, Notes added in proof on p. 121

were considered with the boundary atoms being connected to Bethe lattices. No Coulomb forces were included and so just the susceptibility of the cluster was calculated. The method was applied to the one-mode system Ga/InP and to the two-mode system GaP/As. Good agreement with experiment was claimed.

Other methods that have been applied include the "replica" or "$n \to 0$" trick by Schmeltzer and Beserman (1980) and the recursion method by Herscovici and Fibich (1980). The latter authors used their results to discuss the 1/2 mode question (section 4) and to calculate the density of states for ZnS/Se and ZnSe/Te. Coulomb forces were not included nor were optical properties specifically calculated.

4. *One-/two-mode criteria*

There have been several attempts to develop relatively simple criteria to predict the type of mode behaviour for mixed systems. Most procedures use just the atomic masses which can only be satisfactory when force constant changes are not important. In strongly ionic materials variations of the effective charge can lead to changes in the strong long-range force constants that ought to be taken into account.

Most criteria have been developed with the aim of predicting the occurrence of local and gap modes. The occurrence of such modes for $c \to 0, 1$ suggests a two-mode behaviour (II). Should no gap mode be predicted or if it is known that there is no gap in the phonon density of states but a local mode is predicted, then a partly two-mode behaviour (II′) may occur. This is characterised by a two-mode behaviour for most values of c with the low-frequency impurity mode disappearing well before the dilute limit. The other intermediate situation occurs when the longitudinal modes follow a one-mode pattern and the transverse modes a two-mode pattern (or vice versa). This is the mixed-mode (M) behaviour and has been observed only for KRb/I (see, however, Ga/InP in section 5.3). None of the criteria described here can predict its occurrence. Finally the absence of local and gap modes leads to the expectation of a one-mode behaviour (I).

The predictions of the various procedures to be outlined below are collected in tables 2–4. References for the experimental work can be found in section 5 where each system is discussed in the same order as in these tables.

The simplest criterion to use is to observe whether the optic bands overlap. Usually the band width is taken to be $(\omega_{LO} - \omega_{TO})$ but this is not always the case, GaSb (Farr et al. 1975) being a good example. A strong overlap suggests one-mode behaviour on the basis that no frequencies can occur in the mixed system in frequency ranges where none occur in either pure crystal. This criterion does very well but, of course, it cannot predict II′ behaviour.

Table 2
Mode criteria for I/VII systems.

	Exp.	RO	CM	HN	EKL	GMP	LBB	SH
LiH/D	II	I	1.6				I	I
NH_4Cl/Br	I	I	0.4	(0.7)	I	I	I	I
KRb/Cl	I	I	0.6	(1.0)	I	?	I	I
K/RbBr	II	I	1.1	(1.7)	II	M	I	I
K/RbI	M	II	1.3	2.0	II	II	II	II
KCl/Br	I	I	0.8	(1.3)	I	I	I	I
RbCl/Br	II	I	1.2	(1.9)	II	M	I	I
CuCl/Br	II	II	1.0	(1.8)	II	II	II	II
CuCl/I	II	II	1.2	1.5	II	II	II	II
CuBr/I	I	I	0.5	0.8	I	I	I	I
AgCl/Br	I	I	1.3	(2.9)	II′	II	I	I

However, adding the information that GaAs does not have a gap, leads to the prediction that Ga/InAs should be II′, in agreement with experiment. As Behera et al. (1977) have pointed out, this general difficulty can be removed by adding the condition that the 1D gap mode must lie within the experimental gap. If there is either a gap mode or a local mode then some form of mixed-mode behaviour (M or II′) is predicted. Their results show that under this condition the LBB predictions for Ga/InP, Ga/InAs, Ga/InSb, GaAs/Sb and InAs/Sb change to some form of mixed-mode behaviour.

Using the MREI model, Chang and Mitra (1968), CM, suggest that two-mode behaviour occurs when

$$M_{AC}/M_B = r_{CM} > 1. \tag{4.1}$$

This is based on requiring a local mode to occur and setting $F_{AB} + F_{BC} = F_{AC}$. Gap modes appear automatically under this condition. It is reasonably successful but predicts type I instead of type II′ in the III/V's. Harada and Narita (1971), HN, criticised this approximation and suggested that the force constant approximation should be applied to the short-range forces, not the transverse forces. This led to a new condition

$$M_{AC}R_{AC}/M_B = r_{HN} > 1, \qquad R_{AC} = \left(\epsilon^0_{AC} + 2\right)/\left(\epsilon^\infty_{AC} + 2\right) \tag{4.2}$$

for two-mode behaviour. Noting that a large R_{AC} indicates strong Coulomb forces they also suggested that one-mode behaviour will occur whenever $R_{AC} > 1.55$, irrespective of the value of r_{HN}. Values of r_{CM} and r_{HN} are given in the appropriate columns in tables 2–4. Cases where R_{AC} is greater than 1.55 are indicated by brackets. This extra factor does help in table 3 (II/VI). However, neither criteria are that successful.

Table 3
Mode criteria for II/VI systems.

	Exp.	RO	CM	HN	EKL	LBB	GMP	SH	K	HF
Mg/ZnS	II		0.9	1.3		I		II′	II	II
Mg/ZnTe	II		1.8	2.8	II	II	II	II	II	II
Mg/CdTe	II	II	2.5	3.4	II	II	II	II	II	II
Zn/CdS	I	I	0.4	0.6	I	I	I	I	I	I
Zn/CdSe	I	I	0.7	1.0	II′	I	M	I	II′	II′
Zn/CdTe	II	II	0.9	1.2	II	I	M	I	II′	II′
Zn/HgTe	II	II	1.2	1.6	II	II	II	II	II	II
ZnS/Se	II	II	1.1	1.7	II	II	II	II	II	II
ZnSe/Te	I	II	0.5	0.7	II′	I	I	I	I	I
Cd/HgTe	II	II	0.7	0.9	II	I		I	II′	II′
CdS/Se	II	II	1.5		II	II	II	II	II	II
CdS/Te	II	II	1.9	2.5	II	II	II	II	II	II
CdSe/Te	II	II	0.8	1.0	II	I	I	I	II′	II′
Mn/ZnTe	II		0.8	1.0	I	I		(I)	I	I
Mn/CdTe	II		1.1	1.5	II	II		II	II	II
Mn/HgTe	II		1.4	1.9	II	II		II	II	II

Table 4
Mode criteria for III/V systems.

	Exp.	RO	CM	HN	EKL	LBB	GMP	SH	K	HF
Al/GaP	II	II	0.8	0.9	II	II	II	II′	II′	II
Al/GaAs	II	II	1.3	1.6	II′	I	II	II	II	II
Al/GaSb	II	II	1.6	1.8	II	II	II	II	II	II
Ga/InP	I	I	0.3	0.4	II′	I	I	I	I	I
Ga/InAs	II′	II	0.6	0.8	II	I	I	II′	II′	II′
Ga/InSb	II′	II	0.8	0.9	II	II	M	II′	II′	II′
GaP/As	II	II	1.2	1.4	II	II	II	II	II	II
GaAs/Sb	II′	II	0.6	0.6	II	II	I	II′	I	I
InP/As	II	II	1.5	1.8	II	II	II	II	II	II
InAs/Sb	II′	II	0.8	0.8	II	II	M	II′	II′	II′

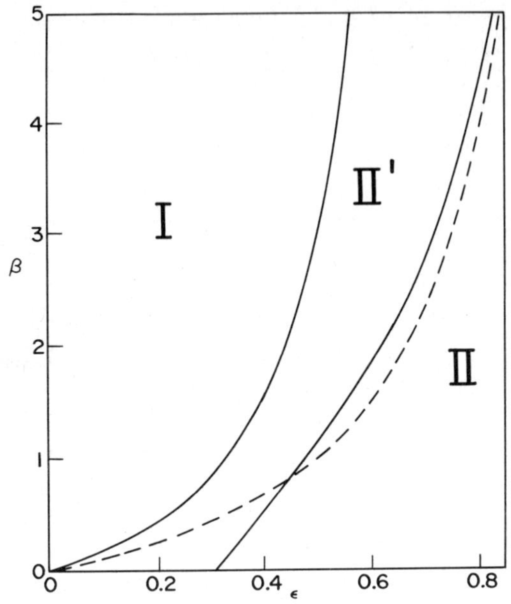

Fig. 15. Phase diagram separating I, II′ and II type regions according to Kleinert (1983). (- - - - - -) is CM result.

The preferable method to calculate local mode frequencies is that using Green functions as outlined in sections 3.1 and 3.2. A full 3D calculation is a laborious excercise, requiring a detailed knowledge of the host crystal phonon eigenfrequencies and eigenvectors. Consequently, simple approximations have been attempted. The calculation is much simpler in 1D but the host crystal phonons are then poorly represented. Lucovsky et al. (1970) (LBB) have overcome this to some extent by adding the extra condition that for a local mode to occur in 3D, the 1D local mode frequency must exceed the ω_{LO} of the host crystal. This turns out to be quite successful although it cannot distinguish between II and II′.

Kleinert (1983) (K) has extended the Green function calculation to 3D using a Bethe lattice approximation. Gap modes do not automatically occur in this approximation so it can distinguish between II and II′. For mass changes only, the method yields just a quadratic equation for the local mode frequency but has poor predictive power. However, the inclusion of a n.n. force constant change of 15% vastly improves the situation. As a five-atom cluster must now be solved the equations are more complicated but Kleinert presented a phase diagram (fig. 15) of $\beta = M_C/M_A$ against $\epsilon = 1 - M_B/M_C$ distinguishing between I, II and II′ types for tetrahedral coordination. This method is very successful in spite of not using band-width information.

Finite concentration theories have also been applied to the problem. Elliott et al. (1974), EKL, have applied a simple CPA model. They used just a semicircular density of states bounded by ω_{LO} and ω_{TO} and then required the existence of a gap for all c for type II to occur. Their condition is

$$|1 - M_{AB}/M_{AC}| > \frac{\omega_{LO}^2/\omega_{TO}^2 - 1}{\omega_{LO}^2/\omega_{TO}^2 + 1} \tag{4.3}$$

with ω_{LO} and ω_{TO} for AB, although they quote it in terms of ϵ^0 and ϵ^∞. There are actually two conditions depending on whether AB or AC is considered the host crystal. Those cases where only one of the conditions is satisfied are indicated by II′. In spite of its simplicity the predictions are almost always correct.

The CPA has also been applied by Sen and Hartmann (1974) (SH) but to a linear chain. Those cases where a gap appeared over only a limited range of c, they called II′ and they further used the Lucovsky et al. (1970) condition that if the local mode frequency is less than ω_{LO}, the system is type I. Their condition was presented as a phase diagram of β against ϵ, fig. 16, that looks somewhat different from that of Kleinert (1983) but is equally successful. Surprisingly, the extension of this calculation to 3D is less successful as pointed out by Grunewald (1982). He found that the II′ region almost disappeared, becoming part of region I.

A rather different approach is that of Herscovici and Fibich (1980), HF, who have employed the recursion method to calculate the spectral function. They compared the second, third and fourth coefficients in the continued

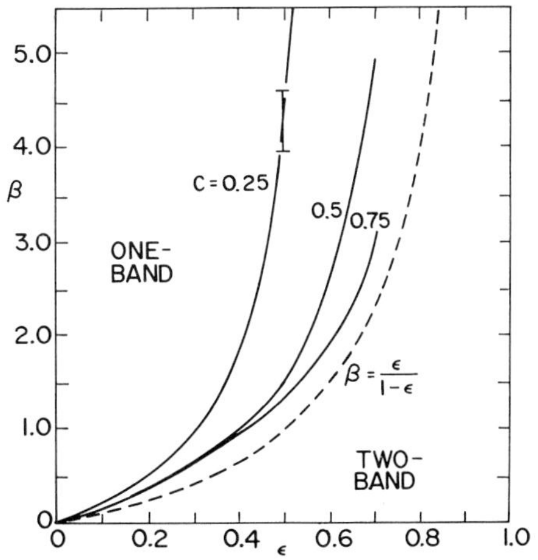

Fig. 16. Phase diagram separating one-mode and two-mode behaviour at various concentrations. From Sen and Hartmann (1974). (- - - - - -) is the CM result.

fraction expression with those of the virtual crystal approximation. They suggested that if all these coefficients differed by more than 10% from the VCA values then the system is type II, whereas if only some differ then it is type II′. The method was applied only to tetragonal systems and a n.n.n. force, set equal to the n.n. force divided by $\sqrt{2}$, was included. It ignored Coulomb forces and so the extension to the alkali halides is not worthwhile. The quantities that have to be less than 0.1 are

$$\delta_1 = (\lambda^2 - 1)^2/4\lambda^2,$$

$$\delta_2 = 2(\lambda - 1)^2/\beta, \tag{4.4}$$

$$\delta_3 = (\lambda/\beta + \delta_2/2)/(1 + 1/\delta_2),$$

where $\lambda^2 = M_C/M_B$. δ_3 differs from the value quoted by Herscovici and Fibich as, on working through their equations, the present author found a factor of 4 missing. This correction does not spoil their general agreement with experiment.

As the simple REI model of Genzel et al. (1974) does not use impurity mode frequencies to fit the parameters it can also be used to predict mode behaviour. This model predicts impurity modes at

$$\omega_{AC,B} = \omega_{TO}(AB)(M_{AB}R_{AB}/M_B)^{1/2} \tag{4.6}$$

and similarly for $\omega_{AB,C}$. Type II behaviour occurs if $\omega_{AC,B} > \omega_{LO}(AC)$ and $\omega_{AB,C} < \omega_{TO}(AB)$. If one of these impurity frequencies is within a host optic band then the model predicts M type. If both impurity frequencies are in-band or below the optic bands then the system is type I. Although Genzel et al. counsel caution as to the use of their model to make such predictions it does very well. However this is not always for the right reasons. For Ga/InP it puts both impurity modes below the reststrahl whereas they are within it experimentally.

Schmeltzer and Beserman (1981a) have applied the renormalisation group method for determining the localisation of states to this 1/2 mode question. Their result is concentration dependent and on their phase diagram for $c = 0.5$, the I/II dividing line lies in between the CM and SH lines. They claim their result explains the II′ behaviour of InAs/Sb.

The mode behaviour of zone boundary (ZB) modes has been addressed by Genzel and Bauhofer (1976) (GB). They pointed out that for transverse and longitudinal modes at the L point in the rocksalt structure and for the longitudinal mode at the X point in the zincblende structure the light ion is at rest in the acoustic mode and the heavy ion is at rest in the optic mode. Consequently, in the absence of significant force constant changes, if $M_A > M_B$,

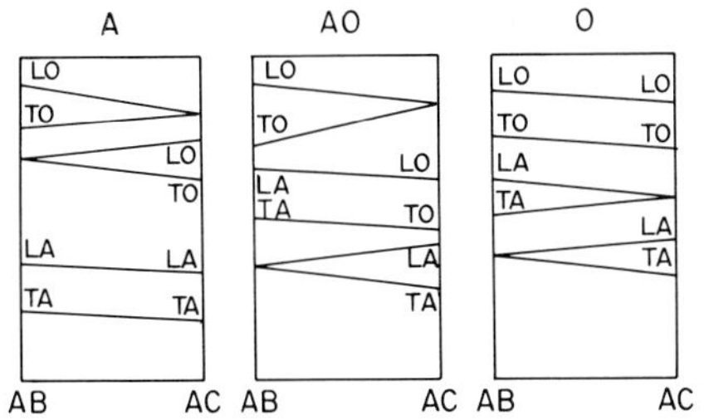

Fig. 17. Sketch of types of ZN behaviour according to Genzel and Bauhofer (1976).

M_C the disorder can be expected to have little effect on these acoustic modes which should then behave in a one-mode manner. If impurity modes exist then the ZB optic modes could be expected to be of type II. In the other limit of $M_A < M_B$, M_C the optic modes should be of type I and the acoustic modes of type II. For the intermediate case, $M_A < M_B < M_C$ the AB acoustic modes could be expected to transform into the AC optic modes with the impurity modes linking up with the corresponding optic and acoustic modes. This is shown schematically in fig. 17 with the above cases being called A, O, AO respectively. This idea has been confirmed experimentally for NH_4Cl/Br, KBr/I, ZnS/Se, CdS/Se and Al/GaSb. Exceptions, AgCl/Br, Ga/InP and GaP/As, can be explained in terms of departures of the pure crystal eigenvectors from the above behaviour.

5. *Mixed crystals systems*

In this section we collect together the experimental and theoretical work on the optical properties of mixed crystals. For the many relatively simple systems that have been investigated in some detail, a summary of the various contributions is given. The more complex systems and/or little investigated systems are collected in tables 5 and 6. It should be noted that this compendium of results is restricted to investigations concerning the straightforward effect of disorder. As mentioned in section 1, it does not include investigations centred on such topics as phase transitions in mixed crystals.

5.1. *Mixed monatomic crystals*

5.1.1. $Si_{1-c}Ge_c$

The mode frequencies in this system have an interesting concentration dependence that is generally interpreted as indicating the presence of short-range order. Contrary to the usual picture it is not the Si local mode in Ge that evolves into the pure Si optic mode, instead it is a higher frequency peak interpreted as being due to Si–Si pairs. In fact the main peaks in the observed

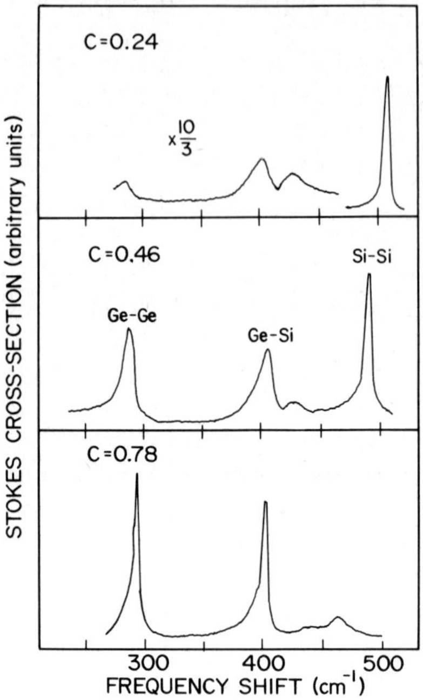

Fig. 18. Raman spectra for $Si_{1-c}Ge_c$ (Renucci et al. 1971a).

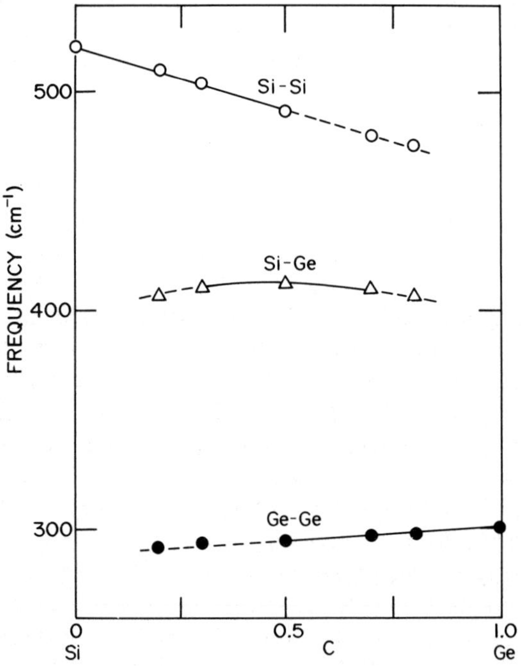

Fig. 19. Concentration dependence of $Si_{1-c}Ge_c$ RS peaks (Ishidate et al. 1984).

spectra are usually labelled as being due to Ge–Ge, Ge–Si and Si–Si pairs. This is done in the RS example shown in fig. 18 taken from Renucci et al. (1971a). As $c \to 1$ the Ge–Si peak becomes the Si local mode in Ge and as $c \to 0$ it may become a Ge resonant mode in Si. However Green function calculations do not produce resonant frequencies at the correct position of 400 cm^{-1}. The overall frequency dependence of these modes is shown in fig. 19.

The above picture has been established by the RS measurements of Feldman et al. (1966) (Si-rich alloys), Renucci et al. (1971a, b) and Byra (1973). The infrared absorption of Cosand and Spitzer (1971) for $c < 0.12$ and $c > 0.88$ shows the Ge–Si mode and the Si–Si mode ($c > 0.88$) as well as several two-phonon peaks. For $c < 0.12$ they also observed a peak at 485 cm^{-1} just below the optic frequency in Si (also seen in RS by Byra) and one at 125 cm^{-1}. It is not clear whether they are defect activated or resonant modes. It should be noted that Braunstein (1963) saw several of these features in his absorption results, but not realising the possibility of disorder-induced dipole moments, interpreted all his peaks as two-phonon in origin.

DATA modes have been clearly identified in RS spectra by Lannin (1977) who identified both the TA two-phonon peaks by analogy with the pure crystal results, and the corresponding first-order peaks in the mixed crystal at half the frequencies. Overall Lannin found that the two-phonon RS spectrum was a simple overtone of the one-phonon spectrum, although a couple of unidentified features did appear. More recently Shen and Cardona (1980) have identified several defect-activated peaks, as well as a resonant mode, in their absorption spectra for $c = 0.84$ using a Green function calculation for the $\langle u^2 \rangle$ of a Si atom in Ge.

Based on the pressure dependences of the RS peaks, Renucci et al. (1971b) questioned whether the main peaks are indeed $\boldsymbol{q} = 0$ optic modes, suggesting they may be density of states features. However their work is not supported by the more extensive pressure studies of Ishidate et al. (1984). Also, Lannin (1977) has made a careful estimate of $\boldsymbol{q} = 0$ and density of states contributions and concluded that $\boldsymbol{q} = 0$ modes usually dominate.

The single impurity and pair approximations have been calculated by Agrawal et al. (1979) for Si in Ge. Although they can reproduce some of the peaks in the infrared absorption, at least one of these peaks may be due to two-phonon absorption (Cosand and Spitzer 1971). Their calculated RS spectra do contain the DATA seen by Lannin (1977) but due to the presence of several two-phonon peaks it is difficult to judge how well they have reproduced the first-order spectrum. Agrawal (1981) has applied a five-atom-cluster Bethe lattice method to this system and he does seem to be able to reproduce the dominant three peaks, if not always of the right intensity. The CPA, being a single-site approximation with no short-range order, is incapable of producing three peaks, as Srivastava and Joshi (1973) have found. However, the CI model of Zinger et al. (1977), being built from three possible unit cells

(Ge–Ge, Ge–Si, Si–Si), does produce three modes. Using mainly pure crystal data they were able to reproduce the experimental results given in fig. 14.

5.1.2. *Se/Te*

The crystal structure of this system is trigonal with three atoms per unit cell. The $\boldsymbol{q}=0$ optic modes have symmetries A_1, A_2 and E with A_1 and E Raman active and A_2 and E infrared active. The IR measured by Keezer et al. (1968) and Geick and Hassler (1969) shows that the low-lying modes (E and A_2) behave in a one-mode manner although Geick and Hassler did find an A_2 local mode in Te that transformed into a gap mode in Se. The higher-frequency modes have weak oscillator strengths so could not be seen. To complete the picture, Geick et al. (1972) measured the RS and found a two-mode behaviour for the high-frequency A_1 and E modes. They also saw an extra E impurity mode that is almost degenerate with the A_2 impurity mode seen in the IR. Geick and Hassler have adapted the MREI to this system and were able to fit the observed behaviour quite well.

5.1.3. $Bi_{1-c}Sb_c$

In the pure crystals there are two Raman frequencies (A_{1g} and E_g, there are no infrared active modes) yet the mixed crystals have five peaks in their RS spectra. However, there is a disagreement as to the concentration dependence of the frequencies of these peaks. Zitter and Watson (1974) using single crystals for $c<0.3$ and $c>0.7$ claim a two-mode behaviour for both the A_{1g} and E_g modes with an extra peak below the highest mode that is not present in the pure crystals. Their results show essentially no concentration dependence for these frequencies. In contrast Lannin (1979a) using single crystals for $c \leqslant 0.17$, and polycrystalline sputtered films otherwise found a much more complicated behaviour for large c (fig. 20). The results suggest a two-mode behaviour plus impurity mode for A_{1g} and what might be called a one-mode plus impurity mode for E_g. However, the nature of the modes for large c can only be surmised by continuity. The extra structure does seem to suggest that there are significant departures from randomicity, a conclusion that Lannin (1979b) also arrived at from his study of the two-phonon RS.

5.2. *Mixed I/VII crystals* *

5.2.1. *Li systems*

The infrared absorption for Li^6/Li^7F and LiH/D has been measured by Montgomery and Hardy (1965) for $0 \leqslant c \leqslant 1$. The first system is, as expected, one-mode whereas the second behaves in a kind of two-mode manner. For D concentrations of less than 5% only the ω_{TO}(LiH) absorption is seen. How-

* See, Notes added in proof on p. 124.

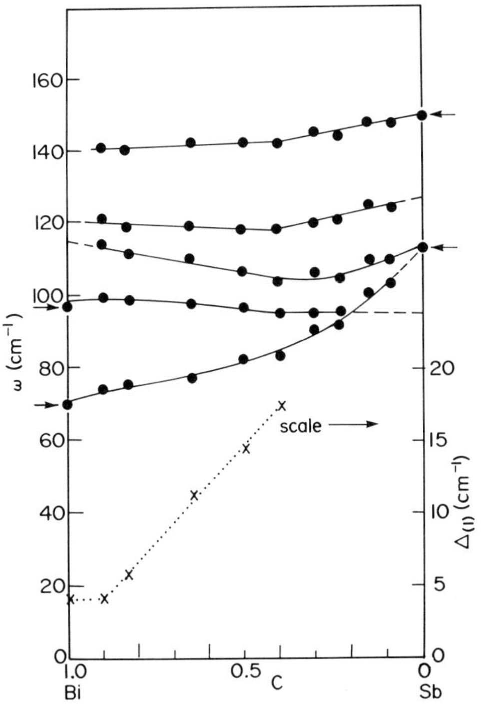

Fig. 20. RS peaks and shoulder positions for $Sb_{1-c}Bi_c$. The dotted line shows the linewidth of the lowest mode (Lannin 1979a).

ever, at greater concentrations of D there is a dramatic switch in intensity to a frequency close to ω_{TO}(LiD), i.e., to the impurity mode. The absorption frequency then remains essentially constant all the way to LiD. The local mode of H in LiD was not seen. Low-concentration theory is capable of producing this kind of behaviour but not the actual frequencies (Elliott and Taylor 1967, Jaswal and Hardy 1968). However, the REI model is not capable of giving even a qualitative fit to the observed behaviour.

5.2.2. $NH_4Cl_{1-c}Br_c$

This system has the special feature that it can exist in 3 phases (β, γ, δ) of different symmetries due to the orientational ordering properties of the NH_4^+ ions. Particularly interesting is the β/γ transition between the CsCl structure (β) and a tetragonal structure (γ) that has two molecules per unit cell yet is otherwise very little different from the β-phase as regards atomic separations. The reduction of the Brillouin zone in going from β to γ causes the M point to be folded back to the zone centre. Bauhofer et al. (1974a) have used this idea to examine the β-phase M-point phonons by doing RS in the γ-phase. They found a one-mode optic/ two-mode acoustic pattern, which is in accord with the ideas of GB. An inelastic neutron scattering experiment by Perry et al. (1978) to test these ideas was inconclusive.

The IR and RS of the β-phase have been measured by Bauhofer et al. (1974b). The IR results indicate that the LO and TO modes have a one-mode behaviour. The RS is due to defect-activated modes of A_{1g}, E_g and F_{2g} symmetry. In the pure materials, modes of the first two symmetries are predominantly ZB TA modes, but there are many modes having F_{2g} symmetry. However, the plot of frequency against concentration again has the form of one-mode optic/two-mode acoustic with the impurity modes (i.e., those at small c, $1-c$) splitting into separate A_{1g}, E_g and F_{2g} peaks at large c, $1-c$. The δ-phase (T_d^1 symmetry) shows a one-mode optic behaviour for the limited range of concentrations for which it exists.

The one-mode prediction of the GMP model indicates that there should be a weak impurity mode within or below the reststrahlen band. Bauhofer et al. (1976a) have in fact identified an impurity mode for $c > 0.54$ well below the reststrahlen band around 110 cm^{-1}. These data were used in fig. 11 to illustrate this kind of one-mode behaviour.

A brief summary of this work is given by Bauhofer et al. (1976b).

5.2.3. Mixed alkali halides

As the rock salt crystal structure is invariant under inversion the $\boldsymbol{q}=0$ optic modes are only infrared active. RS scattering can occur only due to the effects of disorder on the polarisability tensor or to two-phonon processes. Consequently the two kinds of experiment will be discussed separately.

5.2.3.1. Infrared reflectivity

Fertel and Perry (1969) have measured the IR for K/RbI and KCl/Br and extracted $\epsilon(\omega)$, hence obtaining the transverse and longitudinal frequencies. However, only IR curves have been presented for K/RbCl, K/RbBr and RbCl/Br by Angress et al. (1976, 1978) [see also Gledhill and Angress (1977) for the temperature-dependent IR of RbCl/Br] which makes a detailed analysis difficult. KCl/Br and K/RbCl are, without doubt, one-mode systems. The reflectivity curves of K/RbI (fig. 9) strongly suggest a two-mode behaviour. In fact the line shapes of Im $\epsilon(\omega)$ and $\epsilon(\omega)^{-1}$ indicate that this is a mixed-mode system, L/1 mode, T/2 mode. This behaviour has been confirmed by inelastic neutron scattering (Renker et al. 1983). However, both RbCl/Br (fig. 21) and K/RbBr (which have almost identical mass disorder parameters) appear to show a two-mode behaviour for less than 50% of the lighter impurity atom. The IR for RbCl/Br should also be compared with that for Ga/InP (fig. 28) which now appears to be mixed mode. It should be noted that RbCl/Br has been quoted as one-mode, but the identification was based on just one $c=0.5$ IR measurement of Fertel and Perry (1969). Angress et al.

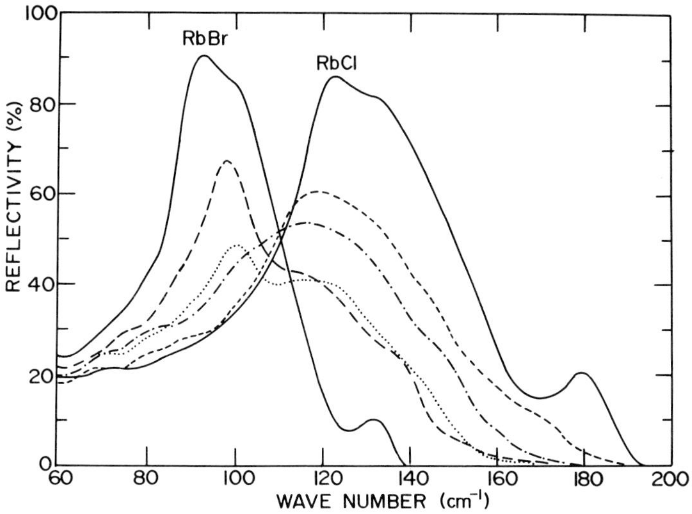

Fig. 21. Reflectivity for $RbCl_{1-c}Br_c$: (- - - - - -) $c = 0.25$, (· — · — ·) $c = 0.5$, (· · · · · ·) $c = 0.65$, (— — —) $c = 0.8$, (———) $c = 0.0$ and 1.0 (Angress et al. 1976).

(1976) were able to reproduce the observed behaviour using a 200 atom chain. Also the simple mass criteria indicates a two-mode behaviour and the GMP model indicates a mixed-mode behaviour. In fact, Srinivasan and Lakshmi (1978) have carried out a pseudo-atom shell-model calculation for K/RbBr, which for $\boldsymbol{q} = 0$ is just an REI calculation. They produced a mixed-mode behaviour, very similar to that fitted to K/RbI, yet apparently only aware of the measurement of Fertel and Perry, did not realise how successful they may have been. Further analysis of these systems would be worthwhile.

The IR has been measured for several mixtures of the form $(KCl)_{1-c}(RbBr)_c$ by Angress et al. (1978) with the results showing a smooth transition between the reflectivities of KCl and RbBr.

5.2.3.2. Raman scattering

The first RS experiments were those of Stekhanov and Eliashberg (1960) for $KCl_{1-c}Br_c$. Although mainly interested in two-phonon processes, they did conclude that defect activated first-order scattering was also occurring. The more recent work is that of Hurrell et al. (1968) for $c = 0.92$, Fertel and Perry (1971) and a particularly complete set of measurements by Nair and Walker (1971). Only A_{1g} and T_{2g} spectra were seen, although E_g is also permitted. In amongst the two-phonon peaks of A_{1g} symmetry they identified two first-order peaks, one at 120 cm^{-1} that exists only for $c < 0.3$ and the other moving linearly from 125 cm^{-1} ($c \approx 0$) to 155 cm^{-1} ($c \approx 1$). They suggested that the former is a Br resonant mode and the latter is a DATA mode. The main peak for the T_{2g} spectra also shows a linear behaviour but they were unable to identify its origin. The RS for $KBr_{1-c}I_c$ and $K_{1-c}Rb_cCl$ have been measured by Nair and Walker (1973). The results for KBr/I were similar to those for KCl/Br, there being a dominant first-order A_{1g} peak moving linearly with concentration (after a correction for small crystallite sizes). This peak was identified with a ZB TO mode, hence showing the one-mode/optic behaviour expected according to GB. Several other weak first-order peaks were observed. No E_g scattering was detected but a first-order T_{2g} peak was observed. Due to it being weak at $c \approx 0.1$ it was not identified.

The behaviour of the RS for K/RbCl was rather different as in this case the dominant first-order scattering was in the E_g and T_{2g} spectra. The E_g peak was clearly identified and behaves in a linear manner with c. It is quite likely due to ZB TO modes although the extrapolation to $c = 0$ did not quite give the pure crystal frequency. Once more the T_{2g} peak was very broad and its origin could not be identified.

5.2.4. Cu halides

The RS in the optic region was measured by Murahashi et al. (1973) for CuCl/Br and by Murahashi and Koda (1976) for CuCl/Br, CuCl/I and CuBr/I. CuBr/I behaves in a one-mode manner. However, the interpretation

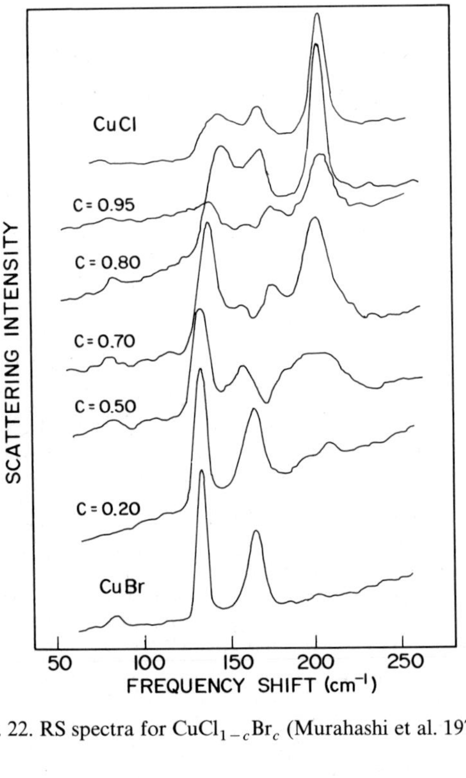

Fig. 22. RS spectra for $CuCl_{1-c}Br_c$ (Murahashi et al. 1973).

of the behaviour of the other two mixtures is complicated by the appearance of a third peak in CuCl (unexpected because of the wurtzite structure). However, both CuCl/Br and CuCl/I behave in a basically two-mode manner as can be seen for CuCl/Br in fig. 22. Murahashi and Koda were able to fit an MREI mode by assuming that the Br or I impurity mode emerges from the lowest of the three CuCl peaks. Livescu et al. (1981) have also measured the RS for CuCl/Br and, as well, have measured the polaritons for this system. They concluded that a three-oscillator model is needed to fit their polariton dispersion curves. Essential to their argument is the suggestion that the third peak in CuCl is due to the possibility that the Cu^+ ion can be off-centre. In the mixed crystal this would happen only for a Cu^+ ion surrounded by a complete tetrahedral of Cl^- ions and so this possibility dies out rather rapidly as the Br concentration is increased, leading to a conventional two-mode behaviour.

The low-frequency RS has been studied by Vardeny and Brafman (1979). A close group of low-frequency (≈ 35 cm^{-1}) DATA peaks and the corresponding two-phonon peaks were identified and showed a weak linear dependence on concentration. An anomalous temperature dependence of these peaks was explained by invoking the off-centre hypothesis. Talwar et al. (1981a) have calculated the RS using the single impurity model and were able to produce peaks at the appropriate frequencies. However, as the RS experiments used unpolarised light, a detailed comparison with experiment was not possible.

5.2.5. $AgCl_{1-c}Br_c$

The IR and RS have been measured by Bootz et al. (1974). Contrary to the simple mass criteria and to the GMP model, their results indicate that the system behaves in a one-mode manner. In fact the reststrahlen bands do overlap so this suggestion is not surprising. The two-phonon peaks in the RS spectra were not analysed in detail. However, the recent resonant RS experiments of Fujii et al. (1983) suggest that this is a type M system. As this is an indirect gap material nonzero $\boldsymbol{q}$ phonons are involved in the resonant scattering. Consequently Fujii et al. were able to measure the concentration dependence of the ZB phonons. They identified a local mode for $c \approx 1$ and an inband mode for $c \approx 0$. The concentration dependence of the ZB phonons is complicated due to the Ag^- vibrating in the optic modes in AgBr and also in the acoustic modes in AgCl. This means that the arguments of GB do not apply.

5.3. Mixed II/VI crystals *

5.3.1. $Mg_{1-c}Zn_cS$

This system crystallises in the wurtzite structure for $c > 0.7$ and the RS has

* See, Notes added in proof on p. 124.

been measured in this range by Novik et al. (1974). It shows a clear two-mode behaviour.

5.3.2. $Mg_{1-c}Zn_cTe$

The RS in the Zn-rich region has been measured by Zigone et al. (1978) and Vodopyanov et al. (1978) with the latter authors also measuring the IR. The RS results appear to be in good agreement whereas the infrared results for the MgTe modes are lower than the Raman results. This system was assumed to have the zincblende structure and is clearly two-mode. There is a peak at $\approx 100\ cm^{-1}$ in the Raman scattering which may be a defect-activated mode. First-order replicas due to resonant effects were seen in both sets of Raman studies.

5.3.3. $Mg_{1-c}Cd_cTe$

The IR and RS has been measured by Nakashima et al. (1972, 1973) for $c > 0.4$ and shows the system to be two-mode. The crystal structure in this region is zincblende but is partly polytype for $c < 0.7$. In the absence of information about MgTe they fitted the MREI to the nonzero c results by adjusting the parameters for MgTe. However, their extrapolated value for ω_{LO} ($320\ cm^{-1}$) departs quite strongly from that measured by Zigone et al. (1978), due, in part, to the wurtzite structure of MgTe. For $c = 0.65$ resonant scattering occurred enhancing a number of two-phonon peaks. However, there can be seen in each of the published figures an unidentified peak at $\approx 200\ cm^{-1}$ that could be a defect-activated mode.

5.3.4. $Zn_{1-c}Cd_cS$

The first measurement of the IR was by Lucovsky et al. (1967) who interpreted their results as indicating this to be a one-mode system. This was criticised by Lisitsa et al. (1969) who claimed from their transmission measurements that the system behaves in a more complicated manner. However, the RS results of Vodopyanov et al. (1971) and Vavilov et al. (1971) and the infrared reflectivity results of Mityagin et al. (1976) support the original suggestion. Note that this system crystallises in the wurtzite structure so Lucovsky et al. saw two sets of modes corresponding to propagation $\perp$ and $\parallel$ to the c direction. Similarly, a noninfrared active mode of symmetry A_2 was seen in RS. Mityagin et al. selected just modes $\perp$ to c and, using temperature studies, identified a small peak between ω_{TO} and ω_{LO} as defect activated. A plot of their frequencies against concentration looks just like a standard one-mode system, as predicted by the REI models, with weak subsidiary modes that are almost degenerate within the main band (see fig. 11). This extra peak has also been detected by RS for $c \approx 0, 1$ by Krol et al. (1977a). They agreed that the peak at $c = 0.05$ is due to impurity modes but noting its large width at $c = 0.94$ suggest that, in this case, there is a resonant interaction of the lattice vibrations. Vinogradov and Mityagin (1978), noting a narrow gap in the calculated density of states of ZnS at $\approx 330\ \mathrm{cm}^{-1}$ (Kunc 1974), suggested that the observed weak mode for small c is in fact a gap mode.

5.3.5. $Zn_{1-c}Cd_cSe$

RS has been used to follow the E_2 mode (Brafman and Mitra 1968) and an LO phonon (Brafman 1972) as a function of concentration. The frequencies change smoothly between the $c = 0$, 1 values and so, along with the detailed spectrum for $c = 0.525$ (Brafman 1972), a one-mode behaviour is indicated. This is supported by the spectra for $c = 0.2$, 0.5 obtained by Valakh et al. (1980). They also saw a strong asymmetry of the ω_{TO} line that they interpreted as being due to a resonant interaction with a lower frequency two-phonon band. It should be noted that the crystal structure is rather varied, ranging through cubic, two-phase or polytype and hexagonal.

5.3.6. $Zn_{1-c}Cd_cTe$

The IR has been measured by Harada and Narita (1971) and Vodopyanov et al. (1972) with both experiments indicating a two-mode behaviour over the concentration range measured, $0.3 < c < 1$. The RS of Vinogradov and Vodopyanov (1976) supports this conclusion as they were able to trace the CdTe transverse mode down to $c = 0.02$ removing any question that this system is partly two-mode. The REI model describes this system well and Harada and Narita show a good fit to their reflectivity data.

5.3.7. $Zn_{1-c}Hg_cTe$

The only reported results are the IR curves of Gebicki et al. (1975) for $c > 0.86$. These results suggest a two-mode behaviour for this system.

5.3.8. $ZnS_{1-c}Se_c$

Brafman et al. (1967, 1968) have presented a comprehensive set of data for this system having measured both the IR and RS at $c = 0.18$, 0.40, 0.67, 0.985. The system is clearly two-mode. This is confirmed by the RS of Schmeltzer et al. (1980) who concentrated on the two-phonon peaks in order to study the zone boundary phonons. They identified a two-mode behaviour for the TA(X) modes (fig. 23), with the ZnS optic modes transforming into the local mode of S in ZnSe, and the TO(X) mode of ZnSe connecting with the gap mode of Se in ZnS. However, contrary to the behaviour at $\boldsymbol{q} = 0$ where the LO mode of ZnSe also connects with this gap mode, they claim their infrared absorption results support the assumption that this mode transforms into the LA(L) mode of ZnS, which is as suggested by GB. There is also the indication of a DATA peak in the RS results. Schmeltzer and Beserman (1981b) have also studied the RS in this system but concentrated on the temperature dependence in order to investigate the possibility of a two-phonon resonant state.

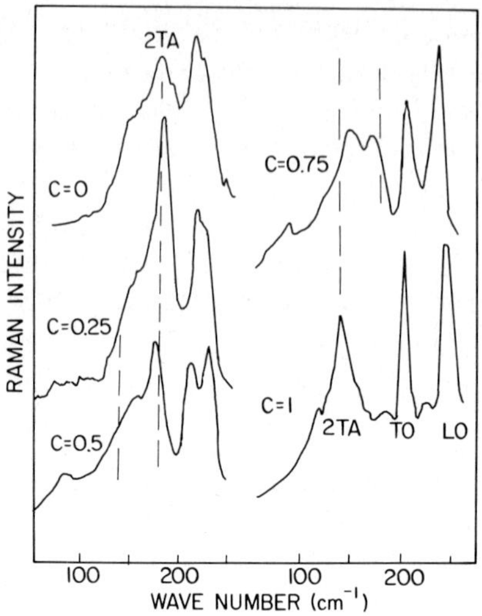

Fig. 23. RS spectra of $ZnS_{1-c}Se_c$ illustrating behaviour of two-phonon peak (2TA) (Schmeltzer et al. 1980).

It is of interest to note that the powder neutron diffraction studies of Beserman et al. (1976c) support a two-mode behaviour for this system (and a one-mode behaviour for ZnSe/Te, see below).

5.3.9. $ZnSe_{1-c}Te_c$

The RS has been measured by Nakashima et al. (1971) and Artamonov et al. (1979) establishing this system to be of the one-mode type. The two results disagree as to the concentration dependence of the TO mode even though both measurements were made at room temperature. Although the values shown by Nakashima et al. lie just under a linear interpolation between the end point frequencies, those of Artamonov et al. are almost constant. These authors discuss this effect in terms of an anharmonic resonance effect. They also suggest that the broad peaks below 150 cm^{-1} are due to two-phonon scattering.

5.3.10. $CdS_{1-c}Se_c$

The crystal structure for this system is wurtzite but there is little anisotropy in the optic frequencies of CdS and CdSe. Consequently, the distinction between modes $\perp$ and $\|$ to the c axis is often ignored in the experiments on the mixed crystals. The IR has been measured by Balkanski et al. (1966) ($E \perp c$), Verleur and Barker (1967a) ($E \perp c$, $E \| c$) and by Parrish et al. (1967). The system is clearly two-mode. RS experiments have been carried out by Parrish et al. (1967) (TO modes not observed) and by Chang et al. (1968). They are in agreement with the IR results. The two-phonon infrared absorption was measured by Balkanski et al. (1966) but not analysed. Parrish et al. (1967) and Beserman (1977) have measured both the two-phonon absorption and RS (see also Hayek and Brafman 1971) and interpret their results as indicating a two-mode behaviour for the optic modes and a one-mode behaviour for the acoustic modes. This is in agreement with the arguments of GB.

Verleur and Barker (1967a) have applied an appropriately generalised version of their CI mode to this system and obtained an excellent fit to the IR provided they allowed for departures from random disorder. The MREI also provides a good description of the concentration dependence of the transverse and longitudinal frequencies, as shown by Chang and Mitra (1968).

5.3.11. $CdS_{1-c}Te_c$

The IR and RS have been measured by Mityagin et al. (1977) for $c > 0.3$. However, they were not able to see a transverse peak in the RS. The various peaks show a smooth two-mode behaviour as a function of concentration in spite of a wurtzite/zincblende phase transition for $c \approx 0.75$. It is interesting to note that their extrapolation of the CdTe modes to $c = 0$ suggests and impurity mode of 157 cm^{-1} which is well below the reported gap in the density of states of CdS (Nusimovici et al. 1970) of 175 cm^{-1} to 195 cm^{-1}. This discrepancy questions the reliability of the model for the phonon dispersion curves used by Nusimovici et al.

5.3.12. $CdSe_{1-c}Te_c$

For $c \lesssim 0.4$ this system has the wurtzite structure, for $c \gtrsim 0.6$ it is zincblende and in between it is a mixture. However, as in similar systems, these phase changes appeared to have no effect on the optic frequencies. The IR has been measured by Vinogradov et al. (1973) and Gorska and Nazarewicz (1973, 1974). It is a two-mode system to which Gorska and Nazarewicz (1974) have successfully fitted the REI model. They ignored the wurtzite modification as there is little evidence of mode splitting for motion $\perp$, $\parallel$ to the c-axis. The RS has been carried out by Plotnichenko et al. (1977) who confirmed the above results for the longitudinal modes. They were also able to observe the E_2 mode, characteristic of the wurtzite structure as far as $c = 0.35$.

The two-phonon absorption has been measured and interpreted by Gorska et al. (1975). Their conclusion is shown in fig. 24, the pattern being rather different from what is to be expected according to GB. The mass ratios are those for the type of behaviour we call AO in fig. 17. It is not clear whether this difference is due to some complexity of the unknown CdSe dispersion curves or to the phase changes, or both.

5.3.13. $Cd_{1-c}Hg_cTe$

A particularly interesting feature of this system is that there is a transition from semiconductor to semimetal at $c = 0.85$. It is also predicted to be a one-mode system yet the experimental results clearly show two well-separated

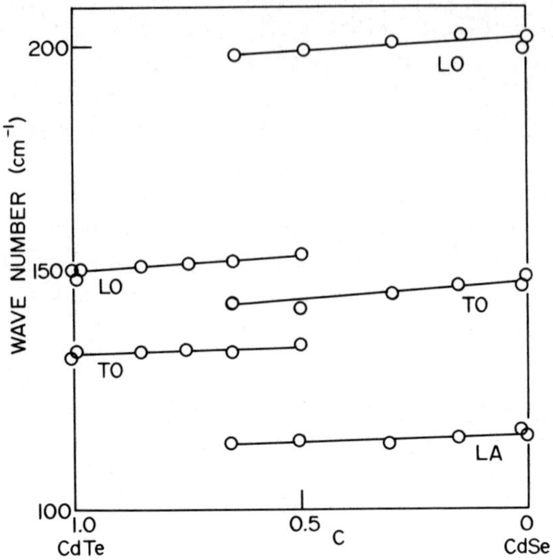

Fig. 24. Concentration behaviour of optic modes in $CdSe_{1-c}Te_c$ (Gorska et al. 1975).

modes (fig. 25). The IR has been measured by Carter et al. (1971), Kim and Narita (1971), Baars and Sorger (1972) and Polian et al. (1976) for large *c* and by Hoclet et al. (1979) for small *c*, with the results being in general agreement.

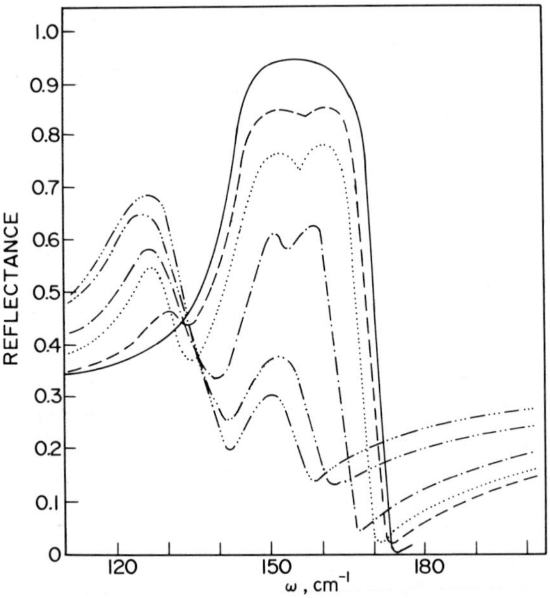

Fig. 25. Reflectivity of $Cd_{1-c}Hg_cTe$; (———) $c = 0$, (– – –) $c = 0.09$, (· · · · · ·) $c = 0.24$, (· – · – ·) $c = 0.52$, (– · · –) $c = 0.71$, (– · · · –) $c = 0.80$ (after Krozyrev et al. 1983).

Recently, Kozyrev et al. (1983) have extended the measurements across the concentration range and, noting the fine structure for $c < 0.5$, fitted a simplified version of the CI model of Verleur and Barker (1966). They found they needed to assume a departure from random disorder in order to obtain a good fit.

The RS for large c has been measured by Mooradian and Harman (1971) but their results are in poor agreement with the IR results. In particular the frequencies of the HgTe modes and the longitudinal CdTe mode show a much weaker concentration dependence than these determined by IR. The effect of the vanishing band gap has been looked for by Mooradian and Harman (1971) and Polian et al. (1976). The first set of authors increased the temperature from 2 K to 300 K for $c = 0.84$ causing the value of the band gap to increase from below to above the phonon frequencies. They saw no evidence of any coupling effects. Polian et al. examined the dielectric function for $c = 0.87$ and showed, using an REI model for the optic modes, the need to include interband effects in order to obtain a good fit to their reflectivity results. There appears to be no evidence for the breakdown of the adiabatic approximation in this situation.

Amirtharaj et al. (1983) and Tiong et al. (1984) have carried out polarisation studies for $c = 0.8$ and, besides the optic modes, have identified an A_1 mode (DA?), a forbidden TO mode and a mode whose origin, they claim, is due to a coupling between an LO mode and the electrons. A low-frequency DATA peak has been identified by Shen and Chu (1983) in their infrared absorption curves.

5.3.14. $Mn_{1-c}Zn_cTe$

Infrared reflectivity and absorption measurements have been made for $c = 0.95$, 0.90, 0.85 by Olszewski et al. (1981). The IR has an extra peak by $c = 0.95$ yet this system would seem to be a prime candidate to have a one-mode behaviour. The Zn and Mn masses are very similar and the values of ω_{TO} and ω_{LO} are very similar, 180 cm^{-1}, 207 cm^{-1} for ZnTe and 187 cm^{-1}, 210 cm^{-1} for hypothetical cubic MnTe as estimated by extrapolating the results for Mn/CdTe (Gebicki and Nazarewicz 1978). However, the Harada and Narita criteria is right on the border line between one- and two-mode behaviour. In fact, on analysis of the IR results Olszewski et al. found that the Mn causes a weak split mode just below ω_{LO}(ZnTe) which would suggest a one-mode behaviour. Valakh and Litvinchuk (1983a) have observed these weak TO and LO modes using RS and confirmed that $\omega_{LO} < \omega_{TO}$, as predicted by the REI model, using resonant RS.

5.3.15. $Mn_{1-c}Cd_cTe$

This is a system where the end point members have different crystal structures. However, all studies have been for $c > 0.3$ where the structure is zincblende. The IR measurements of Gebicki and Nazarewicz (1978) show a well-defined two-mode behaviour. This is confirmed by the RS measurements of Gebicki et al. (1980), Picquart et al. (1980) and Venugopalan et al. (1980, 1982). Several peaks other than the optic modes are present in the Raman spectra. Gebicki et al. and Picquart et al. were able to identify most of them as due to two-phonon scattering but there are a few that must be either DATA or DALA modes. Picquart et al. also measured the infrared absorption, confirming this identification. The lack of a concentration dependence for these frequencies suggests a two-mode behaviour for these zone boundary modes. Venugopalan et al. have carried out polarisation studies for $c = 0.6$ and have shown that the low-frequency (DATA and DALA) peaks have significant A_1 symmetry indicating that the polarisability tensor $\boldsymbol{a}$ is affected by the disorder (section 3.2). They also comment that the low-frequency structure resembles the CdTe density of states estimated by Sennett et al. (1969).

5.3.16. $Mn_{1-c}Hg_cTe$

The IR has been measured by Gebicki and Nazarewicz (1977) for $0.31 < c < 0.83$ in which range the crystal structure is zincblende. There is a clear well-separated two-mode behaviour. For $c > 0.8$ the free carrier concentration becomes very large leading to a strong plasmon/LO mode coupling. McKnight et al. (1978) have examined the IR in this range finding the expected strong dependence of the longitudinal frequencies.

5.4. Mixed III/V crystals *

5.4.1. $Al_{1-c}Ga_cP$

The IR work of Sobotta et al. (1975) shows this to be a two-mode system. However, the REI model does not give a good fit to the concentration dependence of the GaP TO mode (Sobotta and Riede 1978). The model requires that the frequency of this mode should show a maximum near $c = 0.5$ that is not seen experimentally. This was also found to be the case by Lucovsky et al. (1976) when they analysed their infrared results. They were able to obtain a fit by allowing each force constant to vary independently with concentration. Both experiments show subsidiary structure in the GaP reflectivity peak for large c but it was not discussed. It should be noted the two-phonon optical absorption has been measured by Berndt and Kopylov (1978) and their peak assignments confirm the two-mode nature of the system.

The RS has been studied by Bairamov et al. (1980) for $c \geqslant 0.77$. They observed an extra peak at 377 cm^{-1} as well as the expected four optic modes. It would seem reasonable to suppose that it is related to the extra structure seen in the reflectivity. However, there is insufficient experimental evidence to test this supposition.

* See, Notes added in proof on p. 125.

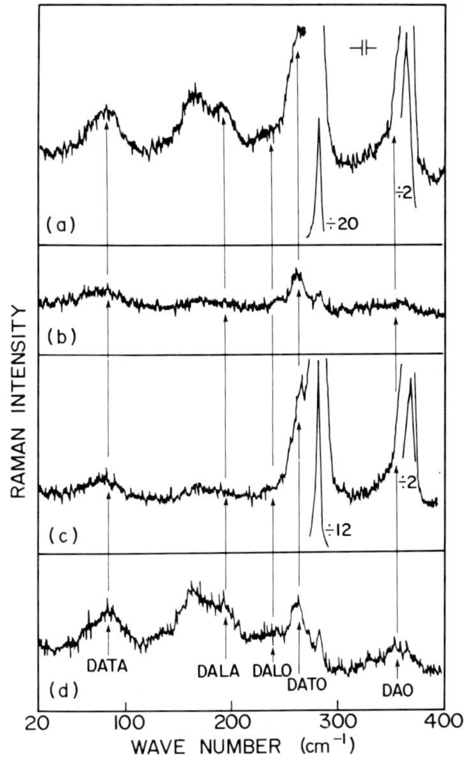

Fig. 26. Raman spectra for $Al_{0.2}Ga_{0.8}As$ for different symmetries: (a) A_1+E+F_2, (b) 3E, (c) F_2, (d) A_1+4E (Saint-Cricq et al. 1981).

5.4.2. $Al_{1-c}Ga_cAs$

The IR was first measured by Ilegems and Pearson (1970) who found this system to be a two-mode system with no extra structure. RS studies (Kawamura et al. 1972, Tsu et al. 1972) do show structure other than what can be identified as the two LO modes and the two TO modes. In the Al rich crystals an impurity mode just below the gap mode is seen plus a broad feature at 200 cm^{-1}. This was interpreted as being due to DALA modes activated by the effects of disorder on the electronic wavefunctions. Kim and Spitzer (1979) have reexamined this system carefully using the same sample for both the infrared and Raman studies. They have confirmed the above observations and, as well, have observed the DALA peak in Ga-rich crystals. Further structure was seen below both the local mode and GaAs LO mode peaks. By using different polarisations Saint-Cricq et al. (1981) have separated the different representations that contribute to the Raman scattering for $c=0.8$ (fig. 26). They also find DATA, DALO peaks whose positions correlate quite well with

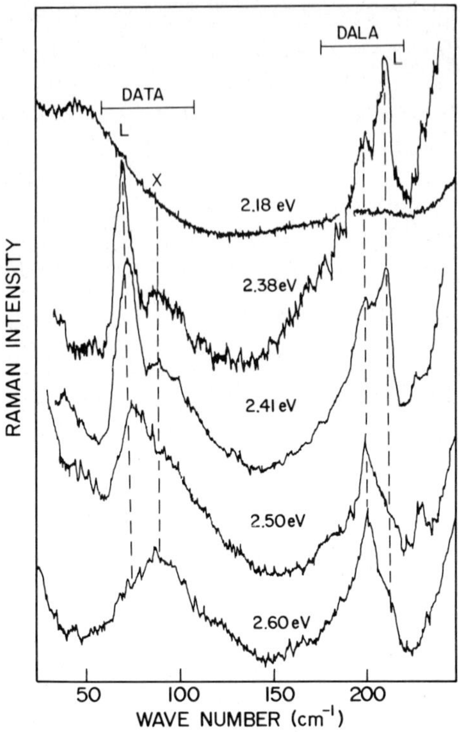

Fig. 27. Resonant RS for $Al_{0.44}Ga_{0.56}As$ (Carles et al. 1983).

the low concentration *t*-matrix calculations of Talwar et al. (1981b). It has also been pointed out (Carles et al. 1982) that the experimental $A_1 + 4E$ component, which does not contain LO or TO contributions, compares well with calculated densities of states. Jusserand et al. (1981) and Jusserand and Sapriel (1981) and Carles et al. (1983) have used the resonant Raman effect to enhance the intensities of the DATA and DALA peaks in order to help with their identification. Figure 27 shows this effect very clearly. Other resonant Raman studies are those of Shah et al. (1973) and Balkanski et al. (1978), the latter observing high-frequency replicas of the first-order spectrum.

The CPA calculation of Bonneville (1984), which includes the Coulomb forces in an approximate manner, reproduces the behaviour of the optic modes quite well. Also, Sobotta and Riede (1974) have used their MREI modes to calculate the IR, reproducing the experimental results reasonably well.

5.4.3. $Al_{1-c}Ga_cSb$

The IR data of Lucovsky et al. (1975) and Ipatova et al. (1976a) show this to be a two-mode system. The MREI approximation fits the data of Lucovsky et al. quite well although they chose to fine tune the approximation by using different scaling parameters [see eq. (3.24)] for the different force constants in order to remove a small systematic deviation. The RS has been measured by Biryulin et al. (1979) and fitted to the CI model of Zinger et al. (1976). However, there are systematic differences between the infrared and Raman results in the local and gap mode regions. The Raman spectra determined by Charfi et al. (1978) show a discrepancy just for $\omega_{\mathrm{AlSb,Ga}}$. They also found a DALA at 140 cm^{-1} for $c < 0.5$. Charfi et al. were able to obtain a very good fit to their data using the version of the REI approximation due to Jahne (1976c).

The two-phonon absorption has been measured by Ipatova et al. (1976b) and the two-phonon Raman scattering by Charfi et al. (1980). The latter authors found a two-mode behaviour for the ZB optic phonons and a one-mode behaviour for the ZB acoustic phonons, in accordance with the GB prediction.

5.4.4. $Ga_{1-c}In_cP$

This was established to be a one-mode system by Lucovsky et al. (1971) using both IR and RS. The IR is shown in fig. 28. However, the pronounced structure did not lead to a type II assignment as can be seen from the dielectric constant in fig. 29. However, subsidiary TO and LO modes were

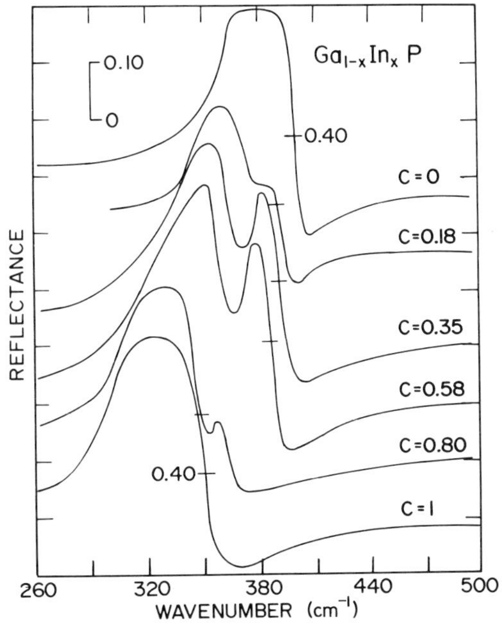

Fig. 28. Reflectivity of $Ga_{1-c}In_cP$ (Lucovsky et al. 1971).

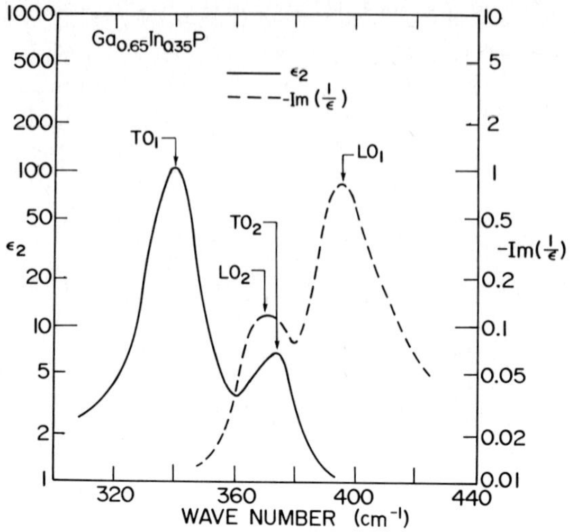

Fig. 29. Im $\epsilon(\omega)$ and $\mathrm{Im}(\epsilon(\omega)^{-1})$ for $Ga_{0.65}In_{0.35}P$ (Lucovsky et al. 1971).

identified at frequencies between the dominant modes at all concentrations, in accord with the REI behaviour shown in fig. 11. Their origin has been the cause of some argument.

Hirlimann et al. (1976a, b) and Beserman et al. (1976a, b) on the basis of their Raman studies, particularly for small c, find just an extra mode that is longitudinal in nature and suggest that it cannot be an impurity mode but is due to an anharmonic coupling between the LO mode and the phonon continuum. Jahne et al. (1979) have also measured the IR and RS, confirming the mainly longitudinal nature of the extra mode, particularly for $c > 0.1$, but explaining its origin as being the extra inband mode expected in a one-mode REI calculation.

A more recent investigation by Jahne et al. (1981) suggests that the impurity mode at $c \sim 1$ is in fact just 6 cm^{-1} above the InP band, thus being a local mode. If this is the case then the system behaves in a mixed-mode manner and there must then be a rapid switch in intensity from the InP LO mode to the longitudinal component of the Ga local mode as $1 - c$ increases. Based on their work for Ga/InP/As, Jusserand and Slempkes (1984) also claim that Ga forms a local mode just 2 cm^{-1} above the InP band [they do not reference Jahne et al. (1981)]. Consequently, they suggest that this is an M-type system. Bedel et al. (1984a) also support the idea that this is not a type I system and present the detailed collection of points shown in fig. 30.

Besides this mode, the Raman studies of Bairamov et al. (1981a, b) have identified a DATA mode at 87 cm^{-1}. The Green function calculation for

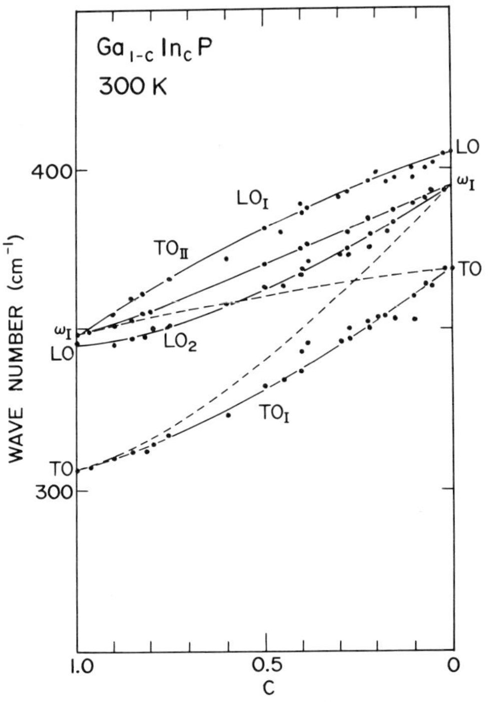

Fig. 30. Optic mode concentration dependence of $Ga_{1-c}In_cP$ (Bedel et al. 1984a). (Full lines are guides for the eye.)

small c of Kleinert and Jahne (1981) appears to explain these observed features. An F_2 component accounts for the DATA mode and the infrared inband mode, whereas the inband mode as seen by RS is dominated by the A_1 component. Two-phonon infrared absorption has been measured by Beserman and Schmeltzer (1977) (see also Beserman et al. 1978) and Ulrici and Jahne (1978) have investigated the two-phonon RS. It was found that the acoustic as well as optic ZB modes also behave in a one-mode manner. This is contrary to the GB prediction and to the ZB REI calculation of Jahne and Ulrici (1980). However, the CPA calculation of Kleinert (1984) does produce a one-mode behaviour for the $\boldsymbol{q} = 0$ and ZB phonons.

In conclusion, though, it should be noted that very recently Galtier et al. (1984) have questioned the existence of $\boldsymbol{q} = 0$ modes in this system. Their argument is based largely on a linear pressure dependence of the LT splitting which, they claim, indicates that the observed LO and TO modes cannot originate from the zone centre. (They found this dependence to be quadratic for Ga/P/As.)

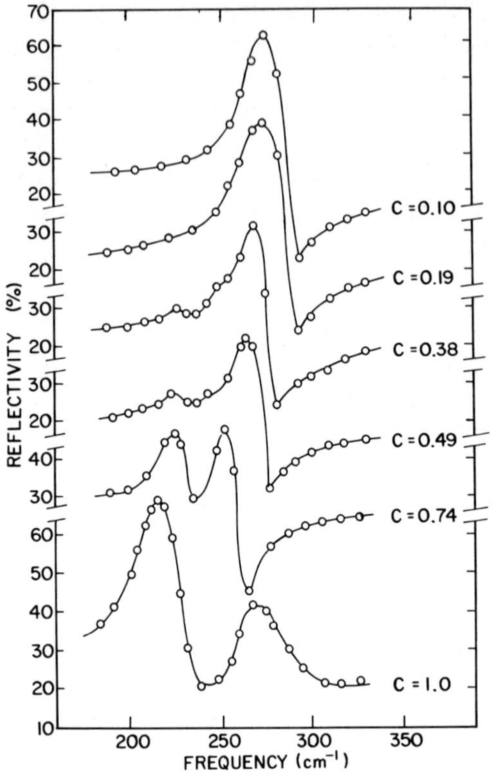

Fig. 31. Reflectivity of $Ga_{1-c}In_cAs$ (Yamazaki et al. 1980). (Structure between 250 and 300 cm^{-1} for $c = 1.0$ is an interference effect due to small sample thickness.)

5.4.5. $Ga_{1-c}In_cAs$

The IR measurements of Brodsky and Lucovsky (1968), Lucovsky and Chen (1970) and Yamazaki et al. (1980) all show partly two-mode behaviour (fig. 31), as the contributions due to the InAs modes disappear below $c = 0.2$. The effect is well reproduced by the Bethe lattice CPA calculation of Kleinert (1982).

Yamazaki et al. (1980) have applied a much simplified version of the CI mode of Verleur and Barker (1966) to this system. They found that the oscillator strength for InAs modes becomes very small as c decreases below 0.2 accounting for the above behaviour. However, they do not show a full comparison between calculation and experiment.

5.4.6. $Ga_{1-c}In_cSb$

The infrared reflectivity studies of both Brodsky et al. (1970) and Gasanly et al. (1971) show this system to be a partly two-mode system rather like

Ga/InAs. Gasanly et al. have fitted an REI model to their experimental results and the agreement seems good, although no direct comparison is shown. However, they did calculate the reflectivity for $c = 0.25$, 0.9, and indeed found that oscillator strength of the InSb mode for $c < 0.25$ is sufficiently small that this mode produces no structure in the reflectivity. The CPA calculations of Kleinert (1982, 1984) reproduce this partly two-mode behaviour as can be seen in figs. 6 and 10, the latter comparing the calculation with the experimental results of Brodsky et al.

5.4.7. $GaP_{1-c}As_c$

The IR results of Verleur and Barker (1966) and Chen et al. (1966) indicate this to be a two-mode system but with extra structure. It was for this reason that Verleur and Barker introduced their CI model and the comparison between that model and experiment is very good (fig. 13), the fit indicating appreciable short-range order. Teicher et al. (1984) offer some support for this conclusion based on their two-phonon RS measurements.

By their choice of polarisation geometries in the RS experiments, Strahm and McWhorter (1969) were able to separate the longitudinal and transverse modes. It is of interest to note that for $c = 0.15$ the GaAs modes do not show dominant longitudinal or transverse behaviour. This is as expected for impurity modes but is contrary to the predictions of isodisplacement models. The actual frequencies compare well with those predicted by the model of Verleur and Barker although the predicted fine structure frequencies tend to be too low. Some of this structure could well be called defect activated as is done in other materials. Hirlimann et al. (1976a) have also presented polarisation studies of the Raman scattering with these results (and those for Ga/InP) being interpreted by Beserman et al. (1976a) as due to anharmonic mode coupling. A small peak just below ω_{LO} for small c has been identified by Schmeltzer and Beserman (1980) as a localised mode on the basis of their application of the "replica" trick.

The two-phonon transmission has been measured by Chen et al. (1966), Osamura and Murakami (1972) and, as noted above, Teicher et al. (1984). Their spectra show a clear two-mode behaviour. The GB criteria indicate a type OA behaviour which includes the LA mode of GaP transforming into the LO mode of GaAs. The data do not support this suggestion. However, the calculation of Jahne and Ulrici (1980), which incorporates the theoretical finding that it is the Ga atoms that move in the LA(X) mode in both GaP and GaAs, do agree with experiment.

5.4.8. $GaAs_{1-c}Sb_c$

The infrared reflectivity work of Lucovsky and Chen (1970) shows this to be a partly two-mode system, with the GaSb modes disappearing for $c \lesssim 0.9$. The early work of Potter and Stierwalt (1964) was confined to smaller values of c

and so missed this extra structure. This behaviour is reproduced by the CPA calculation of Kleinert (1984) which also agrees with the experimental finding that the two-mode behaviour exists over a larger range of c in Ga/InSb as compared to GaAs/Sb.

5.4.9. $InP_{1-c}As_c$

This was established to be a two-mode system by the IR studies of Kekelidze et al. (1973). The earlier work of Oswald (1959) did not extend to low enough frequencies to see the InAs modes. Kekelidze et al. used the version of the REI model that includes local and nonlocal charge (Gasanly et al. 1971) to calculate the reflectivities and found good agreement with experiment. The REI calculation of Talwar et al. (1980) which neglects the local/nonlocal division also agrees well with experiment.

The RS work of Carles et al. (1980) and Bedel et al. (1984b) also shows two-mode behaviour both for the $\boldsymbol{q} = 0$ modes and for the optic zone boundary modes that contribute to the two-phonon peaks. The acoustic ZB modes contributing to the two-phonon peaks behave in a one-mode manner as suggested by the GB criteria. These Raman studies also show the presence of DATA and DALA peaks which behave in a one-mode manner. The low concentration calculation of Talwar et al. (1980) reproduces these peaks but the positions of the peaks are not in good agreement with experiment, probably due to a deficiency in the lattice dynamics model used for InP. Bedel et al. have also examined the resonant Raman scattering in this system.

5.4.10. $InAs_{1-c}Sb_c$

The IR studies of Lucovsky and Chen (1970) show this to be a partly two-mode system. However, only the reflectivities for $c = 0.15$, 0.75 and 0.80 are shown making it difficult to compare with even an REI model.

5.4.11. $Al_{1-x}Ga_xP_yAs_{1-y}$

The IR has been measured by Lucovsky et al. (1974). They were able to interpret the structure in terms of the characteristic ω_{TO} frequencies of the various constituent materials. For instance in GaP-rich alloys there is a band clearly identifiable with GaP and with other peaks that appear at frequencies near the ω_{TO}'s of GaAs and AlP. These were interpreted as impurity modes leading to a three-mode behaviour (the frequency for AlAs is close to that of GaP and so is not expected to appear separately). Sen and Lucovsky (1975) have carried out a 1D ATA calculation that supports this kind of identification.

5.4.12. $Ga_{1-x}In_xP_{1-y}As_y$

The IR for $x = y$ has been reported by Sirota et al. (1977). By comparison with the results for ternary mixtures such as GaP/As etc. they were able to identify

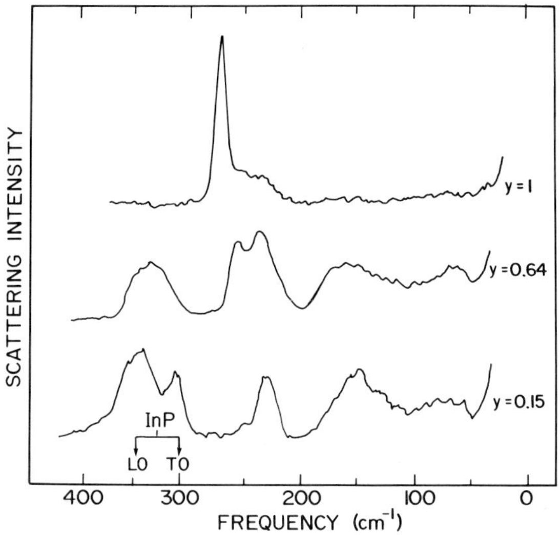

Fig. 32. Raman spectra for $Ga_xIn_{1-x}P_{1-y}As_y$ with $y = 2.2x$ (Pinczuk et al. 1978).

the various peaks in a manner similar to that of Lucovsky et al. (1974) for Al/GaP/As. Zinger et al. (1979) have examined the IR of two different mixtures ($x = 0.72$, $y = 0.58$; $x = 0.89$, $y = 0.65$) and analysed their results using an extension of their CI model. This has four possible cells and consequently four ω_{TO} modes. Their calculated values agree fairly well with experiment with both the experimental and calculated values changing little with concentration and being roughly similar to those of the four constituents. The RS result of Jusserand and Slempkes (1984) support this conclusion.

A special set of mixtures of interest is those with $y/(1-x) \approx 2.2$ as they all have the same lattice constant as InP which can be used as a substrate. The RS has been measured by Pinczuk et al. (1978) (see also Portal et al. 1979) and the IR by Amirtharaj et al. (1980) with generally fair agreement. Typical Raman spectra are shown in fig. 32 with the collected frequencies of both experiments in fig. 33. The high-frequency InP modes agree very well but there are inconsistencies in the GaAs, InAs regions with the infrared results showing more structure. Amirtharaj et al. did report some discrepancies between samples from different sources, as is shown in fig. 33.

5.5. Miscellaneous mixed crystals *

5.5.1. Co/NiO

Due to the very small mass difference between Co and Ni it is not surprising that the IR results of Gielisse et al. (1965) show a one-mode behaviour.

* See, Notes added in proof on p. 126.

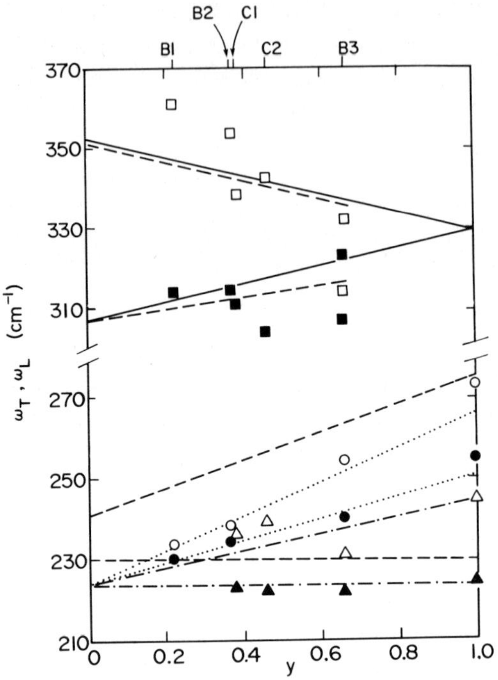

Fig. 33. Optic frequencies in $Ga_xIn_{1-x}P_{1-y}As_y$; □ InP-like, ○ GaAs-like, △ InAs-like (TO solid, LO open). (- - - - - -) results of Pinczuk et al. (1978). B and C refer to samples from different sources. The lines are from a model (Amirtharaj et al. 1980).

5.5.2. $M(H_cF_{1-c})_2$, M = Ca, Sr, Ba

The local modes due to H impurities are usually measured at very low H concentrations. However Vergnat-Grandjean et al. (1979, 1982) have measured the infrared absorption for values of c up to 0.64. The MF_2 peak changes little with c but in CaF_2 develops a low-frequency two-phonon peak. There is evidence of an LT splitting of the H impurity mode at high c and the results also show modes that were attributed to MH_2. The situation is complicated by the presence of several peaks due to H clusters.

5.5.3. Ca/SrF_2, Sr/BaF_2

Verleur and Barker (1967b) have measured the IR spectra for these systems and RS spectra have been measured by Chang et al. (1966) and Lacina and Pershan (1970). Both systems have a one-mode behaviour, although there is a small amount of structure in the reflectivity curves. Verleur and Barker [see also Barker and Verleur (1967)] have fitted their CI model to the reflectivity, successfully reproducing the structure, as well as both the transverse frequency and the Raman frequency, as a function of concentration. The low c theory has been applied to Ca/SrF_2 by Lacina and Pershan. Estimating force constant changes from the pure crystal parameters they were able to obtain the one-mode behaviour for the Raman frequency and calculate the reflectivity for 25% Sr in reasonable agreement with experiment (fig. 5).

5.5.4. Ca / PbF_2

Ciepielewski and Kosacki (1982) have measured the RS spectra of this system for $0.3 < c < 0.7$. They claim a type II behaviour but their peaks are very broad and at most the main peak only develops a shoulder.

5.5.5. Sr / CdF_2

Values for the Raman frequency, ω_{LO} and ω_{TO} have been reported by Gobeau et al. (1978). They have presented an REI calculation for this one-mode system which reproduces the linear concentration dependence of the Raman frequency, but not the reported nonlinear behaviour of ω_{TO} and ω_{LO}.

5.5.6. KMg / NiF_3

The IR has been measured by Perry and Young (1967) and Barker et al. (1968). There are three infrared active modes in the pure crystals and consequently three reststrahl bands that are, in fact, well separated. In the mixed crystals the central band tends to split in a two-mode manner (fig. 34). Barker et al. have fitted an REI model adapted to the larger unit cell of the perovskites. Using relatively few parameters they obtain a good fit to almost all the infrared active transverse and longitudinal modes. As expected, the high- and low-frequency regions show a one-mode behaviour. It is interesting to observe that both the experimental and calculated modes in the middle region show the mixed-mode behaviour seen in KRb/I.

5.5.7. $Ge_{1-c}(GaSb)_c$

This has been studied by RS (Krabach et al. 1983) and appears to show a II′ behaviour as the GaSb peak disappears for $c < 0.27$. This can be explained by the fact that Ge does not have a gap in its density of state. They interpret some low-frequency peaks as being DATA and DALA modes.

5.5.8. PbSe / Te

This is an interesting system in that the simple mass criterion is on the one/two-mode border line, there are no gaps in the pure crystal phonon density of states and yet the IR of Finkenrath et al. (1979) shows what looks like a mixed-mode behaviour. However, both the extrapolated impurity mode frequencies are inband. An examination of the phonon densities of states suggests that a resonant mode due to Te replacing Se could be expected at about the experimental frequency but any resonant frequency of Se in PbTe would be significantly higher than that observed. Both pure crystals have high dielectric constants (PbTe is paraelectric) and a large LT splitting. Conse-

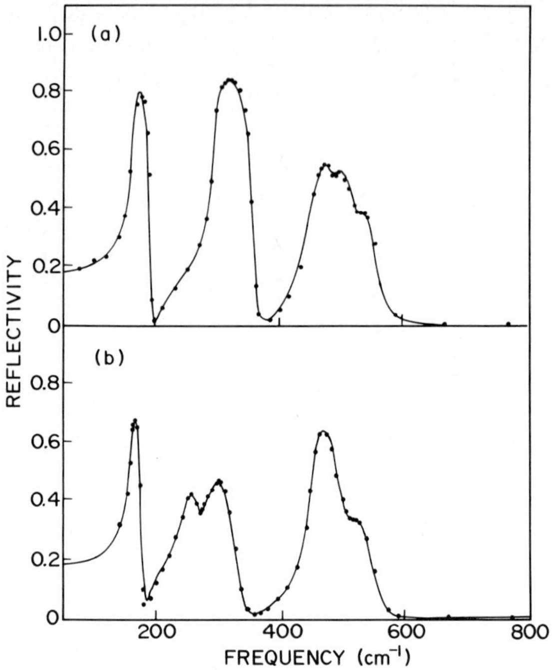

Fig. 34. Reflectivity of (a) $KMgF_3$, (b) $KMg_{0.5}Ni_{0.5}F_2$. Points are experimental results and lines are an oscillator fit (Barker et al. 1968).

quently the usual criteria are probably invalid. Finkenrath et al. note that the simple REI model of Genzel et al. (1974) does not describe the system and the present author has found that the full REI model yields complex frequencies due to a large negative f_{BC}.

5.5.9. Zr, Hf(S / Se)$_3$

These materials have a chain-like structure and have a special mode behaviour. This has been established for $Zn(S/Se)_3$ by the RS studies of Jandl and Provencher (1981), Provencher et al. (1982) and Zwick et al. (1983) and by the IR study of Deslandes and Jandl (1984). $Hf(S/Se)_3$ has been studied by RS by Zwick et al. (1982) and their results are shown in fig. 35. There are three classes of modes, I, II and III. Group I modes are associated with interchain forces, no doubt of long range, and show a type I behaviour. Internal deformation chain modes (Group II) have a type II behaviour. However the Group III modes due to n.n. chalcogen–chalcogen interactions behave in a type II manner but with an extra impurity mode that dies out as c goes to zero and one. It is assigned to a S–Se pair in analogy to the assignment of the extra modes seen in Si/Ge, Se/Te and Sb/Bi. This is sometimes called three-mode behaviour.

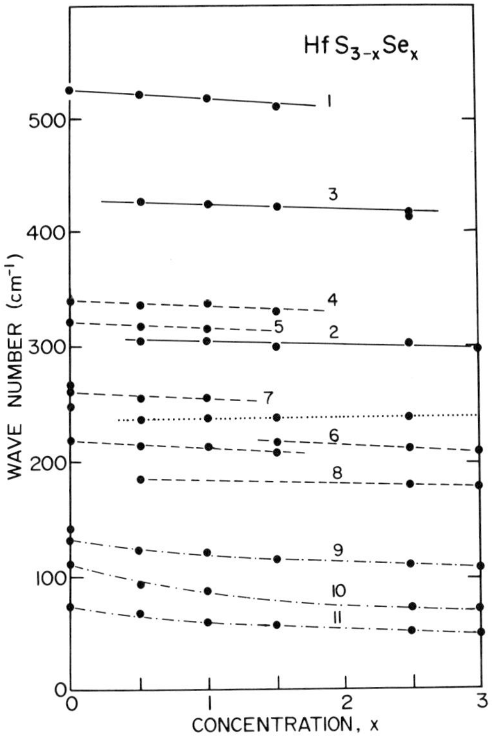

Fig. 35. Mode frequencies for $Hf(S_{1-c}Se_c)_3$. Lines 1, 2, 3 – Group III; lines 4, 5, 6, 7, 8 – Group II; lines, 9, 10, 11 – Group I (Zwick et al. 1982).

5.6. Layer crystals *

5.6.1. GaS / Se

This system has received a lot of attention. The IR has been measured both for $E \perp C$ (Abdullaev et al. 1976, Allakhverdiev et al. 1979) and $E \parallel C$ (Riede et al. 1980). The results show a standard two-mode behaviour to which the MREI and CI(Z) models have been fitted. RS spectra have been obtained by Hayek et al. (1973), Mercier and Voitchovsky (1974), Gamarts et al. (1978) and Abdullaev et al. (1979). Recently, Gasanly et al. (1983) have reported on an extensive study of this system using polarised RS and IR. There are many Raman active modes in the pure crystals and the mixed crystal spectra are similarly complicated. Most modes show a one-mode behaviour with a few apparently two-mode in character. However, the interpretation is complicated by a structural phase transition that appears to take place gradually as c

* See, Notes added in proof on p. 128.

Table 5
Mixed layer crystals.

Ga/InSe	RS, I?	Belenkii et al. (1978)
		Vinogradov et al. (1979)
GaS/Se	see text	
Ge/SnS	RS, I	Golubev and Vodopyanov (1981)
Ge/SnSe	insuff. data	Golubev and Vodopyanov (1981)
GeS/Se	RS, I+II	Vodopyanov et al. (1979a)
SnS/Se	see text	
$Tl(Ga/In)S_2$	RS, I+II	Gasanly et al. (1980)
$TlGa(S/Se)_2$	RS, I+II	Bakhyshov et al. (1979)
	IR	Gasanly et al. (1979)

Table 6
General mixed crystals.

H/DCl	IR, RS, I+l.m.	Bureau and Brunel (1981)
naphthalene h_8/d_8	IR, I+II	Ahlgren and Kopelman (1980)
$(NH_4/K)_2CuCl_4 \cdot 2H_2O$	RS, I+II	Sahni and Bansal (1978)
$Ca(Mo/W)O_4$	RS, I+II	Karapetyan et al. (1976)
		Krol et al. (1977b)
$Ni(S/Se)_2$	RS, I+imp	Lemos et al. (1980)
$Cu(Ga/In)S_2$	IR, I	Bodnar et al. (1978)
$CuGa(S/Se)_2$	RS, I+II	Anedda et al. (1983)
(ZnSe)/(GaAs)	IR, II	Lucovsky and Mikkelsen (1976)
(ZnTe)/(CdSe)	RS, I+II	Valakh et al. (1983b)
$Zn/Cd\ Cr_2Se_4$	IR, I	Wakamura et al. (1976)
$Zn/Hg\ Cr_2Se_4$	IR, I+II	Wakamura et al. (1973)
$(GaAs)/(Ga_2Se_3)$	IR, ?	Lucovsky and Mikkelsen (1976)
$Ga_2(S/Se)_3$	IR, II	Musaeva et al. (1977)
$(Sr/Pb)TiO_3$	RS, I+imp	Burns and Dacol (1981)
Sr/EuS	IR, I	Mutzenich et al. (1983)
Zr/HfO_2	RS, I	Ignatev et al. (1980)
$(Sb/Bi)_2Te_3$	RS, I	Richter et al. (1977)
SbSBr/I	IR, I	Bartzokas and Siapkas (1976)
$Cs(Mg/Co)Cl_3$	RS, I	Johnstone et al. (1981)
$Ba_2(Na/K)Nb_5O_{15}$	RS, I+II	Boudou and Sapriel (1980)
$Ba/PbTiO_3$	RS, I?	Burns (1974)
$Eu/SrAs_3$	RS, I+II	Oles et al. (1981)
$Eu(P/As)_3$	RS, I+II	Oles et al. (1981)
$Tl(InS_2)/(GaSe_2)$	IR, I+II	Allakhverdiev et al. (1982)
TlS/Se	RS, I+II	Allakhverdiev et al. (1980)
	IR, I+II	Aliev et al. (1983)
$Hg(Br/I)_2$	RS, I+II	Nakashima et al. (1974)
$Bi_2(Se/Te)_3$	RS, I+II	Richter et al. (1977)
$Hg_2(Cl/Br)_2$	RS, I, II+imp	Barta et al. (1982)
	IR	Limonov and Markov (1983)

changes. The IR shows no evidence of this change but there are indications of a phase change in the high-frequency absorption spectra of Allakhverdiev and Tagyev (1977). It may be noted that the RS spectra of GaSe/Te (Cerdeira et al. 1977) show large effects due to a structural phase change at finite c.

5.6.2. $Sn(S/Se)_2$

The RS spectra of Jandl et al. (1978), Vodopyanov et al. (1979b) and Walsh et al. (1980) all show a two-mode behaviour. Walsh et al. note that the local and gap modes frequencies do not split into separate A_{1g} and E_g components until the rather large concentration of 25%. They suggest that this reflects the 2D nature of the crystals. The IR has been measured by Harbec and Jandl (1982) and requires a 3 oscillator fit. It also shows a two-mode behaviour with an extra, almost degenerate, impurity mode.

References

Abdullaev, G.B., K.R. Allakhverdiev, G.L. Belenki, R.K. Nani, E.Y. Salaev and M.M. Tagiev, 1976, Sov. Phys.-Semicond. **9**, 1313.

Abdullaev, G.B., K.R. Allakhverdiev, R.K. Nani, E.Y. Salaev and M.M. Tagyev, 1979, Phys. Status Solidi (a) **53**, 549.

Agrawal, B.K., 1969, Phys. Rev. **186**, 712.

Agrawal, B.K., 1981, Solid State Commun. **37**, 271.

Agrawal, B.K., S. Tripathi, A.K. Misra and D.N. Talwar, 1979, Phys. Rev. B **19**, 5277.

Ahlgren, D.C., and R. Kopelman, 1980, Chem. Phys. **48**, 47.

Aliev, R.A., K.R. Allakhverdiev, R.M. Sardarly and N. Safarov, 1983, Sov. Phys.-Solid State **25**, 520.

Allakhverdiev, K.R., and M.M. Tagyev, 1977, Phys. Status Solidi (a) **39**, K111.

Allakhverdiev, K.R., S.S. Babaev and M.M. Tagyev, 1979, Phys. Status Solidi (b) **93**, K67.

Allakhverdiev, K.R., M.A. Nizametdinova, N.Yu. Safarov, L.K. Vodopyanov and L.V. Golubev, 1980, Phys. Status Solidi (b) **102**, K117.

Allakhverdiev, K.R., M.M. Godzhaev, A.I. Nadzhafov and R.M. Sardarly, 1982, Sov. Phys.-Solid State **24**, 1442.

Amirtharaj, P.M., G.D. Holah and S. Perkowitz, 1980, Phys. Rev. B **21**, 5656.

Amirtharaj, P.M., K.K. Tiong and F.H. Pollack, 1983, J. Vac. Sci. Technol. A **1**, 1744.

Anedda, A., G. Bongiovanni, F. Raga, E. Fortin and M. Quintero, 1983, Nuovo Cimento **2 D**, 1950.

Angress, J.F., W.G. Chambers, G.A. Gledhill and W. Smith, 1976, J. Phys. C **9**, 3717.

Angress, J.F., G.A. Gledhill, J.D. Clark, W.G. Chambers and W. Smith, 1978, Infrared Reflection Spectra of Some Mixed Alkali Halides, in: Proc. Int. Conf. Lattice Dynamics, Paris, 1977, ed. M. Balkanski (Flammarion, Paris) p. 422.

Artamonov, V.V., M.Ya. Valakh and N.I. Vitrikhovskii, 1979, Sov. Phys.-Solid State **21**, 1015.

Baars, J., and F. Sorger, 1972, Solid State Commun. **10**, 875.

Bairamov, B.H., V.N. Bessolov, E. Jahne, Y.P. Yakovlev, V.V. Torporov and S.B. Ubaidullaev, 1980, Sov. Tech. Phys. Lett. **6**, 618.

Bairamov, B.H., V.V. Toporov, Sh.B. Ubaidullaev, L. Hildisch and E. Jahne, 1981a, Solid State Commun. **37**, 963.

Bairamov, B.H., V.N. Vishnevskii, M.I. Demchuk, V.V. Toporov, S.B. Ubaidullaev, L. Hilisch and E. Jahne, 1981b, Sov. Phys.-Solid State **23**, 13.
Bakhyshov, N.A., N.M. Gasanly, B.M. Yavadov, V.I. Tagirov and Sh.M. Efendiev, 1979, Phys. Status Solidi (b) **91**, K1.
Balkanski, M., 1980, Optical Properties due to Phonons, in: Optical Properties of Solids, ed. M. Balkanski (North-Holland, Amsterdam) ch. 8.
Balkanski, M., R. Beserman and J.M. Besson, 1966, Solid State Commun. **4**, 201.
Balkanski, M., L.M. Falikov, C. Hirlimann and K.P. Jain, 1978, Solid State Commun. **25**, 261.
Barker, A.S., and A.J. Sievers, 1975, Rev. Mod. Phys. **47**, Suppl. No. 2.
Barker, A.S., and H.W. Verleur, 1967, Solid State Commun. **5**, 695.
Barker, A.S., J.A. Ditzenberger and H.J. Guggenheim, 1968, Phys. Rev. **175**, 1180.
Barta, C., G.F. Dobrzhanskii, G.M. Zinger, M.F. Limonov and Y.F. Markov, 1982, Sov. Phys.-Solid State **24**, 1672.
Bartzokas, A., and D. Siapkas, 1976, Ferroelectrics **12**, 127.
Bauhofer, W., L. Genzel and I.R. Jahn, 1974a, Phys. Status Solidi (b) **63**, 465.
Bauhofer, W., L. Genzel, C.H. Perry and I.R. Jahn, 1974b, Phys. Status Solidi (b) **63**, 385.
Bauhofer, W., L. Genzel and W. Konig, 1976a, Phys. Status Solidi (b) **78**, K121.
Bauhofer, W., L. Genzel, C.H. Perry and I.R. Jahn, 1976b, Optical Phonons and Phase Transitions in $(NH_4)Cl_{(1-x)}Br_{(x)}$ Mixed Crystals, in: Proc. 3rd Int. Conf. Light Scattering Solids, Campinas, 1975, eds M. Balkanski, R.C.C. Leite and S.P.S. Porto (Flammarion, Paris) p. 918.
Bedel, E., R. Carles, G. Landa and J.B. Renucci, 1984a, Rev. Phys. Appl. **19**, 17.
Bedel, E., R. Carles, A. Zwick, J.B. Renucci and M.A. Renucci, 1984b, Phys. Rev. B **30**, 5923.
Behera, S.N., P. Nayak and K. Patnaik, 1977, Pramana **8**, 255.
Belenkii, G.L., L.N. Alieva, R.Kh. Nani, E. Yu. Saleav and V.Yu. Shteinshraiber, 1978, Sov. Phys.-Solid State **20**, 1860.
Benedek, G., and G.F. Nardelli, 1967a, Phys. Rev. **154**, 872.
Benedek, G., and G.F. Nardelli, 1967b, Phys. Rev. **155**, 1004.
Berndt, V., and A.A. Kopylov, 1978, Sov. Phys.-Solid State **20**, 156.
Beserman, R., 1977, Solid State Commun. **23**, 323.
Beserman, R., and D. Schmeltzer, 1977, Solid State Commun. **24**, 793.
Beserman, R., G. Gilat and C. Hirlimann, 1976a, Raman Study of the Coupling Between Disordered Activated One-Phonon Density of States and the Normal Modes of Vibration of Doped GaP, in: Proc. 13th Int. Conf. Physics of Semiconductors, Rome, 1976, ed. F.G. Fumi (North-Holland, Amsterdam) p. 1290.
Beserman, R., C. Hirlimann, M. Balkanski and J. Chevallier, 1976b, Solid State Commun. **20**, 485.
Beserman, R., M. Zigone, W. Drexel and C. Marti, 1976c, Solid State Commun. **18**, 419.
Beserman, R., D. Schmeltzer, C. Herscovici and M. Fibich, 1978, Zone Edge Phonons in Mixed Crystals, in: Proc. Int. Conf. Lattice Dynamics, Paris, 1977, ed. M. Balkanski (Flammarion, Paris) p. 441.
Bilz, H., D. Strauch and R.K. Wehner, 1984, Vibrational Infrared and Raman Spectra of Non-Metals, in: Handbuch der Phys. 25/2d, ed. L. Genzel (Springer Verlag, Berlin) p. 1.
Biryulin, Yu.F., G.M. Zinger, I.P. Ipatova, Yu.E. Pozhidaev and Yu.V. Shmartsev, 1979, Sov. Phys.-Semicond. **13**, 948.
Bodnar, I.V., A.G. Karoza and G.F. Smirnova, 1978, Phys. Status Solidi (b) **86**, K171.
Bonneville, R., 1980, Phys. Rev. B **21**, 368.
Bonneville, R., 1981, Phys. Rev. B **24**, 1987.
Bonneville, R., 1984, Phys. Rev. B **29**, 907.
Bootz, B., W. von der Osten and N. Uhle, 1974, Phys. Status Solidi (b) **66**, 169.
Born, M., and K. Huang, 1954, Dynamical Theory of Crystal Lattices (Oxford University Press, London).
Bottger, H., 1976, Phys. Status Solidi (b) **77**, 561.

Boudou, A., and J. Sapriel, 1980, Ferroelectrics **29**, 37.
Brafman, O., 1972, Solid State Commun. **11**, 447.
Brafman, O., and S.S. Mitra, 1968, Phys. Rev. **171**, 931.
Brafman, O., I.F. Chang, G. Lengyel, S.S. Mitra and E. Carnall, 1967, Phys. Rev. Lett. **19**, 1120.
Brafman, O., I.F. Chang, G. Lengyel, S.S. Mitra and E. Carnall, 1968, Optical Phonons in $ZnS_{(x)}Se_{(1-x)}$ Mixed Crystals, in: Local Excitations in Solids, Irvine, 1967, ed. R.F. Wallis (Plenum, New York) p. 602.
Braunstein, R., 1963, Phys. Rev. **130**, 879.
Brodsky, M.H., and G. Lucovsky, 1968, Phys. Rev. Lett. **21**, 990.
Brodsky, M.H., G. Lucovksy, M.F. Chen and T.S. Plaskett, 1970, Phys. Rev. B **2**, 3303.
Bureau, J.C., and L.C. Brunel, 1981, Chem. Phys. **63**, 293.
Burns, G., 1974, Phys. Rev. B **10**, 1951.
Burns, G., and F.H. Dacol, 1981, J. Raman Spectrosc. **10**, 227.
Byra, W.J., 1973, Solid State Commun. **12**, 253.
Carles, R., N. Saint-Cricq, J.B. Renucci and R.J. Nicholas, 1980, J. Phys. C **13**, 899.
Carles, R., N. Saint-Cricq, A. Zwick, M.A. Renucci and J.B. Renucci, 1983, Nuovo Cimento **2 D**, 1712.
Carles, R.N., A. Zwick, M.A. Renucci and J.B. Renucci, 1982, Solid State Commun. **41**, 557.
Carter, D.L., M.A. Kinch and D.D. Buss, 1971, Optical Phonons and Dielectric Constants in $Hg_{(0.796)}Cd_{(0.204)}Te$, in: Proc. Int. Conf. Physics of Semimetals and Narrow Gap Semiconductors, Dallas, 1970 (Pergamon, Oxford) p. 273.
Cerdeira, F., E.A. Meneses and A. Gouskov, 1977, Phys. Rev. B **16**, 1648.
Chang, I.F., and S.S. Mitra, 1968, Phys. Rev. **172**, 924.
Chang, I.F., and S.S. Mitra, 1971, Adv. Phys. **20**, 359.
Chang, I.F., S.S. Mitra, J.N. Plendl and L.C. Mansur, 1968, Phys. Status Solidi **28**, 663.
Chang, R.K., B. Lacina and P.S. Pershan, 1966, Phys. Rev. Lett. **17**, 755.
Chang, R.K., J.M. Ralston and D.E. Keating, 1968, Temperature Dependence of Raman Linewidth and Intensity of Semiconductors, in: Light Scattering Spectra in Solids, New York, 1968, ed. G.B. Wright (Springer Verlag, New York) p. 369.
Charfi, F., M. Zouaghi, C. Llinares, M. Balkanski, Ch. Hirlimann and A. Joullie, 1978, Small Wave Vector Modes in $Ga_{(1-x)}Al_{(x)}Sb$, in: Proc. Int. Conf. Lattice Dynamics, Paris, 1977, ed. M. Balkanski (Flammarion, Paris) p. 438.
Charfi, F., M. Zouaghi, A. Joullie, M. Balkanski and Ch. Hirlimann, 1980, J. Phys. (France) **41**, 83.
Chen, Y.S., W. Shockley and G.L. Pearson, 1966, Phys. Rev. **151**, 648.
Ciepielewski, P., and I. Kosacki, 1982, Solid State Commun. **44**, 417.
Cosand, A.E., and W.G. Spitzer, 1971, J. Appl. Phys. **42**, 5241.
Deslandes, J., and S. Jandl, 1984, Phys. Rev. B **29**, 2088.
Elliott, R.J., and P.L. Leath, 1975, High Concentration Mixed Crystals and Alloys, in: Dynamical Properties of Solids, Vol. 2, ed. G.H. Horton and A.A. Maradudin (North-Holland, Amsterdam) p. 385.
Elliott, R.J., and D.W. Taylor, 1967, Proc. R. Soc. A **296**, 161.
Elliott, R.J., J.A. Krumhansl and P.L. Leath, 1974, Rev. Mod. Phys. **46**, 465.
Farge, Y., and M.P. Fontana, 1979, Electronic and Vibrational Properties of Point Defect in Ionic Crystals (North-Holland, Amsterdam).
Farr, M.K., J.G. Traylor and S.K. Sinha, 1975, Phys. Rev. B **11**, 1587.
Feldman, D.W., M. Askin and J.H. Parker, 1966, Phys. Rev. Lett. **17**, 1209.
Fertel, J.H., and C.H. Perry, 1969, Phys. Rev. **184**, 874.
Fertel, J.H., and C.H. Perry, 1971, Phys. Lett. A **36**, 315.
Fetter, A.L., and J.D. Walecka, 1971, Quantum Theory of Many Particle Systems (McGraw-Hill, New York).

Finkenrath, H., G. Franz and N. Uhle, 1979, Phys. Status Solidi (b) **95**, 179.
Fujii, A., H. Stolz and W. von der Osten, 1983, J. Phys. C. **16**, 1713.
Galtier, P., J. Chevallier, M. Zigone and G. Martinez, 1984, Phys. Rev. B **30**, 726.
Gamarts, E.M., B.S. Zadokhin and A.N. Starukin, 1978, Sov. Phys.-Solid State **20**, 2153.
Ganguly, B.N., and J.L. Birman, 1968, Microscopy Theory of Lattice Raman Scattering in Crystals containing Impurities, in: Light Scattering Spectra in Solids, New York, 1968, ed. G.B. Wright (Springer Verlag, New York) p. 487.
Gasanly, N.M., V.K. Subashiev, M.I. Aliev, A.A. Kukharaskii and V.M. Evdokimov, 1971, Sov. Phys.-Solid State **13**, 54 .
Gasanly, N.M., B.D. Dzhavadov, V.I. Tagirov and E.A. Vinogradov, 1979, Phys. Status Solidi (b) **95**, K27.
Gasanly, N.M., R.E. Guseinov, A.S. Ragimov and V.I. Tagirov, 1980, Phys. Status Solidi (b) **101**, K121.
Gasanly, N.M., A.F. Goncharov, N.N. Melnik and A.S. Ragimov, 1983, Phys. Status Solidi (b) **120**, 137.
Gebicki, W., and W. Nazarewicz, 1977, Phys. Status Solidi (b) **80**, 307.
Gebicki, W., and W. Nazarewicz, 1978, Phys. Status Solidi (b) **86**, K135.
Gebicki, W., A. Krol and W. Nazarewicz, 1975, Phys. Status Solidi (b) **67**, K89.
Gebicki, W., E. Amzallag, M. Picquart, Ch. Julien and M. Le Postollec 1980, J. Phys. (France) **41**, C5-339.
Geick, R., and J. Hassler, 1969, Phys. Status Solidi **33**, 689.
Geick, R., E.F. Steigmeier and H. Auderset, 1972, Phys. Status Solidi (b) **54**, 623.
Genzel, L., and W. Bauhofer, 1976, Z. Phys. B **25**, 13.
Genzel, L., T.P. Martin and C.H. Perry, 1974, Phys. Status Solidi (b) **62**, 83.
Gielisse, P.J., J.N. Plendl, L.C. Mansur, R. Marshall, S.S. Mitra, R. Mykolajewycz and A. Smakula, 1965, J. Appl. Phys. **36**, 2446.
Gledhill, G.A., and J.F. Angress, 1977, Phys. Status Solidi (b) **79**, K107.
Gobeau, J., M. Heuret and J.P. Mon, 1978, Lattice Dynamics of the Mixed System $Sr_{(1-x)}Cd_{(x)}F_2$, in: Proc. Int. Conf. Lattice Dynamics, Paris, 1977, ed. M. Balkanski (Flammarion, Paris) p. 432.
Golubev, L.V., and L.K. Vodopyanov, 1981, Sov. Phys.-Solid State **23**, 180.
Gorska, M., and W. Nazarewicz, 1973, Phys. Status Solidi (b) **57**, K65.
Gorska, M., and W. Nazarewicz, 1974, Phys. Status Solidi (b) **65**, 193.
Gorska, M., J. Grad and W. Nazarewicz, 1975, Phys. Status Solidi (b) **70**, 299.
Grim, A., A.A. Maradudin, I.P. Ipatova and A.V. Subashiev, 1972, J. Phys. Chem. Solids **33**, 775.
Grunewald, G., 1982, J. Phys. C **15**, 3663.
Haas, M., H.B. Rosenstock and R.E. McGill, 1969, Solid State Commun. **7**, 1.
Harada, H., and S. Narita, 1971, J. Phys. Soc. Jpn. **30**, 1628.
Harbec, J.Y., and S. Jandl, 1982, Phys. Rev. B **25**, 6126.
Harrison, W.A., 1980, Electronic Structure and the Properties of Solids (Freeman, San Francisco).
Hayek, M., and O. Brafman, 1971, Resonance Raman Effect in Solid Solutions, in: Proc. 2nd Int. Conf. Light Scattering Solids, Paris, 1971, ed. M. Balkanski (Flammarion, Paris) p. 76.
Hayek, M., O. Brafman and R.M.A. Leith, 1973, Phys. Rev. B **8**, 2772.
Hayes, W., and R. Loudon, 1978, Scattering of Light by Crystals (Wiley, New York).
Herscovici, C., and M. Fibich, 1980, J. Phys. C **13**, 4463.
Hirlimann, C., R. Beserman and M. Balkanski, 1976a, Raman Lineshape of $Ga_{(x)}In_{(1-x)}P$ and $GaAs_{(1-x)}P_{(x)}$, in: Proc. 13th Int. Conf. Physics of Semiconductors, Rome, 1976, ed. F.G. Fumi (North-Holland, Amsterdam) p. 208.
Hirlimann, C., R. Beserman, M. Balkanski and J. Chevallier, 1976b, Raman Study of One and Two Phonon Coupling in Mixed $Ga_{(x)}In_{(1-x)}P$, in: Proc. 3rd Int. Conf. Light Scattering Solids, Campinas, 1975, eds M. Balkanski, R.C.C. Leite and S.P.S. Porto (Flammarion, Paris) p. 129.

Hoclet, M., P. Plumelle, M. Vandevyver, R. Triboulet and Y. Marfaing, 1979, Phys. Status Solidi (b) **92**, 545.
Hurrell, J.P., S.P.S. Porto, T.C. Damen and S. Mascarenhas, 1968, Phys. Lett. A **26**, 194.
Ignatev, B.V., V.F. Kalabukhova and A.A. Sobol, 1980, Sov. Phys.-Solid State **22**, 890.
Ilegems, and G.L. Pearson, 1970, Phys. Rev. B **1**, 1576.
Ipatova, I.P., R.R. Ichkitidze, I.N. Sochilina, Yu.I. Ukhanov and Yu.V. Shmartsev, 1976a, Sov. Phys.-Solid State **18**, 306.
Ipatova, I.P., R.R. Ichkitidze, I.N. Sochilina, Yu.I. Ukhanov and Yu. V. Shmartsev, 1976b, Sov. Phys.-Solid State **18**, 342.
Ishidate, T., S. Katagiri, K. Inoue, M. Shibuya, K. Tsuji and S. Minomura, 1984, J. Phys. Soc. Jpn. **53**, 2584.
Jahne, E., 1976a, A New Approach to Long Wavelength Optical Phonons in Mixed Crystals, in: Proc. 13th Int. Conf. Physics of Semiconductors, Rome, 1976, ed. F.G. Fumi (North-Holland, Amsterdam) p. 212.
Jahne, E., 1976b, Phys. Status Solidi (b) **74**, 275.
Jahne, E., 1976c, Phys. Status Solidi (b) **75**, 221.
Jahne, E., and B. Ulrici, 1980, Phys. Status Solidi (b) **101**, 169.
Jahne, E., W. Pilz, M. Giehler and L. Hildisch, 1979, Phys. Status Solidi (b) **91**, 155.
Jahne, E., P. Kleinert, B.Th. Bairamov and V.V. Toporov, 1981, Phys. Status Solidi (b) **104**, 531.
Jandl, S., and R. Provencher, 1981, J. Phys. C **14**, L461.
Jandl, S., Y.H. Harbec and C. Carlone, 1978, Solid State Commun. **27**, 1441.
Jaswal, S.S., and J.R. Hardy, 1968, Phys. Rev. **171**, 1090.
Johnstone, I.W., G.D. Jones and D.J. Lockwood, 1981, Solid State Commun. **39**, 395.
Jusserand, B., and J. Sapriel, 1981, Phys. Rev. B **24**, 7194.
Jusserand, B., and S. Slempkes, 1984, Solid State Commun. **49**, 95.
Jusserand, B., J. Sapriel, F. Alexandre and P. Delpech, 1981, J. Phys. (France) **42**, C6-43.
Kaplan, T., and M. Mostoller, 1974, Phys. Rev. B **10**, 3610.
Karapetyan, V.E., A.V. Krol and A.I. Ryskin, 1976, Sov. Phys.-Solid State **18**, 1231.
Kawamura, H., R. Tsu and L. Esaki, 1972, Phys. Rev. Lett. **29**, 1397.
Keezer, R.C., G. Lucovsky and M.L. Slade, 1968, Solid State Commun. **6**, 765.
Kekelidze, N.P., G.P. Kekelidze and Z.D. Makharadze, 1973, J. Phys. Chem. Solids **34**, 2117.
Kim, O.K., and W.G. Spitzer, 1979, J. Appl. Phys. **50**, 4362.
Kim, R., and S. Narita, 1971, J. Phys. Soc. Jpn. **31**, 613.
Kleinert, P., 1979a, Phys. Status Solidi (b) **91**, 455.
Kleinert, P., 1979b, Phys. Status Solidi (b) **95**, 163.
Kleinert, P., 1982, Phys. Status Solidi (b) **114**, 459.
Kleinert, P., 1983, Phys. Status Solidi (b) **118**, 283.
Kleinert, P., 1984, Phys. Status Solidi (b) **122**, 81.
Kleinert, P., and F. Bechstedt, 1978, Phys. Status Solidi (b) **85**, 253.
Kleinert, P., and E. Jahne, 1981, Phys. Status Solidi (b) **107**, 177.
Krabach, T.N., N. Wada, M.V. Klein, K.C. Cadien and J.E. Greene, 1983, Solid State Commun. **45**, 895.
Krol, A.W., N.V. Levichev, A.L. Natadze and A.I. Ryskin, 1977a, Solid State Commun. **24**, 151.
Krol, A.W., N.V. Levichev and A.I. Ryskin, 1977b, Sov. Phys.-Solid State **19**, 452.
Krol, A.W., N.V. Levichev, A.L. Natadze and A.I. Ryskin, 1978, Sov. Phys.-Solid State **20**, 85.
Krozyrev, S.P., L.K. Vodopyanov and R. Triboulet, 1983, Sov. Phys.-Solid State **25**, 361.
Kunc, K., 1974, Ann. Phys. (Paris) **8**, 319.
Kutty, A.P.G., 1974, Solid State Commun. **14**, 213.
Lacina, W.B., and P.S. Pershan, 1970, Phys. Rev. B **1**, 1765.
Lannin, J., 1977, Phys. Rev. B **16**, 1510.
Lannin, J., 1979a, Phys. Rev. B **19**, 2390.

Lannin, J., 1979b, Solid State Commun. **29**, 159.
Lemos, V., G.M. Gualberto, J.B. Salzberg and F. Cerdeira, 1980, Phys. Status Solidi (b) **100**, 755.
Limonov, M.F., and Y.F. Markov, 1983, Sov. Phys.-Solid State **25**, 623.
Lisitsa, M.P., M.Y. Valakh and N.K. Konovets, 1969, Phys. Status Solidi **34**, 269.
Livescu, G., Z. Vardeny and O. Brafman, 1981, Phys. Rev. B **24**, 1952.
Lucovksy, G., and J.C. Mikkelsen, 1976, J. Electron. Mater. **5**, 179.
Lucovksy, G., K.Y. Cheng and G.L. Pearson, 1975, Phys. Rev. B **12**, 4135.
Lucovksy, G., R.D. Burnham and A.S. Alimonda, 1976, Phys. Rev. B **14**, 2503.
Lucovsky, G., and M.F. Chen, 1970, Solid State Commun. **8**, 1397.
Lucovsky, G., E. Lind and E.A. Davies, 1967, Infrared Active Lattice Modes of the Mixed System $Cd_{(1-x)}Zn_{(x)}S$, in: Int. Conf. II–VI Semiconducting Compounds, Providence, 1967, ed. D.G. Thomas (W.A. Benjamin, New York) p. 1150.
Lucovsky, G., M.H. Brodsky and E. Burstein, 1970, Phys. Rev. B **2**, 3295.
Lucovsky, G., M.H. Brodsky, M.F. Chen, R.J. Chicotka and A.T. Ward, 1971, Phys. Rev. B **4**, 1945.
Lucovsky, G., R.D. Burnham, A.S. Alimonda and H.A. Six, 1974, Infrared Reflectance of Quaternary Alloys $Ga_{(1-x)}Al_{(x)}As_{(1-y)}P_{(y)}$, in: 12th Int. Conf. on Physics of Semiconductors, Stuttgart, 1974, ed. M.H. Pilkuhn (Teubner, Stuttgart) p. 326.
Maradudin, A.A., 1965, Rep. Prog. Phys. **28**, 331.
Maradudin, A.A., and J. Oitmaa, 1969, Solid State Commun. **7**, 1143.
Maradudin, A.A., E.W. Montroll, G.H. Weiss and I.P. Ipatova, 1971, Theory of Lattice Dynamics in the Harmonic Approximation, 2nd edition (Academic, New York).
Martin, T.P., 1975, Phys. Status Solidi (b) **67**, 137.
Massa, N.E., J.F. Vetelino and S.S. Mitra, 1982, Phys. Rev. B **26**, 4579, 4606.
Matossi, F., 1951, J. Chem. Phys. **19**, 161.
McKnight, S.W., P.M Amirtharaj and S. Perkowits, 1978, Solid State Commun. **25**, 357.
Mercier, A., and J.P. Voitchovsky, 1974, Solid State Commun. **14**, 757.
Mityagin, Y.A., L.K. Vodopyanov and E.A. Vinogradov, 1976, Sov. Phys.-Solid State **17**, 1341.
Mityagin, Y.A., V.G. Plotnichenko, L.K. Vodopyanov and L.D. Budennaya, 1977, Sov. Phys.-Solid State **19**, 1811.
Montgomery, D.J., and J.R. Hardy, 1965, Effect of Isotopic Composition on Lattice Vibration Absorption of Infrared Radiation in Ionic Crystals, in: Proc. Int. Conf. Lattice Dynamics, Copenhagen, 1963, ed. R.F. Wallis (Pergamon, Oxford) p. 491.
Mooradian, A., and T.C. Harman, 1971, Raman Scattering From $Hg_{(1-x)}Cd_{(x)}Te$, in: Proc. Int. Conf. Physics of Semimetals and Semiconductors, Dallas, 1970 (Pergamon, Oxford) p. 297.
Murahashi, T., and T. Koda, 1976, J. Phys. Soc. Jpn. **40**, 747.
Murahashi, T., T. Koda, Y. Oka and T. Kushida, 1973, Solid State Commun. **13**, 307.
Musaeva, L.G., M.D. Khomutova and N.M. Gasanly, 1977, Sov. Phys.-Solid State **19**, 1030.
Mutzenich, G., R. Faymonville and P. Grosse, 1983, Phys. Status Solidi (b) **119**, 531.
Myles, C.V., 1983, Phys. Rev. B **28**, 4519.
Nair, I., and C.T. Walker, 1971, Phys. Rev. B **3**, 3446.
Nair, I., and C.T. Walker, 1973, Phys. Rev. B **7**, 2740.
Nakashima, S., T. Fukumoto and A. Mitsuishi, 1971, J. Phys. Soc. Jpn. **30**, 1508.
Nakashima, S., T. Fukumoto, A. Mitsuishi and K. Itoh, 1972, J. Phys. Soc. Jpn. **32**, 1438.
Nakashima, S., T. Fukumoto, A. Mitsuishi and K. Itoh, 1973, J. Phys. Soc. Jpn. **35**, 1437.
Nakashima, S., H. Mishima and H. Tai, 1974, J. Phys. Chem. Solids **35**, 531.
Newman, R.C., 1973, Infrared Studies of Crystal Defects (Taylor and Francis, London).
Novik, A.E., I.V. Pevnitskii, A.I. Ryskin, L.A. Sysoev and G.I. Khilko, 1974, Sov. Phys.-Solid State **16**, 1147.
Nusimovici, M.A., M. Balkanski and J.L. Birman, 1970, Phys. Rev. B **1**, 595.
O'Hara, M.J., C.W. Myles, J.D. Dow and R.D. Painter, 1981, J. Phys. Chem. Solids **42**, 1043.

Oles, B., H.J. Stolz and H.G. Von Schnering, 1981, Phys. Status Solidi (b) **106**, 157.
Olszewski, A., W. Wojdowski and W. Nazarewicz, 1981, Phys. Status Solidi (b) **104**, K155.
Osamura, K., and Y. Murakami, 1972, Trans. Jpn. Inst. Met. **13**, 171.
Oswald, F., 1959, Z. Naturforschg. a **14**, 374.
Ovander, L.N., and N.S. Tyu, 1979, Phys. Status Solidi (b) **91**, 763.
Parrish, J.F., C.H. Perry, O. Brafman, I.F. Chang and S.S. Mitra, 1967, Phonons in Mixed Crystal System $CdS_{(x)}Se_{(1-x)}$, in: Int. Conf. II–VI Semiconducting Compounds, Providence, 1967, ed. D.G. Thomas (Benjamin, New York) p. 1164.
Perry, C.H., and E.F. Young, 1967, J. Appl. Phys. **38**, 4616.
Perry, C.H., I.R. Jahn, V. Wagner, W. Bauhofer, L. Genzel and J.B. Sokoloff, 1978, Phonon Dispersion in $NH_4Cl_{(1-x)}Br_{(x)}$ Mixed Crystals, in: Proc. Int. Conf. Lattice Dynamics, Paris, 1977, ed. M. Balkanski (Flammarion, Paris) p. 419.
Pershan, P.S., and W.B. Lacina, 1969, Raman Scattering from Mixed Crystals, in: Int. Conf. Light Scattering Spectra of Solids, New York, 1968, ed. G.B. Wright (Springer Verlag, New York) p. 439.
Picquart, M., E. Amzallag, M. Balkanski, Ch. Julien, W. Gebicki and W. Nazarewicz, 1980, Phys. Status Solidi (b) **99**, 683.
Pinczuk, A., and E. Burstein, 1975, Fundamentals of Inelastic Light Scattering in Semiconductors and Insulators, in: Light Scattering in Solids, ed. M. Cardona (Springer Verlag, Heidelberg) p. 25.
Pinczuk, A., J.M. Worlock, R.E. Nahory and M.A. Pollack, 1978, Appl. Phys. Lett. **33**, 461.
Plotnichenko, V.G., L.V. Golubev and L.K Vodopyanov, 1977, Sov. Phys.-Solid State **19**, 1582.
Polian, A., R. Le Youllec and M. Balkanski, 1976, Phys. Rev. B **13**, 3558.
Portal, J.C., P. Perrier, M.A. Renucci, S. Askenazy, R.J. Nicholas and T. Pearsall, 1979, A Study of the Conduction Band and Phonons of $Ga_{(x)}In_{(1-x)}As_{(y)}P_{(1-y)}$ by the Shubnikov–De Haas Effect, Magnetophonon Resonance and Raman Scattering, in: 14th Int. Conf. Physics of Semiconductors, Edinburgh, 1978, ed. B.L.H. Wilson (Institute of Physics, London) p. 829.
Potter, R.F., and D.L. Stierwalt, 1964, Reststrahlen Frequencies for Mixed $GaAs_ySb_{1-y}$ system, in: 7th Int. Conf. Physics of Semiconductors, Paris 1964, ed. M. Hulin (Dunod, Paris) p. 1111.
Poulet, R., and J.P. Mathieu, 1976, Vibration Spectra and Symmetry of Crystals (Gordon and Breach, New York).
Prasad, R., and A. Bansil, 1980, Phys. Rev. B **21**, 496.
Provencher, R., S. Jandl and C. Carlone, 1982, Phys. Rev. B **26**, 7049.
Renker, B., N.M. Butt and N.E. Massa, 1983, Phys. Rev. B **27**, 1450.
Renucci, M.A., J.B. Renucci and M. Cardona, 1971a, Raman Scattering in Ge–Si Alloys, in: Proc. 2nd Int. Conf. Light Scattering Solids, Paris, 1971, ed. M. Balkanski (Flammarion, Paris) p. 326.
Renucci, M.A., J.B. Renucci and M. Cardona, 1971b, Solid State Commun. **9**, 1651.
Richter, W., H. Kohler and C.R. Becker, 1977, Phys. Status Solidi (b) **84**, 619.
Riede, V., H. Neumann, H. Sobotta and F. Levy, 1980, Solid State Commun. **34**, 229.
Rosenstock, H.B., and R.E. McGill, 1968, Phys. Rev. **176**, 1004.
Sahni, V.C., and M.L. Bansal, 1978, Phys. Status Solidi (b) **90**, 415.
Saint-Cricq, N., R. Carles, J.B. Renucci, A. Zwick and M.A. Renucci, 1981, Solid State Commun. **39**, 1137.
Schmeltzer, D., and R. Beserman, 1980, Phys. Rev. B **22**, 6330.
Schmeltzer, D., and R. Beserman, 1981a, Phys. Rev. Lett. **47**, 860.
Schmeltzer, D., and R. Beserman, 1981b, J. Phys. C **14**, 5003.
Schmeltzer, D., R. Beserman and D. Slamovits, 1980, Phys. Rev. B **22**, 4038.
Sen, P.N., and W.M. Hartmann, 1974, Phys. Rev. B **9**, 367.
Sen, P.N., and G. Lucovsky, 1975, Phys. Rev. B **12**, 2988.
Sennett, C.T., D.R. Bosomworth, W. Hayes and A.R.L. Spray, 1969, J. Phys. C **2**, 1137.

Shah, J., A.E. Digiovanni, T.C. Damen and B.J. Miller, 1973, Phys. Rev. B **7**, 3481.
Shen, S.C., and M. Cardona, 1980, Solid State Commun. **36**, 327.
Shen, S.C., and J.H. Chu, 1983, Solid State Commun. **48**, 1017.
Sirota, N.N., I.V. Bodnar and G.F. Smirnova, 1977, Phys. Status Solidi (a) **41**, 669.
Sobotta, H., and V. Riede, 1974, Phys. Status Solidi (b) **63**, K143.
Sobotta, H., and V. Riede, 1978, Czech. J. Phys. B **28**, 536.
Sobotta, H., V. Riede and K. Buchheiser, 1975, Czech. J. Phys. B **25**, 841.
Srinivasan, R., and G. Lakshmi, 1978, Lattice Dynamics of $K_{(1-x)}Rb_{(x)}Br$ in the Pseudo-Crystal Model, in: Proc. Int. Conf. Lattice Dynamics, Paris, 1977, ed. M. Balkanski (Flammarion, Paris) p. 429.
Srivastava, V., and S.K. Joshi, 1973, Phys. Rev. B **8**, 4671.
Stekhanov, A.I., and M.B. Eliashberg, 1960, Opt. & Spectrosc. **10**, 174.
Strahm, N.D., and A.L. McWhorter, 1969, Raman Scattering from Lattice Vibrations of $GaAs_{(x)}P_{(1-x)}$, in: Light Scattering in Solids, New York, 1968, ed. G.B. Wright (Springer Verlag, New York) p. 455.
Talwar, D.N., M. Vandevyver and M. Zigone, 1980, J. Phys. C **13**, 3775.
Talwar, D.N., M. Vandevyver, K. Kunc and M. Zigone, 1981a, Phys. Status Solidi (b) **103**, 381.
Talwar, D.N., M. Vandevyver and M. Zigone, 1981b, Phys. Rev. B **23**, 1743.
Taylor, D.W., 1973, Solid State Commun. **13**, 117.
Taylor, D.W., 1975, Dynamics of Impurities in Crystals, in: Dynamical Properties of Solids, Vol. 2, eds G.H. Horton and A.A. Maradudin (North-Holland, Amsterdam) p. 285.
Taylor, D.W., 1976, J. Phys. C **9**, 453.
Teicher, M., R. Beserman, M.V. Klein and H. Morkoc, 1984, Phys. Rev. B **29**, 4652.
Tiong, K.K., P.M. Amirtharaj, P. Parayanthal and F.H. Pollack, 1984, Solid State Commun. **50**, 891.
Tsu, R., H. Kawamura and L. Esaki, 1972, Raman Scattering of Local and Collective Phonon Modes in $Ga_{(1-x)}Al_{(x)}As$, in: Proc. 11th Int. Conf. Physics of Semiconductors, Warsaw, 1972 (Elsevier, Amsterdam) p. 1135.
Ulrici, B., and E. Jahne, 1978, Phys. Status Solidi (b) **86**, 517.
Valakh, M.Y., and A.P. Litvinchuk, 1983a, Sov. Phys.-Solid State **25**, 1597.
Valakh, M.Y., L.M. Lisita, V.I. Siderenko and G.N. Polissky, 1980, Phys. Lett. A **78**, 115.
Valakh, M.Y., A.P. Litvinchuk and G.G. Tarasov, 1983b, Sov. Phys.-Solid State **25**, 1752.
Vardeny, Z., and O. Brafman, 1979, Phys. Rev. B **19**, 3290.
Vavilov, V.S., V.S. Vinogradov, L.K Vodopyanov and B.S. Umarov, 1971, Experimental and Theoretical Investigation of Raman Spectra in $Cd_{(1-x)}Zn_{(x)}S$ Solid Solutions, in: Proc. 2nd Int. Conf. Light Scattering Solids, Paris, 1971, ed. M. Balkanski (Flammarion, Paris) p. 338.
Venkataraman, G., L.A. Feldkamp and V.C. Sahni, 1975, Dynamics of Perfect Crystals (M.I.T. Press, Cambridge).
Venugopalan, S., A. Petrou, R.R. Galaska and A.K. Ramdas, 1980, Solid State Commun. **35**, 401.
Venugopalan, S., A. Petrou, R.R. Galaska and A.K. Ramdas, 1982, Phys. Rev. B **25**, 2681.
Vergnat-Grandjean, D., P. Vergnat, J.F. Brice and R. Leveque, 1979, Phys. Status Solidi (b) **96**, 611.
Vergnat-Grandjean, D., P. Vergnat and J.F. Brice, 1982, Phys. Status Solidi (b) **113**, 352.
Verleur, H.W., and A.S. Barker, 1966, Phys. Rev. **149**, 715.
Verleur, H.W., and A.S. Barker, 1967a, Phys. Rev. **155**, 750.
Verleur, H.W., and A.S. Barker, 1967b, Phys. Rev. **164**, 1169.
Vinogradov, E.A., and Yu.A. Mityagin, 1978, Sov. Phys.-Solid State **20**, 1825.
Vinogradov, E.A., and L.K. Vodopyanov, 1976, Sov. Phys.-Solid State **17**, 2088.
Vinogradov, E.A., L.K. Vodopyanov and G.S. Oleinik, 1973, Sov. Phys.-Solid State **15**, 322.
Vinogradov, E.A., N.M. Gasanly, A.F. Goncharov, B.M Dzhavadov and N.N. Melnik, 1979, Sov. Phys.-Solid State **21**, 906.

Vinogradov, V.S., 1970, Sov. Phys.-Solid State **11**, 1666.
Vodopyanov, L.K., B.S. Umarov, L.A. Sysoev and L.A. Sarkisov, 1971, Sov. Phys.-Solid State **13**, 660.
Vodopyanov, L.K., E.A. Vinogradov, A.M. Blinov and V.A Rukavishnikov, 1972, Sov. Phys.-Solid State **14**, 219.
Vodopyanov, L.K., E.A. Vinogradov, N.N. Melnik, V.G. Plonitchenko, J. Chevallier and J.C. Guillaume, 1978, J. Phys. (France) **39**, 627.
Vodopyanov, L.K., L.V. Golubev and D.I. Bletskan, 1979a, Sov. Phys.-Solid State **21**, 1053.
Vodopyanov, L.K., L.V. Golubev and D.I. Bletskan, 1979b, Sov. Phys.-Solid State **21**, 2003.
Wakamura, K., T. Arai, S. Onari, K. Kudo and T. Takahashi, 1973, J. Phys. Soc. Jpn. **35**, 1430.
Wakamura, K., T. Arai and K. Kudo., 1976, J. Phys. Soc. Jpn. **40**, 1118.
Walsh, D., S. Jandl and J.Y. Harbec, 1980, J. Phys. C **13**, L125.
Wanser, K.H., and R.F. Wallis, 1979, Solid State Commun. **32**, 967.
Xinh, N.X., A.A. Maradudin and R.A. Coldwell-Horsfall, 1965, J. Phys. (France) **26**, 717.
Yamazaki, S., A. Ushirokawa and I. Katoda, 1980, J. Appl. Phys. **51**, 3722.
Zigone, M., C. Hirlimann, M. Jouanne, J. Chevallier and M.S. Martin, 1978, Long Wavelength Optical Phonon Replica in $Mg_{(x)}Zn_{(1-x)}Te$ Mixed Crystals, in: Proc. Int. Conf. Lattice Dynamics, Paris, 1977, ed. M. Balkanski (Flammarion, Paris) p. 177.
Zinger, G.M., I.P. Ipatova and A.V. Subashiev, 1976, Sov. Phys.-Semicond. **10**. 286.
Zinger, G.M., I.P. Ipatova and A.V. Subashiev, 1977, Sov. Phys.-Semicond. **11**, 383.
Zinger, G.M., M.A. Ilin, E.P. Rashevskaya and A.I. Ryskin, 1979, Sov. Phys.-Solid State **21**, 1522.
Zitter, R.N., and P.C. Watson, 1974, Phys. Rev. B **10**, 607.
Zwick, A., G. Landa, M.A. Renucci and R. Carles, 1982, Phys. Rev. B **26**, 5694.
Zwick, A., G. Landa, R. Carles, M.A. Renucci and A. Kjekshus, 1983, Solid State Commun. **45**, 889.

Notes added in proof

The following sections, whose organisation follows closely that of the main text, have been added with the intention of bringing the article up to data as of mid 1987.

[Addenda to section 3. Disorder theory]

Recursion method
Until recently the theoretical approximations used for calculating the optical properties of mixed crystals have been the CPA (section 3.3) and various isodisplacement approximations (sections 3.4 and 3.5). The recursion method had only been used in this context to discuss the one-/two-mode question (sections 3.6 and 4). Starting in 1985 several calculations have appeared using this method, along with realistic models for the lattice dynamics, to describe the behaviour of the optical properties.

The first calculations appear to be those of Sinai et al. (1985) and Sinai and Wu (1985). They have reported real space recursion calculations for the density of states for Ga/InP and Ga/InAs. In the latter case they illustrated how the type II′ behaviour of Ga/InAs arises.

Almost simultaneously, Kobayashi et al. (1985a) presented a very detailed set of calculations for the Al/GaAs system (type II) which allowed them to interpret the many DA peaks seen in the experimental Raman spectra (see section 5.4.2). More recently Kobayashi and Roy (1987b) have extended these calculations to all the type II′ III/V materials (see table 4). A further application is that of Kobayashi et al. (1985b) and Newman et al. (1985) who have used the results of their calculations for $(GaSb)/Ge_2$ to discuss whether or not a phase transition occurs as the structure changes from zincblende to diamond (they suggest there is a phase transition). Finally Kobayashi and Roy (1987a) have applied the $\boldsymbol{k} = 0$ recursion method to the superlattice GaAs–Al/GaAs.

All these calculations use upwards of 1000 atoms in the clusters over which they average, in order to produce representative densities of states. Further, the method is confined to short-range forces and so no Coulomb forces are included. This means such properties as the LO–TO splitting can be discussed only in an artificial manner (see, for instance, Kobayashi et al. (1985b)). The inclusion of Coulomb forces seems to require the more direct approach of solving the dynamical equations of motion as has already been done for amorphous systems by Thorpe and de Leeuw (1986).

Coherent potential approximation (CPA)

The CPA continues to be employed with, for instance, Jusserand et al. (1985) using it to examine the behaviour of $Ga_{1-c}In_cP$. Their calculated spectral functions at $c \simeq 1$ are in qualitative agreement with experiment which now seems to indicate that this system is of type M rather than of type I as was thought originally (see section 5.4.4). This is in contrast to the type I behaviour indicated by the CPA calculation of Kleinhert (1984*). Jusserand et al. (1985) also presented results for InP/As and Al/GaAs.

A new application of CPA is its extension to quaternary systems. Essentially the same approximation has been presented by both Kleinhert (1985a) and Gregg and Myles (1985) for 1D chains. Although, as Kleinhert does in fact suggest, the self-energy $\boldsymbol{E}$ (eq. (3.13)) couples both sublattices when they are both disordered, both calculations make the simplifying approximation of setting to zero those elements that are off-diagonal in the sublattice index. However, there is little overlap between their calculations. Gregg and Myles compare CPA densities of states with those calculated for chains of 50000 atoms (see section 3.6) for Al/GaP/As. The five mixtures for which they present results correspond to 10% of AlP, AlAs, GaP and GaAs, respectively, plus AlGaPAs.

In contrast, Kleinhert concentrates on applying an embedded cluster version of the CPA to mixtures of Ga/InP/As that are lattice matched to InP

* References with a starred date refer to the main bibliography.

(see end of section 5.4.12 and below) and presents both densities of states (compared with those calculated for linear chains of 20000 atoms) and susceptibilities. Kleinhert also makes a comparison with experiment (Pickering 1981) which is qualitatively successful.

Besides disorder on both the sublattices, quaternaries can be formed with disorder on only one sublattice (in this case $\boldsymbol{E}$ is non-zero only on the disordered sublattice). Gregg and Myles (1985) give densities of states for several mixtures of Al/Ga/InAs.

The only approach to three dimensions has been the Bethe lattice calculation of Kleinhert (1985b) for Ga/InAs/Sb.

Generating function method

The 1D problem has been treated using the generating function technique by Lemieux et al. (1985). They have produced a diagram distinguishing I, II and II′ mode systems that is very similar to that of Sen and Hartman (1974*) (fig. 16) but with all the curves moved to somewhat smaller ϵ. The calculation was extended to include next nearest neighbour forces by Tremblay and Breton (1984) in an attempt to make K/RbBr and RbCl/Br appear one mode. However, according to experiment (table 2 and fig. 21), they both appear to be two mode. A further application of the method is that of Lopez Castillo and Tremblay (1986a) to the quasi 1D materials $M(S/Se)_3$, see below.

Spectral line shapes

The asymmetric shape of first-order Raman lines in disordered crystals has been receiving particular attention. Initially it was suggested that it arose from an anharmonic coupling between the discrete Raman phonon and the continuum phonon density of states (see, for instance, Beserman et al. (1976b*) for Ga/InP and Hirlimann et al. (1976a*) for GaP/As).

More recently, Jusserand and Sapriel (1981*) have suggested that the asymmetry arises from the activation of $\boldsymbol{q} \neq 0$ modes due to disorder in combination with the frequency dispersion of these modes. They modelled the effect by folding the Raman phonon spectral function with a Lorentzian centred on $\boldsymbol{q} = 0$. However Parayanthal and Pollak (1984) suggest that it is preferable to use a Gaussian spacial correlation function instead of the Lorentzian and claim a more general description of this asymmetry.

A comparison between the two explanations is presented by Olego et al. (1986) for Zn/CdTe. They conclude that the discrete-continuum model gives the better fit to experiment. As well, the asymmetry of the peaks for GaAs/Sb (see below) is not well described by this procedure. Further, Lopez Castillo and Tremblay (1986a) obtained asymmetric Raman peaks in their 1D calculations (see above and $Zr(S/Se)_3$ below) in spite of using very flat dispersion curves.

Finally, Lopez Castillo and Tremblay (1986b) have shown how the disorder-induced linewidth can be calculated using sum rules in the one-mode case.

[Addenda to section 5.2. Mixed I/VII crystals]

K/RbI
Using dispersive Fourier transform spectroscopy Memon and Tanner (1986) have detected two TO modes for Rb concentrations of 25% and 50%. Their values are in agreement with those of Fertel and Perry (1969*) although it should be noted that the REI fit of Fertel and Perry indicates that this system is of type M rather than II.

Cu/AgI
The Raman spectra measured by Livescu and Brafman (1986) show this to be a two-mode system. They also detected a weak DA peak.

[Addenda to section 5.3. Mixed II/VI crystals]

Mn/ZnTe
Both the earlier work of Olszewski et al. (1981*) and the more recent RS and IR of Oles and von Schnering (1985) fitted to an REI model indicate this to be a type I system (section 5.3.14). However these studies were confined to Mn concentrations of less than 15%. More recently, Peterson et al. (1986) has measured the RS of samples with Mn concentrations of up to 70%. They were able to fit an REI model showing type M behaviour and identify a number of two-phonon peaks.

Mn/CdTe
Shen et al. (1985) have reported a low frequency band in the infrared absorption, similar to the one they observed in Cd/HgTe (see also Shen and Chu 1983*).

Cd/HgTe
Kozyrev et al. (1983) and Vodopyanov et al. (1985) have fitted the dielectric constant to the IR and in so doing introduced several weak oscillators. These they interpret as being due to clustering and fitted a version of the isodisplacement model of Verleur and Barker (1966*) (see section 3.5). Vodopyanov et al. (1985) have also used resonant RS to enhance these weak modes, obtaining good agreement with the IR frequencies.

Fine structure outside the reststrahlen band has been observed in both the infrared absorption and reflectivity by Shen et al. (1985) and Shen and Chu (1986). They identify the structure with various two-phonon combinations.

Amirtharaj (1985) have presented further discussions of their polarised RS for $Cd_{0.2}Hg_{0.8}Te$ (and also for $Mn_{0.1}Hg_{0.9}Te$).

[Addenda to section 5.4. Mixed III/V crystals]

Al / GaAs
Further RS studies have been presented by Wang and Zhang (1986) and by Nakahara et al. (1987). Following the earlier work of Jusserand and Sapriel (1981*) they identify and interpret a number of two-phonon peaks. Wang and Zhang observe that the TO(X, L) and LO(X, L) modes have a type II behaviour. Along with the observation by Jusserand and Sapriel that the TA(X, L) modes have type I behaviour, this confirms the type A behaviour (fig. 17) predicted by GB. Nakahara et al. have also measured the pressure dependence of the RS and, hence, of the identified modes.

Kamijoh (1986) have observed that these DA peaks in the RS can be enhanced by doping with Zn.

Ga / InP
Bedel et al. (1985) using resonant RS have found DALA modes and confirmed the type II behaviour suggested earlier (see section 5.4.4 and the theory section above). They have also interpreted a number of two phonon peaks.

Ga / InAs
The RS of Kakimoto and Katoda (1985) supports the conclusions of the IR data (see section 5.4.5) that this is a two-mode system although they show data only for 53% In. Using a form of the REI approximation that includes the possibility of clustering (Yamazaki et al. 1980*) and data from GaAs and InAs, they have extracted a clustering parameter from their experimental results. Further, they used the differences between the REI predictions and their measured values to estimate the bond lengths in the alloys and obtain results in good agreement with the EXAFS results of Mikkelsen and Boyce (1982). Kakimoto and Katoda also report clustering parameters for $Al_{0.5}Ga_{0.5}As$, $Ga_{0.52}In_{0.48}P$ and $GaP_{0.3}As_{0.7}$.

GaAs / Sb
RS has now been reported by Cohen et al. (1985) and by McGlinn et al. (1986) with both these studies indicating a type II behaviour in contrast to the type II′ behaviour seen earlier (section 5.4.8). Cohen et al. also found both a DALA mode and a DATA mode existing over most of the concentration range. Both sets of authors note that the asymmetry of the ‘GasAs’ LO peak is not well described by the spacial correlation models described above.

[Addenda to section 5.5. Miscellaneous mixed crystals]

Fe / ZnF$_2$ and Fe / MnF$_2$

The B_{1g} and A_{1g} modes of both these systems are of type I according to the Raman studies of Vianna et al. (1984). However, whereas the E_g mode of Fe/MnF$_2$ is type I that for Fe/ZnF$_2$ is type II.

Hg$_2$(Cl / Br)$_2$

Limonov and Markov (1985) have identified a number of DA peaks in their Raman spectra.

(GaSb) / Ge$_2$

Beserman et al. (1985) have presented further RS on this system and give a discussion of the line width and its relation to possible phase changes. As mentioned above, Newman et al. (1985) and Kobayashi et al. (1985b) have used the recursion method to calculate the density of states and used their results to explain a number of anomalous effects observed in the RS.

Quasi 1D systems

TlS / Se, TlGa / InTe$_2$, TlIn(Se / Te)$_2$

Gasanly et al. (1983) have measured the polarised IR and RS of these systems. They found that they could fit the concentration dependence of the IR active E_u and A_{2u} modes by using separate REI models. Both TlS/Se and TlGa/InTe$_2$ are basically of type II whereas TlIn(Se/Te)$_2$ shows type M for A_{2u} modes and all three types for E_u modes. All three systems have low frequency interchain modes that are of type I. The Raman spectra exhibit a variety of multimode behaviours. Allakhverdiev et al. (1985) have presented a very similar set of measurements for TlS/Se along with a different form for the REI models.

Experimental results have been reported for Tl(In/Tl)Se$_2$ by Gasanly et al. (1984). The IR modes show both type I and II$'$ behaviour.

Zr / TiS$_3$, Zr / HfS$_3$, Zr(S / Se)$_3$

RS spectra for Zr/HfS$_3$ have been measured by Nouvel et al. (1985) and polarised RS for Zr/TiS$_3$ (Ti concentration $\leq 33\%$) by Gard et al. (1986). In spite of the disorder being on the Zr sublattice in both cases, there are differences in behaviour. For instance, Zr/TiS$_3$ shows a three-mode behaviour for the highest frequency A_g mode. Further, the assignment of mode behaviour for some of the other spectral peaks is not always in agreement although this could be due, in part, to the small concentration range investigated by Gard et al.

Deslandes and Jandl (1984) have continued their study of $Zr(S/Se)_3$ by measuring the infrared reflectivity and find a one- to two-mode change in behaviour at large S concentrations. A theoretical attempt to describe this system has been given by Lopez Castillo and Tremblay (1986a) based on their generating function approach mentioned above. Rather than use a strict 1D chain, they modelled the system by a linear array of 2D unit cells. Although not attempting a quantitative comparison with experiment they were able to give a general picture of how this system behaves.

Defect chalcopyrites

The RS spectra of Razzetti et al. (1983) and Lottici and Razzetti (1984) show $Zn/CdGa_2S_4$ to be of type I with the exception of the lowest frequency mode which is type II. $Cd(Ga/In)_2S_4$ might behave likewise except that there is a phase change as the Ga concentration decreases and only type I behaviour is seen (Razzetti et al. (1983)).

Stronger effects are seen when the anions are disordered. The RS spectra for both $ZnGa_2(S/Se)_4$ (Lottici et al. 1984) and $CdGa_2(S/Se)_4$ (Lottici and Razzetti 1984 and Lottici et al. 1984) are predominantly of type II. More details of the RS spectra for $CdGa_2(S/Se)_4$ are given by Parisini and Lottici (1985) along with an REI calculation for the type II high frequency ZnS like modes.

Quaternary mixtures

ZnS / Se / Te

The CI model fit of Burlakov et al. (1985c) to their IR results suggests that this system behaves as a mixture of both type I and II although not all the predicted modes are seen. Note that the rather limited RS studies of Valakh et al. (1983) were able to detect only a two mode behaviour.

CdS / Se / Te

The IR studies of Burlakov et al. (1985b, c) show that this is a 3 mode system and they were able to fit a CI model to their data. Gupta et al. (1986) have applied the REI model of Kutty (1974*) to this data but the fit is not as good as that of Burlakov et al.

Ga / InP / As

As well as the authors referenced in section 5.4.12, Pickering (1981) has also measured the IR for this system lattice matched to InP and Soni et al. (1986) have measured the resonant RS. Inoshita (1984) has reported RS spectra for this system lattice matched to GaAs. A detailed comparison between experiment and the results of a CI calculation has been given by Zinger et al. (1984) although the agreement is not particularly good. Using the same model but with fewer approximations about the force constants, Inoshita was able to produce an impressive fit to both the quaternary results as well as to those for the various ternary mixtures possible in this system. However this was at the expense of 26 force constants. It should be noted that the 1D embedded cluster calculation of Kleinhert is also in fair agreement with the behaviour of the lower three TO modes.

Ga / InAs / Sb

Pickering (1986) has measured the IR for the GaSb-rich mixtures and for the InAs-rich mixtures, there being a miscibility gap in between. In both these regions the system behaved in a three-mode manner.

[Addenda to section 5.6. Layer crystals]

CdCl / Br and CdBr / I

The Raman spectra of both these layered mixed systems show a complex multimode behaviour. A study of the line intensities suggests that clustering occurs in CdCl/Br (Syme et al. 1986) but not in CdBr/I (Vickers et al. 1985).

GaS / Se

Gasanly and Melnick (1984) and Gasanly et al. (1986) have reported further RS spectra. The latter authors have measured both the temperature and pressure dependences of the first and second order spectra.

$Ti(S/Se)_2$, $Ti(Se/Te)_2$, $Ti/ZrSe_2$

Freund and Kirby (1984) have measured the polarised RS from the A_{1g} and E_g modes of these systems. $Ti(Se/Te)_2$ is clearly of type I. The spectra for

Table 7
General mixed crystals

Zn/HgTe	IR, II(TO)	Kumazaki and Nishiguchi (1986)
ZnSe/Te	IR, M	Burlakov et al. (1985a, c)
Cd/HgSe	IR, I	Kumazaki et al. (1986)
Al/InP	RS, II	Bour et al. (1987)
$Fe/MnCl_2$	RS, I	Mischler et al. (1981)
$Fe/CoCl_2$	RS, I	Lockwood et al. (1982)
Cd/PbF_2	IR, RS, I	Kosacki et al. (1986)
		Valakh et al. (1986)
$Sn(S/Se)_2$	IR, RS, II	Garg (1986)
$Hf(S/Se)_2$	IR, II	Kliche (1986)
$Ti/HfSe_2$	IR, II(TO)	Taguchi et al. (1984)
$Pt(S/Se)_2$	IR, II	Kliche (1985)
$Cd(Cr/In)_2S_4$	RS, I	Watanabe et al. (1986)
$CuAl/GaSe_2$	RS, I+II	Azhnyuk et al. (1986)
$Pb_5(Ge/SiO_4)(VO_4)_2$	IR, I+II	Klanjsek Gunde et al. (1987)

$Ti(S/Se)_2$ show a general two-mode like behaviour but with some oddities such as an A_{1g} mode apparently turning into an E_g mode as the S concentrations increases. Freund and Kirby have developed an REI model for this system which is in some agreement with their measurements. They claim a two-mode behaviour for $Ti/ZrSe_2$ although their spectra do not seem to support this identification.

References added in proof

Allakhverdiev, K.R., U.A. Aleshenko, N.A. Bakyshov, L.P. Vodopyanov, F.M. Gashimzade, R.M. Sardarly and Y.A. Shteinshraiber, 1985, Phys. Status Solidi b **127**, 459.

Amirtharaj, P.M., K.K. Tiong, P. Parayanthal and F.H. Pollak, 1985, J. Vac. Sci. & Technol. A **3**, 226.

Azhnyuk, Y.N., V.V. Artamonov and I.V. Bodnar, 1986, J. Appl. Spectrosc. **43**, 1276.

Bedel, E., R. Carles, A. Zwick, M.A. Renucci and J.B. Renucci, 1985, Phys. Status Solidi b **130**, 467.

Beserman, R., J.E. Greene, M.V. Klein, T.N. Krabach, T.C. Romano and S.I. Shah, 1985, Raman Scattering from Metastable $(GaSb)_{1-x}Ge_{2x}$ Alloys: Theory and Experiment, in: Proc. 17th Int. Conf. on the Physics of Semiconductors, San Francisco, 1984, eds J.D. Chadi and W.A. Harrison (Springer, New York) p. 961.

Bour, D.P., J.R. Shealy, G.W. Wicks and W.J. Schaff, 1987, Appl. Phys. Lett. **50**, 615.

Burlakov, V.M., A.P. Litvinchuk and V.N. Pyrkov, 1985a, Sov. Phys.-Solid State **27**, 131.

Burlakov, V.M., A.P. Litvinchuk and V.N. Pyrkov, 1985b, Sov. Phys.-Solid State **27**, 480.

Burlakov, V.M., A.P. Litvinchuk, V.N. Pyrkov, G.G. Tarasov and N.I. Vitrikhovskii, 1985c, Phys. Status Solidi b **128**, 389.

Cohen, R.M., M.J. Cherng, R.E. Benner and G.B. Stringfellow, 1985, J. Appl. Phys. **57**, 4817.

Deslandes, J., and S. Jandl, 1984, Phys. Rev. B **30**, 6019.

Freund, G.A., and R.D. Kirby, 1984, Phys. Rev. B **30**, 7122.

Gard, P., F. Cruege, C. Sourisseau and O. Gorochov, 1986, J. Raman Spectrosc. **17**, 283.

Garg, A.K., 1986, J. Phys. C **19**, 3949.

Gasanly, N.M., and N.N. Melnik, 1984, Sov. Phys.-Solid State **26**, 913.

Gasanly, N.M., A.S. Ragimov, A.F. Goncharov, N.N. Melnik and E.A. Vinogradov, 1983, Physica B **115**, 381.

Gasanly, N.M., N.N. Melnik, A.S. Ragimov and V.I. Tagirov, 1984, Sov. Phys.-Solid State **26**, 336.

Gasanly, N.M., N.N. Melnik, V.I. Tagirov and A.A. Yushin, 1986, Phys. Status Solidi b **135**, K107.

Gregg, J.R., and C.W. Myles, 1985, J. Phys. Chem. Solids **46**, 1305.

Gupta, H.C., G. Sood, J. Malhotra and B.B. Tripathi, 1986, Phys. Rev. B **34**, 2903.

Inoshita, T., 1984, J. Appl. Phys. **56**, 2056.

Jusserand, B., D. Paquet and K. Kunc, 1985, CPA Lattice Dynamics of III–V Mixed Crystals: Theory and Experiment, in: Proc. 17th Int. Conf. on the Physics of Semiconductors, San Francisco, 1984, eds J.D. Chadi and W.A. Harrison (Springer, New York) p. 1165.

Kakimoto, K., and T. Katoda, 1985, Jpn. J. Appl. Phys. **24**, 1022.

Kamijoh, T., A. Hashimoto, N. Watanabe and M. Sakuta, 1986, Phys. Rev. B **33**, 7281.

Klanjsek Gunde, M., B. Orel and V. Moiseenko, 1987, Phys. Status Solidi b **139**, K75.

Kleinhert, P., 1985a, Phys. Status Solidi b **127**, 109.

Kleinhert, P., 1985b, Phys. Status Solidi b **130**, 489.

Kliche, G., 1985, J. Solid State Chem. **56**, 26.

Kliche, G., 1986, Solid State Commun. **59**, 587.
Kobayashi, A., and A. Roy, 1987a, Phys. Rev. B **35**, 2237.
Kobayashi, A., and A. Roy, 1987b, Phys. Rev. B **35**, 5611.
Kobayashi, A., J.D. Dow and E.P. O'Reilly, 1985a, Superlattices and Microstructures **1**, 471.
Kobayashi, A., K.E. Newman and J.D. Dow, 1985b, Phys. Rev. B **32**, 5312.
Kosacki, I., K. Hibner, A.P. Litvinchuk and M.Ya. Valakh, 1986, Solid State Commun. **57**, 729.
Kozyrev, S.P., L.K. Vodopyanov and R. Triboulet, 1983, Solid State Commun. **45**, 383.
Kumazaki, K., and N. Nishiguchi, 1986, Solid State Commun. **60**, 301.
Kumazaki, K., N. Nishiguchi and M. Cardona, 1986, Solid State Commun. **58**, 425.
Lemieux, M.A., P. Breton and A.-M.S. Tremblay, 1985, J. Phys. Lett. **46**, L1.
Limonov, M.F., and Y.F. Markov, 1985, Sov. Phys.-Solid State **27**, 232.
Livescu, G., and O. Brafman, 1986, J. Phys. C **19**, 2663.
Lockwood, D.J., G. Mischler, A. Zwick, I.W. Johnstone, G.C. Psaltakis, M.G. Cottam, S. Legrand and J. Leotin, 1982, J. Phys. C **15**, 2973.
Lopez Castillo, J.M., and A.-M.S. Tremblay, 1986a, Phys. Rev. B **33**, 6599.
Lopez Castillo, J.M., and A.-M.S. Tremblay, 1986b, Phys. Rev. B **34**, 8482.
Lottici, P.P., and C. Razzetti, 1984, J. Mol. Struct. **115**, 133.
Lottici, P.P., A. Parisini and C. Razzetti, 1984, Prog. Cryst. Growth & Charact. **10**, 289.
McGlinn, T.C., T.N. Krabach, M.V. Klein, G. Bajor, J.E. Greene, B. Kramer, S.A. Barnett, A. Lastras and S. Gorbatkin, 1986, Phys. Rev. B **33**, 8396.
Memon, A., and D.B. Tanner, 1986, Int. J. Infrared & Millim. Waves **7**, 1805.
Mikkelsen, J.C., and J.B. Boyce, 1982, Phys. Rev. Lett. **49**, 1412.
Mischler, G., D. Bertrand, D.J. Lockwood, M.G. Cottam and S. Legrand, 1981, J. Phys. C **14**, 945.
Nakahara, J., T. Ichimori, S. Minomura and H. Kukimoto, 1987, J. Phys. Soc. Jpn. **56**, 1010.
Newman, K.E., J.D. Dow, A. Kobayashi and R. Beserman, 1985, Solid State Commun. **56**, 553.
Nouvel, G., A. Zwick and M.A. Renucci, 1985, Phys. Rev. B **32**, 1165.
Olego, D.J., P.M. Raccah and J.P. Faurie, 1986, Phys. Rev. B **33**, 3819.
Oles, B., and H.G. von Schnering, 1985, J. Phys. C **18**, 6289.
Parayanthal, P., and F.H. Pollak, 1984, Phys. Rev. Lett. **52**, 1822.
Parisini, A., and P.P. Lottici, 1985, Phys. Status Solidi b **129**, 539.
Peterson, D.L., A. Petrou, W. Giriat, A.K. Ramdas and S. Rodriguez, 1986, Phys. Rev. B **33**, 1160.
Pickering, C., 1981, J. Electron. Mater. **10**, 901.
Pickering, C., 1986, J. Electron. Mater. **15**, 51.
Razzetti, C., P.P. Lottici, L. Zanotti and M. Curti, 1983, Phys. Status Solidi b **118**, 743.
Shen, S.C., and J.H. Chu, 1986, Chin. J. Phys. **6**, 294.
Shen, S.C., J.H. Chu and H.J. Ye, 1985, Phonon Spectra of Mixed Crystals $Cd_xHg_{1-x}Te$ and $Cd_{1-x}Mn_xTe$, in: Proc. 17th Int. Conf. on the Physics of Semiconductors, San Francisco, 1984, eds J.D. Chadi and W.A. Harrison (Springer, New York) p. 1189.
Sinai, J.J., and S.Y. Wu, 1985, Phys. Status Solidi b **130**, K91.
Sinai, J.J., S.Y. Wu and Z.-B. Zheng, 1985, Phys. Rev. B **31**, 3721.
Soni, R.K., S.C. Abbi, K.P. Jain, M. Balkanski, S. Slempkes and J.L. Benchimol, 1986, J. Appl. Phys. **59**, 2184.
Syme, R.W.G., D.J. Lockwood, N.L. Rowell and K.S. Chao, 1986, Phys. Rev. B **34**, 8906.
Taguchi, I., H.P. Vaterlaus and F. Levy, 1984, Solid State Commun. **49**, 79.
Thorpe, M.F., and S.W. de Leeuw, 1986, Phys. Rev. B **33**, 8490.
Tremblay, A.M.S., and P. Breton, 1984, J. Appl. Phys. **55**, 2389.
Valakh, M.Y., A.P. Litvinchuk, V.I. Sidorenko and N.I. Vitrikhovskii, 1983, Sov. Phys.-Solid State **25**, 1112.
Valakh, M.Y., I. Kosacki and A.P. Litvinchuk, 1986, Sov. Phys.-Solid State **28**, 362.
Vianna, S., C.B. de Araujo and S.M. Rezende, 1984, Phys. Rev. B **30**, 3516.
Vickers, R.E.M., R.W.G. Syme and D.J. Lockwood, 1985, J. Phys. C **18**, 2419.

Vodopyanov, L.K., S.P. Kozyrev, Y.A. Aleshchenko, R. Triboulet and Y. Marfaing, 1985, Optical Observation of Clusters in Distribution of Cd and Hg Ions in Cation Sublattice of $Cd_{1-x}Hg_xTe$, in: Proc. 17th Int. Conf. on the Physics of Semiconductors, San Francisco, 1984, eds J.D. Chadi and W.A. Harrison (Springer, New York) p. 947.

Wang, X.-J., and X.-Y. Zhang, 1986, Solid State Commun. **59**, 869.

Watanabe, J., M. Udagawa, T. Kamigaichi and K. Ohbayashi, 1986, J. Phys. C **19**, 2351.

Zinger, G.M., I.P. Ipatova and A.I. Ryskin, 1984, Sov. Phys.-Semicond. **18**, 13.

CHAPTER 3

Effect of Composition Disorder on the Electronic Properties of Semiconducting Mixed Crystals

A.L. EFROS and M.E. RAIKH

A.F. Ioffe Physico-Technical Institute
Academy of Science of the USSR
Polytechnicheskaya 26, Leningrad, USSR

Optical Properties of Mixed Crystals
Edited by
R.J. Elliott and I.P. Ipatova

Contents

1. Introduction 135
2. Density of states in mixed crystals 136
  2.1. The interaction of electrons with composition fluctuations 136
  2.2. Density of states (qualitative discussion) 139
  2.3. Band-edge and impurity-level shifts 141
  2.4. Formulation of the optimum fluctuation method 146
  2.5. The form of the prefactor 149
3. Effect of composition disorder on the fundamental absorption edge 152
  3.1. Excitons localized by composition fluctuations 152
  3.2. Estimate of the exciton absorption linewidth 157
  3.3. Localization threshold 161
4. Effect of composition fluctuations on kinetic phenomena 164
  4.1. Composition fluctuations and electron mobility 164
  4.2. Effect of composition fluctuations on the hopping conduction activation energy ... 166
Appendix I 169
Appendix II 171
Appendix III 173
References 173

1. Introduction

Mixed crystals of semiconductors are materials of great promise in modern electronics because their properties can be changed smoothly as the composition is varied.

The most characteristic feature of mixed crystals is that various types of atoms are distributed randomly over lattice sites. It turns out that for many purposes this randomness may, in a sense, be neglected and the electronic spectra of these systems can be described quite satisfactorily by ordinary band theory. For this reason, mixed crystals are often treated in the "virtual-crystal" approximation in which the real solid is replaced by a certain ideal virtual crystal. The periodic potential of such a crystal is obtained by averaging the true potential produced by substitutional atoms; this procedure should be carried out separately for each sublattice. This approach has been found to describe quite well the changes in the properties of mixed crystals as their composition varies (Cohen and Bergstresser 1966). Apparently this comes about because both substitutional and host atoms belong to the same group of the periodic system * and have similar properties.

However, in a number of optical and electrical experiments this randomness is of major importance. For example, it strongly affects the interband light absorption (Dean et al. 1969), broadens the exciton reflection line (Suslina et al. 1978) and the intra-impurity absorption line (Berndt et al. 1977) and is responsible for changes in electronic mobility (Greene et al. 1979).

This chapter is devoted to a discussion of such phenomena. But, in contrast to the excellent reviews already published (Elliott et al. 1974, Pikhtin 1977, Nelson 1982), we shall be mainly concerned with the theoretical aspects of the problem. It must be admitted that, at present, the discussion of these phenomena does not go much beyond purely qualitative arguments. We hope, nevertheless, that a detailed review of relevant theoretical ideas may be of some interest to the reader.

* This is not the case in recently obtained alloys $(GaAs)_xGe_{(1-x)}$ (Alferov et al. 1982), where the group IV atoms (Ge) are substituted by atoms of groups III and V (Ga and As). It has been pointed out by D'yakonov and Raikh (1982) that these materials exhibit a phase transition as a function of x, associated with the fact that Ge possesses a center of inversion whereas GaAs does not.

We shall restrict our considerations to the phenomena associated with long-range disorder in the sense that the characteristic length is large compared to the lattice spacing. The disorder may then be treated as large-scale fluctuations in the fractional concentration of the mixed crystal's components. In the following we shall call them "composition fluctuations". The characteristic length for studying these fluctuations is, most commonly, the wavelength of an electron. The condition for the validity of the method to be discussed is that this length should be larger than the average distance between the substitutional atoms.

The chapter is organized in the following way.

In section 2 we obtain the form of the interaction potential between electrons and composition fluctuations and discuss the density of states in mixed crystals. This problem is discussed in the most detail. The density of states is the most important characteristic of disordered systems and its behavior is a subject of some controversy in the literature. In this section – for the first time, to our knowledge – a detailed discussion is given of the optimum fluctuation method as applied to the "white-noise" potential. The specific feature of this case is that the interaction leads to an infinite energy shift and a renormalization of energy is necessary.

Section 3 is concerned with the effect of composition fluctuations on the exciton absorption and luminescence. The composition fluctuations give rise to the localization of the center-of-mass of the exciton. Recent experiments on luminescence make it possible to detect the localization threshold. The exciton absorption linewidth associated with the composition fluctuations is also discussed.

Finally, section 4 examines the effect of composition fluctuations on the mobility of band electrons as well as on the activation energy of hopping conduction.

2. Density of states in mixed crystals

2.1. The interaction of electrons with composition fluctuations

As an example, we study the behavior of a conduction band electron. We assume that, in a given sublattice, an A atom may be substituted by a B atom, so that the composition of the sublattice may be written as A_xB_{1-x}, where x is the fractional concentration of A atoms ($0 \leqslant x \leqslant 1$; in what follows x will usually be referred to simply as the "composition"). Let $E_c(x)$ be the conduction band minimum for the crystal with composition x. The composition of the sample is supposed to vary from point to point, with mean value x_0. Then, for a given volume with a large number of substitutional atoms, one writes $x = x_0 + \Delta x$. Assuming $\Delta x \ll x_0$, the energy at the bottom of the conduction band is

$$E_c(x_0) + \alpha \Delta x, \tag{2.1}$$

where

$$\alpha = \left. \frac{\mathrm{d}E_c}{\mathrm{d}x} \right|_{x=x_0}. \tag{2.2}$$

The deviation of the fractional concentration of A atoms from its mean value can be conveniently written in the form

$$\Delta x = \xi(r)/N, \tag{2.3}$$

where $\xi(r)$ is the deviation of the (absolute) concentration of A atoms from its mean value and H is the total concentration of (sub)lattice sites occupied by either A or B atoms. The function $\xi(r)$ is the concentration averaged over a volume with a large number of both A and B atoms, centered at the point r. This function may be used when the fluctuations of interest are much larger in space than the distance between substitutional atoms.

The potential energy of an electron at the bottom of the conduction band is

$$V(r) = \frac{\alpha}{N}\xi(r). \tag{2.4}$$

Equation (2.4) describes the interaction of an electron with the composition fluctuations. Although qualitatively derived, this result is quite accurate if one neglects the deformation associated with the fluctuations of concentration. It is also supposed that there are no long-range forces between the electrons and substitutional atoms. A derivation of eq. (2.4) based on the Schrödinger equation is given in Appendix I.

The statistical distribution of ξ may be described by the correlation function $\langle \xi(r)\xi(r') \rangle$, $\langle \xi(r)\xi(r')\xi(r'') \rangle$, etc., where $\langle \cdots \rangle$ denotes averaging over different configurations of atoms. The problem of finding these functions is the most complicated and least studied problem in the theory of mixed crystals. The reason is that they are determined by interactions between atoms during sample preparation. In theoretical studies these interactions are usually ignored or, in other words, the positions of A and B atoms are assumed to be random. The two-point correlation is then approximated by

$$\langle \xi(r)\xi(r') \rangle = x(1-x)N\delta(r-r'). \tag{2.5}$$

Here the factor xN gives the concentration of A atoms and determines the fluctuations if $x \ll 1$. In the opposite limit $1 - x \ll 1$ the fluctuations are determined by B-type atoms.

Sometimes (Shlimak et al. 1977), in order to achieve agreement between experimental and theoretical results, the assumption of uncorrelated distribution has to be given up. The simplest alternative is to assume that A atoms form clusters whose positions are (again) uncorrelated. If the concentration of clusters is low and the size of the fluctuations is large compared with the average distance between the clusters, the correlation function can be written, by analogy with (2.5), in the form

$$\langle \xi(r)\xi(r') \rangle = QNx\delta(r - r'), \tag{2.6}$$

where Q is a typical number of atoms in a cluster. It may be seen from eq. (2.6) that fluctuations are enhanced by clusterization.

An interesting case of one-dimensional composition fluctuations occurs in ZnS crystals (Maslov and Suslina 1982, Maslov et al. 1983). Normally, both the cubic form of zinc sulphide (sphalerite) and its hexagonal form (wurtzite) may be described as fixed sequences of alternating close-packing planes. However, as a result of crystal growth conditions, the actual sequence may differ from the periodic one, thus leading to the formation of the so-called packing faults. These are distributed randomly in the crystal and their role is that of substitutional atoms. The correlation function for this case is a one-dimensional version of eq. (2.5). The concentration of the packing faults varies along the axis normal to the close-packing layers and is, therefore, a function of only one coordinate.

In three dimensions, the random potential correlation function is obtained by substituting (2.4) into (2.5):

$$\langle V(r)V(r') \rangle = \gamma\delta(r - r'), \tag{2.7}$$

where

$$\gamma = \frac{\alpha^2}{N} x(1 - x), \tag{2.8}$$

One can see from eq. (2.2) that the parameter α characterizes the shift of the bottom of the conduction band with composition. If $E_c(x)$ is assumed to be a linear function, the quantity α is independent of x and is given by an energy difference between the conduction band minima in materials with $x = 0$ and $x = 1$. This provides us with a simple method to determine its value. Namely, if two such semiconductors are brought into contact, then band discontinuities occur at the boundary and for the conduction band (ΔE_c) the band discontinuity is just the parameter α.

The band discontinuities for various pairs of semiconductors have recently been measured by Kowalczyk et al. (1982) and by Katnani et al. (1982). The values of the discontinuities were determined from the spectra of photoelectrons produced by X-rays in a heterojunction.

So far we have discussed the influence of composition fluctuations on the motion of a conduction electron. The valence band holes are affected in exactly the same way. The parameter α in this case is given by ΔE_v, i.e., by the valence-band discontinuity at the heterojunction. Since the sum $\Delta E_v + \Delta E_c$ of the discontinuities is the difference between the band gaps of the two constituent semiconductors (i.e., with $x = 0$ and $x = 1$), it is sufficient to measure the discontinuities for one band only. According to Kowalczyk et al. (1982), at the GaAs–InAs heterojunction $\Delta E_v = 0.17$ eV and $\Delta E_c = 0.9$ eV. It is shown by Katnani et al. (1982) and Katnani and Margaritondo (1983) that the experimental values of the band discontinuities agree quite well with those calculated by Harrison (1977).

A rough estimate for the value of α can be obtained from a comparison of work functions in the constituent semiconductors or from the photoelectron emission threshold.

The above arguments leading to eqs. (2.2)–(2.4) should be accepted only with some serious reservations. Indeed, it has been taken for granted that the position E_c of the bottom of the conduction band is entirely determined by the value of x at a given point of the sample and coincides with the average position of the bottom of the band in a crystal with composition x. This is not the case, however, if one takes into account the fact that the composition fluctuations are accompanied by deformations of the lattice. It then turns out that the deformation at point r is determined not only by the value of the function ξ at point r but also by its values in the neighboring region. As a result, the interaction of an electron with fluctuations becomes nonlocal. This effect may be ignored if the quantity $(1/a)(\mathrm{d}a/\mathrm{d}x)$, where a is the lattice constant, is sufficiently small. The nonlocality also occurs when the potential of the atoms is of the long-range Coulomb nature. In the present paper these effects are neglected.

2.2. Density of states (qualitative discussion)

The effect of composition fluctuations on the density of states manifests itself, first of all, in the fact that, at energies below the bottom of the conduction band, the density of states does not vanish, but forms an exponentially decreasing "tail". These states are, actually, the lowest bound states in potential wells produced by the composition fluctuations. For energies low enough in the band gap, the density of states is proportional to $\exp[-F(E)]$, where E is the energy of an electron measured from the bottom of the conduction band. Our first concern will be to find the shape of the function

$F(E)$ in the region where the density of states is exponentially small, that is, $F(E) \gg 1$ ("the low-energy tail"). This can be done with the help of the optimum fluctuation method developed by Halperin and Lax (1966) and by Zittartz and Langer (1966). To find the density of states, one should, in principle, take into account all the fluctuations which produce the electron level with energy E. The probabilities of such fluctuations are exponentially small and differ from each other considerably. Therefore, the main contribution to the density of states comes from fluctuations close to the so-called optimum fluctuation which is the most probable among all the fluctuations producing the level E. The exponential in the density of states is identical to the exponential in the probability of the optimum fluctuation. As to the prefactor, it can be found by taking into account all the fluctuations close to the optimum one.

In this subsection the optimum fluctuation method is given in its simplified version which only gives $F(E)$ to within a constant factor. The fluctuations in this version are characterized by two parameters: the size R and the excess number of atoms Z.

Let us consider, for simplicity, the case $x \ll 1$, when the composition fluctuations are determined entirely by the fluctuations in the concentration of A atoms. If in a certain volume, say in a block $R \times R \times R$, the excess number of A atoms is Z, then the fluctuation in the concentration is $\xi = Z/R^3$. According to eq. (2.4), this fluctuation is felt by an electron as a potential well with depth $V = \alpha Z/NR^3$. If we require that this well has a level with energy E, then its depth should be of the order of $|E|$ and its size should be equal to (or larger than) the de Broglie wavelength of the electron, $\hbar/\sqrt{m|E|}$, where m is the conduction-band effective mass. Clearly, for a given excess concentration, the probability of a fluctuation falls off exponentially with size R, so that for the optimum fluctuations we have $R = \hbar/\sqrt{m|E|}$. We thus find that the binding energy E of an electron to a fluctuation-induced potential well, and Z_E, the excess number of A atoms required for producing this well, are related by

$$Z_E = N\hbar^3/\alpha m^{3/2}|E|^{1/2}. \tag{2.9}$$

The density of states ρ is proportional to the probability of finding an excess number Z_E of A atoms in a region with linear dimension $R_E = \hbar/\sqrt{m|E|}$. If Z_E is much smaller than xNR_E^3, which is the average concentration of A atoms in such a region, then Gaussian statistics are applicable, so that the probability in question is $\exp(-Z_E^2/xNR_E^3)$. Using eq. (2.9) and introducing $E_0 = \alpha^4 x^2 m^3/\hbar^6 N^2$, one finds

$$F(E) = \sqrt{|E|/E_0}\,. \tag{2.10}$$

The quantity E_0 is thus a characteristic decay energy for the density of states in the band gap. We obtained it here by closely following the original arguments of Alferov et al. (1968). Equation (2.10) is valid for $|E| \gg E_0$. On the other hand, for Gaussian statistics to be valid, we require $Z_E \ll xNR_E^3$, which means that $|E| \ll \alpha x$. It follows that eq. (2.10) is only true for $E_0 \ll \alpha x$, a condition which is generally fulfilled.

So far the concentration of A atoms has been considered to be low. If concentrations of A and B atoms are comparable, the composition fluctuations are determined by atoms of both types. Then, in analogy with eq. (2.5), we shall write $x(1-x)$ instead of x, and the composition dependence of the characteristic energy E_0 will be of the form $x^2(1-x)^2$. Therefore, for any x we obtain

$$E_0 = \frac{x^2(1-x)^2 \alpha^4 m^3}{\hbar^6 N^2} = \frac{\gamma^2 m^3}{\hbar^6}. \tag{2.11}$$

Thus, the states in the low-energy tail are discrete levels in potential wells produced by the composition fluctuations. If energy increases from the bottom of the conduction band, the states become progressively plane-wave-like. The role of the composition fluctuations in this region is reduced to that of scattering centers which limit the electron mobility. For large positive energies ($E \gg E_0$), the density of states tends to its unperturbed value

$$\rho(E) = \frac{(2m)^{3/2}}{4\pi^2\hbar^2}\sqrt{E}. \tag{2.12}$$

Thus, at energies of the order of E_0 above the bottom of the conduction band, the density of states is considerably affected by the composition fluctuations. It should be emphasized that the quantity E_0 is the only parameter to determine the behavior of the density of states in a random potential obeying eq. (2.6). In other words, the density of states is a universal function of the ratio E/E_0.

In sections 2.3 and 2.4 a quantitative theory is presented for calculating the exponential and prefactor of the density of states in a region where it is exponentially small. The main result is that a numerical factor neglected in eq. (2.10) is actually of crucial importance. In fact, $F(E) = 13.2\sqrt{E/E_0}$ (Baranovskii and Efros 1978), which makes it quite natural that the band-tailing in mixed crystals is generally small.

2.3. Band-edge and impurity-level shifts

It should be remembered that eq. (2.4) for the interaction of electrons with fluctuations is obtained under the assumption that the size of the fluctuation under study is larger than the lattice spacing. However, $\xi(r)$ as defined by eq. (2.5) fails to meet this requirement. In fact, eq. (2.5) describes a random function containing arbitrarily small spatial harmonics ("white noise"). This function, although computationally convenient, gives rise to some difficulties in three-dimensional models. We will show in this section that these short-wavelength fluctuations lead to a divergent second-order energy correction. It may be argued that the only effect of this divergency is that the density of states, as a function of energy, is shifted uniformly towards negative energies by a certain amount Γ. It is impossible to calculate Γ by the method under discussion. In the first place, eqs. (2.4) and (2.5) are not applicable to small-scale fluctuations. To make things worse, the effective-mass approximation breaks down. The only result obtainable is, therefore, a rough estimate of the shift, with the delta-function in eq. (2.5) "smeared" over a length of the order of the lattice constant a. It turns out that the quantity Γ is considerably larger than the characteristic energy for the density of states, E_0.

The existence of the shift gives rise to the question – what precisely is the physical meaning of $E_c(x_0)$, a quantity introduced in section 2.1 as the conduction band minimum for $x = x_0$. The potential energy of an electron has the form $E_c(x_0) + V(r)$. By definition, the spatial average of $V(r)$ is zero. Hence, $E_c(x_0)$ is the expectation value for the potential energy of an electron. This is just the way the conduction band minimum is defined in the virtual-crystal approximation (see Appendix I). The quantity Γ is the shift of the bottom of the conduction band due to the composition fluctuations. A similar shift occurs in the valence band, but in the opposite direction, so that the composition fluctuations lead to narrowing of the band gap in semiconducting mixed crystals (see Baldareschi and Maschke 1975).

Although $E_c(x)$ is not an experimentally observable minimum of the band eq. (2.2) can be reasonably used for estimating α from experimental data; this is because in general the correction to eq. (2.2) due to the shift Γ is small since $d\Gamma/dx \ll \alpha$. This means that, in estimating α from the displacement of bands with composition, one may neglect the fluctuation-induced shift of the bottom band.

Apart from the band-edge shift, the composition fluctuations also produce the shift of impurity levels relative to the new band edge. Suppose a mixed crystal is doped with nonisoelectronic impurities which produce localized electronic states. The short-range fluctuations lower an impurity level in the same manner as the band edge, but fluctuations larger than the localization radius produce only a classical shift whose expectation value is zero. Thus, the shift of the impurity level differs from that of the band edge, and this difference is determined by fluctuations of the order of the localization radius. Therefore, in contrast to the band-edge shift Γ, the impurity level shift can be exactly calculated within our method.

We now turn to the mathematical treatment of the energy shift problem.

The behavior of an electron in a mixed crystal is described by the Schrödinger equation with a random potential obeying eq. (2.7). The second-order term of the perturbation theory for the electron energy may be written in the form

$$E_p^{(2)} = \left\langle \sum_{p'} \frac{\left| \int \mathrm{d}^3R\, \Psi_{p'}^* V(R) \Psi_p \right|^2}{E_p - E_{p'}} \right\rangle, \tag{2.13}$$

where ψ_p is a plane wave normalized to the volume and $E_p = p^2/2m$. Using eq. (2.7), one obtains

$$E_p^{(2)} = \frac{\gamma}{(2\pi\hbar)^3} \int \frac{\mathrm{d}^3p'}{E_p - E_{p'}}. \tag{2.14}$$

The integral (2.14) understood in the sense of a principal value, diverges linearly at large p'. This divergence results from the delta-function approximation in eq. (2.7). Halperin and Lax (1967), in whose paper this divergency seems to have been discussed first (see also Zinger et al. 1977), point out that it results in the shift of the band edge. This can be demonstrated by means of a procedure similar to mass renormalization in quantum electrodynamics. In our case, a "bare" energy is measured from the level $E_c(x_0)$ and, after renormalization, all observable quantities are functions of energy measured from the shifted band edge. Introducing a diagrammatical notation, we denote the unperturbed Green function $G^{(0)} = (E - p^2/2m + \mathrm{i}\delta)^{-1}$ by a full line. The interaction with the random potential is denoted by a broken line associated with factors γ and $\int \mathrm{d}^3q/(2\pi\hbar)^3$, where q is a momentum corresponding to the broken line. The only divergent irreducible diagram for the self-energy part is shown in fig. 1. It is associated with the quantity

$$\Sigma^{(2)} = \frac{\gamma}{(2\pi\hbar)^3} \int \frac{\mathrm{d}^3q}{E - q^2/2m + \mathrm{i}\delta}. \tag{2.15}$$

Suppose that an extra factor $(1 + q^2a^2/\hbar^2)^{-1}$ corresponds to a broken line.

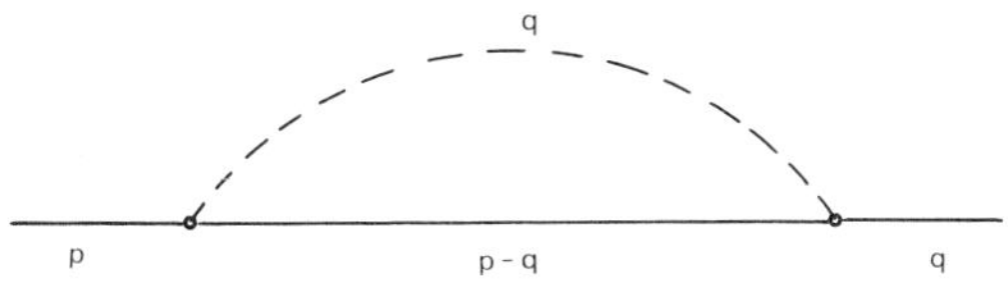

Fig. 1. Divergent irreducible diagram for a self-energy part.

This means that, in eq. (2.7), the delta function is "smeared" over a distance of the order of the lattice spacing a or, in other words, it is replaced by $(8\pi a^3)^{-1}\exp(-|r|/a)$. Let us introduce a quantity Γ defined by

$$\Gamma = -\frac{\gamma}{(2\pi\hbar)^3}\int \frac{d^3q}{(q^2/2m)(1+q^2a^2/\hbar^2)}. \tag{2.16}$$

We may now add and subtract this quantity from the energy E in the Schrödinger equation for an electron in a random field. The difference $E-\Gamma=E'$ we shall call the "physical energy" whereas the sum of the remaining Γ-term with the random potential will be treated as a perturbation. Then, in the unperturbed Green function, the energy E should be replaced by physical energy E'. In any diagram each irreducible self-energy part $\Sigma^{(2)}$ should be supplemented by the term $-\Gamma$, so that it takes the form

$$\tilde{\Sigma}^{(2)} = \frac{\gamma}{(2\pi\hbar)^3}\int \frac{d^3q}{(1+q^2a^2/\hbar^2)(E-q^2/2m+i\delta)} - \Gamma \tag{2.17}$$

which tends to a finite value at $a\to 0$. It follows that this limit can be taken in all diagrams. Then the shift Γ becomes the only a-dependent quantity in the theory. Evaluating the integral (2.16), one finds

$$\Gamma = -\frac{1}{2\pi}\frac{\gamma m}{\hbar^2 a} = -\frac{\alpha^2 m x(1-x)}{2\pi N\hbar^2 a}. \tag{2.18}$$

After renormalization the Schrödinger equation reads

$$-\frac{\hbar^2}{2m}\Delta\Psi + (V(R)-\Gamma)\Psi = E'\Psi. \tag{2.19}$$

Introducing dimensionless variables

$$R' = \frac{\sqrt{2mE_0}}{\hbar}R, \quad V' = \frac{V}{E_0}, \quad \varepsilon' = \frac{E'}{E}, \quad \Gamma' = \frac{\Gamma}{E_0} \tag{2.20}$$

eqs. (2.19) and (2.17) can be rewritten as

$$-\Delta\Psi + (V'-\Gamma')\Psi = \varepsilon'\Psi, \tag{2.21}$$

$$\langle V'(R_1')V'(R_2')\rangle = 2^{3/2}\delta(R_1'-R_2'). \tag{2.22}$$

As already mentioned, in all the diagrams one may let the quantity Γ tend to infinity because as can be seen from eqs. (2.21) and (2.22), all the physical quantities are functions of the parameter ε' only. The density of states, in particular, will be determined for all energies only by the principal energy $E - \Gamma = E'$ related to E_0.

We thus conclude that the composition fluctuations produce the narrowing of the band gap, with a characteristic composition dependence of the form $x(1-x)$ [see eq. (2.18)]. This is just the behavior generally observed, but unfortunately, it is not yet clear whether it can be explained by composition fluctuations or whether one should ascribe it to the monotonic change in semiconductor parameters. As remarked earlier, eq. (2.18) should only be considered as a rough estimate, since the method does not claim to describe fluctuations of the order of interatomic distances. In particular, it would be entirely inconsistent to substitute into this formula the band minimum effective mass. More detailed discussion of the fluctuation-induced shift is given by Van Vechten and Bergstresser (1970), Berolo et al. (1973), Baldareschi and Maschke (1975).

Let us now discuss the shift of an impurity level. If an impurity atom with a local state (for example, a donor) is introduced into a mixed crystal, then fluctuations in the local state energy will respond to the configuration of substitutional atoms in the vicinity of the impurity. This results in nonhomogeneous broadening of the level, as indicated in section 4.2. Besides, there exists a shift of the impurity level relative to the renormalized band edge. The shift of the impurity level relative to the unrenormalized band edge, as given by second-order perturbation theory, is

$$\left\langle \sum_{\nu' \neq \nu} |V_{\nu\nu'}|^2/(\varepsilon_\nu - \varepsilon_{\nu'}) \right\rangle,$$

where $V_{\nu\nu'} = \int d^3r\ \psi_\nu^* V(r) \psi_{\nu'}$. The energy levels and wave functions of an electron in the impurity potential are ε_ν and ψ_ν, respectively. Once again, the smeared delta function is needed, used in the procedure leading to eq. (2.16). The level shift ΔE_ν, relative to the renormalized band edge is obtained by subtracting the shift Γ from the above expression. Finally,

$$\Delta E_\nu = \sum_{\nu' \neq \nu} \frac{\iint d^3r\, d^3r'\ \psi_\nu(r)\psi_\nu^*(r')\psi_{\nu'}(r)\psi_{\nu'}^*(r')}{\varepsilon_\nu - \varepsilon_{\nu'}}$$

$$\sum \times \frac{\exp\left[-|r-r'|/a\right]}{8\pi a^3} - \Gamma. \tag{2.23}$$

Even though both terms in eq. (2.23) diverge at $a \to 0$, their difference remains finite and is independent of a. Mathematically, the shift we obtained is similar to the Lamb shift in quantum electrodynamics. The order of magnitude of the shift is given by eq. (2.18) with a replaced by the impurity state radius. It should be noted that the shift is small compared with the impurity level broadening discussed in section 4.2.

2.4. Formulation of the optimum fluctuation method

In what follows we consider the states which are deep in the forbidden band ($\varepsilon' < 0$, $|\varepsilon'| \gg 1$). For a three-dimensional system with uncorrelated potential, the main difficulty in applying the optimum fluctuation method is to remove the divergency associated with the shift Γ. This problem has been resolved by Brezin and Parisi (1978). Although the scheme we outline here differs from theirs, the final results are the same. To remove the divergency it is necessary from the very beginning to introduce the physical energy ε' measured from the shifted band edge. The density of states can be written as a ratio of two functional integrals

$$\rho(\varepsilon') = \frac{1}{\Omega E_0} \frac{\int \sum_{\nu} \delta(\varepsilon' - \varepsilon_\nu - \Gamma') W(V') \mathrm{D}V'}{\int W(V') \mathrm{D}V'}, \tag{2.24}$$

where Ω is the volume of the system and ε_ν is the eigenvalue of the equation

$$-\Delta\psi_\nu + V'\psi_\nu = \varepsilon_\nu \psi_\nu. \tag{2.25}$$

The quantity W is the probability of finding the fluctuation $V'(R')$. It may be readily seen from eq. (2.22) that $W(V') = \exp[-2^{-5/2}\int \mathrm{d}^3R' V'^2(R')]$. In the low-energy tail the only contribution to the density of states comes from the ground state with $\nu = 0$ and energy ε_0. Equation (2.24) for the density of states can be conveniently rewritten in new variables defined as

$$r = \sqrt{|\varepsilon'|}\, R', \quad \varepsilon = \varepsilon_0/|\varepsilon'|, \quad U = V'/|\varepsilon'|, \quad \tau = \Gamma'/|\varepsilon'|. \tag{2.26}$$

It gives

$$\rho(\varepsilon') = \frac{1}{\Omega E_0 |\varepsilon'|} \frac{\int \delta(-1 - \varepsilon - \tau) \exp\left[\left(-\sqrt{|\varepsilon'|}/2^{5/2}\right) \int U^2(r)\, \mathrm{d}^3 r\right] \mathrm{D}U}{\int \exp\left[\left(-\sqrt{|\varepsilon'|}/2^{5/2}\right) \int U^2(r)\, \mathrm{d}^3 r\right] \mathrm{D}U}. \tag{2.27}$$

In the new variables the Schrödinger equation retains its form, replacing V' by U and ε_ν by ε.

The optimum fluctuation is a function $U_0(r)$ which minimizes the argument of the exponential in eq. (2.27) provided that, in the potential well $U_0(r)$, the ground state energy is equal to -1. The standard procedure (Halperin and Lax 1966) leads to the equations

$$U_0 = -\varphi^2, \qquad \Delta\varphi + \varphi^3 = \varphi. \tag{2.28, 2.29}$$

It should be emphasized that the argument of the delta function is not reduced to zero by the optimum fluctuation. The dimensionless shift τ is compensated by a change in energy ε due to small-scale corrections to the optimum fluctuation potential. It is at this point that our calculation scheme differs from those proposed earlier.

To evaluate the prefactor, fluctuations close to the optimum fluctuation should be taken into account. For this purpose we expand the difference $U(r) - U_0(r)$ as a sum over a complete set of functions $w_n(r)$:

$$U(r) - U_0(r) = \sum_n a_n w_n(r). \tag{2.30}$$

The functional integral (2.27) reduces then to integration over the coefficients a_n. It is convenient to choose $w_0(r)$ in the form

$$w_0 = U_0/J_0^{1/2}, \qquad J_0 = \int U_0^2 \, \mathrm{d}^3r = \int \varphi^4 \, \mathrm{d}^3r. \tag{2.31}$$

With second-order perturbation terms taken into account, the ground-state energy in the potential $U(r)$ is

$$\varepsilon = -1 + a_0 \frac{J_0^{1/2}}{I} + \sum_{m,n\neq 0} a_m a_n \left(w_m \left(-1 - \hat{H}_1 \right)^{-1} w_n \right)_{00}, \tag{2.32}$$

where

$$I = \int \varphi^2 \, \mathrm{d}^3r, \qquad \hat{H}_1 = -\Delta - \varphi^2. \tag{2.33}$$

All the diagonal matrix elements $(w_n)_{00}$ are zero for $n \neq 0$.

As will be seen from the following, at $|\varepsilon'| \gg 1$ the integral in the numerator of eq. (2.27) is determined by $a_n \ll 1$ and $a_0 \sim a_n^2$. For this reason the terms in a_0^2 and $a_0 a_n$ are omitted in eq. (2.32). Substituting eq. (2.30) into the exponential in the numerator of eq. (2.27) one should also omit the term of order a_0^2. In this approximation one gets

$$\rho(\varepsilon') = \frac{I \exp\left[-\sqrt{|\varepsilon'|}\,J_0/2^{5/2} + \left(\sqrt{|\varepsilon'|}/2^{3/2}\right)\tau I\right]}{\Omega E_0 |\varepsilon'| J_0^{1/2}}$$

$$\times \frac{\int \prod_{i=1}^{\infty} \mathrm{d}a_i \exp\left[-\left(\sqrt{|\varepsilon'|}/2^{5/2}\right) \sum_{m,n\neq 0} a_m a_n K_{mn}\right]}{\int \prod_{i=0}^{\infty} \mathrm{d}a_i \exp\left[-\left(\sqrt{|\varepsilon'|}/2^{5/2}\right)\left(\left(J_0^{1/2} - a_0\right)^2 + \sum_{n\neq 0} a_n^2\right)\right]} \tag{2.34}$$

where

$$K_{mn} = \delta_{mn} - 2I\left(w^{(m)}\left(1 + \hat{H}_1\right)^{-1} w^{(n)}\right)_{00} \tag{2.35}$$

and a_0 integration in the numerator has been carried out with the help of the delta function. Note that the matrix K_{mn} may be regarded as a matrix element of the operator

$$\boldsymbol{K} = 1 - 2\varphi\left(1 + \hat{H}_1\right)^{-1}\varphi \tag{2.36}$$

with the w_n as basis functions. If the w_n are eigenfunctions of $\boldsymbol{K}$, the quadratic form in the upper exponential of eq. (2.34) is diagonalized and the integrals over a_i are reduced to Gaussian ones.

As shown by Houghton and Schäfer (1979), for the eigenfunctions $J_1^{-1/2}\partial\varphi^2/\partial r_i$ ($i = x, y, z$; $J_1 = \int \mathrm{d}^3r\,(\partial\varphi^2/\partial x)^2$) of the operator $\boldsymbol{K}$, the corresponding eigenvalues are zero. Integration over the corresponding expansion coefficients a_{1x}, a_{1y}, a_{1z} is equivalent, to within a factor $J_1^{3/2}$, to integration over the coordinates of the center of the optimum fluctuation and yields $(2mE_0|\varepsilon'|/\hbar^2)^{3/2}\Omega$, which is the dimensionless volume of the system. As a result, the density of states takes the form

$$\rho(\varepsilon') = \frac{IJ_1^{3/2}}{2^5\pi^2 J_0^{1/2} E_0}\left(\frac{2mE_0|\varepsilon'|}{\hbar^2}\right)^{3/2} \exp\left[-\frac{\sqrt{|\varepsilon'|}\,J_0}{2^{5/2}} + \frac{\sqrt{|\varepsilon'|}\,\tau}{2^{3/2}} I\right]$$
$$\times (\det{}'\boldsymbol{K})^{-1/2}. \tag{2.37}$$

Here $\det'\boldsymbol{K}$ is the product of all the eigenvalues of the operator $\boldsymbol{K}$ with the exception of those equal to zero and of the one which is equal to infinity and corresponds to the eigenfunction $J_0^{-1/2}\varphi^2(r)$.

Next we turn to the question of the energy renormalization. The argument of the exponential in eq. (2.37), when written in dimension variables, is

$$-\frac{\sqrt{|E-\Gamma|}}{2^{5/2}\sqrt{E_0}}J_0 + \frac{I\Gamma}{2^{3/2}\sqrt{E_0\,|E-\Gamma|}}. \tag{2.38}$$

As shown in section 2.5, det$'\boldsymbol{K}$ contains a diverging factor which, in the limit $a \to 0$, exactly cancels the second term in eq. (2.38) with the result that the quantity Γ, diverging at $a \to 0$, enters into eq. (2.37) only as an energy shift.

The problem, however, needs further discussion. Our procedure starts with adding and subtracting Γ in the Schrödinger equation. Suppose we replace Γ by $\Gamma + A\sqrt{E_0\,|E-\Gamma|}$, where A is a numerical factor (correction terms of just the same kind enter into the quantity Re Σ at negative energies). The term $A\sqrt{E_0\,|E-\Gamma|}$ is small compared with Γ, but, when added to the quantity Γ, it gives rise, in eq. (2.38), to the term $A(I - J_0/4)/2^{3/2}$. In calculating the density-of-states prefactor one should take into account the terms of this order. Adding or subtracting Γ in the Schrödinger equation is an identical transformation. Therefore, $A(I - J_0/4)/2^{3/2}$ must be zero for any A provided that, in the procedure leading to eq. (2.37), the terms of order unity are properly taken into account. It then follows that the proposed renormalization procedure is self-consistent only if $J_0 = 4I$ or, in other words, if this relation is an inherent property of eq. (2.29). This is indeed the case (see Rashba 1982).

2.5. *The form of the prefactor*

As shown in Appendix II, the determinant on the right-hand side of eq. (2.37) can be written in the form

$$\det{}'\boldsymbol{K} = -\left(\frac{J_1}{4I}\right)^3 \frac{\det{}'\left(\hat{H}_3 + 1\right)}{\det{}'\left(\hat{H}_1 + 1\right)}, \tag{2.39}$$

where

$$\hat{H}_3 = -\Delta - 3\phi^2. \tag{2.40}$$

The prime in the lower determinant of eq. (2.39) indicates that, in evaluating the product of the eigenvalues of $\hat{H}_1 + 1$, one should exclude the zero eigenvalue related to the function φ. Similarly, in the upper determinant three zero eigenvalues of the operator $\hat{H}_3 + 1$ should be excluded, which correspond to the functions $\partial\varphi/\partial x$, $\partial\varphi/\partial y$, $\partial\varphi/\partial z$. Each of these determinants may be considered as the product of two factors, one of which corresponds to the discrete spectrum of an operator and the other to the continuous spectrum.

Let D_d be the ratio of the "discrete" factors and D_c the ratio of the "continuous" factors. Then

$$\det{}'\boldsymbol{K} = -\left(\frac{J_1}{4I}\right)^3 D_c D_d. \tag{2.41}$$

Let us first evaluate the quantity D_c. At large distances the eigenfunctions of operators $\hat{H}_1$ and $\hat{H}_3$ are proportional to $\sin(kr - \pi l/2 + \delta_l^{(1,3)})$, where l is the orbital quantum number, k is the momentum and $\delta_l^{(1)}$, $\delta_l^{(3)}$, are k-dependent scattering phases for potentials $-\varphi^2$ and $-3\varphi^2$, respectively. As a boundary condition, we require that wave functions vanish at $r = L$. If the eigenvalues of $\hat{H}_1$ and $\hat{H}_3$, in continuous parts of their spectra, are, respectively, $K_{n,l}^{(1)2}$ and $K_{n,l}^{(3)2}$, then

$$K_{n,l}^{(1,3)} L = \pi(n + l/2) - \delta_l^{(1,3)}. \tag{2.42}$$

Taking into account $2l+1$-fold degeneration of the eigenvalues, one gets

$$D_c = \prod_{n,l} \frac{K_{n,l}^{(3)2} + 1}{K_{n,l}^{(1)2} + 1}$$

$$= \exp\left[\sum_{l,n} (2l+1)\left(K_{n,l}^{(3)} - K_{n,l}^{(1)}\right) \frac{2K_{n,l}^{(1)}}{K_{n,l}^{(1)2} + 1}\right]. \tag{2.43}$$

From the condition (2.42) it follows that $K_{n,l}^{(3)} - K_{n,l}^{(1)} = (\delta_l^{(3)} - \delta_l^{(1)})/L$. We now change in eq. (2.43) from summation to integration. This gives

$$-\tfrac{1}{2} \ln D_c = \frac{1}{\pi} \sum_{l=0}^{\infty} (2l+1) \int_0^\infty \mathrm{d}k \frac{k}{k^2+1} \left(\delta_l^{(3)} - \delta_l^{(1)}\right). \tag{2.44}$$

As indicated earlier, the quantity $\ln D_c$ diverges. We can now single out the diverging term in eq. (2.44). For this purpose we transform the right-hand side of it by adding and subtracting the quantity $\frac{1}{2} \ln D_c$ given by eq. (2.44) with the exact scattering phases replaced by phases $\tilde{\delta}_l^{(1)}$ and $\tilde{\delta}_l^{(3)}$ calculated in the Born approximation (Landau and Lifshitz 1959):

$$\tilde{\delta}_l^{(3)} - \tilde{\delta}_l^{(1)} = \pi \int_0^\infty \mathrm{d}r\, \varphi^2(r) J_{l+1/2}^2(kr) r, \tag{2.45}$$

where $J_{l+1/2}$ is the Bessel function. One finds

$$-\tfrac{1}{2} \ln D_c = -\tfrac{1}{2} \ln \tilde{D}_c + \frac{1}{\pi} \sum_{l=0}^{\infty} (2l+1) \int_0^\infty \frac{\mathrm{d}k\, k}{k^2+1}$$

$$\times\left[\left(\delta_l^{(3)} - \delta_l^{(1)}\right) - \left(\tilde{\delta}_l^{(3)} - \tilde{\delta}_l^{(1)}\right)\right] \tag{2.46}$$

in which the second term is convergent. Substituting (2.45) into eq. (2.44) and using the summation formula for Bessel functions

$$\sum_{l=0}^{\infty} (2l+1) J_{l+1/2}^2(z) = \frac{2}{\pi} z \tag{2.47}$$

(see Gradstein and Ryzhik 1966) one obtains

$$\tfrac{1}{2} \ln \tilde{D}_c = -\frac{I}{2\pi^2} \int_0^{k_{max}} dk \frac{k^2}{k^2+1}, \tag{2.48}$$

where the cutoff parameter should be considered much larger than unity: $k_{max} \gg 1$.

According to eqs. (2.37), (2.41) and (2.46), the quantity $-\frac{1}{2} \ln \tilde{D}_c$, divergent at $k_{max} \to \infty$, enters in the argument of the density-of-states exponential. As already mentioned, this quantity should be combined with the second (diverging at $a \to 0$) term of eq. (2.38). The sum of these two terms is

$$\frac{I\Gamma}{2^{3/2}\sqrt{E_0 |E-\Gamma|}} + \frac{I}{2\pi^2} \int_0^{k_{max}} dk \frac{k^2}{k^2+1}. \tag{2.49}$$

Introducing a dimensionless variable $k = q\hbar/\sqrt{2m|E-\Gamma|}$ and substituting it into eq. (2.16) for the quantity Γ, eq. (2.49) can be written in the form

$$\frac{I}{2\pi^2}\left[\int_0^{k_{max}} dk \frac{k^2}{k^2+1} - \int_0^{\infty} \frac{dk}{1+(k^2a^2/\hbar^2)2m|E-\Gamma|}\right]. \tag{2.50}$$

Now if k_{max} tends to infinity and a to zero, the sum (2.50) becomes

$$-\frac{I}{2\pi^2} \int_0^{\infty} \frac{dk}{k^2+1} = -\frac{I}{4\pi}. \tag{2.51}$$

Thus, in eq. (2.37) for the density of states, the divergent terms cancel out. Finally, in the low-energy tail,

$$\rho(E') = \frac{C}{E_0}\left(\frac{m|E'|}{\hbar^2}\right)^{3/2} \exp\left[-\sqrt{|E'|/\theta E_0}\right], \tag{2.52}$$

where

$$\theta = 2^5/J_0^2 \tag{2.53}$$

and

$$C=\frac{I^2}{2^{3/2}\pi^2\,|D_{\rm d}|^{1/2}}$$

$$\times\exp\left[-\frac{I}{4\pi}+\frac{1}{\pi}\sum_{l=0}^{\infty}(2l+1)\left[\left(\delta_l^{(3)}-\delta_l^{(1)}\right)-\left(\tilde{\delta}_l^{(3)}-\tilde{\delta}_l^{(1)}\right)\right]\right]. \quad (2.54)$$

The quantities C and θ are dimensionless parameters whose values should be calculated on a computer. The parameter θ is found to be $1/178$ (Baranovskii and Efros 1978), whereas C, according to Brezin and Parisi (1980), equals 17.5. Alternatively, the parameter θ, together with the function φ, a solution for eq. (2.29), can be estimated variationally (see Appendix III). For the case of degenerate band structure, the argument of the density-of-states exponential has been calculated by Kusmartsev and Rashba (1983).

3. *Effect of composition disorder on the fundamental absorption edge*

3.1. *Excitons localized by composition fluctuations*

As may be seen from eq. (2.52), the characteristic energy for the density of states in the low-energy tail is $E_0/178$, that is, very much smaller than a qualitative estimate. Typically, absorption spectra of direct-band A^3B^5 materials are not much affected by the composition fluctuations. For example, in $In_xGa_{1-x}As$ mixed crystals, given the values of α from section 2.1, the quantity E_0 at $x=0$ does not exceed 3×10^{-2} meV.

As for A^2B^6 compounds, the fluctuations result in the broadening of exciton lines. Goede et al. (1978) and Suslina et al. (1978) have studied the composition dependence of the exciton reflection spectra for the mixed crystals $Zn_xCd_{1-x}S$ and $Zn_xCd_{1-x}Te$. The exciton reflection line is found to reach a maximum at $x=0.5$. In $Zn_xCd_{1-x}S$ the maximum linewidth was about 9 meV and in $Zn_xCd_{1-x}Te$ it was about 3.5 meV. Lai and Klein (1980), Permogorov et al. (1981a, b, 1982), Areshkin et al. (1982) and Cohen and Sturge (1982) have investigated the effect of the fluctuations on the photoluminescence of $GaAs_{1-x}P_x$, $CdS_{1-x}Se_x$ and $Zn_xCd_{1-x}S$. The main result is that, at helium temperatures, a wide and bright luminescence line is observed, thought to be due to radiation of excitons trapped by the fluctuation-induced potential wells.

Baranovskii and Efros (1978) have suggested that the composition fluctuations might result in a localization of excitons. We reproduce here their estimation for the binding energy and characteristic length of such a state.

We examine the behavior of an exciton in a crystal with fluctuating composition. Let m_e, r_e, m_h, r_h be the masses and positions of an electron and a hole, respectively, and let $V_e = (\alpha_e/N)\xi(r_e)$, $V_h = (\alpha_h/N)\xi(r_h)$ be the potential fluctuations as seen by them. Let us introduce a coordinate of the relative motion of the two carriers $r = r_e - r_h$ and a coordinate of the exciton center-of-mass $R = (m_e r_e + m_h r_h)/M$, where $M = m_e + m_h$. Then the Schrödinger equation for the exciton reads:

$$\left[-\frac{\hbar^2}{2M}\Delta_R - \frac{\hbar^2}{2\mu}\Delta_r - \frac{e^2}{r} + V_e\left(R + \frac{m_h}{M}r\right) - V_h\left(R - \frac{m_e}{M}r\right)\right]\Psi(r, R)$$

$$= E\Psi(r, R), \tag{3.1}$$

where $\mu = m_e m_h/M$ is the reduced mass of the electron–hole pair. Now, we suppose that the exciton is broadened only slightly, that is, the linewidth is smaller than the exciton energy E_{ex}. It is then appropriate to look for the solution in the form $\psi = \Phi(r)\chi(R)$, where the function Φ satisfies the equation

$$\left(-\frac{\hbar^2}{2\mu}\Delta_r - \frac{e^2}{r}\right)\Phi = -E_{ex}\Phi \tag{3.2}$$

for relative motion of the two carriers. Substituting this into (3.1), multiplying (3.1) by $\Phi(r)$ and integrating over the volume, one obtains, using (3.2), an equation for the function $\chi(R)$

$$\left[-\frac{\hbar^2}{2M}\Delta_R + \tilde{V}_e(R) - \tilde{V}_h(R)\right]X = (E + E_{ex})X, \tag{3.3}$$

where

$$\tilde{V}_e(R) = \int d^3r\, V_e\left(R + \frac{m_h}{M}r\right)\Phi^2(r), \tag{3.4}$$

$$\tilde{V}_h(R) = \int d^3r\, V_h\left(R - \frac{m_e}{M}r\right)\Phi^2(r). \tag{3.5}$$

We thus have come to an equation for a particle of mass M in a random field. The characteristic length of the relative motion of an electron and hole is known to be $a_B = \hbar^2\kappa/\mu e^2$, where e is the electron charge and κ is the dielectric constant. Let us assume a_B to be small compared with R_L, a characteristic length for localization of the exciton as a whole. Then, in the integrands of eqs. (3.4) and (3.5), one can drop the terms $(m_h/M)r$ and

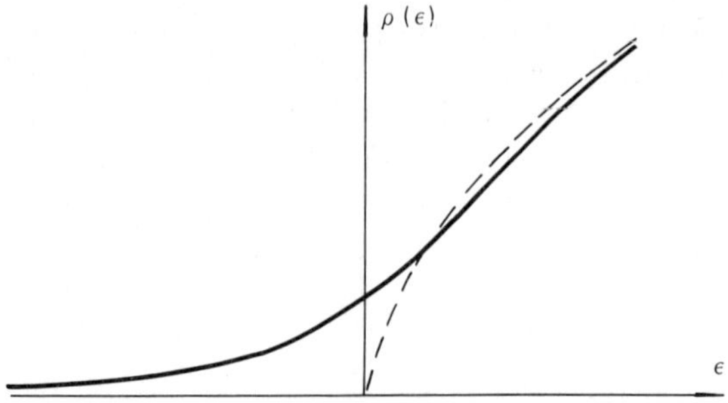

Fig. 2. Schematic representation of the density of excitonic states in a mixed crystal as a function of center-of-mass energy (solid curve). The dashed curve is the unperturbed "free-carrier" density.

$(m_e/M)r$ in the arguments of V_e and V_h, respectively. Integration over r is then readily done and we obtain

$$\tilde{V}_e(R) - \tilde{V}_h(R) = \frac{\alpha}{N}\xi(R), \tag{3.6}$$

where

$$\alpha = \alpha_e - \alpha_h = \frac{dE_g}{dx}. \tag{3.7}$$

Thus, the motion of the center-of-mass of the exciton is described by the same equation as the motion of an electron (see discussion in section 2). Let $\varepsilon = E + E_{ex}$ be the energy of the exciton center-of-mass motion. The corresponding density of states is shown in fig. 2. At large positive energies one applies the free-carrier density given by eq. (2.1), while for $\varepsilon < 0$, $|\varepsilon| \gg E_0/178$, eq. (2.52) is valid, which relates to excitons localized by the fluctuations. It is worth reminding ourselves that, for the case of an exciton, the quantity α in eq. (2.11) should be replaced by dE_g/dx while instead of m the total mass of the exciton M should appear.

Let us discuss in more detail the condition $R_L \gg a_B$. It follows from section 2 that $R_L = \hbar/\sqrt{M|\varepsilon|}$. In the theory of the exciton absorption the $|\varepsilon|$ of interest are of the order of the exciton absorption linewidth Δ. If the mass of an electron and a hole are the same order of magnitude, then the condition $R_L = \hbar/\sqrt{M\Delta} \gg a_B$ is equivalent to the condition $\Delta \ll E_{ex}$. For comparable values of m_e and m_h, the violation of this condition implies that the exciton is destroyed by the effect of the fluctuations.

A somewhat different situation arises in the limit $m_h \gg m_e$. In this case the condition $R_L \ll a_B$ means that the wave function of an electron is much more spread out than that of a hole, so that the exciton is now very much like a donor atom. There is, however, a substantial difference between the two systems. Variations of the donor level are brought about by the effect of the composition fluctuations on the electron *only*, while the exciton energy is also shifted due to the interaction of fluctuations with the hole. Since the hole is much more localized than the electron, it must be much more sensitive to the fluctuations. This is indeed the case if α_e and α_h are the same order of magnitude. The above theory is then applicable at $R_L \ll a_B$, but the parameter α in eq. (3.6) should be replaced by α_h, that is, one should only take into account the potential acting on the hole. The density of states and exciton linewidth will be determined by the energy E_0^h, where the index h means that, in eq. (2.11), one should replace α by α_h and m by m_h. It should be noted that in this case the exciton linewidth Δ may happen to exceed the binding energy E_{ex}, and this does not necessarily imply disintegration of the exciton, provided the interaction of an electron with fluctuations is weak enough.

Let us now suppose that the quantity α_h is so small that the interaction between the hole and the fluctuations may be ignored (a corresponding limitation on the magnitude of α_h will be obtained shortly). In the Schrödinger equation for the exciton center-of-mass motion only the potential energy $\tilde{V}_e(R)$, as defined by eq. (3.4) is involved. If $R_L \gg a_B$, then $\tilde{V}_e(R) = V_e(R)$ and the above theory is valid, with $\alpha = \alpha_e$. At $R_L < a_B$, however, as one passes over from $V_e(R)$ to $\tilde{V}_e(R)$, the smoothing of small-scale fluctuations becomes quite important. The exciton moves in the potential whose minimum spatial parameter is of the order of a_B. The mean-square-root fluctuation for this potential can be calculated in a hydrogenic approximation for the relative motion:

$$\phi = \frac{1}{\sqrt{\pi a_B^3}} \exp[-r/a_B]. \tag{3.8}$$

With the help of eq. (3.4) we find

$$\begin{aligned} W^2 &= \langle \tilde{V}_e^2(R) \rangle \\ &= \frac{\alpha_e^2}{N} \int d^3r\, d^3r' \left\langle \xi\left(R + \frac{m_h}{M} r\right) \xi\left(R + \frac{m_h}{M} r'\right) \right\rangle \phi^2(r) \phi^2(r'). \end{aligned} \tag{3.9}$$

Using eq. (2.5), one finds

$$W^2 = \frac{\alpha_e^2}{N} x(1-x) \int \phi^4\, d^3r = \frac{\alpha_e^2 x(1-x)}{8\pi N a_B^3}. \tag{3.10}$$

From the above inequalities it follows that $\hbar^2/Ma_B^2 \ll W$, i.e., a typical potential well contains many levels, so that the potential may be considered to be classical. The density of states for this case has been found to be of the form (Shklovskii and Efros 1979)

$$\rho(\varepsilon) = \frac{(2m)^{3/2}}{2\pi\hbar^3} \int_{-\infty}^{\varepsilon} \sqrt{\varepsilon - V}\, F(V)\, \mathrm{d}V. \tag{3.11}$$

Since the potential $\tilde{V}_e(R)$ is proportional to the Gaussian random function $\xi(R)$, the potential distribution function must be Gaussian

$$F(V) = \frac{1}{\sqrt{2\pi}\, W} \exp\left[-V^2/2W^2\right]. \tag{3.12}$$

In this case the tail in the density of excitonic states is described by the Gaussian law $\rho(\varepsilon) \sim \exp(-\varepsilon^2/2W^2)$. This behavior differs substantially from that found in section 1. The specific feature of this case is that the composition fluctuations "catch" the exciton by the electron rather than by the hole. The latter is only affected by the fluctuations through its interaction with the electron. The electron wave function smoothes the potential off, so that the motion of the heavy hole becomes quasi-classical.

Let us now discuss the validity of eqs. (3.10)–(3.12). The interaction of the hole with the fluctuations may be ignored if the broadening E_0^h due to this interaction is much less than W. This condition can be written in the form

$$\alpha_h \ll \alpha_e \left(\hbar^2/m_h a_B^2 W\right)^{3/4}. \tag{3.13}$$

Another necessary condition, $m_h a_B^2 W/h^2 \gg 1$, reflects the quasi-classical character of the motion. In this case the exciton broadening up to the composition fluctuations is similar to the donor-level broadening, so that, in the remainder of this paper, we shall refer to it as the "donor-like case".

Another situation worth mentioning is an exciton bound by an impurity. The line-broadening of such an exciton is quite analogous to that in the donor atom. Since an impurity-bound state is, generally, more confined in space than a state localized by the fluctuations, the broadening of an impurity-bound exciton is more pronounced. This phenomenon has been experimentally observed by Suslina et al. (1979).

Up to now the difference $\alpha_e - \alpha_h = \alpha$ has been assumed to be of the same order of magnitude as α_e and α_h themselves. The alternative situation $\alpha \ll \alpha_e$, α_h has been examined by Ipatova et al. (1983). In this case the bands are shifted uniformly as the composition changes, so that the bound gap remains constant. This reduces to zero the effective potential energy (3.6) for the interaction of an exciton as a whole with fluctuations. The reason is that, in eqs. (3.4) and (3.5), one neglects nonlocality. Actually, an exciton interacts effectively with the fluctuations whose size is of the order of a_B, the radius of the relative motion of the electron and hole. Thus, the optimum fluctuation turns out to have two characteristic lengths: a small one of the order of a_B and a large one of the order of the localization radius of an exciton as a whole.

The question of the density of excitonic states has been discussed by Lai and Klein (1980) and Cohen and Sturge (1982). In both papers the behavior obtained is of the form $\rho \sim \exp(-|\varepsilon|/W')$, where W' is of the same order of magnitude as W [eq. (3.10)]. This result disagrees with ours.

3.2. Estimate of the exciton absorption linewidth

For an exciton produced by light, the energy ε of its center-of-mass motion and the light frequency ω are related by

$$\varepsilon = \hbar\omega - E_g + E_{ex}. \qquad (3.14)$$

It is known that, due to the law of conservation of momentum, the exciton has a very low energy ε, so that its behavior, as discussed earlier, is very sensitive to composition fluctuations. The density of excitonic states is plotted in fig. 2. In the region $\varepsilon \gg E_0$ the wave function of the exciton center-of-mass motion is very nearly a plane wave. As energy increases, the conservation of momentum law becomes increasingly effective in every single act of the exciton absorption, which is the reason for the short-wave absorption cutoff.

As to the long-wave cutoff, this is explained by the fact that at $\varepsilon < 0$, $|\varepsilon| \gg E_0/178$ the number of excitonic states decreases exponentially with $|\varepsilon|$. As a result, the exciton absorption line is bell-shaped, with width Δ. The shape of the line can be obtained analytically only in the donor-like case. When the fluctuations are of importance for both the electron and hole, analytical results are only obtainable for short- and long-wave wings of the line (Ablyazov et al. 1983). The linewidth can be estimated from an interpolation formula which satisfies the normalization condition and, on the wings, agrees with the corresponding analytical expressions. Let us discuss this procedure in detail.

The absorption coefficient $\alpha(\omega)$ may be written as $\alpha(\omega) = \alpha_0 A(\varepsilon)$, where α_0 is a slowly varying function of ω and

$$A(\varepsilon) = \frac{1}{\Omega}\left\langle \sum_i \left| \int d^3R\, \Psi_i(R) \right|^2 \delta(\varepsilon - \varepsilon_i) \right\rangle. \qquad (3.15)$$

Here $\psi_i(R)$ is the wave function for the exciton center-of-mass motion, ε_i is the corresponding energy and Ω is the normalization volume. It can be readily seen that

$$\int_{-\infty}^{\infty} A(\varepsilon)\, \mathrm{d}\varepsilon = 1. \tag{3.16}$$

Let us calculate $A(\varepsilon)$ at large positive ε. In this region the wave functions ψ are close to plane waves so that perturbation theory is applicable. One obtains

$$A(\varepsilon) = \frac{1}{\Omega} \int \frac{\mathrm{d}^3 k'}{(2\pi)^3} \frac{\left\langle \left| \int V(R) \exp[\mathrm{i}k'R]\, \mathrm{d}^3 R \right|^2 \right\rangle}{\varepsilon_{k'}^2} \delta(\varepsilon - \varepsilon_{k'}), \tag{3.17}$$

where $\varepsilon_{k'} = \hbar^2 k'^2 / 2M$. Using the correlation function (2.7), we find

$$\left\langle \left| \int V(R) \exp[\mathrm{i}kR]\, \mathrm{d}^3 R \right|^2 \right\rangle = \int \mathrm{d}^3 R\, \mathrm{d}^3 R' \exp[\mathrm{i}k(R - R')] \langle V(R) V(R') \rangle$$

$$= \frac{\alpha^2 x(1-x)}{N} \Omega. \tag{3.18}$$

Finally,

$$A(\varepsilon) = \frac{0.072}{\varepsilon} \left(\frac{E_0}{\varepsilon} \right)^{1/2}, \qquad \varepsilon \gg E_0/178. \tag{3.19}$$

A similar result has been obtained by Zinger et al. (1977) for infrared lattice absorption.

We now consider the region of negative energies $\varepsilon < 0$, $|\varepsilon| \gg E_0/178$. The absorption coefficient is then exponentially small so that the optimum fluctuation method may be applied. We rewrite (3.15) in the form of a functional integral

$$A(\varepsilon) = \frac{1}{\Omega} \int \mathrm{D}V\, W(V) \sum_i \left| \int \mathrm{d}^3 R\, \psi_i(R) \right|^2 \delta(\varepsilon - \varepsilon_i), \tag{3.20}$$

where the integration is carried out over all possible fluctuations $V(R)$. $W(V)$ stands for the probability of the fluctuation and is exponentially small in the region considered. In calculating the integral (3.20) by the optimum fluctuation method, only the ground state in the potential well $V(R)$ should be taken into account. The main contribution to the integral comes from the fluctuations close to the optimum one $V_0(R)$, so that the factor $|\int \psi(R)\, \mathrm{d}^3 R|^2$ has to be taken out of the functional integral at $\psi_i = \psi_0$, where ψ_0 is the wave function of the ground state of the exciton in the potential $V_0(R)$. The remaining functional integral is, by definition, the density of states $\rho(\varepsilon)$, given by eq. (2.52). Thus,

$$A(\varepsilon) = \left| \int \Psi_0(R)\, \mathrm{d}^3 R \right|^2 \rho(\varepsilon). \tag{3.21}$$

It is now expedient to write $\psi_0(R)$ in terms of the dimensionless function φ, which obeys eq. (2.29):

$$\psi_0(R) = \left(\frac{2M|\varepsilon|}{\hbar^2} \right)^{3/4} \frac{\varphi\left(R\sqrt{2M|\varepsilon|/\hbar^2}\right)}{\left(\int \mathrm{d}^3 r\, \varphi^2(r)\right)^{1/2}}. \tag{3.22}$$

A solution of eq. (2.29), obtained variationally in Appendix III, has the form $\varphi(r) = 4\sqrt{2}\exp(-\sqrt{3}\,r)$. Substituting it into eq. (3.2) and using eqs. (2.52) and (3.21), one gets

$$A(\varepsilon) = \frac{242}{E_0} \exp\left[-13.2\sqrt{|\varepsilon|/E_0}\right]; \qquad \varepsilon < 0,\ |\varepsilon| \gg E_0/178. \tag{3.23}$$

The next step is to match eqs. (3.19) and 3.23), taking into account the normalization condition (3.16). The interpolation formula may be written in the form

$$A(\varepsilon) = \frac{178}{E_0} \begin{cases} 0.95 \dfrac{(q+1)^{1/2}}{(q+5.4)^2}, & q > 0 \\ 1.36 \exp\left[-\sqrt{13.84 - q}\right], & q < 0 \end{cases} \tag{3.24}$$

where $q = 178\,\varepsilon/E_0$. $A(\varepsilon)$ defined in this way satisfies condition (3.16) and at the point $q = 0$ both the function and its first derivative are continuous. The function (3.24) is plotted in fig. 3. The resulting linewidth is

$$\Delta = 0.08 E_0 = 0.08 \frac{\alpha^4 x^2 (1-x)^2 M^3}{\hbar^6 N^2}. \tag{3.25}$$

We now turn our attention to the donor-like situation. In this case a typical potential well has a depth W and a linear dimension a_B. Such a well contains many levels related to the motion of the exciton center-of-mass ($W \gg \hbar^2/Ma_\mathrm{B}^2$). In other words, the exciton moves in a smooth quasi-classical potential. As a first approximation, the exciton may be treated as a free

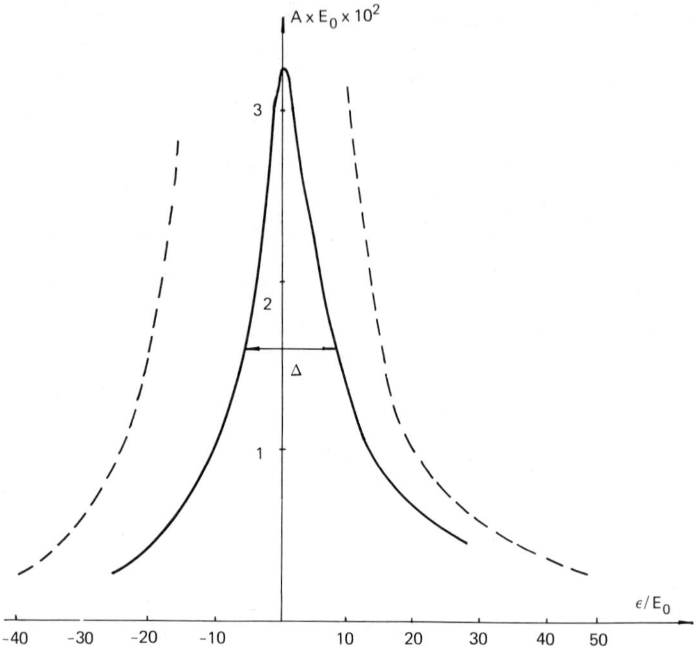

Fig. 3. Exciton absorption line for a mixed crystal, calculated from the interpolation formula (3.24) (solid curve). The dashed curve show the asymptotic behavior of the absorption coefficient at large positive and large negative energies.

particle and the random potential $V(R)$ as a local displacement of the exciton band edge. Then, in eq. (3.15) one can use

$$\Psi_k = \frac{1}{\sqrt{\Omega}} \exp[\mathrm{i}kR] \tag{3.26}$$

and

$$\varepsilon_k = \frac{\hbar^2 k^2}{2M} + V. \tag{3.27}$$

One then obtains

$$A(\varepsilon) = \langle \delta(\varepsilon - V) \rangle. \tag{3.28}$$

Using the distribution function (3.12), we find

$$A(\varepsilon) = \frac{1}{\sqrt{2\pi}} \exp[-\varepsilon^2/W^2]. \tag{3.29}$$

Thus, in the donor-like case, the shape of the exciton line is Gaussian.

One faces two difficulties in comparing the obtained linewidth with experiment. First, the experimental data available are related to reflection and luminescence rather than absorption. Secondly, the values of α_e and α_h are unknown for A^2B^6 mixed crystals (one only knows the difference $\alpha_e - \alpha_h = dE_g/dx$). Besides, the exciton masses are not known with sufficient accuracy. A quantitative attempt to compare the theory with experiment has been undertaken by Suslina et al. (1978).

The width of the exciton absorption line in strong magnetic fields has been calculated by Raikh and Efros (1984). The linewidth is shown to increase with magnetic field. The reason for this is twofold. First, in a strong magnetic field the wave function of the relative motion of an exciton shrinks in the transverse direction. This results, as eq. (2.10) shows, in an increase of the mean-square-fluctuation in the potential felt by the exciton as a whole. On the other hand, the translational mass for the transverse motion of the exciton as a whole increases with magnetic field, and this effect favors the localization of the exciton.

3.3. Localization threshold

If recombination processes are neglected, that is, the exciton lifetime is assumed to be infinite, then excitonic states can be classified as either delocalized or localized. The wave functions are, respectively, either "smeared" throughout the crystal or confined to a limited region in space. According to the modern theory (Mott and Davis 1979, Shklovskii and Efros 1979), there is an energy E_{th} which separates these two types of states. When this energy is approached from the localization side, the localization radius goes to infinity. It should be noted, however, that in the vicinity of this threshold the exciton wave function cannot be considered as a one-parameter exponential, but, as fig. 4 shows, it is only modulated by such an exponential while containing small-scale harmonics of the order of $\hbar/\sqrt{ME_0}$. The wave functions of delocalized states also contain in this vicinity small-scale harmonics and a large-wavelength envelope function. It should be emphasized that, in passing through the localization threshold, it is only the wave function's large-scale structure that changes. The shape of the function in a small region of space does not allow us to draw any conclusion whatsoever as to the type of state it describes. Characteristics such as the density of states or the light absorption

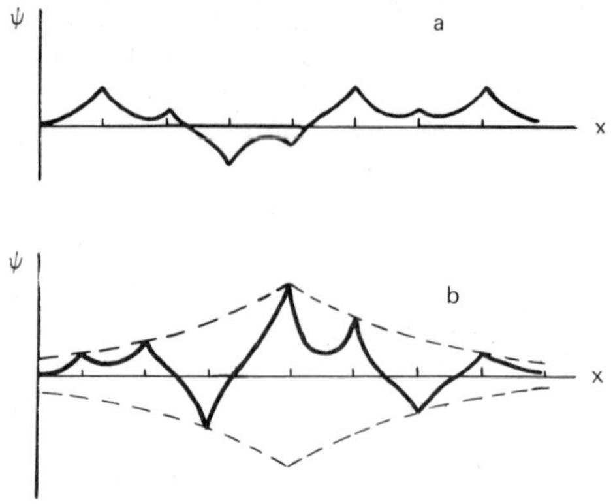

Fig. 4. Wave function in the neighborhood of the localization threshold: (a) delocalized state, (b) localized state.

coefficient are insensitive to the large-scale structure of the wave function and cannot therefore be used for studying the localization threshold.

In electronic systems the localization threshold can be studied from the zero-temperature conductivity (Rosenbaum et al. 1980) which vanishes below it. For excitons, the threshold may be thought to be observable in experiments on luminescence.

This conclusion was drawn by Cohen and Sturge (1982) and Permogorov et al. (1982). Both groups report a detailed study of low-temperature exciton luminescence of $CdS_{1-x}Se_x$ mixed crystals under monochromatic excitation in various regions of the absorption spectrum. The typical data are shown in fig. 5 for $CdS_{0.97}Se_{0.03}$ at $T = 2$ K. The dashed line is the spectrum of photoluminescence when the excitation is well into the exciton band. The weak maximum at $E = 2.46$ eV is a phonon replica. The full lines show the spectra of photoluminescence for different excitation energies. The spectra are detected through their LO_1 phonon replica. All the lines in fig. 5 are shifted to shorter wavelengths by the phonon energy so that the sharp peak corresponds to the line of monochromatic excitation.

The main idea of Cohen and Sturge (1982) and Permogorov et al. (1982) is that the sharp peaks in photoluminescence spectra appear when the excitation energy is in the localized region. If an excitation energy is above the localization threshold, there are no sharp peaks at the excitation energy because the lifetime is long enough for thermalization of the exciton to the localized region, which makes the exciton unobservable at the excitation frequency.

Open circles and dotted lines show the ratio of the integral intensity of the sharp peaks to the integral intensity of the wings at different excitation energies. Extrapolation of the dotted line to the zero ratio gives the mobility threshold. It is important that the short-wavelength edge of luminescence for

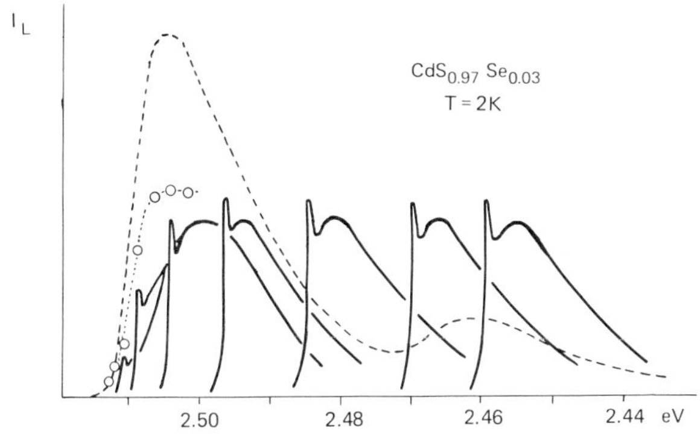

Fig. 5. First phonon sidebands of the luminescent line for various (monochromatic) excitation frequencies (solid curves). The curves are shifted to higher frequencies by the LO_1 phonon energy and normalized on the ordinate to have a common wing maximum. The peak-to-wing integrated intensity ratio is shown for the sidebands as a function of excitation frequency (open circles and dotted curves). The dashed curve is the luminescence spectrum for excitations high into the conduction band. After Permogorov et al. (1982).

excitations well into the band coincides with the mobility threshold obtained by extrapolation. This is additional evidence for the fact that, at the mobility threshold, the excitation energy migration is drastically slowed down.

It has also been suggested (Permogorov et al. 1983) that the mobility threshold may be investigated through energy dependence of the luminescence polarization. It turns out that, for localized excitons produced by linearly polarized light, the fluorescence is also linearly polarized. This is believed to be due to anisotropy in the wave functions of localized excitons. Because of the anisotropy, states with different dipole moments are also different in energy. The excited dipoles are oriented by light. Unless the orientation is destroyed during the exciton lifetime, we shall have polarized emission. As may be seen from fig. 6, the mobility threshold obtained this way coincides with that from the monochromatic excitation method described above.

More evidence comes from the work by Ovsyankin and Fedorov (1982), who have investigated the mobility threshold of small excitons in disordered tysonite structure $S_2Yb_2F_8$. At monochromatic excitation the luminescence quantum efficiency was measured as a function of the light frequency. At high enough light intensities the quantum efficiency of excitation into localized states is much greater than that for delocalized states. The explanation proposed is that delocalized excitons are intensively quenched by defects. Summing up, one may expect that experiments on luminescence could indeed reveal the mobility threshold.

In our opinion, an experimental study of Anderson's localization in the excitonic system is a significant problem. This phenomenon is now being

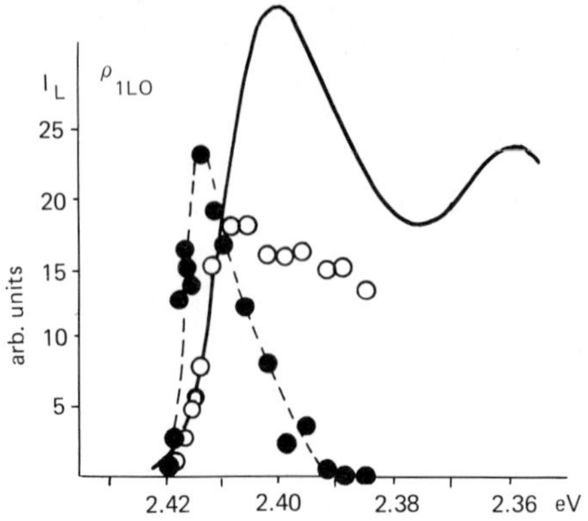

Fig. 6. Luminescence spectrum for monochromatic excitations high into the conduction band (solid curve). Open circles – the same as in fig. 5. Closed circles show the frequency dependence of the degree of polarization ρ_{1LO} for the first sidebands of the luminescence line (in relative units). After Permogorov et al. (1983).

widely investigated on electrons (for an updated review see the English edition of the monograph by Shklovskii and Efros, 1979). However, in electronic systems the localization is greatly complicated by Coulomb interactions. In particular, the dielectric constant goes to infinity at the localization threshold. That is why the behavior of electron wave functions, in this vicinity, may differ significantly from that predicted by Anderson's model (Anderson 1958) with electron–electron interactions completely ignored. On the contrary, an excitonic system is expected to be well-described in terms of the Anderson model. When compared, the results obtained for both systems might elucidate the role of Coulomb interactions at the localization threshold. Theoretical study of this problem is at present in its initial stage and faces serious difficulties (Altshuler and Aronov 1983).

4. *Effect of composition fluctuations on kinetic phenomena*

4.1. Composition fluctuations and electron mobility

In mixed crystals, charged carriers are scattered by composition fluctuations. Predominantly, the scattering is by the fluctuations of the order of the electron wavelength. In semiconductors this length is, generally, much larger than the lattice constant, so that the above scheme may well be used for describing the scattering.

For simplicity, we consider the case of an isotropic quadratic energy spectrum (Shlimak et al. 1977) The collision frequency of an electron with momentum p is

$$\nu_p = \frac{2\pi}{\hbar} \sum_q \langle |V_q|^2 \rangle (1 - \cos\theta_q)\delta(\varepsilon_p - \varepsilon_{p-q}), \tag{4.1}$$

where θ_q is the scattering angle, $\varepsilon_p = p^2/2m$ and

$$\langle |V_q|^2 \rangle = \frac{1}{\Omega} \int d^3r \exp[iqr] \langle V(r)V(0) \rangle. \tag{4.2}$$

Using the correlation function (2.7), one obtains

$$\langle |V_q|^2 \rangle = \frac{\alpha^2 x(1-x)}{\Omega N}. \tag{4.3}$$

From this it may be seen that the scattering by composition fluctuations is equivalent to that by a short-range potential. Substituting eq. (4.3) into eq. (4.1) we find

$$\nu_p = \frac{\alpha^2 x(1-x) mp}{\pi \hbar^4 N}. \tag{4.4}$$

From eq. (4.4) the Hall mobility in a weak magnetic field may be easily shown to have the form

$$\mu_H = \frac{\pi^{3/2}}{2\sqrt{2}} \frac{e\hbar^4}{\alpha^2 x(1-x) m^{5/2} (kT)^{1/2}}. \tag{4.5}$$

Shlimak et al. (1977) have generalized this expression to the case of the complicated Ge valence band and compared it with the hole mobility data for p-type $Ge_{0.92}Si_{0.08}$. Their conclusion is that at nitrogen temperatures the scattering from the fluctuations greatly affects the hole mobility, although there is no wide enough energy range where the mobility is due to this mechanism alone.

The effect of the fluctuations on the electron mobility has been demonstrated most clearly by Greene et al. (1979). They studied mixed crystals of $In_{1-x}Ga_xAs_yP_{1-y}$. The parameter x was chosen in such a way that the lattice constant coincided with that of the InP substrate. The y-dependence of the Hall mobility was found to display a minimum at $y \approx 0.5$. The temperature dependence of the electron mobility was also investigated and, in the range 77–300 K was found to be of the form $\mu_H \sim T^{-1/2}$. According to the authors' analysis, these y- and T-dependencies cannot be accounted for by any other relevant scattering mechanisms (polar optical phonons and ionized impurities).

4.2. *Effect of composition fluctuations on the hopping conduction activation energy*

At low temperatures the electrical conduction of doped semiconductors is, generally, via the hopping mechanism, that is, the charge transfer is due to tunneling of the electrons from one donor to another. Since the impurity levels differ in energy, these transitions are also accompanied by phonon absorption or emission. For this reason the conductivity depends exponentially on the temperature

$$\sigma = \sigma_0 \exp[-\varepsilon_A/kT], \tag{4.6}$$

where ε_A is an activation energy for the hopping conduction and σ_0, while a slowly varying function of T, depends exponentially on the impurity concentration.

Let us consider a mixed crystal doped with electrically active impurities, say, with donors. The composition fluctuations in the vicinity of an impurity center produce an energy spread of impurity states in addition to that caused by the Coulomb interaction between the charged impurities. As a result, the activation energy ε_A depends on the composition of the mixed crystal. Figure 7 shows the experimental composition dependence of the activation energy for gallium-doped Si_xGe_{1-x} at $0 < x < 0.06$ (Gel'mont et al. 1974). It can be seen

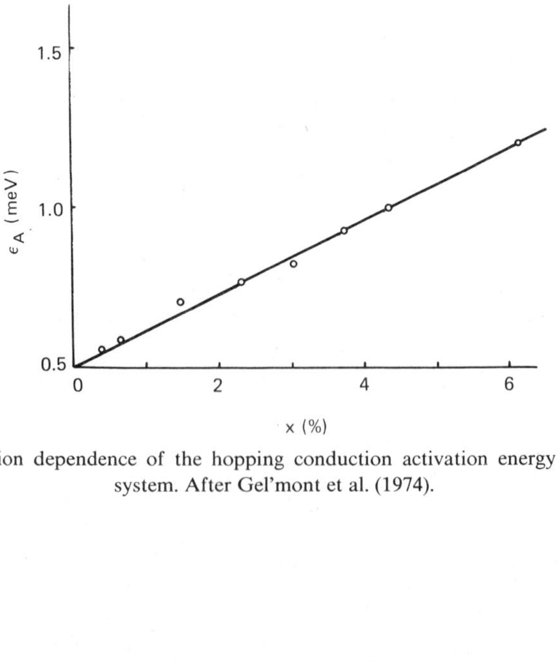

Fig. 7. Composition dependence of the hopping conduction activation energy for a Si_xGe_{1-x} system. After Gel'mont et al. (1974).

that ε_A increases significantly with x. At $x \sim 0.04$–0.06 it may be argued that ε_A is largely determined by the composition fluctuations.

In our view, mixed crystals are very promising objects for studying hopping transport. The point is that usually, with the energy spread due to the Coulomb interaction, occupation of donor levels is characterized by long-range correlations. The occupation of a given donor is determined by the potentials at this point due to all other donors, while their own charges are, in turn, determined by the occupation of the given donor. This is a source of great difficulty in the theory of hopping conduction (Shklovskii and Efros, 1979). A mixed crystal is, in this respect, a simpler system. If the energy spread is due to composition fluctuations, then for donors with a small overlap, the donor levels are statistically quite independent of each other. One therefore expects mixed crystals to be a good tool for checking the hopping conduction theory free from complications due to the Coulomb interaction.

Generally, the calculation of the activation energy reduces to a complicated percolation problem which has to be solved by computer modelling (Shklovskii and Efros 1980). If the Coulomb interaction may be ignored, a simple expression for the activation energy is obtained by applying perturbation theory to the percolation problem (Skal et al. 1975). It turns out that ε_A may be expressed in terms of the distribution function of the levels over energy $F(\varepsilon)$. Namely,

$$\varepsilon_A = \int_{-\infty}^{\infty} d\varepsilon_i \int_{-\infty}^{\infty} d\varepsilon_j \; \varepsilon_{ij} F(\varepsilon_i) F(\varepsilon_j), \tag{4.7}$$

where

$$\varepsilon_{ij} = \tfrac{1}{2}\{ |\varepsilon_i - \mu| + |\varepsilon_j - \mu| + |\varepsilon_i - \varepsilon_j| \}. \tag{4.8}$$

ε_i is the fluctuation-induced energy shift at the ith impurity and μ is the chemical potential measured from the average impurity level. For a given degree of compensation K the chemical potential is defined by the condition

$$1 - K = \int_{-\infty}^{\mu} F(\varepsilon)\, d\varepsilon. \tag{4.9}$$

The shift of an impurity level may result from either the composition fluctuations in the region of the impurity state or fluctuations in the potential of the central cell. The former mechanism can be described within the scheme we are discussing. The first-order energy correction ε for the impurity level is

$$\varepsilon = \frac{\alpha}{N} \int d^3r \, \phi^2(r) \xi(r), \tag{4.10}$$

where $\phi(r)$ is the wave function of the impurity center. Using the same argument that led to eq. (3.12), we obtain

$$F(\varepsilon)=\frac{1}{\sqrt{2\pi}\,W}\exp[-\varepsilon^2/2W^2], \tag{4.11}$$

where W is defined by eqs. (3.8) and (3.10).

Using eqs. (4.7) and (4.9), the activation energy is

$$\varepsilon_{\mathrm{A}}=\frac{W}{\sqrt{\pi}}\left[1+\frac{\sqrt{\pi}\,\mu}{W}(1-2K)+\sqrt{2}\exp[-\mu^2/2W^2]\right], \tag{4.12}$$

where μ is related to K by eq. (4.9).

Figure 7 shows the experimental data for $K=0.4$. In this case

$$\varepsilon_{\mathrm{A}}=\frac{0.98}{2\sqrt{\pi}}\,\frac{\alpha(xN)^{1/2}}{Na_{\mathrm{B}}^{3/2}}. \tag{4.13}$$

Equation (4.12) is obtained for a hydrogen-like impurity center. In the above experiment the conduction was due to hopping by the Ga acceptor levels, so that, in their calculations, Gel'mont et al. (1974) took into account the 4-fold degeneration of the valence band. The result obtained differs from eq. (4.13) by replacing the factor 0.98 by 0.54.

Finally we discuss the role of the central cell. At small x the fluctuations in the central cell potential can be described as follows. Consider a mixed crystal Si_xGe_{1-x} doped with gallium. The contribution of the central cell potential into the Ga ionization energy is about 1 meV in germanium and about 30 meV in silicon. Therefore, if one of the four nearest neighbours of the Ga atom is the silicon atom, the Ga level could well be shifted by several meV. Let Δ be the magnitude of the shift. At small x the configurations with two or more neighbouring Si atoms are too rare to be taken into account. From eq. (4.7) one then easily obtains $\varepsilon_{\mathrm{A}}=8x\Delta$. If this mechanism is assumed to be the only one to explain the activation energy concentration dependence (see fig. 7), then Δ will be equal to 1.7 meV, which we consider quite reasonable.

The only way to explain the experimental data by large-scale fluctuations [eq. (4.13)] is to assume that, in the germanium crystal, the silicon atoms form clusters of several tens of atoms. This, as discussed in section 2.1, enhances the fluctuations.

It should be remembered that the more significant the fluctuations in the central cell potential the deeper the impurity level under study. Owing to the central cell potential, the ground-state wave function, at small r has a term in $1/r$. This term leads to a divergency of the integral $\int d^3r\,\phi^4(r)$ which is involved in eq. (3.10) for the quantity W. Thus the contribution from small distances can only be ignored at very low central cell potentials.

The energy spread of impurity states has also been estimated from the intra-impurity absorption in $Ga_xIn_{1-x}P$ mixed crystals doped with silicon and tellurium (Berndt et al. 1977, 1978). The results were again interpreted in terms of the fluctuations in the central cell potential. We believe that measurements of the hopping conductivity and the intra-impurity absorption on the same material could permit a thorough verification of the hopping conduction theory.

Appendix I

Potential energy for interaction of an electron with composition fluctuations

The true potential in a mixed crystal may be written as

$$V(r) = \sum_i (c_i V_A(r - R_i) + (1 - c_i) V_B(r - R_i)), \tag{AI.1}$$

where V_A and V_B are the potentials of A and B atoms, respectively. The coefficients c_i are unity or zero according to whether there is an A atom or a B atom present at site i. (The sites are assumed to form a perfect lattice. We thus neglect the displacements of the atoms due to the random nature of their neighbourhood.)

To a first approximation the potential $V(R)$ may be replaced by $V_0(r)$, the periodic potential of the virtual crystal, which actually means that all the c_i are replaced by their mean value x

$$V_0(r) = \sum_i (x V_A(r - R_i) + (1 - x) V_B(r - R_i)). \tag{AI.2}$$

The perturbing potential $W(r) = V(r) - V_0(r)$ has the form

$$W(r) = \sum_i (c_i - x)(V_A(r - R_i) - V_B(r - R_i)). \tag{AI.3}$$

Let us introduce the wave functions

$$\psi_R = \frac{1}{\sqrt{\Omega}} \exp[ikr] U_k(r), \tag{AI.4}$$

where Ω is the normalization volume and $u_k(r)$ is a periodic function of r, obeying the Schrödinger equation in the virtual-crystal potential:

$$\hat{H}_0\Psi_k = \left(-\frac{\hbar^2}{2m_0}\Delta + V_0(r)\right)\Psi_k = E_k\Psi_k. \tag{AI.5}$$

It is appropriate to seek a solution of the equation

$$(\hat{H}_0 + W(r))\Psi = E\Psi \tag{AI.6}$$

in the form

$$\Psi = \sum_k B_k\Psi_k(r). \tag{AI.7}$$

Substituting eq. (AI.7) into eq. (AI.6), multiplying eq. (AI.6) by ψ_k^* and integrating over the volume, one gets

$$(E_k - E)B_k + \frac{1}{\Omega}\sum_{k_1} B_{k_1}\int d^3r \exp[i(k_1 - k)r]\, U_{k_1}U_k^* \times\left[\sum_i (c_i - x)(V_A(r - R_i) - V_B(r - R_i))\right] = 0. \tag{AI.8}$$

Interchanging the summation and integration and using the periodicity of $u_k(r)$, we find

$$(E_k - E)B_k + \frac{1}{\Omega N}\sum_{k_1} B_{k_1}\alpha_{kk_1}\sum_i (c_i - x)\exp[i(k_1 - k)R_i] = 0, \tag{AI.9}$$

where

$$\alpha_{kk_1} = N\int d^3r\, U_{k_1}U_k^*(V_A(r) - V_B(r))\exp[i(k_1 - k)r]. \tag{AI.10}$$

For large-scale fluctuations, the function B_k should only be considered to be nonzero in a small region about the point $k = 0$, so that in eq. (AI.9) we may write $E_k = \hbar^2k^2/2m$ and approximate α_{k,k_1} by

$$\alpha = N\int d^3r |U_0(r)|^2(V_A(r) - V_B(r)). \tag{AI.11}$$

(In going from eq. (AI.10) to eq. (AI.11) the difference $V_A(r) - V_B(r)$ is assumed to fall off over the distance of a few lattice constants. If it contains a long-range component, one cannot replace $\exp[i(k_1 - k)r]$ in eq. (AI.10) by unity, even for small k, k_1'. In this case the interaction of an electron with the composition fluctuations is nonlocal.) The right-hand side of eq. (AI.11) indeed gives the same quantity α that is defined by eq. (2.2). This is really seen from the identity

$$\frac{\partial E_k}{\partial x} = \int d^3r \, \Psi_k^* \frac{\partial \hat{H}_0}{\partial x} \Psi_k. \tag{AI.12}$$

In eq. (AI.9) the sum over the lattice sites can be expressed in terms of $\xi(r)$, a deviation of the concentration of A atoms from its mean value xN

$$\sum_i (c_i - x) \exp[i(k_1 - k)R_i] = \int d^3r \, \xi(r) \exp[i(k_1 - k)r]. \tag{AI.13}$$

The equation for B_k may then be rewritten in the form

$$\left(\frac{\hbar^2 k^2}{2m} - E\right) B_k + \frac{\alpha}{\Omega N} \sum_{k_1} B_{k_1} \int d^3r \, \xi(r) \exp[i(k_1 - k)r] = 0. \tag{AI.14}$$

Transforming now to the spatial wave function

$$\phi(r) = \sum_k B_k \exp(ikr), \tag{AI.15}$$

multiplying eq. (AI.14) by $\exp(ikr)$ and summing over k, we obtain the Schrödinger equation for the function $\phi(r)$

$$-\frac{\hbar^2}{2m} \Delta\phi + \frac{\alpha}{N} \xi\phi = E\phi. \tag{AI.16}$$

The potential energy in this equation is identical to that of eq. (2.4).

Appendix II

Proof of eq. (2.39)

We employ the regularization procedure of Houghton and Schafer (1979). We add the quantity $-\delta\varphi^2$ to the Hamiltonian (2.33) and introduce the operator

$$\boldsymbol{K}_\delta = 1 - 2\varphi\left(\hat{H}_\delta + 1\right)^{-1}\varphi, \qquad \hat{H}_\delta = -\Delta - \varphi^2(1 + \delta). \tag{AII.1}$$

Using the properties of this operator

$$\boldsymbol{K}_{\delta}\varphi^{2}=\left(1+\frac{2}{\delta}\right)\varphi^{2};\qquad \boldsymbol{K}_{\delta}\frac{\partial\varphi^{2}}{\partial x}=-\frac{\delta}{2-\delta}\frac{\partial\varphi^{2}}{\partial x}\tag{AII.2}$$

it can easily be shown that

$$\det{}'\boldsymbol{K}=-4\lim_{\delta\to 0}\frac{1}{\delta^{2}}\det\boldsymbol{K}_{\delta}.\tag{AII.3}$$

On the other hand, the determinant of $\boldsymbol{K}_{\delta}$ may be written in the form

$$\det\boldsymbol{K}_{\delta}=\frac{\det\left(\hat{H}_{\delta}-2\varphi^{2}+1\right)}{\det\left(\hat{H}_{\delta}+1\right)}.\tag{AII.4}$$

The operator in the numerator has a 3-fold degenerate eigenvalue vanishing at $\delta\to 0$. The corresponding eigenfunctions are $\partial\varphi/\partial x$, $\partial\varphi/\partial y$, $\partial\varphi/\partial z$. For small δ the term $-\delta\varphi^{2}$ in the Hamiltonian may be treated as a perturbation; calculated to the first order, the eigenvalue sought is

$$\delta\left(\int \mathrm{d}^{3}r\,\varphi^{2}\left(\frac{\partial\varphi}{\partial x}\right)^{2}\Big/\int \mathrm{d}^{3}r\left(\frac{\partial\varphi}{\partial x}^{2}\right)\right).$$

Thus the numerator of eq. (AII.4) is of order δ^{3}. As to the denominator, it is proportional to δ, since the eigenvalue corresponding to the function φ goes to zero at $\delta\to 0$ and at small δ is equal to $\delta J_0/I$. Substituting eq. (AII.4) into eq. (AII.3) and taking the limit $\delta\to 0$, we obtain

$$\det{}'\boldsymbol{K}=-\frac{4I}{J_0}\left(\frac{\int \mathrm{d}^{3}r\,\varphi^{2}(\partial\varphi/\partial x)^{2}}{\int \mathrm{d}^{3}r(\partial\varphi/\partial x)^{2}}\right)^{3}\frac{\det{}'(-\Delta-3\varphi^{2}+1)}{\det{}'(-\Delta-\varphi^{2}+1)}.\tag{AII.5}$$

The primes indicate that, in evaluating the determinants, the zero eigenvalues should be excluded. The integrals involved in the right-hand side of eq. (AII.5) are easily expressed in terms of the quantities I and J_1 introduced in the text. Namely, $\int \mathrm{d}^{3}r(\partial\varphi/\partial x)^{2}=I$, $\int \mathrm{d}^{3}r\,\varphi^{2}(\partial\varphi/\partial x)^{2}=J_1/4$ (the first relationship is obtained using $J_0=4I$). Equation (2.39) follows immediately.

Appendix III

Variational solution of eq. (2.29) and coefficient θ

The function $\varphi(r)$ obeying eq. (2.29) realizes an extremum of the functional

$$G(\varphi) = \int d^3r\left[(\nabla\varphi)^2 + \varphi^2 - \tfrac{1}{2}\varphi^4\right]. \tag{AIII.1}$$

Using the trial function $B\exp(-\eta r) = \varphi$ and minimizing the functional with respect to B and η, one finds $B = 4\sqrt{2}$, $\eta = \sqrt{3}$. Thus, a variational estimate for the solution of eq. (2.29) is

$$\varphi(r) = 4\sqrt{2}\exp[-\sqrt{3}\,r]. \tag{AIII.2}$$

When substituted into eq. (2.53), this function gives $\theta = 2^5/J_0^2 = 27/2^9\pi^2 = 1/187$, in reasonable agreement with the computer-calculated value $\theta = 1/178$.

References

Ablyazov, N.N., M.E. Raikh and A.L. Efros, 1983, Fiz. Tverd. Tela **25**, 353 [Sov. Phys.-Solid State **25**, 199].
Alferov, Zh.I., E.L. Portnoy and A.A. Rogachev, 1968, Fiz. Tekh. Poluprovodn. **2**, 1194 [Sov. Phys.-Semicond. **2**, 1001].
Alferov, Zh.I., M.Z. Zhingarev, S.G. Konnikov, I.I. Mokan, V.P. Ulin, V.E. Umanskii and B.S. Yavich, 1982, Fiz. Tekh. Poluprovodn. **16**, 831 [Sov. Phys.-Semicond. **16**, 532].
Altshuler, B.L., and A.G. Aronov, 1983, in: Electron–Electron Interactions in Disordered Systems, eds A.L. Efros and M. Pollak (North-Holland, Amsterdam) p. 1.
Anderson, P.W., 1958, Phys. Rev. **109**, 1492.
Areshkin, A.G., L.G. Suslina and D.L. Fedorov, 1982, Pis'ma v Zh. Eksp. & Teor. Fiz. **35**, 427.
Baldareschi, A., and K. Maschke, 1975, Solid State Commun. **16**, 99.
Baranovskii, S.D., and A.L. Efros, 1978, Fiz. Tekh. Poluprovodn. **12**, 2233 [Sov. Phys.-Semicond. **12**, 1328].
Berndt, V., A.A. Kopilov and A.N. Pikhtin, 1977, Fiz. Tekh. Poluprovodn. **11**, 2206 [Sov. Phys.-Semicond. **11**, 1296].
Berndt, V., A.A. Kopilov and A.N. Pikhtin, 1978, Fiz. Tekh. Poluprovodn. **12**, 1628 [Sov. Phys.-Semicond. **12**, 964].
Berolo, O., J.C. Wolley and J.A. Van Vechten, 1973, Phys. Rev. B **8**, 3794.
Brezin, E., and G. Parisi, 1978, J. Stat. Phys. **19**, 269.
Brezin, E., and G. Parisi, 1980, J. Phys. C: Solid State Phys. **13**, L307.
Cohen, E., and M.D. Sturge, 1982, Phys. Rev. B **25**, 3828.
Cohen, M.L., and T.K. Bergstresser, 1966, Phys. Rev. **141**, 789.
Dean, P.J., G. Kaminsky and R.B. Zetterstrom, 1969, Phys. Rev. **181**, 1149.
D'yakonov, M.I., and M.E. Raikh, 1982, Fiz. Tekh. Poluprovodn. **16**, 890 [Sov. Phys.-Semicond. **16**, 570].
Elliott, R.J., J.A. Krumhansl and P.L. Leath, 1974, Rev. Mod. Phys. **46**, 465.

Gel'mont, B.L., A.R. Gadgiev, B.I. Shklovskii, I.S. Shlimak and A.L. Efros, 1974, Fiz. Tekh. Poluprovodn. **8**, 2377 [1975, Sov. Phys.-Semicond. **8**, 1549].
Goede, O., L. John and D. Henning, 1978, Phys. Status Solidi (b) **89**, K183.
Gradstein, I.S., and I.M. Ryzhik, 1966, Tables of Integrals, Sums, Series and Products (Academic Press, New York).
Greene, P.D., S.A. Wheller, A.R. Adams, A.N. El-Sabbahy and C.N. Ahmad, 1979, Appl. Phys. Lett. **35**, 78.
Halperin, B.I., and M. Lax, 1966, Phys. Rev. **148**, 722.
Halperin, B.I., and M. Lax, 1967, Phys. Rev. **153**, 802.
Harrison, W., 1977, J. Vac. Sci. & Technol. **14**, 1016.
Houghton, A., and L. Schäfer, 1979, J. Phys. A **12**, 1309.
Ipatova, I.P., A.U. Maslov and A.V. Subashiev, 1983, Fiz. Tverd. Tela **25**, 2051.
Katnani, A.D., and G. Margaritondo, 1983, J. Appl. Phys. **54**, 2522.
Katnani, A.D., P.R. Daniels, Fe-Xio Zhao and G. Margaritondo, 1982, J. Vac. Sci. & Technol. **20**, 662.
Kowalczyk, S.P., W.J. Schaffer, E.A. Kraut and R.W. Grant, 1982, J. Vac. Sci. & Technol. **20**, 705.
Kusmartsev, F.V., and E.I. Rashba, 1983, Pis'ma v Zh. Eksp. & Teor. Fiz. **37**, 106 [JETP Lett. **37**, 130].
Lai, S., and M.V. Klein, 1980, Phys. Rev. Lett. **44**, 1087.
Landau, L.D., and E.M. Lifshitz, 1959, Quantum Mechanics (Pergamon, New York).
Maslov, A.U., and L.G. Suslina, 1982, Fiz. Tverd. Tela **24**, 3394 [Sov. Phys.-Solid State **24**, 1928].
Maslov, A.U., D.L. Fedorov, L.G. Suslina, V.G. Melekhin and A.G. Areshkin, 1983, Fiz. Tverd. Tela **25**, 1408 [Sov. Phys.-Solid State **25**, 809].
Mott, N.F., and E.A. Davis, 1979, Electronic Processes in Non-crystalline Materials (Clarendon Press, Oxford).
Nelson, R.J., 1982, in: Excitons, eds E.I. Rashba and M.D. Sturge (North-Holland, Amsterdam) p. 319.
Ovsyankin, V.V., and A.A. Fedorov, 1982, Pis'ma v Zh. Eksp. & Teor. Fiz. **35**, 199 [JETP Lett. **35**].
Permogorov, S., A. Reznitsky, V. Travnikov, S. Verbin, G.O. Müller, P. Flögel and M. Nikiforova, 1981a, Phys. Status Solidi (b) **106**, K57.
Permogorov, S., A. Reznitsky, V. Travnikov, S. Verbin, G.O. Müller and M. Nikiforova, 1981b, J. Lumin. **24/25**, 409.
Permogorov, S., A. Reznitsky, S. Verbin, G.O. Müller, P. Flögel and M. Nikiforova, 1982, Phys. Status Solidi (b) **113**, 589.
Permogorov, S., A. Reznitsky, S. Verbin and V. Lysenko, 1983, Solid State Commun. **47**, 5.
Pikhtin, A.N., 1977, Fiz. Tekh. Poluprovodn. **11**, 425 [Sov. Phys.-Semicond. **11**, 245].
Raikh, M.E., and Al.L. Efros, 1984, Fiz. Tverd. Tela **26**, 106 [Sov. Phys.-Solid State **26**, 61].
Rashba, E.I., 1982, in: Excitons, eds E.I. Rashba and M.D. Sturge (North-Holland, Amsterdam) p. 543.
Rosenbaum, T.F., K. Andres, G. Thomas and R.N. Bhatt, 1980, Phys. Rev. Lett. **45**, 1723.
Shklovskii, B.I., and A.L. Efros, 1979, Electronic Properties of Doped Semiconductors (Nauka, Moscow) (English Translation Springer, 1984).
Shklovskii, B.I., and A.L. Efros, 1980, Fiz. Tekh. Poluprovodn. **14**, 825 [Sov. Phys.-Semicond. **14**, 487].
Shlimak, I.S., A.L. Efros and I.V. Yanchev, 1977, Fiz. Tekh. Poluprovodn. **11**, 257 [Sov. Phys.-Semicond. **11**, 149].
Skal, A.S., B.I. Shklovskii and A.L. Efros, 1975, Fiz. Tverd. Tela **17**, 506 [Sov. Phys.-Solid State **17**, 316].
Suslina, L.G., A.G Plyukhin, D.L. Fedorov and A.G. Areshkin, 1978, Fiz. Tekh. Poluprovodn. **12**, 2238 [Sov. Phys.-Semicond. **12**, 1331].

Suslina, L.G., A.G. Plyukhin, O. Goede and D. Henning, 1979, Phys. Status Solidi (b) **94**, K185.
Van Vechten, J.A., and T.K. Bergstresser, 1970, Phys. Rev. B **1**, 3351.
Zinger, G.M., I.P. Ipatova, A.V. Subashiev, 1977, Fiz. Tverd. Tela **19**, 2258 [Sov. Phys.-Solid State **19**, 1322].
Zittartz, J., and J.S. Langer, 1966, Phys. Rev. **148**, 741.

CHAPTER 4

Infrared and Raman Studies of Disordered Magnetic Insulators

W. HAYES

Clarendon Laboratory
Oxford, England

and

M.C.K. WILTSHIRE

Department of Solid State Physics
The Australian National University, Canberra
and Institute of Physical Sciences
CSIRO, Canberra, Australia

Optical Properties of Mixed Crystals
Edited by
R.J. Elliott and I.P. Ipatova

Contents

1. Introduction 179
2. Theoretical considerations 182
3. Experimental systems 188
   3.1. Transition metal oxides 188
   3.2. Transition metal fluorides with the rutile structure 190
      3.2.1. Concentrated mixtures 190
      3.2.2. Dilute mixtures 194
   3.3 Transition metal compounds with the cubic perovskite structure 199
      3.3.1. Mixtures of magnetic ions 199
      3.3.2. Dilute mixtures 200
   3.4. Transition metal compounds with the layer perovskite structure 202
   3.5. Transition metal halides with the cadmium chloride structure 204
      3.5.1. Dilute mixtures 204
      3.5.2. Mixtures of magnetic ions 210
4. Conclusions 212
References 212

1. Introduction

The techniques of infrared spectroscopy and of Raman scattering provide a convenient and valuable approach to the study of magnons in magnetic insulators (Tinkham 1970, Fleury and Loudon 1968). Because the wavevectors of the photons used in these studies are very small, the wavevectors $\boldsymbol{q}$ of the magnons excited have the value $\boldsymbol{q} \simeq 0$. Since there are very few ferromagnets transparent to electromagnetic radiation, infrared and Raman studies have been largely confined to antiferromagnets. In antiferromagnets the magnon energy at $\boldsymbol{q} = 0$ has the finite value

$$\nu_{\mathrm{afmr}} = [2H_{\mathrm{A}}(H_{\mathrm{E}} + H_{\mathrm{A}})]^{1/2}, \tag{1}$$

where H_{E} and H_{A} are the exchange and anisotropy fields respectively. For relatively small H_{A} the value of the antiferromagnetic resonance frequency (1) becomes

$$\nu_{\mathrm{afmr}} = [2H_{\mathrm{E}}H_{\mathrm{A}}]^{1/2}. \tag{2}$$

In addition to ν_{afmr} it is possible to excite two-magnon excitations with $\boldsymbol{q}_1 + \boldsymbol{q}_2 = 0$ using infrared and Raman techniques. These two-magnon bands are dominated by magnons near the edge of the Brillouin zone where the density of magnon states is relatively high.

The experimental techniques used in the study of Raman scattering by magnons and the theory of the scattering cross section for one-magnon and two-magnon excitations have been described by Hayes and Loudon (1978) (see also Lockwood (1982)). In materials for which $H_{\mathrm{E}} \gg H_{\mathrm{A}}$ both the one-magnon frequency (cf. eq. (2)) and its Raman cross section are proportional to $H_{\mathrm{A}}^{1/2}$. Hence antiferromagnets with small magnon frequencies at $\boldsymbol{q} = 0$ tend to be poor one-magnon light scatterers. For example, the $\boldsymbol{q} = 0$ one-magnon frequency for MnF_2 is 8.7 cm^{-1} and for FeF_2 is 52 cm^{-1}; the latter is much easier to observe in light scattering than the former.

The two-magnon scattering in the case of MnF_2, for example, peaks at ~ 100 cm^{-1}. The intensity of this scattering is generally comparable to or greater than that of one-magnon scattering, indicating that the scattering mechanism is not the one-magnon scattering mechanism taken to second order. The short-range nature of the exchange-scattering mechanism actually involved restricts excitation to nearest-neighbour spin pairs on opposite sub-lattices (Fleury and Loudon 1968, Hayes and Loudon 1978). The theory takes account of the fact that the two magnons involved in the scattering event interact so that the energy of the two-magnon excitation is not quite the same as the sum of the individual magnon energies.

Single-magnon and two-magnon excitations may also be observed using the techniques of far-infrared absorption (Martin 1967) and the results for pure crystals have been reviewed by Bloor and Copland (1972). More recently resonance techniques have been employed using monochromatic radiation obtained from carcinotrons at energies up to 20 cm^{-1} and from far-infrared lasers (for example, the H_2O laser emits at 40 and 127.5 cm^{-1} and D_2O at 118.65 cm^{-1}). The sample is placed in a magnetic field which is varied until resonant absorption of the monochromatic radiation takes place. With this technique it is necessary to know the g-value of the mode or to make measurements at a number of frequencies in order to extrapolate to the zero-field position.

Effects of isolated impurities on the vibrational properties of crystals have been extensively studied (Barker and Sievers 1975). Depending on the mass and force-constant changes associated with the impurity one may get localised vibrational modes above the highest band modes in the crystal or gap modes, which are localised, if there is a gap in the density of single-phonon states between the acoustic and optic modes. Resonance in the band-mode regions may also be observed. Analogous effects occur with magnetic excitations. Since in an antiferromagnet the pure-crystal magnon bands have a finite lower frequency (eqs. (1) and (2)), as well as a finite upper frequency, magnetic impurity modes occur below as well as above the magnon bands; the location depends on the relative magnitude of the impurity–host exchange compared with the host–host exchange.

Anisotropy energy also plays a significant role in determining the location of impurity modes. For example, the Mn impurity mode in FeF_2 lies below the FeF_2 band partly because the Fe–Mn exchange is weaker than the Fe–Fe exchange, and also because the Mn anisotropy energy is extremely small, whereas the Fe impurity mode in MnF_2 lies above the MnF_2 magnon band because Fe^{2+} has a large anisotropy energy (Enders et al. 1972).

In principle, the energy of the impurity mode, combined with a knowledge of the pure crystal exchange and anisotropy energy, makes a determination of the host–impurity exchange possible. For an impurity A replacing a host atom B the empirical relationship

$$J_{AB} = \sqrt{J_{AA} J_{BB}} \qquad (3)$$

generally holds quite well. By means of cluster models or Green-function theories the impurity magnetic modes can be calculated and the exchange parameter J_{AB} refined. A review of magnetic impurity excitations has been given by Cowley and Buyers (1972).

With increasing concentration of impurities we gradually come into the mixed crystal regime. Here, as in the isolated impurity case, wavevector conservation is no longer required. However, the infrared and Raman experiments still measure the response of the crystal at $\boldsymbol{q} \simeq 0$. This is no longer ideally a δ-function for a one-magnon excitation, as with a perfect crystal. As the disorder increases the response peak shifts and broadens while a background response with structure grows, covering the entire range of frequency of the one-magnon density of states.

If a crystal is diluted with a non-magnetic constituent, e.g., MnF_2 with ZnF_2 forming $Mn_{1-x}Zn_xF_2$, the single-magnon response remains concentrated at a single peak whose intensity and energy decrease with increasing x. For a crystal of mixed magnetic ions a single peak occurs if the constituents are sufficiently similar magnetically and two peaks if they are different. This classification into one- and two-mode behaviour is also a property of phonons in mixed crystals (Elliott et al. 1974). Examples of single-band magnetic systems are mixtures of transition metal oxides (Becker et al. 1975) and diluted magnets (Coombs et al. 1976) while mixtures of transition metal fluorides give rise to two-band behaviour (Enders et al. 1972). Such behaviour is observed not only in infrared and Raman studies but often, in more detail, in neutron scattering (Cowley et al. 1977a).

In crystals such as $Mn_{1-x}Zn_xF_2$ the antiferromagnetic transition temperature $T_N(x)$ decreases with increasing x until at a critical concentration x_p there is no longer any long-range magnetic order even at $T = 0$. For $x < x_p$ the magnetic ions form finite clusters together with a single infinitely-connected cluster. For $x > x_p$ only finite clusters occur. Although the study of magnon excitations for $x \lesssim x_p$ is of interest, optical studies in this region have not proved very fruitful. This is due to the large width and low intensity of the magnon excitations at $x \sim x_p$ and also interference from phonon excitations. The neutron scattering technique has proved more suitable (Cowley et al. 1980).

The theoretical difficulties associated with excitations in mixed crystals have been discussed by Elliott et al. (1974). Even with a knowledge of exchange interactions in the mixture the calculation of the excitation spectrum is still a formidable problem. The coherent potential approximation (CPA) and its derivatives are widely used. Here one endeavours to set up an effective medium in which the magnons are not scattered on the average either by host atoms or impurity atoms. An alternative approach is to use exchange parameters directly in a computer simulation of the mixture and to extract the excitation spectrum from an analysis of the model's equation of motion.

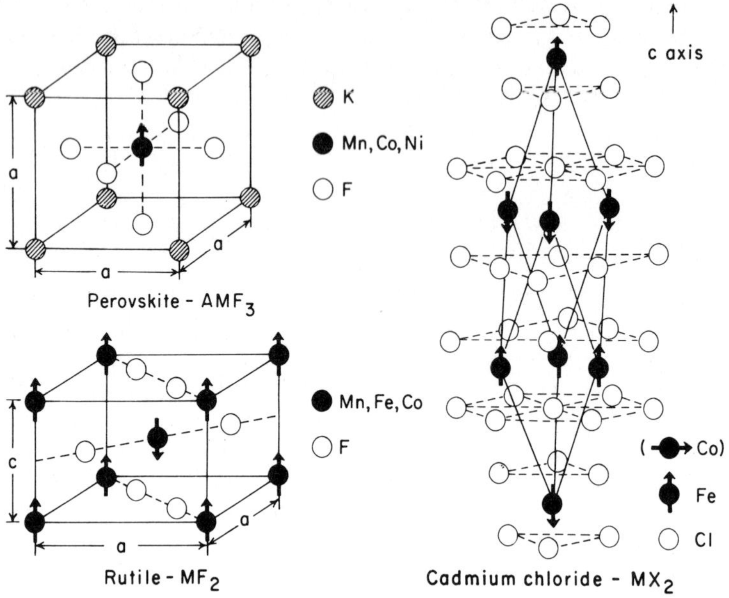

Fig. 1. Schematic representation of the perovskite, rutile and cadmium chloride structures (Lockwood 1982).

Results obtained in this way have been compared with experimental data (Cowley et al. 1977a) and with CPA calculations (Huber 1974). Finally there are cluster models which work reasonable well because of the short range of exchange interactions. These various approaches are reviewed in section 2.

Experimental results are collected in section 3. Most work has been carried out on materials with the sodium chloride, rutile, perovskite and cadmium chloride structures and the latter three structures are represented in fig. 1.

2. *Theoretical considerations*

The simplest way of discussing impurity modes in a magnet is to use the Ising cluster model (see, for example, Cowley and Buyers 1972). In this approximation the pure crystal Hamiltonian is written as

$$\mathcal{H} = \tfrac{1}{2}\sum_{i,j} I_{ij} S_i^z S_j^z - \sum_i H_A \sigma_i S_i^z, \tag{4}$$

where I_{ij} is the (longitudinal) exchange between spins on sites i and j and is usually non-zero only when i and j are nearest or next-nearest neighbours. σ_i

is $+1$ or -1 for the up or down sublattice while H_A is the sum of the single-ion anisotropy and magnetic dipole–dipole interactions. If a magnetic impurity with spin S', exchange I' and anisotropy H'_A is introduced at site 0 the Hamiltonian becomes

$$\mathscr{H}=\tfrac{1}{2}\sum_{i,j\neq 0} I_{ij}S_i^zS_j^z - \sum_{i\neq 0} H_A\sigma_iS_i^z + \sum_{j\neq 0} I'_{0j}S_0'^zS_j^z - H'_A\sigma_0S_0'^z. \tag{5}$$

The excitations described by this Hamiltonian are localised on the individual sites and the impurity frequencies may be readily calculated giving, for example,

$$\omega_i = H'_A + zI'S, \tag{6}$$

where interactions with z nearest neighbours only are considered.

This very simple approximation gives a reasonable estimate of the energy of highly localised modes and also of excitations with wavevectors at the Brillouin zone boundary. However, in the majority of cases of interest there is a transverse part of the exchange that must be included. Then the Hamiltonian of the pure system is written as

$$\mathscr{H}=\tfrac{1}{2}\sum_{ij}\{J_{ij}\mathbf{S}_i\cdot\mathbf{S}_j + K_{ij}S_i^zS_j^z\} - \sum_i H_A\sigma_iS_i^z \tag{7a}$$

$$=\tfrac{1}{4}\sum_{ij}\{J_{ij}(S_i^+S_j^- + S_i^-S_j^+) + 2(J_{ij}+K_{ij})S_i^zS_j^z\} - \sum_i H_A\sigma_iS_i^z. \tag{7b}$$

We can identify $I_{ij}=J_{ij}+K_{ij}$ as the longitudinal exchange. The transverse exchange, given by the first term in eq. (7b), permits the transfer of a spin deviation from one site to another. It is therefore necessary to include those sites in the neighbourhood of the defect in the calculation, and Cowley and Buyers (1972) discuss this and the simplification introduced by taking account of local symmetry. This type of cluster model is very useful for carrying out approximate calculations of impurity-mode energies and gives particularly good results when the impurity modes are highly localised. However, the cluster model is less useful for impurity modes that lie close to the host magnon band.

The most powerful way of taking the host into account as well as the impurity is through the Green-function formalism (Lovesey 1968, Tonegawa 1968). We define the pure-crystal Green function

$$P_{\alpha\beta}(i,\ j;\ \omega)=\frac{(-1)^{\alpha+1}}{2S}\langle\langle S_{i\alpha}^+;\ S_{j\beta}^-\rangle\rangle_{\omega+\mathrm{i}\epsilon}, \tag{8}$$

where α, $\beta = 1$ or 2 specify the sublattice of the antiferromagnet and $\langle\langle A;\, B\rangle\rangle_\omega$ is the Fourier transform of

$$\langle\langle A(t);\, B(0)\rangle\rangle = -\mathrm{i}\theta(t)\langle[A(t),\, B(0)]\rangle. \tag{9}$$

The Green function is usually evaluated by obtaining its equation of motion

$$\omega\langle\langle A;\, B\rangle\rangle_\omega = \langle[A,\, B]\rangle + \langle\langle[A,\, \mathscr{H}];\, B\rangle\rangle_\omega \tag{10}$$

and by using the random phase approximation to decouple the higher-order Green functions that occur on the right-hand side of this equation, e.g.,

$$\langle\langle X_i^+ S_i^z;\, S_j^-\rangle\rangle \simeq \langle S_i^z\rangle\langle\langle S_i^+;\, S_j^-\rangle\rangle \simeq \pm S\langle\langle S_i^+;\, S_j^-\rangle\rangle. \tag{11}$$

Making use of translational invariance, we can obtain the transform

$$P_{\alpha\beta}(\boldsymbol{k},\, \omega) = \sum_{ij} P_{\alpha\beta}(i,\, j;\, \omega)\exp(\mathrm{i}\boldsymbol{k}\cdot\boldsymbol{R}_{ij})\exp(\mathrm{i}\boldsymbol{k}\cdot\boldsymbol{\Delta}_{\alpha\beta}), \tag{12}$$

where $\boldsymbol{\Delta}$ is the vector linking an atom to its neighbour on the opposite sublattice, and this may be written in matrix form as

$$\mathbf{P}(\boldsymbol{k},\, \omega) = \begin{pmatrix} P_{11}(\boldsymbol{k},\, \omega) & P_{12}(\boldsymbol{k},\, \omega) \\ P_{21}(\boldsymbol{k},\, \omega) & P_{22}(\boldsymbol{k},\, \omega) \end{pmatrix}. \tag{13}$$

From this Green-function matrix $\mathbf{P}(\boldsymbol{k},\, \omega)$, which describes the excitations of the pure crystal, we can obtain (Cowley and Buyers 1972, Elliott et al. 1974) experimentally observed quantities such as the neutron scattering cross section, the far infrared absorption coefficient and the Raman scattering intensity.

When there are impurities in the system, we define a Green function for the impure sample as

$$G_{\alpha\beta}(i,\, j;\, \omega) = \frac{(-1)^{\alpha+1}}{2(S_i S_j)^{1/2}}\langle\langle S_{i\alpha}^+;\, S_{j\beta}^-\rangle\rangle_\omega \tag{14}$$

and write the change in the Hamiltonian from the pure crystal $\mathscr{H}_0$ to the impure crystal $\mathscr{H}'$ as

$$\mathscr{H}' = \mathscr{H}_0 + V. \tag{15}$$

Thus in the spectral representation where

$$\mathbf{P} = (E\mathbf{I} - \mathscr{H}_0)^{-1} \tag{16}$$

we have

$$\mathbf{G} = (E\mathbf{I} - \mathscr{H}')^{-1} \tag{17}$$

and the Dyson equation

$$\mathbf{G} = \mathbf{P} + \mathbf{PVG}. \tag{18}$$

It is also convenient to define the T-matrix by

$$\mathbf{G} = \mathbf{P} + \mathbf{PTP}, \tag{19}$$

where

$$\mathbf{T} = \mathbf{V}(\mathbf{I} - \mathbf{PV})^{-1}. \tag{20}$$

Either of these expressions can be expanded to give the perturbation series (Elliott et al. 1974)

$$\mathbf{G} = \mathbf{P} + \mathbf{PVP} + \mathbf{PVPVP} + \ldots \tag{21}$$

These relationships have been written so far with the implicit assumption of a single isolated defect. If there are many impurities, as in a mixed crystal, the equations need to be modified. It must be recalled that the Green functions above have translation symmetry written into them. In a mixed crystal, there is no translational symmetry. Nevertheless, all the observed experimental quantities involve the sum over all sites in the sample, and this is equivalent to taking the configurational average. In a homogeneously random system, this configurational average $\langle \mathbf{G} \rangle$ depends only on the relative positions of the two sites so that $\langle \mathbf{G} \rangle$ has average translational symmetry, like $\mathbf{P}$, and can be labelled with a wavevector $\boldsymbol{k}$. Then the expansion (21) can be rewritten as

$$\langle \mathbf{G} \rangle = \mathbf{P} + \mathbf{P}\Sigma\langle \mathbf{G} \rangle, \tag{22}$$

where Σ is known as the self-energy and is in general a function of $\boldsymbol{k}$ and ω. The different approximations used in treating mixed crystals reflect different ways of treating Σ, and three approximations that are most commonly used will be briefly described here (for a more detailed review see Elliott et al. 1974).

In the virtual crystal approximation (VCA) we write

$$\Sigma = \langle \mathbf{V} \rangle. \tag{23}$$

This approximation to Σ is linear in the impurity concentration, and is real, so the lifetimes of the states are infinite. It is, however, a useful approximation close to the band edges and for small perturbations $\boldsymbol{V}$.

The next simplest approximation is the average T-matrix approximation (ATA). Here the self-energy is taken as

$$\Sigma = \langle \boldsymbol{T} \rangle / \{ \boldsymbol{I} + \langle \boldsymbol{T} \rangle \boldsymbol{P}(0) \}, \tag{24}$$

where $\boldsymbol{P}(0)$ is the host-lattice Green function. This self-energy is complex so that the energy levels are shifted and broadened from those of the pure crystal.

The final level of approximation is the CPA (Soven 1967, Taylor 1967). Here one assumes an effective medium with a Green function $\boldsymbol{G}_e$ such that

$$\boldsymbol{G}_e = \boldsymbol{P} + \boldsymbol{P} \Sigma \boldsymbol{G}_e . \tag{25}$$

Now the true Green function for the medium $\boldsymbol{G}$ must still satisfy the Dyson equation (18) so that, eliminating $\boldsymbol{P}$, we obtain

$$\boldsymbol{G} = \boldsymbol{G}_e + \boldsymbol{G}_e (\boldsymbol{V} - \Sigma) \boldsymbol{G}, \tag{26}$$

We now require that Σ be such that there is no net scattering of the magnons described by $\boldsymbol{G}_e$, that is that the average T-matrix for this scattering be zero. Then since $\boldsymbol{V}$ for a host site is zero and for an impurity site is $\boldsymbol{V}$, we find

$$\boldsymbol{T} = \frac{(1-x)(-\Sigma)}{\boldsymbol{I} + \Sigma \boldsymbol{G}_e(0)} + \frac{x(\boldsymbol{V} - \Sigma)}{\boldsymbol{I} - (\boldsymbol{V} - \Sigma)\boldsymbol{G}_e(0)} = 0 \tag{27}$$

or (see Elliott et al. 1974)

$$\Sigma = x\boldsymbol{V} / \{ \boldsymbol{I} - (\boldsymbol{V} - \Sigma) \boldsymbol{G}_e(0) \}, \tag{28}$$

where $\boldsymbol{G}_e(0)$ is the Green function for the effective medium.

It should be noted that all of the above are single-site approximations, and we have already emphasised the limitations of this approach for systems with significant transverse exchange interactions (i.e., obeying the Hamiltonian (7)). There have been various attempts to deal with this problem. Buyers et al. (1972, 1973) used the Ising model to describe the single-site frequencies, and applied the CPA to these. The transverse interactions were included by assuming that they have a similar frequency and wavevector dependence to the correctly treated longitudinal terms. This approach was very successful in treating (Mn, Co)F_2 (Buyers et al. 1972) where the transverse interactions are by chance independent of composition. It was also applied to the dilute antiferromagnet (Mn, Zn)F_2 (Buyers et al. 1973) by placing a fictitious spin on the non-magnetic zinc sites and introducing a large potential to keep the magnons off these sites. This, again, was reasonably successful (Coombs et al. 1976). These successes are due largely to the fact that the model treats the Ising interactions correctly and includes clustering effects; it is found that the clusters dominate the response of the mixed crystals.

An alternative approach has been developed by Holcomb (1974) which deals with the transverse parts accurately but treats the longitudinal contributions all in the same average way. This model does not predict structure due to cluster effects. For intermediate concentrations it has been successful in describing (Mn, Zn)F_2. Attempts have also been made to treat (Mn, Zn)F_2 using a bond-type CPA (Tahir-Kheli 1972) but this method does not appear to agree as well with the experimental results. Other approaches have been developed by Dzyub (1974a, b) who makes use of a cluster theory and a classical mean-field theory and treats the site-randomness by means of one-body distribution functions. In all these calculations, reasonable agreement with some experimental data has been obtained. However, work on two-dimensional systems, particularly Rb_2(Co, Mg)F_4 (Cowley et al. 1980) and Rb_2(Mn, Mg)F_4 (Cowley et al. 1977a) has shown fine structure that is not well explained by these models.

A more direct approach was developed by Alben and Thorpe (1975) and Thorpe and Alben (1976) who consider the spectral Green function

$$G_{ij}(t) = \frac{-\mathrm{i}\theta(t)}{2\sqrt{S_i S_j}} \langle S_i^+(t) S_j^-(0) \rangle \tag{29}$$

in real space. Making use of the pseudo-translational symmetry resulting from configurational averaging, a partial transform is made to the form

$$G_{ik}(t) = \frac{1}{\sqrt{N}} \sum_j G_{ij}(t) \mathrm{e}^{\mathrm{i}\boldsymbol{k}\cdot\boldsymbol{R}_j}, \tag{30}$$

and the equation of motion of this function is obtained from

$$\mathrm{i}\hbar \frac{\mathrm{d}G_{ik}(t)}{\mathrm{d}t} = [G_{ik}(t), \mathscr{H}], \tag{31}$$

where $\mathscr{H}$ is the Hamiltonian of the mixture (of the form (4) or (7)). Assuming that we know the ground state, i.e., the state at time zero, $G_{ij}(0)$, we can then calculate the time evolution of $G_{ik}(t)$ using

$$G_{ik}(t+\delta t) = G_{ik}(t-\delta t) + 2\delta t \frac{\mathrm{d}G_{ik}(t)}{\mathrm{d}t}. \tag{32}$$

After a finite time T this evolution is truncated, the transform (30) is completed and the time Fourier transform performed, including a damping constant λ to remove termination ripples. This gives

$$G(\boldsymbol{k}, \omega) = \frac{1}{\sqrt{2\pi N}} \int_0^T \sum_i G_{ik}(t) \mathrm{e}^{\mathrm{i}\boldsymbol{k}\cdot\boldsymbol{R}_i} \mathrm{e}^{\mathrm{i}\omega t} \mathrm{e}^{-\lambda t^2}\, \mathrm{d}t, \tag{33}$$

the spectral response function whose imaginary part gives directly the far-infrared absorption or the inelastic neutron scattering cross section. The calculation is performed by setting up a mesh of lattice points occupied with spins at an appropriate concentration, initialising $\boldsymbol{G}_{ik}(0)$, and allowing the evolution to proceed. The size and number of the time steps are determined by the excitation band width and the desired spectral resolution respectively. This technique has been applied to several two-dimensional systems, for example $Rb_2(Ni, Mn)F_4$ (Thorpe and Alben 1976), $Rb_2(Co, Mg)F_4$ (Cowley et al. 1980) and $Rb_2(Mn, Mg)F_4$ (Cowley et al. 1977a), with remarkably good results. It therefore seems that linear spin-wave theory is satisfactory for describing mixed magnets and the simulation method provides a convenient means for testing model Hamiltonians and their parameters against experimental results.

3. Experimental systems

3.1. Transition metal oxides

A small amount of work has been carried out on the transition metal oxides. At first sight these materials should be very suitable for study. They have the rocksalt crystal structure and order as type-2 antiferromagnets for the fcc structure, that is with the moments lying in the (111) plane; all the moments in a given (111) plane are antiparallel to those in nearest (111) planes. However, stoichiometric samples of MnO, NiO, CoO and FeO are difficult to produce. Moreover, their magnetism, particularly that of CoO, is not simple. In all these compounds, besides the strong dipolar anisotropy that confines the moments to the (111) plane, there is an in-plane anisotropy that differs from compound to compound. The Co^2 and Fe^{2+} ions in these materials have particularly strong anisotropies associated with them because their orbital angular momenta are not fully quenched, and CoO undergoes a tetragonal distortion at T_N.

The properties of mixtures of NiO or MnO with CoO or FeO are dominated by the large single-ion anisotropy associated with the Co^{2+} and Fe^{2+} ions. Therefore measurements of antiferromagnetic resonance (AFMR) in the far-infrared have been carried out more with a view to investigating this anisotropy rather than as an investigation of mixed crystals.

Far-infrared spectra have been measured for MnO : Co (Hughes 1971), NiO : Co (Becker et al. 1975) and CoO : Ni (Geis et al. 1976, 1977). In all cases

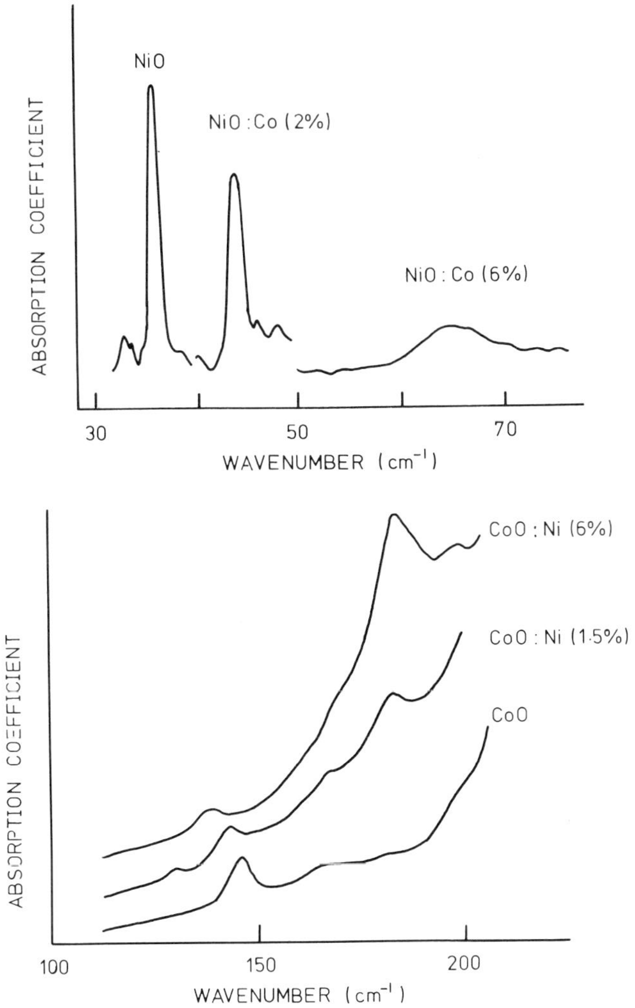

Fig. 2. Concentration dependence of the magnetic-impurity-induced far-infrared absorption for (a) NiO:Co (Becker et al. 1975) and (b) CoO:Ni (Geis et al. 1976).

the AFMR shifts rapidly and broadens with increasing impurity concentration until at $\gtrsim 6\%$ impurity the AFMR becomes too broad to detect (see fig. 2). The variation of the AFMR frequency with composition has been treated using an average crystal model, including an effective anisotropy field, and satisfactory agreement with the experimental data has been obtained (fig. 3); here, both far-infrared and neutron scattering (Wagner et al. 1974) results are shown and are compared with an average crystal model. In this model, the

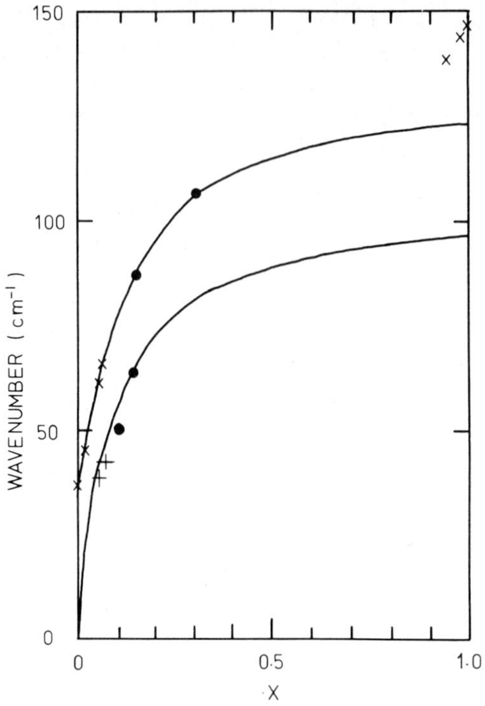

Fig. 3. Comparison of measured and calculated AFMR frequencies in $Ni_{1-x}Co_xO$. × = upper branch, + = lower branch from far-infrared absorption (Becker et al. 1975, Geis et al. 1976, 1977); ● = from neutron scattering (Wagner et al. 1974). The full lines are calculated results from Geis et al. (1977).

anisotropy field of the Ni ions is described by a trigonal term while the Co anisotropy field may contain terms of trigonal, cubic and tetragonal symmetry. The equilibrium direction of the ions in the mixture was calculated and both the in-plane and out-of-plane resonance frequencies were obtained by considering small oscillatory deviations from the equilibrium position; these are shown by the solid line in fig. 3 (Geis et al. 1977). However, the model is not good for the Co-rich mixtures (see Geis et al. 1976) and it is clear that a more detailed treatment of the Co^{2+} ion anisotropy is required.

3.2. *Transition metal fluorides with the rutile structure*

3.2.1. *Concentrated mixtures*

The early work on impurity modes in the antiferromagnetic transition metal fluorides with the rutile structure (fig. 1) was reviewed by Cowley and Buyers (1972). This review was largely concerned with the properties of isolated impurities, and included a comparison of the spectroscopic data then available on MnF_2 doped with Ni, Fe (Oseroff and Pershan 1969, Moch et al. 1969) and Co (Moch et al. 1971) with the results of both cluster and Green-function calculations. These systems all showed localised impurity modes well above the host magnon band. Cowley and Buyers (1972) also predicted, using both the Ising model and Green-function methods, that a local mode should exist in $CoF_2:Mn$ at $\sim 30\ cm^{-1}$, well below the host magnon band.

Localised modes below the host magnons have been found (Enders et al. 1972) in $CoF_2:Mn$ at $28.5\ cm^{-1}$ and in $FeF_2:Mn$ at $50\ cm^{-1}$ using far-infrared Fourier transform spectroscopy. These modes have also been studied using laser methods by Prokhorov and Rudashevskii (1975), Sanders et al. (1980) and Belanger et al. (1982). Manganese impurities in FeF_2 produce a localised mode at $50\ cm^{-1}$, $2.6\ cm^{-1}$ below the bottom of the magnon band, while Co impurities generate a mode at $85.5\ cm^{-1}$, $6.5\ cm^{-1}$ above the magnon band. Both these impurity modes are quite close to the host magnon band, and so are not highly localised. Moreover, since the modes can be shifted relative to the magnon bands by the application of an external magnetic field, the degree of localisation can be changed. Durr and Uwira (1979) have contrasted the effects of the lack of localisation on the modes above and below the band and, using the semiclassical picture of Weber (1969), have examined the eigenvectors of the spins in the region of the defect

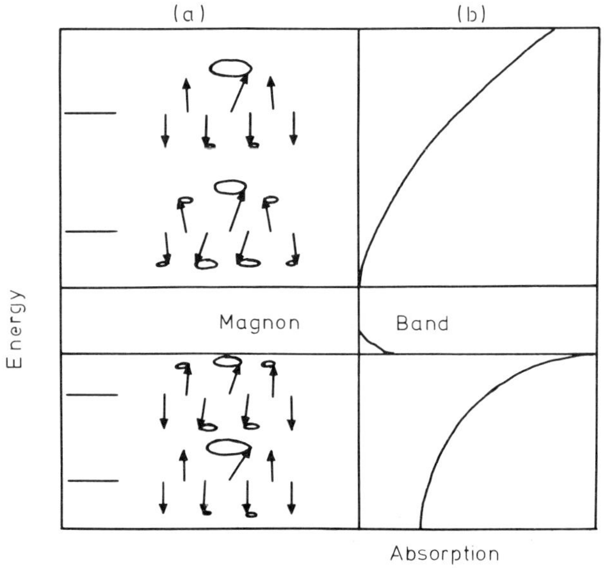

Fig. 4. (a) Schematic behaviour of the spin eigenvectors of a magnetic impurity mode at different energies with respect to the magnon band and (b) the qualitative variation of the far-infrared absorption intensity (after Durr and Uwira 1979).

(fig. 4). For impurity modes below the band, the eigenvectors resemble those of the uniform $k=0$ magnon and all the spin vectors precess in phase. For modes above the host magnon band, the eigenvectors resemble those of the Brillouin zone-boundary magnon, and nearest-neighbouring spins precess in antiphase. Since the far-infrared absorption is proportional to the net transverse magnetic moment of the excitation, we see (Durr and Uwira 1979, Weber 1969) that as a mode below the magnon band rises towards the bottom of the band it will grow in intensity while a mode above the band will be reduced in intensity as it falls towards the top of the magnon band. It should also be pointed out that when an impurity mode lies close to the host magnon band, the influence of the impurity spreads over several lattice sites so that pair effects occur at rather low concentrations. For example, Belanger et al. (1982) detected Mn pair modes in FeF_2 with impurity concentrations as low as 0.04 at%.

Impurity pair modes were first identified by Enders et al. (1972) in $CoF_2:Mn$ at 32 cm^{-1}, between the single impurity line at 28.5 cm^{-1} and the AFMR at 35 cm^{-1}. Using a crude mean-field approach, they calculated an energy of $\sim$ 31 cm^{-1}, close to the observed value. There are, however, several problems that have to be considered when an attempt is made to deal accurately with these pair modes. First, because they are close to the host magnon band, this band must be taken into account. Second, although one can deduce from the frequency of the single impurity line what the host–impurity exchange interactions are, it is not, in general, realistic to assume that the impurity–impurity exchange interactions are simply those of the corresponding pure crystal. Thus in $CoF_2:Mn$, the Mn–Mn interactions are not necessarily those of pure MnF_2, because they may be affected by small changes in the lattice parameter.

Belanger et al. (1982) have made a detailed study of $FeF_2:Mn$, and have considered these problems. They use both a mean-field model and a Green-function calculation to obtain the positions of the pair lines; the Green-function method takes into account interactions with the host magnon band. Constraining their calculations to reproduce both the FeF_2 AFMR and the single Mn impurity mode, they found that it was difficult to obtain precise values for the Fe–Mn interactions because of the distortion of the lattice in the neighbourhood of the impurities. However, it was shown that the nearest-neighbour Mn–Fe exchange was quite large and antiferromagnetic compared to the small ferromagnetic nearest-neighbour exchange in MnF_2 and FeF_2.

The only study in the far-infrared of mixed (as opposed to dilute) magnetic rutiles has been that of Enders et al. (1972). These authors investigated the spectra of powder samples of $Co_{1-x}Mn_xF_2$ and $Fe_{1-x}Mn_xF_2$ for the full range $0<x<1$. For values of x close to 0 or 1, they observe the localised modes and pair lines mentioned earlier. In $Co_{1-x}Mn_xF_2$, the Mn impurity mode at 28.5 cm^{-1} (fig. 5) moves downwards in energy with increasing x to

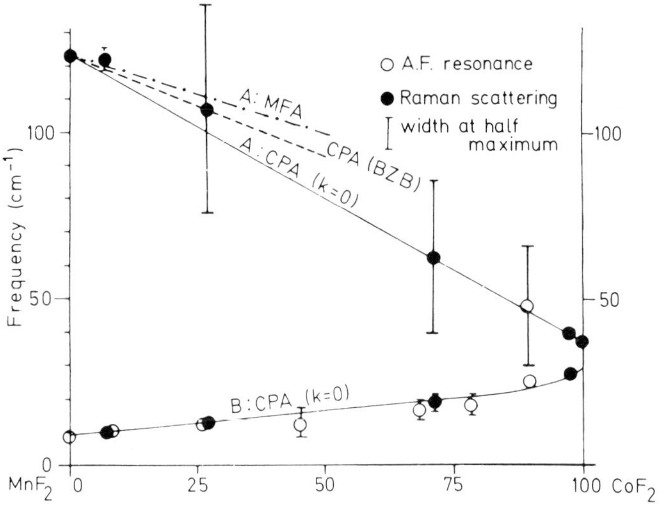

Fig. 5. Single-magnon Raman scattering in $Mn_{1-x}Co_xF_2$ (Gosso and Moch 1977). The curves marked A are associated with cobalt excitations and B with manganese excitations. The solid lines are predictions of CPA (Buyers et al. 1972). (MFA: mean-field approximation, BZB Brillouin zone boundary.)

become the AFMR of pure MnF_2 at 8.7 cm^{-1}. The CoF_2 AFMR at 36 cm^{-1} moves steeply upwards and broadens rapidly and presumably becomes the Co impurity mode at 124 cm^{-1} in MnF_2 (although Enders et al. were unable to follow the mode above ~ 80 cm^{-1}).

Raman studies have also been carried out on the mixed (Co, Mn)F_2 system (Gosso and Moch 1976, 1977). Two one-magnon lines are again found, the higher-frequency line (A in fig. 5) corresponding to the cobalt impurity mode and the lower frequency line (B in fig. 5) to the manganese impurity mode. The $q = 0$ magnon in pure MnF_2 is not observable (see section 1) but the manganese mode increases in intensity with increasing cobalt concentration because of the admixture of orbital angular momentum into the magnon states caused by the Mn–Co interaction. For nearly pure CoF_2 line A corresponds to the $q = 0$ magnon, readily observed in pure CoF_2, and line B becomes the Mn impurity gap mode.

The infrared and Raman data for $Co_{1-x}Mn_xF_2$ are consistent with the results of inelastic neutron scattering experiments (Buyers et al. 1972, Svensson et al. 1975, Cowley et al. 1980). These show a two-band behaviour, the lower band being predominantly Mn spin excitations and the upper band predominantly Co. As the concentration of Mn is increased, the Mn local mode acquires dispersion and the Co-like band moves to higher energy and shows a considerable width. The far-infrared and Raman spectra pick out the long wavelength portions of these dispersion curves and the results agree well with the neutron data.

Buyers et al. (1972) developed a CPA theory to describe the mixed (Mn, Co)F_2 system and compared their results with neutron scattering data. In these calculations the longitudinal part of the exchange was treated by the CPA while the transverse part, which fortuitously remains essentially constant in (Mn, Co)F_2, was treated more approximately. Nevertheless good agreement was obtained between the calculated magnon frequencies and linewidths and the neutron results. This theory has also been used (Gosso and Moch 1976, 1977) to predict the frequency dependence of optical excitations in the mixed crystals and gives good agreement with experiment (fig. 5).

3.2.2. Dilute mixtures

It has been pointed out earlier that one of the major problems in calculating the spectra of mixed magnets is knowing the host–impurity and impurity–impurity interactions and their concentration dependence. If the impurity is non-magnetic, all these interactions are zero. Thus the dilute magnet has no free parameters and so represents a rather severe test of theory. (Mn, Zn)F_2 has been the subject of several investigations, both experimental and theoretical. The far-infrared absorption spectrum of pure MnF_2 contains features due to one-magnon (Richards 1963) and two-magnon (Allen et al. 1966) processes. Both of these are sensitive to dilution and have been studied in some detail. The two-magnon spectrum occurs at a more accessible energy ($\sim 110\ cm^{-1}$) than the single-magnon AFMR at 8.7 cm^{-1}, and dilution effects were first studied in the two-magnon spectrum. Mitlehner et al. (1971) observed the two-magnon spectrum in the infrared in $Mn_{1-x}Zn_xF_2$ for $x < 0.06$ while Buchanan et al. (1972) followed this spectrum up to $x = 0.25$. However, the two-magnon contribution to the spectrum is difficult to determine accurately because of the existence of impurity-induced phonon absorption in the same spectral region. To some extent this can be estimated by also measuring the spectra at a temperature sufficiently above T_N to make the magnon contribution negligible (fig. 6a). This permits the two-magnon peak position to be estimated (Buchanan et al. 1972). However, the Raman spectra give a much clearer measurement of the two-magnon spectrum as a function of dilution and we shall concentrate on the Raman data here.

The two-magnon Raman spectra of $Mn_{1-x}Zn_xF_2$ measured with xy and xz polarisation (Buchanan et al. 1972) are shown in fig. 6b (z coincides with the tetragonal axis). The line shapes are asymmetrical, particularly in xy polarisation. For finite x the peaks move to lower energy and are broadened and the asymmetry is reduced.

The peak position of the two-magnon bands decreases linearly with x up to $x \simeq 0.25$ and then decreases more rapidly. At $x = 0.5$ the transition temperature $T_N(x)$ is ~ 20 K so that the temperature of measurement of the

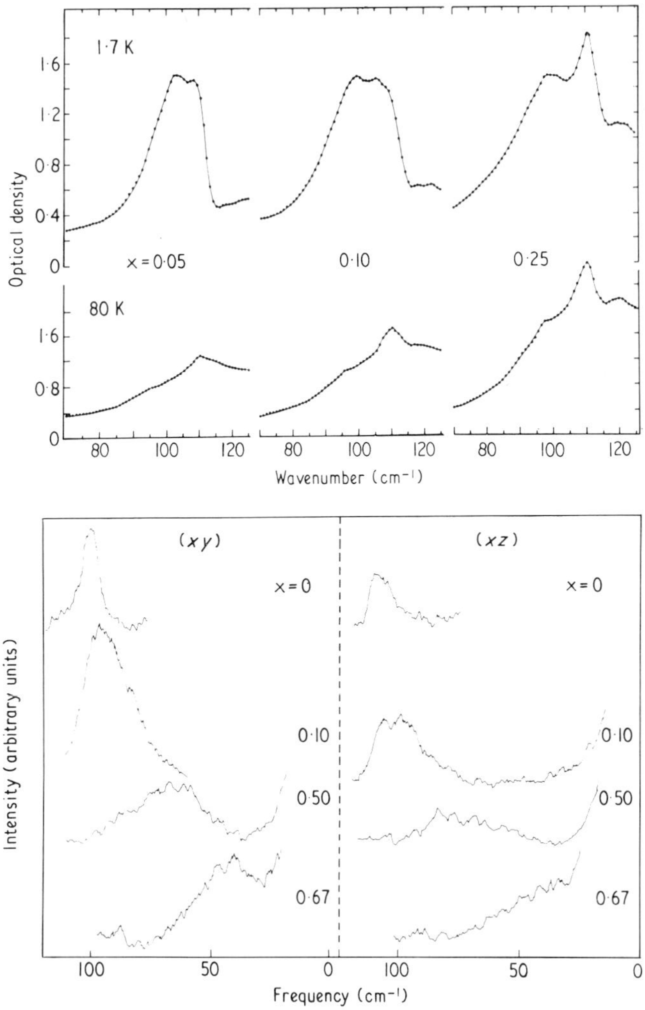

Fig. 6. Two-magnon Raman spectrum of $Mn_{1-x}Zn_xF_2$ obtained by (a) infrared absorption with $\epsilon \perp c$ at 1.7 and 80 K and (b) Raman scattering at ~ 4 K (Buchanan et al. 1972).

scattering (~ 5 K) is effectively zero. For larger values of x the temperature of measurement is an appreciable fraction of $T_n(x)$ and the measured peak positions are lower in the energy than the values appropriate to 0 K.

The Ising model of antiferromagnets is both simple and useful in the present context. It predicts an energy for reversal of a single spin at a single frequency equal to that of the sharp peak at the spin-wave density of states. It may be adapted to two-magnon excitations in mixed crystals by considering the dependence of the energy of a spin on the configuration of the surrounding ions. If we neglect spin–spin interaction and require that at least one nearest neighbour of Mn must be another Mn to give two-magnon scattering we obtain the mean frequency of the two-magnon peak to be

$$\langle \omega \rangle = 2A(1 - 7x/8), \tag{34}$$

where $A = 2SJ$. This does not fall to zero as $x \to 1$ since we have insisted that pairs of Mn atoms must occur as neighbours to contribute to the scattering. It is also possible using this approach to calculate the shape of the two-magnon scattering by calculating a suitable weighted histogram of frequencies for the various possible clusters of nearest neighbours. Such calculated shapes are compared with experimental results in fig. 7; included in this figure also are calculated shapes using the CPA (Buyers et al. 1975).

In general we find with two-magnon Raman studies that excitations are broad and too weak in the vicinity of the percolation threshold x_p (0.75 for the rutile structure) to reveal any unusual behaviour. There is no critical behaviour observable in two-magnon Raman scattering as $x \to x_p$ since light

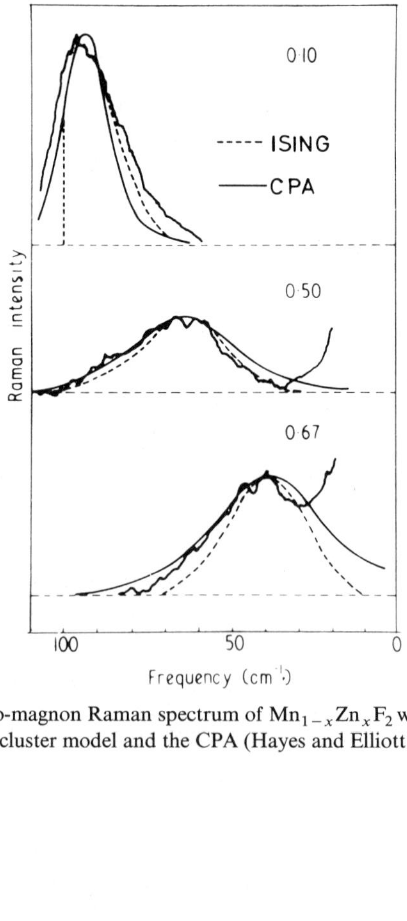

Fig. 7. Comparison of two-magnon Raman spectrum of $Mn_{1-x}Zn_xF_2$ with predictions of both the Ising cluster model and the CPA (Hayes and Elliott 1975).

scattering probes only the short wavelength zone boundary modes which are largely determined by interactions between magnetic nearest neighbours. However, critical scattering near x_p in (Mn, Zn)F_2 has been studied using neutron scattering (Cowley et al. 1977b), supporting the view that the point $(x = x_p, T \to 0)$ may be regarded as a multicritical point, terminating the line of continuous phase transitions of the infinitely connected cluster.

The concentration dependence of one-magnon and two-magnon Raman scattering has been investigated by Montarroyos et al. (1979) for $Fe_{1-x}Zn_xF_2$ and has been explained quite well with a cluster model.

We now turn to the one-magnon spectrum. Because the AFMR frequency in pure MnF_2 is only 8.7 cm^{-1}, and falls with dilution, it is experimentally difficult to observe one-magnon Raman scattering. Infrared measurements although not straightforward, are tractable and two sets of measurements have been reported. Wiltshire (1977) observed the AFMR in single-crystal samples of $Mn_{1-x}Zn_xF_2$ with $x \leqslant 0.5$ while Tennant and Richards (1977) used powder samples in the same concentration range. The results are essentially identical (fig. 8) showing an approximately linear decrease in peak position with zinc concentration that extrapolates to zero frequency at a concentration close to the percolation limit of $x = 0.76$ (Elliott and Heap 1962). This behaviour is

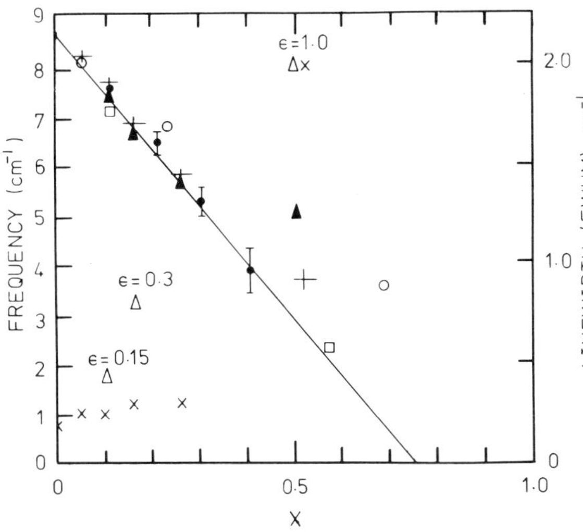

Fig. 8. Concentration dependence of the AFMR frequency and line width in $Mn_{1-x}Zn_xF_2$. + = frequency with experimental errors indicated and × = line width measured by far-infrared absorption on single crystal samples (Wiltshire 1977); ● = frequency measured by far-infrared absorption on powder samples (Tennant and Richards 1977); ○ = frequency measured by neutron scattering (Coombs et al. 1976); □ = frequency from microwave data (Foner 1964); ▲ = frequency and △ = line width calculated using the indicated values of the damping parameter ε in the CPA (Wiltshire 1977). The full line is a linear decrease of the frequency to zero at the percolation limit.

similar to that found by Baker et al. (1966) for the variation of the Néel temperature in $Mn_{1-x}Zn_xF_2$. These results are consistent with Foner's microwave data (1964), but the inelastic neutron scattering results of Coombs et al. (1976) lie somewhat higher in energy. This may be due in part to the difficulty of extracting accurate peak positions at such low frequencies at the zone centre with a Bragg peak also present.

Wiltshire (1977) compared his measurements of AFMR of $Mn_{1-x}Zn_xF_2$ with the CPA calculations of Buyers et al. (1973). A smoothing or damping parameter ϵ has to be included in these calculations and Buyers et al. (1973) set this at 0.1 THz (which gives rise to a linewidth of $\sim 6.7\ cm^{-1}$ fwhm), a value comparable to the frequencies of interest. By increasing the number of points in the lattice sum of the CPA, ϵ can be reduced and Wiltshire (1977) was able to obtain quite good agreement with both the line widths and peak positions for $x < 0.5$. The limiting value of the damping parameter ϵ is set by the existence of branch cuts off the real frequency axis (Nickel and Butler 1973) which move further off the axis as x is increased and are characteristic of this type of CPA. Thus for the more dilute samples a detailed comparison between the CPA (Buyers et al. 1973) and experiments is impossible. However, it should be emphasised that the fairly simple single-site CPA (Buyers et al. 1973) does give a remarkable good description both of the neutron results and the infrared spectra.

McGurn and Tahir-Kheli (1978) have used the far-infrared results to determine the antiferromagnetic-spin-flop critical field in $Mn_{1-x}Zn_xF_2$. In an earlier paper (McGurn and Tahir-Kheli 1977), they developed a path-type CPA for treating the dilute antiferromagnet, and compared the results of this model both with Holcomb's CPA (1974) and the Monte Carlo simulations of Holcomb and Harris (1975). In calculating the spin-flop field McGurn and Tahir-Kheli (1977, 1978) made use of a perturbation theory treatment and also the CPA. They found good agreement between Wiltshire's results (1977) and both their calculations. Also, by making use of a band-CPA rather than a site-CPA (Tahir-Kheli et al. 1978, Tahir-Kheli and McGurn 1978a, b), an estimate of the Néel temperature of (Mn, Zn)F_2 could be obtained (McGurn and Tahir-Kheli 1978) which was in good agreement with the observations of Baker et al. (1966). Moreover, the Néel temperature in the isomorphous system (Fe, Zn)F_2 was also calculated using this model (Tahir-Kheli and McGurn 1978a, b) and, although the measurements of Wertheim et al. (1966) showed distinctly different behaviour for (Fe, Zn)F_2 from that of (Mn, Zn)F_2, good agreement was again obtained. The difference between the two systems was found to be due to the size and nature of the anisotropy. The anisotropy in $Mn_{1-x}Zn_xF_2$ is quite small and mostly dipolar in origin whilst that in $Fe_{1-x}Zn_xF_2$ is large and due mostly to crystal-field effects, i.e., it is a single-ion anisotropy. The difference between these two types of anisotropy certainly accounts for the difference in behaviour of T_N with dilution and may also be responsible for some of the problems that have arisen in the discussion of the $MnF_2:Fe$ and $FeF_2:Mn$ systems (see section 3.2.1). However, as long as the anisotropy is adequately treated, it seems that the CPA in its various forms gives a good description of these mixtures.

3.3. Transition metal compounds with the cubic perovskite structure

3.3.1. Mixtures of magnetic ions

Some results of Raman scattering studies of magnons in $Rb(Mn, Co)F_3$ and $K(Mn, Co)F_3$ have been reported by Gosso and Moch (1976, 1977). However, the most detailed optical study of magnons in a mixed cubic perovskite has been made on $KMn_{1-x}Ni_xF_3$ by Lockwood et al. (1979) (Raman (Parisot et al. 1971) and infrared (Barker et al. 1968) studies of impurity magnetic modes in $KMnF_3$ associated with isolated Ni^{2+} ions were reported earlier).

At low temperature there are three two-magnon bands in the Raman spectrum of $KMn_{1-x}Ni_xF_3$ corresponding to excitations on Ni–Ni, Ni–Mn and Mn–Mn pairs. The Raman spectrum for $x = 0.54$ is shown in fig. 9 for diagonal polarisation. The band at $\gtrsim 600\ cm^{-1}$ is due to Ni–Ni pairs, at $\sim 400\ cm^{-1}$ is due to Mn–Ni pairs and at $\leqslant 200\ cm^{-1}$ is due to Mn–Mn pairs. The peak at $\sim 550\ cm^{-1}$ is a strong two-phonon band. The intensity of the Mn–Mn band is much weaker compared with the Mn–Ni band than would be expected on the basis of simple theory. It is suggested (Lockwood et al. 1979) that this is due to near resonance of the laser light with electronic transitions of Ni^{2+}.

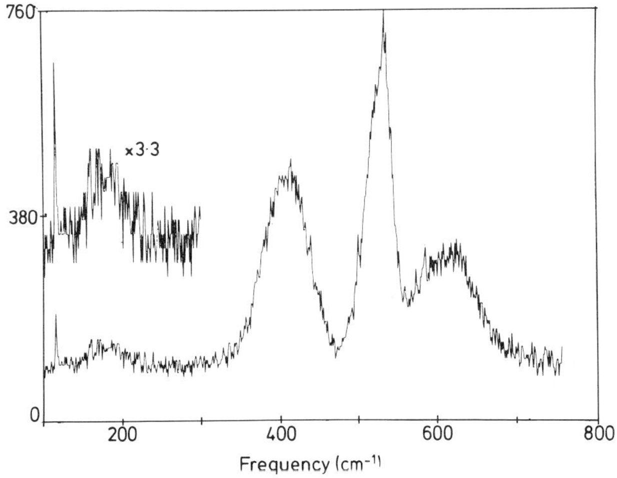

Fig. 9. Two-magnon Raman spectrum of $KMn_{1-x}Ni_xF_3$ with diagonal polarisation for $x = 0.54$ (Lockwood et al. 1979).

The frequency dependence of the two-magnon peaks on x (crystals with $x = 0.84$, 0.73, 0.54 and 0.15 were used) agrees with the predictions of an average cluster model if allowance is made for change in exchange constants with lattice parameter.

3.3.2. Dilute mixtures

A Raman study of the two-magnon spectrum of $KNi_{1-x}Mg_xF_3$ has been carried out by Fleury et al. (1975). The two-magnon Raman spectrum of $KNiF_3$ is again dominated by zone-edge magnons, giving a well-defined band (fig. 10) slightly below twice the zone-edge frequency. This band is known to renormalise with increasing temperature much less rapidly than the zone-centre magnons or the magnetisation. Indeed the two-magnon Raman modes persist to temperatures well above T_N, where long-range magnetic order disappears (Fleury 1969).

Fig. 10 shows the low-temperature magnon pair spectra of $KNi_{1-x}Mg_xF_3$ for $x = 0$, 0.15, 0.44, 0.68 and 0.75. The variation of peak frequency with x is

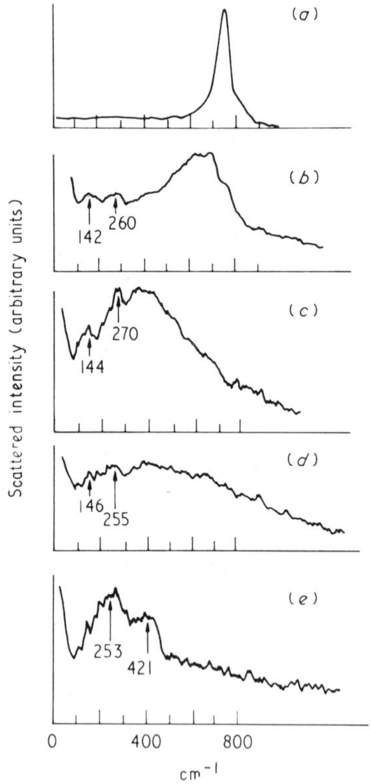

Fig. 10. Two-magnon Raman spectrum of $KNi_{1-x}Mg_xF_3$ for x = (a) 0, (b) 0.15, (c) 0.44, (d) 0.68, and (e) 0.75 at helium temperature (Fleury et al. 1975).

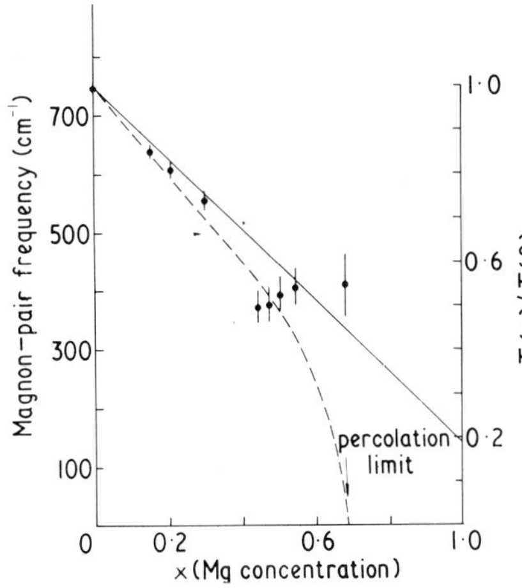

Fig. 11. Variation of two-magnon peak of $KNi_{1-x}Mg_xF_3$ with x at low temperature. The full line is the prediction of an Ising cluster model and the dashed line is a prediction of CPA (Fleury et al. 1975).

shown in fig. 11. The variation is linear in the range $x = 0.0–0.5$. Above $x = 0.5$ the two-magnon frequency is apparently independent of x, at least up to $x = 0.68$. Although some magnetic scattering could be determined up to $x = 0.75$ (the percolation limit for the cubic perovskite structure is $x_p = 0.68$) the uncertainties caused by phonon scattering preclude quantitative frequency determination.

In the perovskite lattice a simple cluster model (Fleury et al. 1975) gives a mean frequency

$$\langle \omega \rangle = 2A[1 - x(z-1)/z] \tag{35}$$

for the magnon pair in dilute crystals; here $A = 2SJz$, $S = 1$ for Ni^{2+} and $z = 6$ is the number of nearest magnetic sites. Equation (35) is plotted as the full curve in fig. 11. Reasonable agreement is obtained with experiment up to $x \simeq 0.5$. The apparent independence of the two-magnon frequency on x for $x > 0.5$ may be due to preferential clustering of Ni^{2+} ions. The broken curve in fig. 10 represents a qualitative estimate of the concentration dependence of long-range order; it is an extrapolation of low-x values obtained using the CPA (Holcomb 1974).

The temperature dependence of the two-magnon spectrum in pure $KNiF_3$ is shown in fig. 12 (Fleury et al. 1975). A broadening and decrease in energy in the two-magnon spectrum is apparent. The mode persists in the paramagnetic region as a slightly underdamped resonance whose frequency at T_N is $\langle \omega \rangle_{T_N} = 0.7\langle \omega \rangle_{T=0}$. The qualitative correspondence between increasing temperature

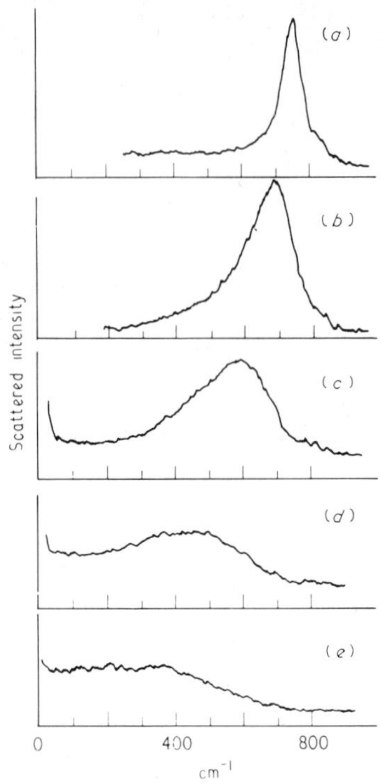

Fig. 12. Two-magnon spectra of pure $KNiF_3$ at (a) 97 K, (b) 181 K, (c) 220 K, (d) 256 K, and (e) 300 K. $T_N = 250$ K (Fleury et al. 1975).

in the pure antiferromagnet (fig. 12) and increasing dilution for $T < T_N(x)$ (fig. 10) is evident. A satisfactory theoretical account of this correspondence is at present lacking (for references to the use of the method of moments for the calculation of the temperature dependence of peak positions of two-magnon excitations in dilute magnetic crystals see Hayes and Elliott (1975)).

3.4. Transition metal compounds with the layer perovskite structure

Two-dimensional (2D) magnets have also been of great interest over the years, first because 2D critical behaviour differs from 3D behaviour and is, in general easier to study, and secondly because the smaller coordination number in 2D systems emphasises the effect of impurities on the energy of a particular site. Materials with the K_2NiF_4 structure are particularly well suited for these investigations and inelastic neutron scattering measurements have been carried out on a wide variety of these compounds. Far infrared studies have been confined to K_2CoF_4 and Rb_2CoF_4 which are good examples of 2D Ising antiferromagnets. The magnon band in these materials lies at quite high

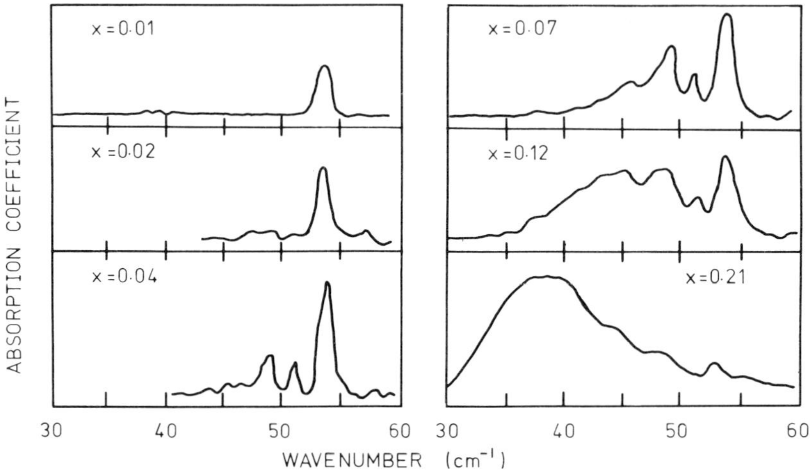

Fig. 13. Concentration dependence of the magnetic-impurity-induced far-infrared absorption in $K_2Co_{1-x}Mn_xF_4$ (Betz et al. 1981).

energies ($\sim 200\ cm^{-1}$) (Ikeda and Hutchings 1978) so that the impurity-induced features below the band should be easy to observe. Betz et al. (1981) have carried out far-infrared studies of K_2CoF_4 and Rb_2CoF_4 containing Fe, Mn or Ni. The Ni impurity mode was observed only in infrared fluorescence and lies quite close to the magnon band. The modes due to Mn and Fe were observed for low impurity concentrations at $54.1\ cm^{-1}$ and $55.4\ cm^{-1}$ respectively; in K_2CoF_4 the Mn mode in Rb_2CoF_4 occurs at $49.7\ cm^{-1}$. The results of this investigation were analysed using a mean-field model and the host–impurity Ising exchange parameters were extracted and were found to obey the empirical relation $J_{AB} = \sqrt{J_{AA}J_{BB}}$ rather well.

At higher concentrations in $K_2Co_{1-x}Mn_xF_4$, additional levels develop in the spectrum and merge into a broad impurity band by $x = 0.2$ (fig. 13). Neither the mean-field Ising model nor Dzyub's cluster model (1974a, b) could explain these levels adequately with reasonable values of the exchange parameters.

The mixture $K_2Co_{1-x}Fe_xF_4$ is of particular interest because there is competition for the direction of order between K_2CoF_4 (parallel to the *c*-axis) and K_2FeF_4 (perpendicular to the *c*-axis). For low concentrations of Fe impurities Betz et al. (1981) found that the isolated Fe^{2+} ions had their moments pulled parallel to the *c*-axis. When the Fe concentration is increased, so that clusters form, the moment tilts away from this axis. Qualitatively similar behaviour has been observed by neutron scattering in other systems with competing anisotropy, for example $K_2(Mn, Fe)F_4$ (Bevaart et al. 1978) and $(Fe, Co)Cl_2$ (Wong et al. 1980). However, no quantitative experimental data are available about the ordering in the mixed phase of these systems.

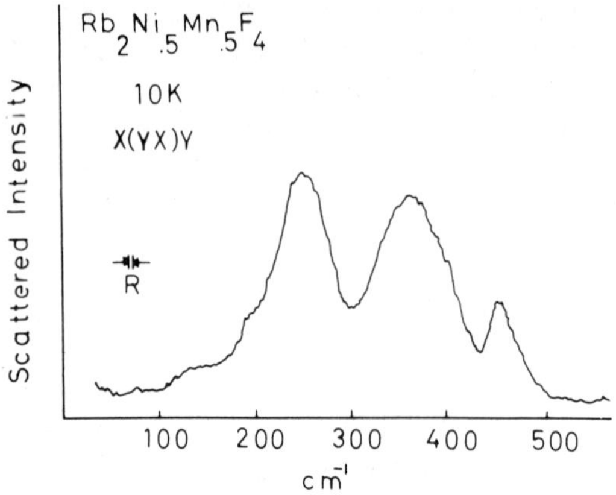

Fig. 14. Two-magnon Raman spectrum of $Rh_2Ni_{0.5}Mn_{0.5}F_4$ (Fleury and Guggenheim 1975).

Neutron scattering measurements (Birgeneau et al. 1975) show that the mixed crystal $Rb_2Mn_{0.5}Ni_{0.5}F_4$ is a two-mode magnon system at low temperatures, giving two well-defined doubly-degenerate magnon branches throughout the Brillouin zone, and Raman scattering measurements (Fleury and Guggenheim 1975) reveal two two-magnon modes due to Mn–Ni (248 cm^{-1}) and Ni–Ni (372 cm^{-1}) pairs (fig. 14). The other peaks in fig. 14 are caused by phonons. Although the absence of a Mn–Mn pair mode in fig. 14 is not unexpected, the intensity of the Mn–Ni mode relative to that of Ni–Ni again seems anomalously large (see section 3.3.1).

3.5. Transition metal halides with the cadmium chloride structure

3.5.1. Dilute mixtures

The transition metal chlorides with the cadmium chloride structure (space group $R\bar{3}m$) form another series of isomorphs with widely differing magnetic properties. For example, $FeCl_2$ has long been known for its so-called metamagnetism, and, more recently, has been extensively studied because it exhibits a tricritical point (Jacobs and Lawrence 1967, Birgeneau et al. 1974). $FeCl_2$ orders antiferromagnetically below 23.5 K. In the ordered state, the ferrous magnetic moments are aligned parallel to the *c*-axis. These moments order ferromagnetically within the layers of the basal plane, and these layers are antiferromagnetically coupled with a much weaker out-of-plane exchange. It is this form of magnetic order that gives rise to the metamagnetism. Thus $FeCl_2$ can be treated to a first approximation as a 2D ferromagnet with quite large anisotropy (Birgeneau et al. 1972). $CoCl_2$, on the other hand, although the overall form of the order is the same and the Néel temperature is comparable to that of $FeCl_2$, has its magnetic moments lying in the basal plane (Wilkinson et al. 1959, Hutchings 1973) and is better described as an X–Y system. The magnetic order of $MnCl_2$ is more complex than either of these (Wilkinson et al. 1957). Its Néel temperature is ~ 2 K but the exchange interactions are not known. All of these magnetic materials are miscible in all proportions and may also be diluted with $MgCl_2$. Moreover, the lattice parameters of $FeCl_2$, $CoCl_2$ and $MgCl_2$ are the same to within 2% so it is expected that the size of the exchange interactions should remain largely unaffected by mixing.

The far-infrared properties of pure $FeCl_2$ and $CoCl_2$ were investigated by Jacobs et al. (1965), who found that the AFMR occurred at 16.5 cm^{-1} and 19.2 cm^{-1} in the two materials respectively. Inelastic neutron scattering measurements (Birgeneau et al. (1972) on $FeCl_2$, Hutchings (1973) on $CoCl_2$) have provided values for the exchange parameters in the pure materials.

Because their lattice parameters are nearly identical to those of $MgCl_2$ and their AFMR frequencies are not too low, $FeCl_2$ and $CoCl_2$ are ideal candidates for dilution studies in the far-infrared. Hayes et al. (1976) studied the far-infrared absorption of $Fe_{1-x}Mg_xCl_2$ for $0 \leqslant x \leqslant 0.95$. The AFMR was found to fall approximately linear in frequency and to broaden with increasing dilution (fig. 15). At the percolation limit, one expects the long wavelength AFMR to vanish, as the infinite cluster also disappears. However, finite clusters still exist and give rise to far-infrared absorption, so the observed spectrum is a combination of the finite cluster response and the long wave-

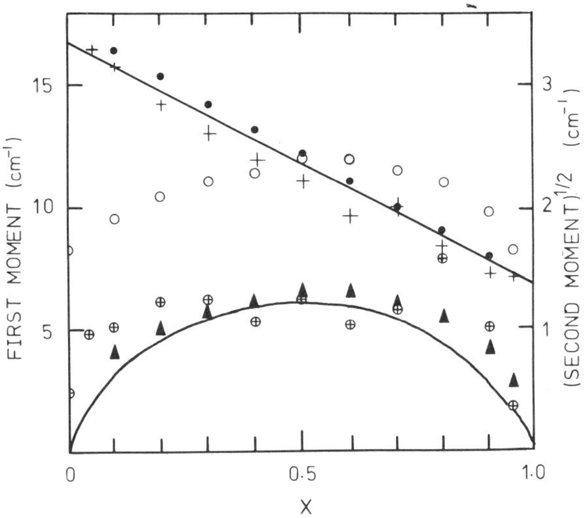

Fig. 15. Concentration dependence of the first and second moments of the AFMR in $Fe_{1-x}Mg_xCl_2$. Left-hand ordinate for first moments: + = measured with the experimental errors indicated: ● = calculated; ——— = mean-field calculation. Right-hand ordinate for second moments: ⊕ = measured; ○ = calculated; ▲ = calculated with the effects of damping removed (Hayes et al. 1976, Wiltshire and Hayes 1978); ——— = Fujiwara's calculation (1977).

length magnon contribution. Because of this the peak frequency of the far-infrared absorption does not fall to zero, but the absorption band persists to high dilution as a cluster mode, eventually becoming an electronic transition at 6.3 cm^{-1} of Fe^{2+} isolated in $MgCl_2$.

The most appropriate value of the percolation limit in these materials is not clear. Whereas the overall magnetism is three-dimensional and antiferromagnetic, the dominant ferromagnetic interactions lie within the basal plane and the system can be approximated as a two-dimensional ferromagnet. In the latter case, the percolation limit $x_p = 0.50$ for the two-dimensional triangular lattice (Shante and Kirkpatrick 1971) is appropriate. However, if the effects of neighbours beyond the first in the plane are considered, the percolation limit for this lattice becomes $x_p = 0.7050$ (including second neighbours) or $x_p = 0.775$ (including third neighbours) (Shante and Kirkpatrick 1971). Moreover, the antiferromagnetic properties derive from the three-dimensional ordering of the two-dimensional layers. In three dimensions, the percolation limit for the hexagonal close-packed structure is $x = 0.80$ (Shante and Kirkpatrick 1971) for nearest neighbour interactions. Other estimates have been given of $x_p = 0.67$ (Hayes et al. 1976, Wada and Ishikawa 1979) and of $x_p = 0.8$ (Ziman and Elliott 1980). It is clear that the three-dimensional percolation limit should lie between 0.67 and 0.8, while that appropriate to the two-dimensional interactions may be as low as $x_p = 0.5$.

A mean-field model provides the simplest way of describing the far-infrared behaviour. Here the exchange energy, written as $2J_{eff}S^2$, is scaled linearly with concentration and the single-ion anisotropy D is held fixed at 6.8 cm^{-1} for $MgCl_2$: Fe. We then write

$$\bar{\nu} = 2(1-x)\cdot J_{eff}S^2 + D,$$

and, choosing $J_{eff} = 4.9$ cm^{-1}, this relation gives reasonable agreement with the observed peak positions (see fig. 15). The mean-field model does not give any description of the variation of the width or shape of the absorption, and a simple cluster model gives linewidths up to a factor of two too large (Hayes et al. 1976).

CPA calculations on $Fe_{1-x}Mg_xCl_2$ have not been published, although Fujiwara (1977) reports a private communication from D.W. Taylor. Fujiwara (1977) suggests that the CPA is not entirely satisfactory because it gives a sharp cut-off at the low frequency band edge, which is precisely the region monitored by far-infrared absorption. Also McGurn and Tahir-Kheli (1978), in discussing the (Mn, Zn)F_2 system, observe that in some unpublished work the two-site CPA when applied to this system gave a line width about an order of magnitude too small for the $k = 0$ magnon. It should also be pointed out that the CPA has not been found to be successful in dealing with two-dimensional systems. Cowley et al. (1977a) found in their neutron study of the dynamics of $Rb_2(Mn, Mg)F_4$ that the CPA gave poor results. They pointed out that the two most common approaches, namely those of Buyers et al. (1972) and of Holcomb (1974), each had their drawbacks. The CPA approach of Buyers et al. (1972) treats the Ising energies (i.e., the longitudinal exchange) exactly and includes the transverse exchange interactions only approximately. Holcomb's method (1974) treats the transverse exchange correctly, but makes all magnetic sites equivalent, thus losing the fine structure in the scattering cross section associated with the Ising energies of the various cluster configurations. The deficiencies of these approaches are exacerbated in two dimensions where the coordination numbers are smaller.

Fujiwara (1977), in an alternative approach, adopted the moment expansion method to obtain the spectral function of a dilute two-dimensional anisotropic ferromagnet approximating to $Fe_{1-x}Mg_xCl_2$. Here, the response from clusters containing four or less spins was calculated independently while the remainder of the spectrum was obtained using Padé Approximants. The calculated peak positions and the second moments were compared with those obtained by Hayes et al. (1976) and reasonable agreement was obtained (fig. 15).

The most successful approach for obtaining calculated spectra has been the equation-of-motion method of Alben and Thorpe (1975) and Thorpe and Alben (1976). This method was found by Cowley et al. (1977a) to give extremely good agreement with the neutron scattering results on $Rb_2(Mn, Mg)F_4$, where the CPA had failed. Wiltshire and Hayes (1978) applied this technique to the $Fe_{1-x}Mg_xCl_2$ series, treating the system as a dilute, two-dimensional ferromagnet and including the out-of-plane interactions as a mean-field term in the single-ion anisotropy. The results of this calculation (fig. 15) agreed quite well with the experimental data of Hayes et al. (1976) and were consistent with the results of Fujiwara's calculation (1977) and, for the very dilute case, with the predictions of a cluster model. There are no adjustable parameters in the calculation as the single-ion anisotropy was obtained from the far-infrared data (Hayes et al. 1976), while the exchange parameters were derived from the neutron results (Birgeneau et al. 1972). It is therefore clear that, at least for those systems where the parameters do not change with composition, linear spin wave theory as applied via the equation-of-motion simulation technique gives a satisfactory description of dilute systems.

Single-magnon Raman scattering has also been studied in $Fe_{1-x}Cd_xCl_2$ for $x < 0.3$ by Mischler et al. (1982). The single-magnon peak for $FeCl_2$ occurs at 17.1 cm^{-1} at 5.8 K and drops rapidly with increasing x. The results are discussed in terms of an Ising cluster model.

The isomorphous system $Co_{1-x}Mg_xCl_2$ was studied by Wiltshire (1979). In this series of mixtures, the AFMR, initially at 19.0 cm^{-1} for $x = 0$ (Jacobs et al. 1965), falls approximately linearly (fig. 16) for $x \leqslant 0.5$. This line extrapolates to zero frequency between $x = 0.65$ and 0.8, as expected on the basis of

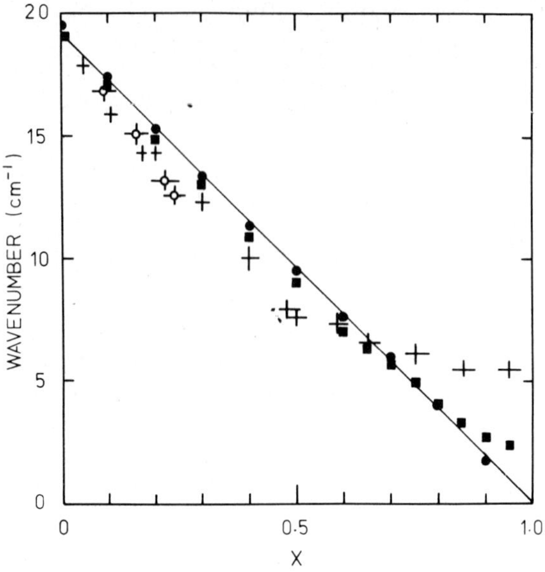

Fig. 16a. Concentration dependence of the AFMR frequency in $Co_{1-x}M_xCl_2$. + = measured, M = Mg, ○ = measured, M = Cd, both with experimental errors indicated; ● = 2D calculation; ■ = 3D calculation. The full line, $\nu = \nu_0(1-x)$, is a guide to the eye.

the three-dimensional percolation limits quoted above (Shante and Kirkpatrick 1971, Hayes et al. 1976, Wada and Ishikawa 1979, Ziman and Elliott 1980). At higher x, however, the falling trend flattens off until the spectrum becomes

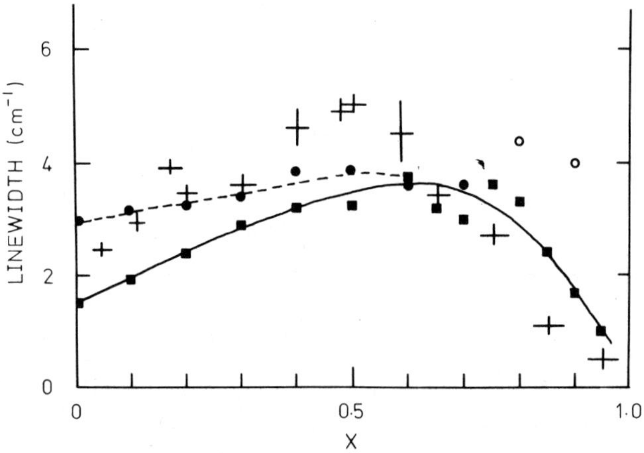

Fig. 16b. Concentration dependence of the full width at half height of the AFMR in $Co_{1-x}Mg_xCl_2$. + = measured with experimental errors indicated. ● = 2D calculations; ○ = 2D calculation, overdamped; dashed line = guide to the eye for the 2D calculation. ■ = 3D calculation; full line = guide to the eye for the 3D calculation.

that of isolated Co^{2+} pairs in $MgCl_2$ with a single absorption peak at 5.5 cm^{-1}. The line width follows a similar pattern to that observed in $Fe_{1-x}Mg_xCl_2$. With increasing x the line broadens, reaching a maximum width around $x = 0.7$, and then it narrows rapidly in the clusters regime (fig. 16b).

Wiltshire (1979) attempted to explain this behaviour using the simulation techniques described earlier. Following the example of $Fe_{1-x}Mg_xCl_2$, a two-dimensional model was used initially, treating the out-of-plane interactions in a mean-field manner to give rise to an effective single-ion term. This approach was not satisfactory because the AFMR fell linearly to zero frequency at $x = 1$, driven by the mean-field term, and the cluster modes at large x were incorrectly calculated. A slight improvement was obtained using a three-dimensional model with Hutchings' parameters (1973); the AFMR then tended to a finite frequency at $x = 1$. However, the cluster modes were still incorrect. The reason for this appeared to be the assumption that the classical Néel state was the true ground state, which is certainly not the case in the cluster regime. However, whilst long-range order persists the simulation appears to be adequate, but considerable care is obviously necessary in the choice of the ground state.

Although critical phenomena in disordered magnetic materials have not been investigated by infrared and Raman techniques (see section 1) they have proved amenable to investigation by more conventional optical methods (Wood and Day 1977). For $FeCl_2$, $T_N = 23.5$ K but below 21 K the application of a magnetic field H parallel to the c-axis causes the lattice to undergo a phase transition from an antiferromagnetic state (low H) to a saturated

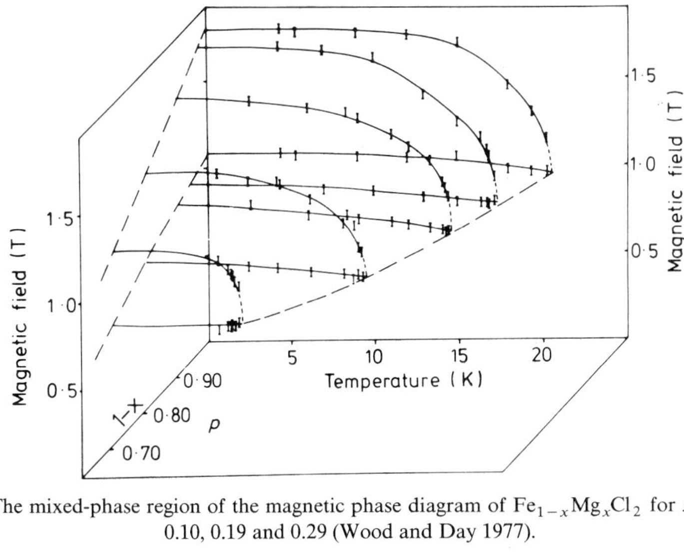

Fig. 17. The mixed-phase region of the magnetic phase diagram of $Fe_{1-x}Mg_xCl_2$ for $x = 0$, 0.05, 0.10, 0.19 and 0.29 (Wood and Day 1977).

paramagnetic state (high H) via a mixed-phase region. The boundaries of the mixed-phase region converge at a critical point (H_t, T_t) which for pure $FeCl_2$ occurs at (0.9 T, 21 K) (fig. 17). This point is readily established by optical means since visible light ($\lambda \sim 600$ nm) is elastically scattered by the crystal in the mixed-phase region, reducing the intensity of a transmitted light beam. In $Fe_{1-x}Mg_xCl_2$, the value of T_t falls smoothly with increasing dilution, tending to zero at $x \simeq 0.4$ (fig. 17). This is closer to the value of 0.5 (Shante and Kirkpatrick 1971) calculated for the site percolation limit of the plane triangular lattice than the three-dimensional percolation limits $0.67 < x_p < 0.8$ (see above).

3.5.2. Mixtures of magnetic ions

Investigations of mixtures of the magnetic transition metal chlorides and bromides have not been as extensive as investigations of the dilute systems. Microwave absorption and resonance techniques have been used to study both manganese and iron impurities. For example, Tuchendler et al. (1978) studied $FeBr_2$ doped with 1% $MnBr_2$ and observed both the manganese impurity modes at low frequency (<100 GHz, $\leqslant 3$ cm^{-1}) and the $FeBr_2$ magnons at higher frequencies (~ 500 GHz, ~ 15 cm^{-1} in zero magnetic field); they also studied (Tuchendler et al. 1980) $Fe_{1-x}Mn_xCl_2$ for $0 \leqslant x \leqslant 0.25$. They were able to deduce values of $2J_1' = 0.26$ cm^{-1} and $2J_2' = -0.18$ cm^{-1} for the inter- and intra-layer Mn–Fe exchange parameters and to estimate the Mn–Mn exchange anisotropy. Moreover, they found that the Fe–Mn exchange parameters were strongly dependent on the Mn concentration although Bertrand et al. (1980) in a similar analysis of $Fe_{1-x}Mn_xCl_2$ did not find it necessary to suppose a concentration-dependent exchange. It would not be surprising if the exchange parameters were concentration dependent in view of the significant difference in lattice parameter between $FeCl_2$ and $MnCl_2$.

The single-magnon excitation in $Fe_{1-x}Mn_xCl_2$ has been studied by Raman scattering over the range $x = 0$ to $x = 0.25$ for $T < T_N$ (Mischler et al. 1981). The magnon frequency for $x = 0$ is 17.1 cm^{-1} at 5.8 K. The fall of this frequency with increasing x, to 10 cm^{-1} at $x = 0.25$, is discussed in terms of a cluster model.

Far-infrared spectra have been reported very recently by Wiltshire and Burton (1982) for the $Fe_{1-x}Co_xCl_2$ system with $0 \leqslant x \leqslant 0.05$ and $0.95 \leqslant x \leqslant 1$. It was found that the FeCl : Co system displayed a localised mode due to the Co impurities about 5.6 cm^{-1} below the $FeCl_2$ AFMR, whilst in $CoCl_2$: Fe only an asymmetric broadening of the AFMR to higher energy was observed. It was pointed out above that $FeCl_2$ orders with the Fe^{2+} spins aligned along the crystalline c-axis whilst $CoCl_2$ has the Co^{2+} spins lying in the basal plane. In mixtures of these materials there is therefore competition between the two types of anisotropy which leads to some very intriguing critical properties (Wong et al. 1980, Aharony and Fishman 1976, Fishman and Aharony 1978,

Wood and Day 1980). Wiltshire (1981) obtained a set of parameters to describe the Fe–Co exchange and Wiltshire and Burton (1982) applied them to calculate the far-infrared spectra. It was found by adjusting the anisotropic NNN exchange, the only parameter not previously determined, that reasonably good agreement between the simulated and measured spectra was obtained.

The Raman spectrum of $Fe_{1-x}Co_xCl_2$ has been studied by Lockwood et al. (1981). The single-magnon Raman peak in $FeCl_2$ occurs at 16.3 cm^{-1} and in $CoCl_2$ at 19.3 cm^{-1}. This peak could be observed in the mixed crystals only for x values of 0.0, 0.026 0.906 and 0.970, because of a very sharp decrease of magnon intensity with increasing disorder (fig. 18).

The tricritical phase diagram of the mixtures $Fe_{1-x}Mn_xCl_2$ and $Fe_{1-x}Co_xCl_2$ have been studied by elastic light scattering (Wood and Day 1980, Haywood et al. 1981). In both cases, behaviour indistinguishable from that of the dilute mixture $Fe_{1-x}Mg_xCl_2$ (Wood and Day 1977) is found. This is not surprising in the case of $Fe_{1-x}Mn_xCl_2$ because the exchange interactions involving the manganese impurities are extremely weak. In $Fe_{1-x}Co_xCl_2$, on the other hand, it appears that the different anisotropies of the Fe^{2+} and Co^{2+} ions dominate the Fe–Co exchange interaction so that, to a first approximation, the two types of spin are uncoupled. This independence of the ordering of the components of the spins parallel and perpendicular to the c-axis is also observed in the neutron scattering results of Wong et al. (1980).

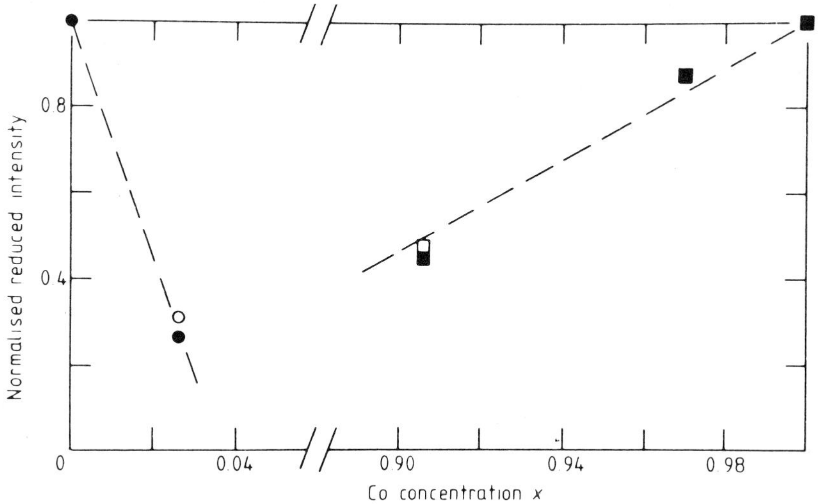

Fig. 18. Variation of intensity with x of the single-magnon Raman peak in $Fe_{1-x}Co_xCl_2$ (Lockwood et al. 1981).

4. Conclusions

There is continuing interest in the use of optical techniques to study magnetic excitations in disordered magnetic insulators. Some work on a magnetic semiconductor has recently been carried out by Grynberg and Picquart (1981) who studied light scattering in $Cd_{1-x}Mn_xTe$ for $0 \leqslant x \leqslant 0.7$ *. The optical studies have pointed out the need for a more detailed understanding of anisotropic exchange and of the variation of exchange interactions with composition, for example, via changes in the lattice parameter. Simulation calculations have highlighted the problems that arise in mixtures where the ordering in the ground state is unknown, as, for example, in competing anisotropy systems. However, despite these deficiencies in our understanding, it has been possible to provide reasonable interpretations of experimental data with simple cluster models, and, on a more fundamental level, using the CPA and simulation methods.

References

Aharony, A., and S. Fishman, 1976, Phys. Rev. Lett. **37**, 1587.
Alben, R., and M.F. Thorpe, 1975, J. Phys. C **8**, L275.
Allen, S.J., R. Loudon and P.L. Richards, 1966, Phys. Rev. Lett. **16**, 463.
Baker, J.M., J.A.J. Laurens and R.W.H. Stevenson, 1966, Proc. Phys. Soc. **77**, 1038.
Barker, A.S., and A.J. Sievers, 1975, Rev. Mod. Phys. **47**, Suppl. 2.
Barker, A.S., J.A. Ditzenberger and H.J. Guggenheim, 1968, Phys. Rev. **175**, 1180.
Becker, C.R., Ph. Lau, R. Geick and V. Wagner, 1975, Phys. Status Solidi (b) **67**, 653.
Belanger, R.M., D. None and M. Motokawa, 1982, Phys. Rev. B **25**, 3186.
Bertrand, D., A.R. Fert, J. Magarino, J. Tuchendler and S. Legrand, 1980, J. Phys. C **13**, 5165.
Betz, E., W. Knierim and U. Durr, 1981, J. Magn. & Magn. Mater. **25**, 61.
Bevaart, L., E. Frikee, J.V. Lebesque and L.J. de Jongh, 1978, Phys. Rev. B **18**, 3376.
Birgeneau, R.J., W.B. Yelon, E. Cohen and J. Makovsky, 1972, Phys. Rev. B **5**, 2607.
Birgeneau, R.J., G. Shirane, M. Blume and W.C. Koehler, 1974, Phys. Rev. Lett. **33**, 1098.
Birgeneau, R.J., L.R. Walker, H.J. Guggenheim, J. Als-Nielsen and G. Shirane, 1975, Phys. Rev. Lett.
Bloor, D., and G.M. Copland, 1972, Rep. Prog. Phys. **35**, 1173.
Buchanan, M., W.J.L. Buyers, R.J. Elliott, R.T. Harley, W. Hayes, A.M. Perry and I.D. Saville, 1972, J. Phys. C **5**, 2011.
Buyers, W.J.L., D.E. Pepper and R.J. Elliott, 1972, J. Phys. C **5**, 2611.
Buyers, W.J.L., D.E. Pepper and R.J. Elliott, 1973, J. Phys. C **6**, 1933.
Buyers, W.J.L., D.E. Pepper and R.J. Elliott, 1975, J. Phys. C **8**, 2183.
Coombs, G.J., R.A. Cowley, W.J.L. Buyers, E.C. Svensson, T.M. Holden and D.A. Jones, 1976, J. Phys. C **9**, 2167.
Cowley, R.A., and W.J.L. Buyers, 1972, Rev. Mod. Phys. **44**, 406.
Cowley, R.A., G. Shirane, R.J. Birgeneau and H.J. Guggenheim, 1977a, Phys. Rev. B **15**, 4292.

* **Note added in proof**: For a recent review of Raman scattering in diluted magnetic semiconductors, see Ramdas and Rodriguez (1987).

Cowley, R.A., G. Shirane, R.J. Birgeneau and E.C. Svensson, 1977b, Phys. Rev. Lett. **39**, 894.
Cowley, R.A., R.J. Birgeneau and G. Shirane, 1980, in: Ordering in Strongly Fluctuation Condensed Matter Systems, ed. T. Riste (Plenum Press, New York) p. 157.
Durr, U., and B. Uwira, 1979, J. Phys. C **12**, L793.
Dzyub, I.P., 1974a, Phys. Status Solidi (b) **61**, 383.
Dzyub, I.P., 1974b, Phys. Status Solidi (b) **66**, 339.
Elliott, R.J., and B.R. Heap, 1962, Proc. R. Soc. A **265**, 264.
Elliott, R.J., J.A. Krumhansl and P.L. Leath, 1974, Rev. Mod. Phys. **46**, 465.
Enders, B., P.L. Richards, W.E. Tennant and E. Catalano, 1972, 18th AIP Conf. (Denver) p. 179.
Fishman, S., and A. Aharony, 1978, Phys. Rev. B **18**, 3507.
Fleury, P.A., 1969, Phys. Rev. **180**, 591.
Fleury, P.A., and H.J. Guggenheim, 1975, Phys. Rev. B **12**, 985.
Fleury, P.A., and R. Loudon, 1968, Phys. Rev. **166**, 514.
Fleury, P.A., W. Hayes and H.J. Guggenheim, 1975, J. Phys. C **8**, 2183.
Foner, S., 1964, Proc. Int. Conf. Mag. 438.
Fujiwara, T., 1977, J. Phys. C **10**, 1039.
Geis, G., R. Geick and C.R. Becker, 1976, J. Magn. & Magn. Mater. **4**, 356.
Geis, G., R. Geick, C.R. Becker and V. Wagner, 1977, Physica B **86–88**, 1257.
Gosso, J.P., and P. Moch, 1976, in: Light Scattering in Solids, eds M. Balkanski, R.C.C. Leite and S.P.S. Porto (Flammarion, Paris) p. 214.
Gosso, J.P., and P. Moch, 1977, Physica B **89**, 209.
Grynberg, M., and M. Picquart, 1981, J. Phys. C **14**, 4677.
Hayes, W., and R.J. Elliott, 1975, in: Light Scattering in Solids, eds M. Balkanski, R.C.C. Leite and S.P.S. Porto (Flammarion, Paris) p. 203.
Hayes, W., and R. Loudon, 1978, Scattering of Light by Crystals (John Wiley and Sons, New York).
Hayes, W., P.J. Walker and M.C.K. Wiltshire, 1976, J. Phys. C **9**, L255.
Haywood, S.K., T.E. Wood and P. Day, 1981, J. Phys. C **14**, 2697.
Holcomb, W.K., 1974, J. Phys. C **7**, 4299.
Holcomb, W.K., and A.B. Harris, 1975, AIP Conf. Proc. **24**, 102.
Huber, D.L., 1974, Phys. Rev. B **10**, 4621.
Hughes, A.E., 1971, Phys. Rev. B **3**, 877.
Hutchings, M.T., 1973, J. Phys. C **6**, 3143.
Ikeda, H., and M.T. Hutchings, 1978, J. Phys. C **11**, L529.
Jacobs, I.S., and P.E. Lawrence, 1967, Phys. Rev. **164**, 866.
Jacobs, I.S., S. Roberts and P.E. Lawrence, 1965, J. Appl. Phys. **36**, 1197.
Lockwood, D.J., 1982, Light Scattering in Solids III, eds M. Cardona and G. Güntherodt (Springer-Verlag, Berlin) p. 59.
Lockwood, D.J., G.J. Coombs and R.A. Cowley, 1979, J. Phys. C **12**, 4611.
Lockwood, D.J., G. Mischler, A. Zwick, I.W. Johnstone, G.C. Psaltakis, M.G. Cottam, S. Legrand and J. Leotin, 1981, J. Phys. C **14**, 2697.
Lovesey, S.W., 1968, J. Phys. C **1**, 102, 118.
Martin, D.H., 1967, Spectroscopic Techniques for Far Infrared, Submillimetre and Millimetre Waves (Interscience, New York).
McGurn, A.R., and R.A. Tahir-Kheli, 1977, J. Phys. C **10**, 4385.
McGurn, A.R., and R.A. Tahir-Kheli, 1978, J. Phys. C **11**, L927.
Mischler, G., D. Bertrand, D.J. Lockwood, M.G. Cottam and S. Legrand, 1981, J. Phys. C **14**, 945.
Mischler, G., A. Zwick, D.J. Lockwood and S. Legrand, 1982, J. Phys. C **15**, L18.
Mitlehner, H., R. Geick, W. Lehmann, R. Weber, G. Dietrich and H. Schoenherr, 1971, Solid State Commun. **9**, 2059.
Moch, P., G. Parisot, R.E. Dietz and H.J. Guggenheim, 1969, in: Light Scattering Spectra of Solids, ed. G.B. Wright (Springer-Verlag, New York) p. 231.

Moch, P., J.P. Gosso and C. Dugautier, 1971, in: Light Scattering in Solids, ed. M. Balkanski (Flammarion, Paris) p. 138.
Montarroyos, E., Cid B de Araujo and S.M. Rezende, 1979, J. Appl. Phys. **50**, 2033.
Nickel, B.G., and W.H. Butler, 1973, Phys. Rev. Lett. **30**, 373.
Oseroff, A., and P.S. Pershan, 1969, in: Light Scattering Spectra of Solids, ed. G.B. Wright (Springer-Verlag, New York) p. 223.
Parisot, G., R.E. Dietz, H.J. Guggenheim, P. Moch and C. Dugautier, 1971, J. Phys. (France) **32**, C1–803.
Prokhorov, A.S., and E.G. Rudashevskii, 1975, JETP Lett. **22**, 99.
Ramdas, A.K., and S. Rodriguez, 1987, Semiconductors and Semimetals (Academic Press, New York) to be published.
Richards, P.L., 1963, J. Appl. Phys. **34**, 1237.
Sanders, R.W., S.M. Rezende, M. Motokawa, R.M. Belanger and V. Jaccarino, 1980, J. Magn. & Magn. Mater. **15–18**, 725.
Shante, V.K.S., and S. Kirkpatrick, 1971, Adv. Phys. **20**, 325.
Soven, P., 1967, Phys. Rev. **156**, 809.
Svensson, E.C., S.M. Kim, W.J.L. Buyers, S. Rolandson, R.A. Cowley and D.A. Jones, 1975, AIP Conf. Proc. **24**, 161.
Tahir-Kheli, R.A., 1972, Phys. Rev. B **6**, 2808.
Tahir-Kheli, R.A., and A.R. McGurn, 1978, Phys. Rev. B **18**, 503; 1978, J. Phys. C **11**, 1413.
Tahir-Kheli, R.A., T. Fujiwara and R.J. Elliott, 1978, J. Phys. C **11**, 497.
Taylor, D.W., 1967, Phys. Rev. **156**, 1017.
Tennant, W.E., and P.L. Richards, 1977, J. Phys. C **10**, L365.
Thorpe, M.F., and R. Alben, 1976, J. Phys. C **9**, 2555.
Tinkham, M., 1970, in: Far Infrared Properties of Solids, eds S.S. Mitra and S. Nudleman (Plenum Press, New York) p. 196.
Tonegawa, T., 1968, Prog. Theor. Phys. **40**, 1195.
Tuchendler, J., J. Magarino, A.R. Fert and D. Bertrand, 1978, Solid State Commun. **27**, 1123.
Tuchendler, J., J. Magarino, D. Bertrand and A.R. Fert, 1980, J. Phys. C **13**, 233.
Wada, K., and T. Ishikawa, 1979, J. Phys. Soc. Jpn. **47**, 95.
Wagner, V., D. Tocchetti and B. Hennion, 1974, Verh. DP.G.(VI) **9**, 645.
Weber, R., 1969, Z. Phys. **223**, 299.
Werthein, G.K., D.N.E. Buchanan and H.J. Guggenheim, 1966, Phys. Rev. **152**, 527.
Wilkinson, M.K., J.W. Cable, E.O. Wolland and W.C. Koehler, 1957, Oak Ridge National Lab. Report ORNL **2430**, 65.
Wilkinson, M.K., J.W. Cable, E.C. Wollan and W.C. Koehler, 1959, Phys. Rev. **113**, 497.
Wiltshire, M.C.K., 1977, J. Phys. C **10**, L37.
Wiltshire, M.C.K., 1979, J. Phys. C **12**, 3571.
Wiltshire, M.C.K., 1981, Solid State Commun. **38**, 803.
Wiltshire, M.C.K., and C.H. Burton, 1982, J. Phys. C **15**, 5649.
Wiltshire, M.C.K., and W. Hayes, 1978, J. Phys. C **11**, 3701.
Wong, P., P.M. Horn, R.J. Birgeneau, C.R. Safinya and G. Shirane, 1980, Phys. Rev. Lett. **45**, 1974.
Wood, T.E., and P. Day, 1977, J. Phys. C **10**, L333.
Wood, T.E., and P. Day, 1980, J. Magn. & Magn. Mater. **15–18**, 782.
Ziman, T.A.L., and R.J. Elliott, 1980, J. Phys. C **13**, 845.

CHAPTER 5

Spectroscopy of Excitons in Disordered Molecular Crystals

Emmanuel I. RASHBA

L.D. Landau Institute for Theoretical Physics
of the Academy of Sciences of the USSR
117334 Moscow, USSR

In memory of
Acad. I.V. Obreimov (1894–1981)
who pioneered this field

Optical Properties of Mixed Crystals
Edited by
R.J. Elliott and I.P. Ipatova

Contents

1. Introduction 217
2. Perfect molecular crystals 218
   2.1. Excitons in a rigid lattice 219
   2.2. Energy bands of excitons and spectroscopy of excitons in perfect crystals 222
   2.3. Excitons: how do we discern them? 228
3. Exciton spectra of doped crystals 228
   3.1. Monomeric isotopic centres 229
   3.2. Aggregate impurity centres 236
   3.3. Vibronic spectra of monomer centres 238
4. Exciton spectra of mixed crystals 242
   4.1. Stability of multiplet structure in the spectra of mixed crystals 242
   4.2. Approximation of average amplitudes and classification of spectra 243
   4.3. Optical and energy spectra of mixed crystals 249
   4.4. Vibronic spectra of mixed crystals 252
5. Reconstruction of exciton band structure. Band-to-band transitions (BTBT) 254
   5.1. Vibronic band-to-band transitions (VBTBT) 254
   5.2. Spin band-to-band transitions (SBTBT) 257
6. Two-level systems. Coherence and dephasing 260
   6.1. Dynamics of quasi-spin 261
   6.2. Optical detection of coherence 263
   6.3. Quenching of the influence of inhomogeneous broadening 265
   6.4. Dimers or "mini-excitons" 268
   6.5. Trapping and inter-impurity transitions 271
7. Band states: transport, localization, coherence 273
   7.1. General definitions 273
   7.2. Critical phenomena in exciton luminescence 276
   7.3. Characteristic times 282
8. Vibrational excitons 286
9. Conclusions 288
List of abbreviations 290
References 291

1. Introduction

This chapter is entirely devoted to molecular excitons in organic crystals. It is well known that there are three basic exciton models: the Frenkel exciton, the Wannier–Mott exciton, and the charge transfer exciton. By the Frenkel exciton (1931) is meant an intra-atomic or intramolecular excitation propagating through the crystal as a wave. The Wannier–Mott exciton is a hydrogen-like formation built-up from an electron and a hole whose radius considerably exceeds the lattice spacing. The charge transfer exciton constitutes an intermediate between these two limiting cases, in that the states with an electron and a hole localized at neighbouring sites make a substantial contribution to its wave function.

The Frenkel model describes lower electron-excited states of crystals consisting of large organic molecules (benzene, naphthalene, etc.) perfectly well, much better than it describes the excitons in solid rare gases for which it was originally formulated. It turned out afterwards that in the latter case the ground state of excitons is better described by the charge transfer model. Therefore the Frenkel exciton has gradually become a synonym of the molecular exciton. The low-temperature spectroscopy of organic molecular crystals grew out of classical works of Obreimov and de Haas (Obreimov 1927, Obreimov and de Haas 1928) and Pringsheim and Kronenberger (Kronenberger and Pringsheim 1926, Kronenberger 1930). Partly in these papers, and to a large extent in studies carried out in the following 20 years by Obreimov, Prikhot'ko and their collaborators: (i) a genetic connection between the spectra of molecules and corresponding molecular crystals was established, (ii) in the spectra of crystals a large number of narrow lines inexplicable within the scope of the Bloch band scheme was found and investigated and (iii) the existence of the bands polarized parallel to the crystal axes and not to the molecule axes was discovered. The first two discoveries stimulated Frenkel's studies (1931), and the third one, Davydov's proposal of exciton multiplets (Davydov 1948).

It should be emphasized from the very beginning that the study of disordered systems plays a special, extremely important role in the physics of molecular excitons. To elucidate this, we would like to point out that in the physics of semiconductors the band structure is usually determined for perfect crystals, and it is at the subsequent stage that one establishes the changes in the energy spectrum resulting from the presence of impurities or defects: impurity levels, impurity bands, tails of the density of states, etc. This can be done owing to the availability of such powerful methods for determining the band structure such as the cyclotron resonance, the oscillatory effects in transport phenomena, etc. The physics of molecular excitons does not have such methods at its disposal. Here, the investigation of the optical spectra of imperfect molecular crystals is a major instrument in reconstructing the band spectrum of excitons in perfect molecular crystals. Below we shall briefly show how this is done practically.

This review is closely related to the book on spectroscopy of molecular excitons by Broude et al. (1981) which also gives much attention to the properties of disordered crystals. A number of problems presented in sections 3 and 4 of this review are treated in that book in considerably more detail. In this connection, it should be noted right from the start that I have endeavoured to write this review in quite another style from that in the book by Broude et al. (1981). That book was addressed to experts in the field of molecular excitons and therefore comprised a large body of specific information (experimental as well as theoretical). Besides, it was deliberately limited to a range of questions most closely connected with the band structure of the singlet exciton spectrum. The present review is addressed mainly to those physicists who are involved in studies of disordered systems, who are not experts in the field of molecular excitons, but in other branches. That is why I have tried to avoid details and have tried from a single point of view to create the most generalized picture of progress in the domain of the spectroscopy of molecular excitons in imperfect crystals. Also for this purpose, the range of questions under discussion has been considerably extended; in particular, triplet excitons and the problem of coherence and dephasing are introduced and mention is made of some of the latest advances in the physics of molecular excitons.

The review is written so that one can read it without consulting other literature extensively. Nevertheless, it would be useful to point out that general information on molecular excitons can be found in the well-known book by Davydov (1968) and more specific information in numerous reviews. Many up-to-date reviews can be found in volume 4 (edited by Agranovich and Hochstrasser 1983) of the present series.

2. *Perfect molecular crystals*

This section contains basic information on the optical and energy spectra of electronic molecular excitons in perfect crystals, the main features permitting the exciton absorption to be recognized, as well as some experimental data providing a means for evaluating the quality of crystals under study and the reliability of the techniques employed.

2.1. Excitons in a rigid lattice

Here we shall limit ourselves to electronic excitons in a rigid lattice, i.e. we shall consider only electronic (but not vibronic) intramolecular excitations, omitting lattice (external) phonons (LP) related to the degrees of freedom of the motion of molecules as a whole (translations and rotations). In such a manner we shall briefly set forth Davydov's theory (1948) in the form best suited for our purpose and at the same time we will introduce the necessary definitions. The Hamiltonian in the second quantization representation is written as

$$H = H_{\mathrm{e}} + H_{\mathrm{ex}}, \tag{2.1}$$

where

$$H_{\mathrm{e}} = E_{\rho} \sum_{\boldsymbol{n}\alpha} a^{+}_{\boldsymbol{n}\alpha} a_{\boldsymbol{n}\alpha}, \qquad E_{\rho} = E_{\mathrm{mol}} + D, \tag{2.2}$$

$$H_{\mathrm{ex}} = \sum_{\boldsymbol{n}\alpha \neq \boldsymbol{m}\beta} M_{\boldsymbol{n}\alpha \boldsymbol{m}\beta} a^{+}_{\boldsymbol{n}\alpha} a_{\boldsymbol{m}\beta}, \quad M_{\boldsymbol{n}\alpha \boldsymbol{m}\beta} = M_{\alpha\beta}(\boldsymbol{n} - \boldsymbol{m}). \tag{2.3}$$

Here H_{e} is the intramolecular Hamiltonian, E_{mol} is the excitation energy of the molecule in the gas, D is the gas-to-crystal energy shift, and H_{ex} is the Hamiltonian describing excitation transfer between the lattice sites. Index $\boldsymbol{n}$ numbers primitive cells, and α numbers different positions within a primitive cell. The molecular states are assumed to be nondegenerate. Small changes which should be introduced for triplet excitons are not essential, for the time being. The molecules $\boldsymbol{n}\alpha$ and $\boldsymbol{m}\alpha$ are termed translationally equivalent. The molecules $\boldsymbol{n}\alpha$ and $\boldsymbol{m}\beta$ with $\alpha \neq \beta$ are termed symmetrically equivalent if they can be interconverted with one of the transformations of the crystal space group. They are shown schematically in fig. 1 for a naphthalene crystal. For simplicity, we assume all molecules in the crystal to be symmetrically equivalent. Hamiltonian (2.1) is diagonalized by transformation to the momentum representation:

$$|\mu \boldsymbol{k}\rangle = \sum_{\boldsymbol{n}\alpha} \psi_{\mu \boldsymbol{k}}(\boldsymbol{n}\alpha) \, |\boldsymbol{n}\alpha\rangle, \qquad \psi_{\mu \boldsymbol{k}}(\boldsymbol{n}\alpha) = N^{-1/2} B_{\alpha}(\mu \boldsymbol{k}) \, \mathrm{e}^{\mathrm{i}\boldsymbol{k}\boldsymbol{n}_{\alpha}}. \tag{2.4}$$

The index μ numbers exciton bands, N is the number of cells in the normalization volume. The Schrödinger equation, determining $B_{\alpha}(\mu \boldsymbol{k})$ is reduced to

$$\sum_{\beta} L_{\alpha\beta}(\boldsymbol{k}) B_{\beta}(\boldsymbol{k}) = \epsilon_{\mu}(\boldsymbol{k}) B_{\alpha}(\boldsymbol{k}), \tag{2.5}$$

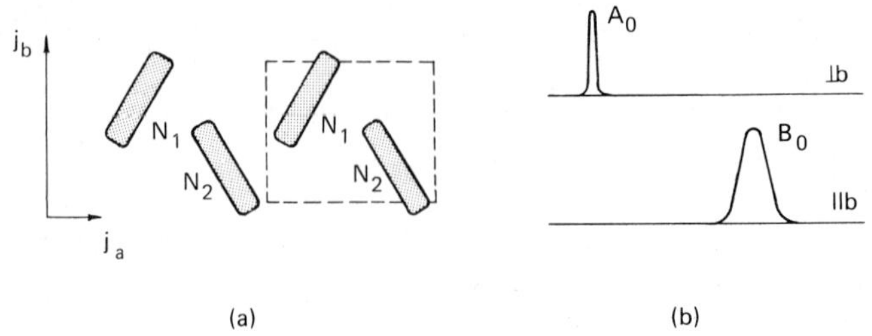

Fig. 1. Schematic arrangement of molecules in crystals of the naphthalene type (a) and the exciton absorption singlet spectrum of a naphthalene crystal (b). The molecules are shown as a projection on a plane *ab*. The dashed line shows a primitive cell. Currents j_a and j_b can be considered as corresponding to normal modes of the elementary cell. The absorption bands A_0 and B_0 form a singlet exciton doublet, the distance between them is the Davydov splitting. All N_1 (or N_2) molecules are translationally equivalent. The sublattices of the N_1 and N_2 molecules are symmetrically equivalent.

where

$$L_{\alpha\beta}(\boldsymbol{k}) = \sum_{\boldsymbol{n}-\boldsymbol{m}} M_{\boldsymbol{n}\alpha\boldsymbol{m}\beta} \exp\left[-\mathrm{i}\boldsymbol{k}(\boldsymbol{n}_\alpha - \boldsymbol{m}_\beta)\right], \tag{2.6}$$

and the total energy $E_\mu(\boldsymbol{k})$ is expressed in terms of $\epsilon_\mu(\boldsymbol{k})$ as

$$E_\mu(\boldsymbol{k}) = E_\rho + \epsilon_\mu(\boldsymbol{k}). \tag{2.7}$$

Equation (2.5) determines Z exciton bands, i.e. Z dispersion law branches $\epsilon_\mu(\boldsymbol{k})$, from the number of molecules in the primitive cell. Since, according to (2.3), the matrix elements of excitation transfer $M_{\boldsymbol{n}\alpha\boldsymbol{m}\beta}$ are zero when $\boldsymbol{n}\alpha = \boldsymbol{m}\beta$, then the trace of this matrix is zero. Consequently,

$$\sum_\mu \int \frac{\mathrm{d}^3k}{(2\pi)^3} \epsilon_\mu(\boldsymbol{k}) = 0 \tag{2.8}$$

and E_ρ is the centre of gravity of the exciton energy spectrum.

The perfect crystal Green function G^0 in the momentum representation is

$$G^0_{\alpha\beta}(\omega\boldsymbol{k}) = \sum_\mu \frac{B_\alpha(\mu\boldsymbol{k}) B^*_\beta(\mu\boldsymbol{k})}{\omega - \epsilon_\mu(\boldsymbol{k}) + \mathrm{i}0} \tag{2.9}$$

and in the site representation

$$G^0_{\boldsymbol{n}\alpha\boldsymbol{m}\beta}(\omega) = \frac{1}{N} \sum_{\boldsymbol{k}} G^0_{\alpha\beta}(\omega\boldsymbol{k}) \exp\left[\mathrm{i}\boldsymbol{k}(\boldsymbol{n}_\alpha - \boldsymbol{m}_\beta)\right]. \tag{2.10}$$

Here and everywhere below, the frequency is measured in energy units ($\hbar = 1$). By analogy with (2.7), ω is related to the total frequency Ω by the relation $\Omega = E_\rho + \omega$. The function

$$S_{\alpha\beta}(\omega \boldsymbol{k}) = -\operatorname{Im} G^0_{\alpha\beta}(\omega \boldsymbol{k}) \tag{2.11}$$

which is called the spectral function, or spectral density, is determined through G^0. In terms of this function we can express the density of states (DOS) in the energy spectrum of the exciton

$$\rho_0(\omega) = \frac{v}{\pi Z} \sum_\alpha \int \frac{\mathrm{d}^3 k}{(2\pi)^3} S_{\alpha\alpha}(\omega \boldsymbol{k})$$

$$= \frac{v}{Z} \sum_\mu \int \frac{\mathrm{d}^3 k}{(2\pi)^3} \delta(\omega - \epsilon_\mu(\boldsymbol{k})). \tag{2.12}$$

Here v is the primitive cell volume.

If light absorption is weak and can be described in terms of perturbation theory, then the tensor of complex conductivity is

$$\hat{\sigma}_c(\omega) = -\frac{1}{\Omega v} \operatorname{Im} \sum_{\substack{\alpha\beta \\ \boldsymbol{n}-\boldsymbol{m}}} G^0_{\boldsymbol{n}\alpha\boldsymbol{m}\beta}(\omega) \boldsymbol{j}_\alpha \otimes \boldsymbol{j}^*_\beta$$

$$= -\frac{1}{\Omega v} \operatorname{Im} \sum_{\alpha\beta} G^0_{\alpha\beta}(\omega, \boldsymbol{k} = 0) \boldsymbol{j}_\alpha \otimes \boldsymbol{j}^*_\beta . \tag{2.13}$$

The real part of the conductivity tensor is

$$\hat{\sigma}(\omega) = \frac{\pi}{\Omega v} \sum_\mu \boldsymbol{j}_\mu \otimes \boldsymbol{j}^*_\mu \delta(\omega - \epsilon_\mu), \qquad \epsilon_\mu \equiv \epsilon_\mu(\boldsymbol{k} = 0). \tag{2.14}$$

Here

$$\boldsymbol{j}_\mu = \sum_\alpha B_\alpha(\mu, \boldsymbol{k} = 0) \boldsymbol{j}_\alpha, \tag{2.15}$$

$\boldsymbol{j}_\alpha$ are matrix elements of the current operator for the intramolecular transition and the symbol $\otimes$, as usual, denotes the direct (the Kronecker) product of vectors. In writing eqs. (2.13) and (2.14) we have accounted for the fact that due to momentum conservation we can take $\boldsymbol{k} = 0$ in $\hat{\sigma}_c$, since the momentum $\boldsymbol{Q}$ of the photon creating an exciton is small compared with the Brillouin momentum $k_{\mathrm{Br}} \sim \pi/a$ (a is the lattice constant) which determines the scale of Brillouin zones.

The basic conclusions of Davydov's theory follow from the above:

(a) Each nondegenerate intramolecular level in the crystal gives rise to Z levels with $\boldsymbol{k} = 0$, forming an exciton multiplet.

(b) The polarization of those optical transitions to these levels which are allowed is determined by vectors $\boldsymbol{j}_\mu$. Therefore the transitions are polarized along the crystallographic and not the molecular axes.

(c) The allowed transitions can be determined by group theory methods.

This phenomenon is called the Davydov splitting.

A two-dimensional scheme of crystals with the naphthalene ($C_{10}H_8$) lattice may serve as the simplest illustration. This crystal belongs to a monoclinic crystal system with two molecules in a primitive cell. Their projections onto the cleavage plane ab of the crystal (b is the monoclinic axis) are shown schematically in fig. 1. The same figure shows the exciton absorption spectrum: A_0–B_0 is the doublet of sharply polarized bands discovered by Prikhot'ko (1944, 1949). Individual components of an exciton multiplet are usually denoted by capital letters corresponding to the crystallographic axes along which they are polarized. The bands polarized along the crystal axes are called sharply polarized, and the bands present in all the components of the spectrum are called weakly polarized.

2.2. *Energy bands of excitons and spectroscopy of excitons in perfect crystals*

The energy bands of excitons can be easily determined from eqs. (2.5) and (2.6), if the matrix elements of excitation transfer $M_{\boldsymbol{n}\alpha\boldsymbol{m}\beta}$ are known. It is the task of finding the set $M_{\boldsymbol{n}\alpha\boldsymbol{m}\beta}$ that involves the main difficulties. For singlet excitons, the excitation transfer from site to site is not, as a rule, associated with the electron transfer. For strong allowed transitions, the main contribution to M must be made by the dipole–dipole interaction. It can be estimated as $M \sim \boldsymbol{d}^2/a^3 \sim \boldsymbol{d}^2/v$, where $\boldsymbol{d}$ is the dipole moment of the transition. The value of $\boldsymbol{d}$ can be conveniently determined from the oscillator strength f; $f = 2m\Omega\boldsymbol{d}^2/\hbar e^2$, where e and m are the electron charge and mass. For strong transitions $f \sim 1$; assuming $\Omega \approx 4$ eV, we obtain $d \approx f^{1/2}e$ Å. If we take $a^3 \approx 500$ Å^3 which is a usual estimate for organic molecules of the naphthalene type, then $M \approx 0.03f$ eV $\approx 250\ f$ cm^{-1}. For the naphthalene lattice the number of nearest neighbours $Z = 4$, and the exciton band width $2\mathfrak{M} \approx 2000f$ cm^{-1}. For instance, $f \approx 0.3$ and $2\mathfrak{M} \approx 600$ cm^{-1} for anthracene ($C_{14}H_{10}$) possessing a similar lattice. This estimate agrees with recent experimental results ($2\mathfrak{M} \approx 350$–$700$ cm^{-1}). For naphthalene $f \approx 4 \times 10^{-3}$, but $2\mathfrak{M} \approx 160$ cm^{-1}, i.e. not much less than in anthracene. In naphthalene, as in other crystals with small f, the bands are formed due to higher multipoles, e.g. octupole–octupole interaction. Calculations of bands from first principles do not prove satisfactory. All the existing results have been obtained by reconstructing the bands from experimental data. The mean band width is $2\mathfrak{M} \sim 100$

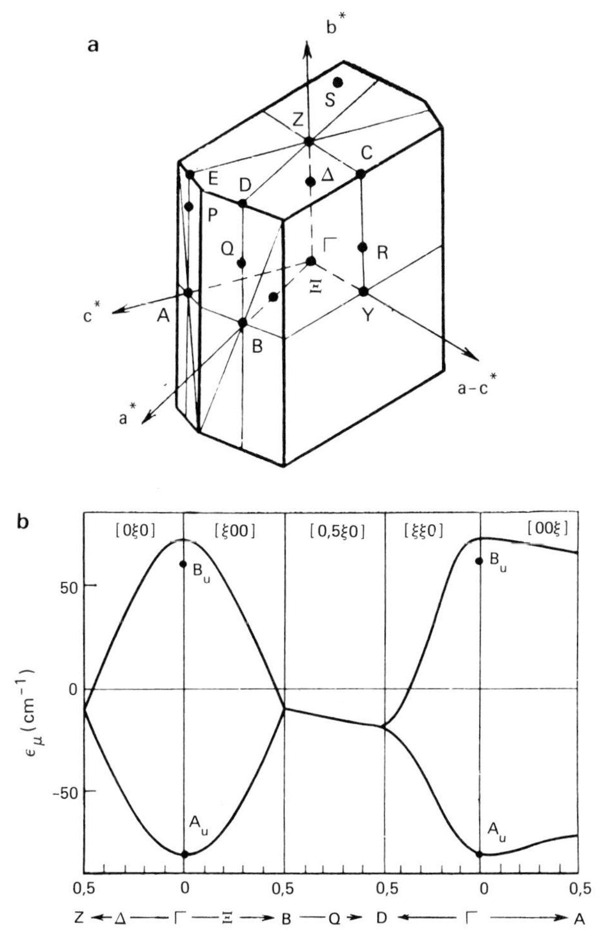

Fig. 2. Brillouin zone of crystals of the naphthalene type (a) and dispersion law of singlet excitons in crystalline naphthalene (b). Points A_u and B_u show the position of centres of gravity of the A_0 and B_0 bands (Broude et al. 1981).

cm^{-1}. By way of example, fig. 2 shows the energy bands of singlet excitons in naphthalene.

The energy bands of triplet excitons result from exchange interactions. Consequently, they are proportional to the overlap integrals of molecular wave functions and, hence, relatively small. The band widths measured experimentally are $2\mathfrak{M} \approx 1–30\ cm^{-1}$. Figure 3 shows the exciton bands of triplet excitons in anthracene.

Homogeneous exterior fields affect the energy spectrum of excitons. The changes are particularly large when the symmetric equivalence of molecules

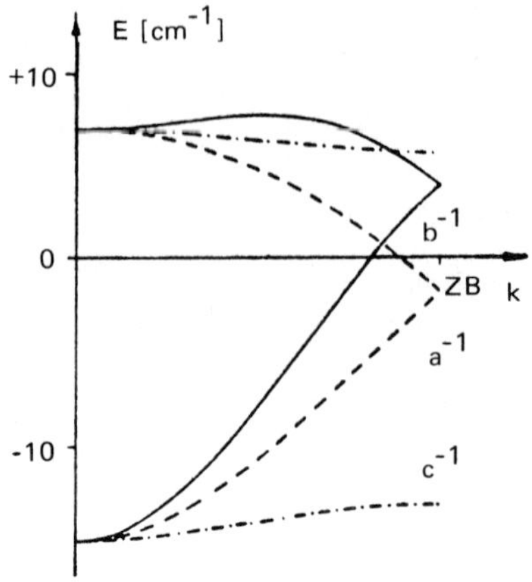

Fig. 3. Dispersion law for triplet excitons in anthracene (Port et al. 1981).

belonging to different sublattices is violated. Under the conditions where the level shift in the outer field exceeds the magnitude of the matrix elements of the excitation transfer, the energy spectrum can undergo qualitative changes. For instance, the exciton dispersion law may convert from a quasi-two-dimensional to a quasi-one-dimensional one, and this, in its turn, may lead to dramatic changes in transport phenomena. The role of such external fields can be played by either directional homogeneous deformations or electric fields. Since both these problems have been treated in recent reviews (Sugakov 1982, Hanson 1983a, b), we do not dwell on them at greater length. Note only, that in spatially inhomogeneous fields there appears an additional mechanism of exciton transport: along with diffusion, a drift arises in the direction of the field gradient (Gribnikov and Rashba 1958).

Formulae (2.5), (2.6) for $\epsilon_\mu(\boldsymbol{k})$ do not take into account the retardation effects which manifest themselves predominantly in strong exciton transitions. As is well known, the retardation of the electromagnetic interactions leads to a qualitative change in the energy spectrum of excitons at $k \sim Q \ll k_{\mathrm{Br}}$ and to formation of polaritons. The polaritons (Pekar 1957, Hopfield 1958, Ginzburg 1958, Agranovich 1959; in ionic lattice dynamics Tolpygo 1950, Huang 1951) are mixed waves with dispersion law $\kappa(\Omega) = (ck/\Omega)^2$, where $\kappa(\Omega)$ is the dielectric susceptibility near exciton resonance. They do not decay if scattering by impurities, defects, and phonons does not occur, that is, they can be observed in perfect crystals at low temperature. The absence of the decay of polaritons means the absence of light absorption. The polaritons, however, make a conventional contribution to the refraction index depending on the oscillator strength of the exciton transition. This suggests that for polaritons the Kramers–Kronig dispersion relations are violated. The validity of these relations is gradually re-established as the scattering of polaritons by phonons

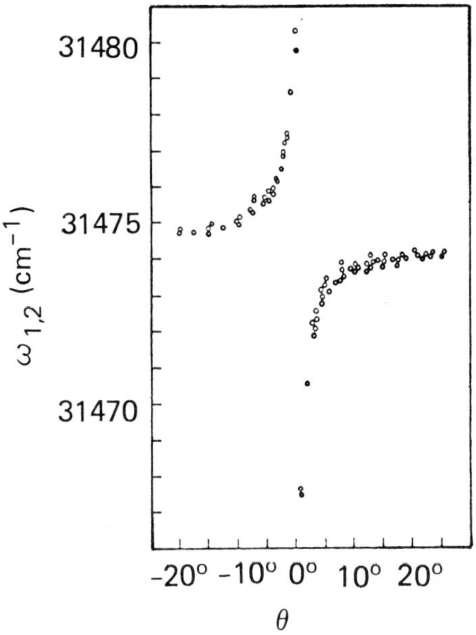

Fig. 4. Two polariton branches of the singlet A_0 exciton in naphthalene, The quantity θ is proportional to the wave vector of the polariton (Stevenson and Small 1983).

and lattice defects increases. These points are discussed at length in books by Agranovich and Ginzburg (1979) and Pekar (1982).

Here, we restrict ourselves to presenting the experimental data recently obtained on pure unstressed naphthalene crystals in which polariton phenomena appear in the vicinity of a very weak ($f \sim 10^{-5}$) singlet A_0 band. They were first observed by Hochstrasser and Meredith (1977) in the three-wave process where, under conditions of phase matching, two low-frequency polaritons fuse into one polariton with frequency close to Ω_{A_0}. Two branches of the dispersion law for the polariton presented in fig. 4 are obtained in different ways: the upper branch from the generation of the second harmonic, the lower one, from the two-photon excitation of fluorescence (Stevenson and Small 1983). The temperature dependence of the signals corresponding to the upper branch is shown in fig. 5. They characterize the temperature dependence of the relaxation rate of a high-frequency polariton produced by fusion of two low-frequency polaritons: the second-harmonic generation decreases monotonically with increasing temperature T, while the intensity of fluorescence increases at first. With a further increase in T both the curves smear out rapidly. Another manifestation of the polariton mechanism of absorption is the drastic growth of the integrated intensity of the A_0 band observed by Robinette and Small (1976) when the temperature is increased from 4 to 20 K.

The degree of sample perfection and spectral resolution attained at present

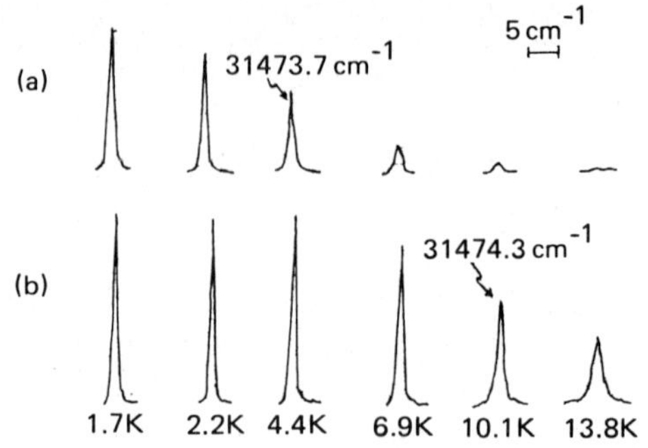

Fig. 5. Temperature dependence of the second harmonic generation (a) and two-photon fluorescence (b) for the upper polariton branch of the singlet A_0 exciton in naphthalene. The intensities are arbitrary, but their general scale is maintained (Stevenson et al. 1981).

can be seen in fig. 6 which presents the low-frequency ${}^{T}B_0$ band of a triplet exciton in naphthalene (Doberer et al. 1982). The width of this band is exceedingly small in N-h_8. The introduction of a small amount of an isotopic impurity results in a strong broadening of the band, and the considerably larger band width of perdeuterated naphthalene N-d_8 can be explained by the presence of trace amounts of lighter isotopic molecules.

The last point to be briefly discussed here is the shape of the exciton light absorption curve. In perfect crystals this absorption is due to an exciton–pho-

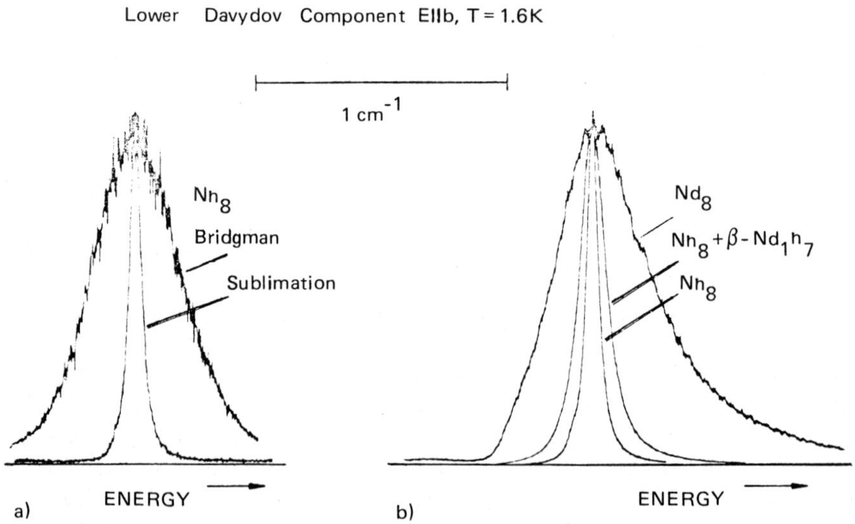

Fig. 6. Comparison of the shape of the long-wave component ${}^{T}B_0$ of the triplet exciton in naphthalene in different samples: (a) in sublimation and Bridgman samples of N-h_8, (b) in sublimation samples of N-h_8, isotopically doped by impurity β-N-d_1h_7 (2%) and in N-d_8 samples free from special doping. The results are reduced to a unified energy scale (Doberer et al. 1982).

non interaction. The interaction, linear in phonon amplitudes, can be written in the general case as

$$H_{\text{ex-ph}} = \frac{1}{V^{1/2}} \sum_{k\mu\mu'} \sum_{qs} g_{\mu\mu' s}(\boldsymbol{k}, \boldsymbol{q}) a^{+}_{k\mu} a_{k-q\mu'} \left(b_{qs} + b^{+}_{-qs} \right), \tag{2.16}$$

where b_{qs} are operators of phonon annihilation, s numbers the phonon branches, V is the normalization volume. Equation (2.16) is obtained by expanding D and $M_{n\alpha m\beta}$ of eqs. (2.2) and (2.3) into molecular displacements, so that g is the sum of corresponding contributions:

$$g_{\mu\mu' s}(\boldsymbol{k}, \boldsymbol{q}) = g^{D}_{\mu\mu' s}(\boldsymbol{q}) + g^{M}_{\mu\mu' s}(\boldsymbol{k}, \boldsymbol{q}). \tag{2.17}$$

It is easily seen that the first term which normally dominates is only dependent on the momentum transfer $\boldsymbol{q}$, but not on the exciton momentum $\boldsymbol{k}$. Under weak exciton–phonon interaction conditions (i.e. neglecting the polariton effects) the theory of such absorption is very simple: in eq. (2.13) G^0 should be substituted by the exact retarded Green function G:

$$G^{-1} = (G^0)^{-1} - \Sigma(\omega, \boldsymbol{k} = 0), \tag{2.18}$$

where $\Sigma(\omega, \boldsymbol{k})$ is the mass operator (or self-energy) of the exciton, calculated in the lowest order of perturbation theory. For a weak exciton–phonon coupling, the same equation can also be deduced from the Weisskopf–Wigner method (1930) if the frequency dependence of the decay constant $\Gamma(\omega)$ is taken into account. Although such formulae for vibrational and electronic excitons have long been known (e.g. Blackman 1933, Lysenko 1936, Davydov and Rashba 1957, Toyozawa 1958, Pekar 1959), they are, as a rule, hardly effective when applied to molecular crystals. Indeed, the only general conclusion is to the effect that, if the point $\boldsymbol{k} = 0$ lies in the middle part of an exciton band, the absorption band is an asymmetrical Lorentz curve. All the other conclusions are dependent upon the dimensionality of the system and the details of the interactions. Their application is hampered by the presence of a large number of vibrational modes and by the fact that the coefficients $g_{\mu\mu' s}$ are practically unknown. Some deformation potentials for naphthalene have been determined only recently (Ostapenko et al. 1978, 1979, Sugakov 1982). Besides, eq. (2.18) has been used for the reconstruction of Im $\Sigma(\omega, \boldsymbol{k} = 0)$ from the absorption measured experimentally (Sheka 1975). We shall revert to this question in section 7.3. For present purposes, suffice it to say that, as is evidenced by the available data, the interaction of excitons with lattice phonons, while being of perceptible magnitude, can still be considered a weak one. Therefore, when studying imperfect crystals, we are going to neglect this interaction whenever possible.

2.3. Excitons: how do we discern them?

The spectra of molecular crystals are very complicated in the exciton region. They comprise a large number of absorption bands of different nature: exciton bands, their satellites which may be assigned to low-dispersion LP groups, the bands of different guests as well as those host molecules which are distorted due to their vicinity to a guest molecule, etc. What criteria can be used then for discerning the exciton absorption bands?

We formulate here the basic criteria:

(1) Polarization along the crystallographic axes.

(2) The presence of Davydov splitting.

(3) The shift of impurity levels near the edge of exciton energy bands ("repulsion") and "disappearance" of the impurity bands falling within the exciton continuum.

(4) The giant anomaly of the polarization and the intensity of the impurity absorption near exciton band edges: the Rashba effect.

(5) Evolution of an exciton multiplet from an impurity band in mixed crystals, or the continuous dependence of the exciton band frequency on composition.

The choice of a criterion depends on the specific properties of the crystal involved. In some cases, one or another of the criteria cannot be successfully applied. The assignment is the more convincing, the larger the number of criteria one manages to use. For instance, the first exciton ever identified corresponds to a narrow line discovered by Scheibe (for a review see Scheibe 1941) in the absorption and fluorescence spectra of an organic polymer, ψ-isocyanine. To recognize the exciton, Franck and Teller (1938) used the polarization of the line along the polymer axis. They also argued in favour of this assignment because in mixed polymers the line moves continuously when the composition is changed (criterion 5, cf. section 4). The Davydov splitting was missing, as $Z=1$. In the case of naphthalene, the first two criteria are met for a A_0–B_0 doublet (singlet transition), but the intensity ratio of these bands differs by a factor of ≈ 400 from that to be expected, considering the direction of optical transition in a free molecule and the orientation of the molecule in a crystal. This anomaly is associated with the effect of the crystal field on a weak intramolecular transition. Up to now, a convincing identification of the A_0–B_0 doublet has been achieved on the strength of criteria 3–5 (see sections 3 and 4).

3. Exciton spectra of doped crystals

According to whether the impurity concentration is high or low, imperfect crystals are classified as mixed or doped crystals. By doped crystals we understand the crystals in which a single guest molecule or several guest molecules (a cluster) are, practically, completely separated by host molecules from other guest molecules (or clusters).

3.1. Monomeric isotopic centres

The simplest impurity centre is an isotopic monomer: it is a center in which one molecule ($\boldsymbol{n}_0\alpha_0$) of the host is substituted by another molecule differing only in isotopic composition (a monomeric centre). In this case, it can be assumed that only one diagonal matrix element is changed in Hamiltonian (2.1)–(2.3). In other words, the perturbation Hamiltonian H_{imp} contains only one term:

$$H_{\text{imp}} = \Delta a^+_{\boldsymbol{n}_0\alpha_0} a_{\boldsymbol{n}_0\alpha_0}, \qquad \Delta = E^{\text{g}}_{\rho} - E^{\text{h}}_{\rho}. \tag{3.1}$$

All $M_{\boldsymbol{n}\alpha\boldsymbol{m}\beta}$ and $\boldsymbol{j}_{\boldsymbol{n}\alpha} = \boldsymbol{j}_\alpha$ remain unchanged, and the function G^0 in (2.13) must be substituted by the Green function G of the perturbed Hamiltonian. It can be written down at once by the method of degenerate perturbations (Lifshitz 1947, Koster and Slater 1954):

$$G_{\boldsymbol{n}\alpha\boldsymbol{m}\beta} = G^0_{\boldsymbol{n}\alpha\boldsymbol{m}\beta} + \frac{\Delta}{1 - \Delta G_0(\omega)} G^0_{\boldsymbol{n}\alpha\boldsymbol{n}_0\alpha_0} G^0_{\boldsymbol{n}_0\alpha_0\boldsymbol{m}\beta}, \tag{3.2}$$

where

$$G_0(\omega) = G^0_{\boldsymbol{n}\alpha\boldsymbol{n}\alpha}(\omega) = \int \frac{\rho_0(\epsilon)}{\omega - \epsilon + \mathrm{i}0} \mathrm{d}\epsilon. \tag{3.3}$$

If Δ is sufficiently large, then a local level splits off the exciton band. Such states are called local or impurity excitons. According to eq. (3.2), the position of the local level, ϵ_i, is determined by the equation

$$\Delta \int \frac{\rho_0(\epsilon)}{\epsilon_i - \epsilon} \mathrm{d}\epsilon = 1. \tag{3.4}$$

In all instances $|\Delta/\epsilon_i| < 1$. This indicates that the level is "repelled" off the band edge. This level always occurs in one- and two-dimensional (1D and 2D) systems and occurs in three-dimensional (3D) systems only when $|\Delta|$ is larger than the critical values $\Delta^{\pm}_{\text{cr}}$ for the splitting of the level off the top or the bottom of the band, respectively. When using a model 3D density of states

$$\rho(\epsilon) = \frac{2}{\pi \mathfrak{M}^2} \sqrt{\mathfrak{M}^2 - \epsilon^2}, \tag{3.5}$$

then $\Delta^{\pm}_{\text{cr}} = \pm \mathfrak{M}/2$.

The local level wave function in site representation is

$$\psi_i(\boldsymbol{n}\alpha) = G^0_{\boldsymbol{n}\alpha\,\boldsymbol{n}_0\alpha_0}(\epsilon_i) \Big/ \left| \frac{\mathrm{d}G_0}{\mathrm{d}\epsilon_i} \right|^{1/2}. \tag{3.6}$$

Its square at an impurity site

$$a^2(\epsilon_i) = |\psi_i(\boldsymbol{n}_0\alpha_0)|^2 = \frac{1}{\Delta^2} \left| \frac{\mathrm{d}G_0}{\mathrm{d}\epsilon_i} \right|^{-1} = \frac{\mathrm{d}\epsilon_i}{\mathrm{d}\Delta}. \tag{3.7}$$

The wave function in momentum representation is

$$\overline{|\psi_i(\mu\boldsymbol{k})|^2} = \frac{1}{NZ} \frac{\{G^0_\mu(\epsilon_i\boldsymbol{k})\}^2}{|\mathrm{d}G_0/\mathrm{d}\epsilon_i|}, \qquad G^0_\mu(\epsilon\boldsymbol{k}) = (\epsilon - \epsilon_\mu(\boldsymbol{k}) + \mathrm{i}0)^{-1}. \tag{3.8}$$

The overbar on the left-hand side of eq. (3.8) denotes the averaging over the distribution of guest molecules in different positions α_0.

Using eqs. (2.13), (3.2) and (3.8) one can find an expression for the conductivity tensor describing the impurity absorption band:

$$\hat{\sigma}_{\mathrm{imp}}(\omega) = \frac{\pi C}{E_\rho v} NZ \sum_\mu \overline{|\psi_i(\mu)|^2} \boldsymbol{j}_\mu \otimes \boldsymbol{j}^*_\mu \delta(\epsilon - \epsilon_i). \tag{3.9}$$

Here $\psi_i(\mu) \equiv \psi_i(\mu, \boldsymbol{k} = 0)$, and C is the fraction of guest-occupied sites. It is convenient to re-write eq. (3.9) in terms of oscillator strengths. Introduce the quantity $\psi_{\mathrm{d}}^2 = (NZ)^{-1}$ which is the limit $|\psi_i(\mu)|^2$ for deep levels, i.e., at $|\Delta| \gg \mathfrak{M}$, as well as oscillator strengths $(f_i)_\mu$ and f_μ for the guest- and host-exciton absorption. The latter quantities are both determined per one (respective) molecule. Then from a comparison of eqs. (2.14) and (3.9) it follows that

$$(f_i)_\mu = \frac{\overline{|\psi_i(\mu)|^2}}{\psi_{\mathrm{d}}^2} f_\mu. \tag{3.10}$$

Finally, introduce the polarization ratio for an impurity band:

$$P^i_{\mu_1/\mu_2} = (f_i)_{\mu_1}/(f_i)_{\mu_2}. \tag{3.11}$$

Then from eqs. (3.8)–(3.10) it follows that

$$P^i_{\mu_1/\mu_2} = \left(\frac{\epsilon_i - \epsilon_{\mu_2}(0)}{\epsilon_i - \epsilon_{\mu_1}(0)} \right)^2 P^0_{\mu_1/\mu_2}, \qquad P^0_{\mu_1/\mu_2} = \frac{f_{\mu_1}}{f_{\mu_2}}. \tag{3.12}$$

Let us now investigate the behaviour of the obtained expressions near the exciton band edge.

For crystals with a 3D spectrum, $\rho_0(\epsilon)$ changes near the band edges ϵ_{ed} as $\rho_0(\epsilon) \propto |\epsilon - \epsilon_{ed}|^{1/2}$, and $G_0(\epsilon)$, according to eq. (3.3), as $G_0(\epsilon) \approx b_1 + b_2 \times |\epsilon - \epsilon_{ed}|^{1/2}$. It follows from eq. (3.7) that near the band edge $a^2(\epsilon_i) \propto |\epsilon - \epsilon_{ed}|^{1/2}$. Naturally, at $|\epsilon_i/\mathfrak{M}| \gg 1$ the value of $a^2(\epsilon) \to 1$; it can be shown that this limit is rapidly reached. The decrease of $a^2(\epsilon_i)$ at $\epsilon_i \to \epsilon_{ed}$ reflects delocalization of the exciton wave function: a large-radius state arises with its centre at the $\boldsymbol{n}_0\alpha_0$ site. From eqs. (3.8) and (3.10) one can see a strong change in $(f_i)_\mu$ in the same limit: the absorption dies down as $|\epsilon - \epsilon_{ed}|^{1/2}$ if ϵ_i approaches the edge of an exciton band to which the transition is forbidden, and flares up as $|\epsilon - \epsilon_{ed}|^{-3/2}$ if ϵ_i approaches that edge of the band to which the transition is allowed.

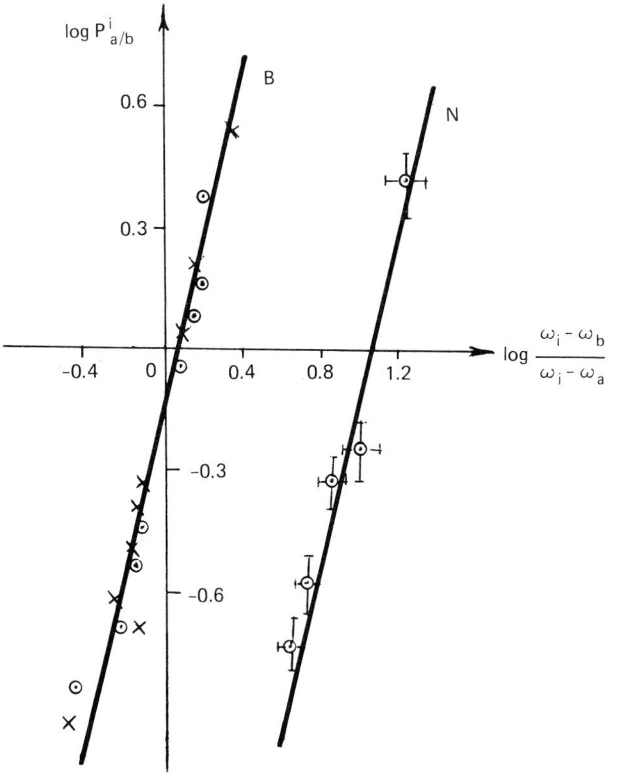

Fig. 7. Polarization ratio of impurity absorption versus impurity absorption frequency ω_i for singlet impurity excitons in benzene and naphthalene. Here ω_a and ω_b are the frequencies of exciton multiplet absorption bands: the A_0 and C_0 bands in benzene and the A_0 and B_0 bands in naphthalene (Sheka 1972).

For the crystals with a 1D energy spectrum, the difference is that $\rho_0(\epsilon) \propto |\epsilon - \epsilon_{ed}|^{-1/2}$. Hence $|dG_0/d\epsilon| \propto |\epsilon - \epsilon_{ed}|^{-3/2}$, i.e. the diminishing of the impurity absorption goes much faster. The suppression of intensity adheres to the law $(f_i)_\mu \propto |\epsilon - \epsilon_{ed}|^{3/2}$ and its enhancement obeys the law $(f_i)_\mu \propto |\epsilon - \epsilon_{ed}|^{-1/2}$.

Equation (3.12) is valid for any dimensionality.

The enormous change in the impurity absorption intensity described above (Rashba 1957, 1962) is known as the Rashba effect. It was first observed by Broude et al. (1961) for isotopic guests in naphthalene. It can be shown, however (Rashba 1957), that this phenomenon is of a general nature and is valid for impurity centres of a general type. A number of examples will be given in section 3.2, while here we shall only briefly describe some experimental data on monomeric centres.

The possibility of experimentally testing the theory implies the possibility of a continuous or quasi-continuous change in Δ, the only parameter of monomeric centres. That the latter possibility actually exists is due to the fact that in organic molecules consisting of a large number of atoms, isotope

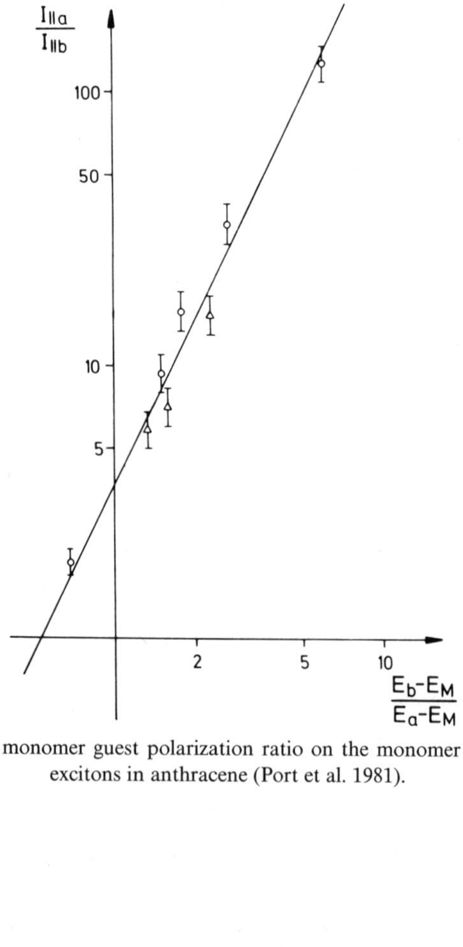

Fig. 8. Dependence of monomer guest polarization ratio on the monomer energy E_M for triplet excitons in anthracene (Port et al. 1981).

substitution is possible for a different number of atoms which may occupy, moreover, different positions. As a result, the isotope shift can be varied by close spacing. It is also essential that the range of Δ variation should be comparable with $\mathfrak{M}$. In this respect, the singlet excitons in naphthalene are a happy choice. The mean isotope shift by one deuterium atom is $\Delta_1 \approx 15\ \text{cm}^{-1}$, the number of hydrogen atoms in the molecule equals 8, so that the maximum shift $\Delta_{max} = 8\Delta_1 \approx 120\ \text{cm}^{-1}$. Hence $\Delta_1 < \mathfrak{M} < \Delta_{max}$.

The square law (3.12) has been tested for singlet excitons in benzene (Dolganov and Sheka 1971) and naphthalene (Sheka 1963a) as well as for triplet excitons in anthracene (Port et al. 1981). Their results are presented in figs. 7 and 8; the agreement between theory and experiment appears convincing. It is interesting that for $\epsilon_{ed} - \epsilon_i = 5\ \text{cm}^{-1}$ in anthracene they observed $P^i = 140$ with $P^0 = 3.6$.

To find the position of impurity bands $\epsilon_i = \epsilon_i(\Delta)$, it is necessary, according to eq. (3.4), to know DOS, $\rho_0(\epsilon)$, for a perfect crystal. Rabin'kina et al. (1970) have chosen the function $\rho_0(\epsilon)$ for the singlet excitons in naphthalene so as to get the best fit with the experimentally determined dependence $\epsilon_i = \epsilon_i(\Delta)$. In so doing, the form of $\rho_0(\epsilon)$ which was obtained (see section 5.1) by Colson et al. (1968) was taken into account. The results are shown in fig. 9.

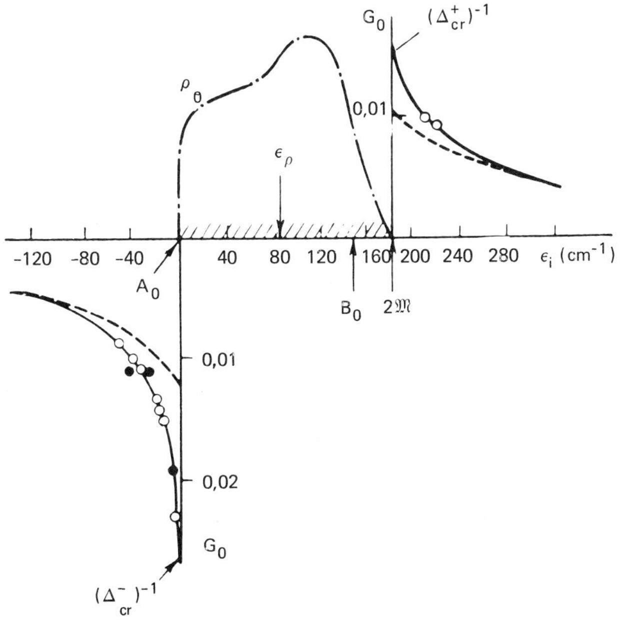

Fig. 9. The experimental (open circles: Sheka 1970; filled circles: Ochs et al. 1974a) and theoretical (solid curve) dependence of Δ^{-1} on the position ϵ_i of the monomer band with respect to the bottom of the exciton band. ρ_0 in DOS, ϵ_p is its centre of gravity. A_0 and B_0 are an exciton doublet (Broude et al. 1981).

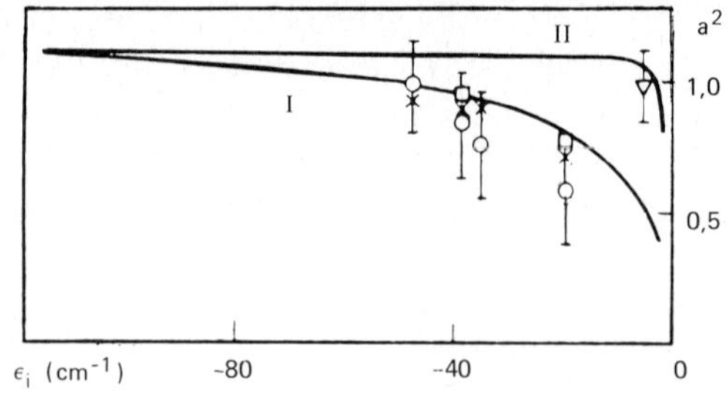

Fig. 10. a^2 as a function of the distance ϵ_i of the local exciton level to the bottom of the singlet exciton band in naphthalene (I) and benzene (II). ○ and × are the experimental data of Broude et al. (1965) and Ochs et al. (1974a), respectively. The full curve for naphthalene is obtained from ρ_0, shown in fig. 9. The points □ are obtained by graphic differentiation of the experimental dependence $\epsilon_i(\Delta)$ (see eq. (3.7)) (Broude et al. 1981).

The dependence of a^2 and ϵ_i which can also be measured experimentally may serve as an independent test for the $\rho_0(\epsilon)$ thus obtained. Due to the partial delocalization of the impurity exciton, its vibronic-fluorescence spectrum consists of two transitions. They differ from each other by their final state, i.e. by whether the intramolecular vibration involved in the vibronic fluorescence is localized at the site occupied by the guest molecule, or at a host-occupied site. Because of the difference in vibrational frequencies of the host and guest molecules, a doublet of bands is seen in the fluorescence spectrum. The intensity ratio of its components determines $a^2(\epsilon_i)$. The experimental data for singlet excitons in naphthalene are shown in fig. 10.

The partial delocalization of the exciton manifests itself also in that part of the impurity fluorescence spectrum which corresponds to transitions involving lattice phonons. A shallow local exciton interacts predominantly with long-wave phonons whose wave vector is of the order of the inverse radius of the centre. Therefore, when $\epsilon_{ed} - \epsilon_i$ decreases, the phonon satellites must narrow, and the long-wave phonons must dominate in them. This phenomenon has been observed by Meletov et al. (1979, 1980) in the singlet transition in naphthalene (fig. 11). For a deep centre (curve A) the shape of a satellite is dictated by the "weighted" density of states in the phonon spectrum. The shape of the satellite can be found from the equations given in section 7.3 if the exciton band width is put equal to zero ($\epsilon(\boldsymbol{q}) = 0$). For a shallow centre, the maximum at 48 cm^{-1} corresponding to short-wave phonons disappears, and a new low-frequency maximum corresponding to long-wave acoustic phonons is sharply visible instead. The calculation of fluorescence spectra for local excitons was also carried out by Ochs et al. (1974b) and Osad'ko and Chigirev (1981).

It should be noted that one and the same effect—partial delocalization of the exciton—manifests itself to a very different extent in different phenom-

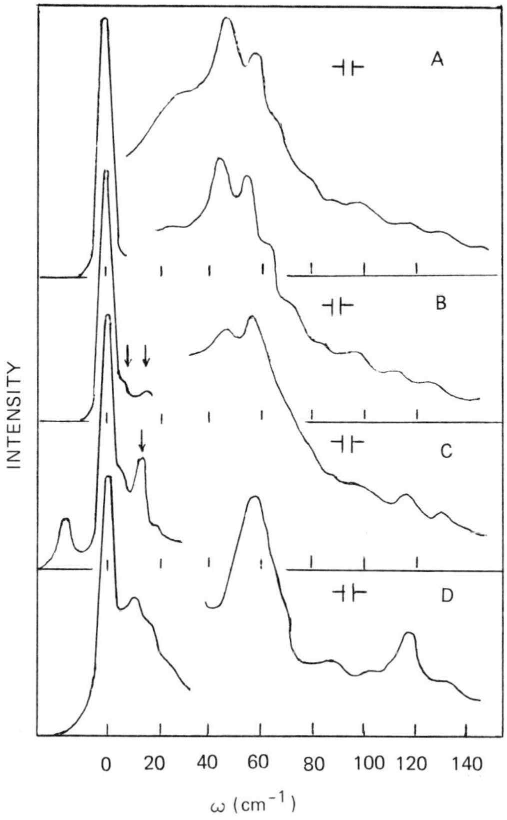

Fig. 11. The change in shape of the phonon satellite with changing position of the energy level of a singlet monomer centre in a naphthalene crystal. The depth of the local level decreases from A to D. The frequency ω is reckoned from the electron transition. Details labelled with arrows correspond to traces of extraneous impurities (Meletov et al. 1980).

ena: the weak dependence of a^2 on ϵ_i (fig. 10) in the systems with a 3D spectrum is not comparable in order of magnitude with the strong change in P^i (fig. 7) in the same range of ϵ_i values.

Thus, the theory of monomeric isotope centres agrees entirely with experiment, and the corresponding experimental data may be used for reconstruction of DOS $\rho_0(\epsilon)$.

The theory of monomeric centres can be constructed not only for isotope impurities but also for chemical ones. It has been developed for electron transitions (Rashba 1957) and generalized to vibronic transitions (Philpott 1968). The difficulty is that there is a large number of unknown matrix elements corresponding to excitation transfer from an impurity to the host lattice. Therefore the analysis of experimental data (Hochstrasser and Small

1966, Talapatra and Misra 1980) is much more laborious and less informative than is the case for isotope substitution.

It was supposed in the foregoing that only one exciton had been trapped by an impurity centre. However, in spite of the fact that eq. (3.4) near each region of the continuum $\rho_0(\epsilon)$ has no more than one solution, a single level can trap more than one exciton. The reason is that the excitons obey the Pauli statistics: in terms of the sites they behave as fermions when filling one site and as bosons when filling different sites. At a large radius of the ψ_i function, i.e. $a^2(\epsilon_i) \ll 1$, the kinematic interaction of excitons (which forbids their coexistence in one site) is rather small, and the impurity centre can bind two excitons. The restriction on the binding of two excitons becomes significantly less rigorous when more complicated centres are considered. And, finally, this must be prompted by the interaction between the excitons if its sign corresponds with attraction. An experiment interpreted in terms of a local biexciton is described at the end of section 3.2.

3.2. Aggregate impurity centres

A more difficult but very fascinating problem is the investigation of more complicated centres. Two types of these centres have been studied experimentally. The first type represents aggregate centres consisting of several molecules with changed isotope composition: dimers, trimers, etc. The second type is an aggregate composed of one alien molecule and several neighbouring host molecules whose electron level is shifted significantly due to the presence of the impurity.

The calculation of all these centres presents no special problems, and there exists a technique for calculating complicated models (see, for instance, Benk et al. 1982). The difficulties involved are due to the Hamiltonian of the crystal, i.e. the set of integrals for excitation transfer, $M_{\boldsymbol{n}\alpha\boldsymbol{m}\beta}$, being unknown. A method for determining $\rho_0(\epsilon)$ based on the analysis of the monomer spectrum was considered in section 3.1. The investigation of monomer spectra, however, cannot yield information on individual integrals $M_{\boldsymbol{n}\alpha\boldsymbol{m}\beta}$. To this end, dimers could be used. Let the guest molecules be located a distance $\boldsymbol{R}$ apart at the sites $\boldsymbol{n}\alpha$ and $\boldsymbol{n}\beta$, then the position of levels is determined by the equation

$$G_0(\epsilon) \pm G_{\boldsymbol{R}}^0(\epsilon) = \Delta^{-1}, \qquad G_{\boldsymbol{R}}^0 \equiv G_{\boldsymbol{n}\alpha\boldsymbol{m}\beta}^0. \tag{3.13}$$

For instance, for deep impurities $|\Delta| \gg \mathfrak{M}$ this equation can be simplified:

$$\epsilon_{\boldsymbol{R}}^{\pm} \approx \Delta + M_{\boldsymbol{n}\alpha\boldsymbol{m}\beta}. \tag{3.14}$$

Thus the spectra of dimers permit $M_{\boldsymbol{n}\alpha\boldsymbol{m}\beta}$ to be determined. The main obstacle lies in the uncertainty with which one may assign one or another of the bands in the dimer spectrum to the particular configurations of isotopic pairs.

This method of determining $M_{n\alpha m\beta}$ was first employed by Hanson (1970) and improved by Hong and Kopelman (1970, 1971a). They proposed three sets of $M_{n\alpha m\beta}$ for singlet excitons in naphthalene. Up to the present, however, the efforts to choose between them and to prove the validity of one of them have failed. It is only known for certain that the largest is the integral $M_{12} \approx 20$ cm^{-1}, relating two neighbouring translationally nonequivalent molecules to each other (fig. 1), and that there is a marked asymmetry of $\rho_0(\epsilon)$ (fig. 9). The latter fact shows that among the other transfer integrals there are some that, although being less than M_{12}, are comparable to it.

In the same way, Doberer et al. (1982) have recently determined the parameters of a triplet exciton band in naphthalene and have used the results for interpreting the complex structure of the high-frequency $^{T}A_0$ component of an exciton doublet (fig. 12). The intensive absorption located on the high-frequency side of the $^{T}A_0$ band is due to the presence of an exceedingly small number of molecules ($\sim 1\%$) isotopically substituted in $^{12}C \rightarrow {}^{13}C$, and is enhanced as a result of the Rashba effect.

By the same means, Hochstrasser and Whiteman (1972) have constructed the exciton band of 1,4-dibromonaphthalene (DBN). DBN consists of stacks of molecules arranged like packs of cards with considerable interaction $M \approx 7.4$ cm^{-1} only inside a stack ($2\mathfrak{M} = 29.4$ cm^{-1}). The Davydov splitting which would reveal an interaction between different stacks has not been observed. The point $\boldsymbol{k} = 0$ lies at the bottom of the band.

The change in absorption and fluorescence spectra as the isotopic cluster grows presents an interesting picture. It reflects a gradual transition from the spectra of doped crystals to the spectra of mixed crystals. For the singlet spectra of naphthalene such investigations were carried out by Broude and Leiderman (1971), Broude et al. (1971) and Port et al. (1975). The connection to the spectra of mixed crystals was traced in great detail by Broude et al. (1981).

The spectra of the defect centres which comprise host molecules perturbed through the presence of a chemical impurity, also exhibit features characteristic for shallow impurity levels (Broude et al. 1959). The spectra of singlet local excitons bound by several types of defect centres in benzene and naphthalene have been studied comprehensively by Ostapenko et al. and described in their review (Ostapenko et al. 1973). Under strong pulsed pumping, a new lower-frequency band has been revealed (Broude et al. 1977) in the fluorescence spectrum of a defect centre in anthracene previously investigated by Glockner and Wolf (1969). Based on the time dependence of its increase and decrease as well as on its dependence on excitation intensity, it has been interpreted as a local-biexciton band. Since the defect centre is a deep one (the depth of the level $|\epsilon_i| = 938$ cm$^{-1} > 2\mathfrak{M}$), the biexciton may arise only due to the attrac-

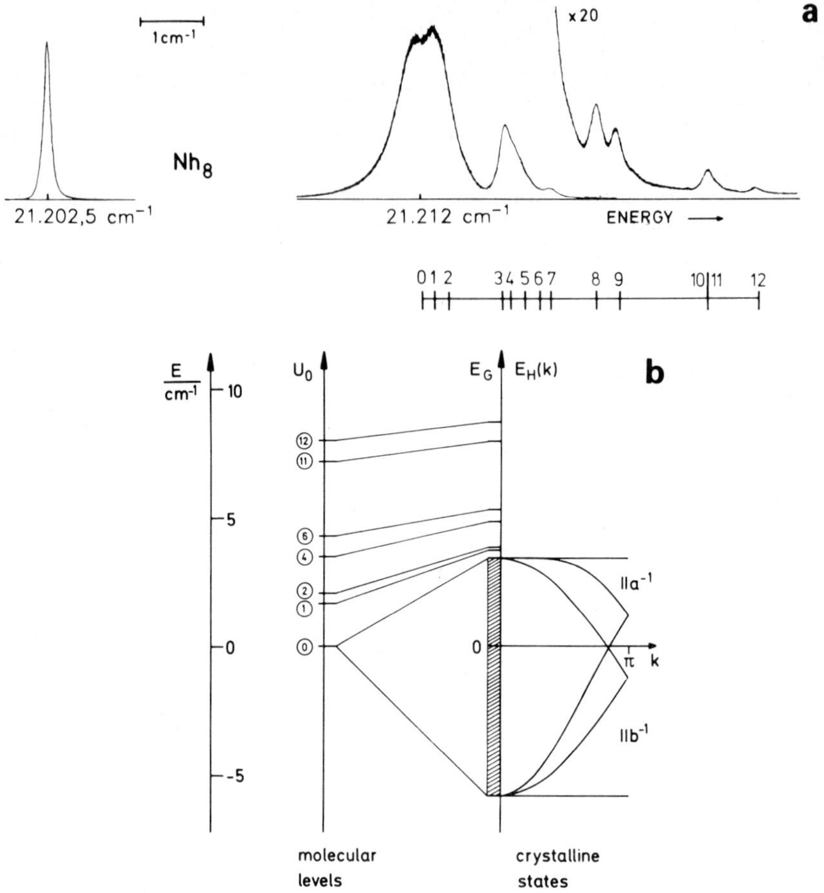

Fig. 12. Triplet exciton absorption in naphthalene. (a) The absorption spectrum of nominally "pure" samples of N-h_8; left curve: b-component; right curve: a-component. (b) Exciton energy bands of naphthalene and the position of the local exciton levels (a calculated one). (1–6), (8–9): the levels of monomers containing one or two atoms of ^{13}C; (10, 11): monomers N-d_1h_7; (7) a dimer of two molecules containing one atom of ^{13}C each. The amount of ^{13}C and d corresponds to the natural abundance of 1% and 10^{-2}%, respectively (Doberer et al. 1982).

tion between excitons. This statement agrees with some other data on the spectra of the strongly pumped anthracene (Broude 1980, Benderskii et al. 1980).

3.3. Vibronic spectra of monomer centres

In the preceding (sub)sections we have excluded intramolecular vibrations from our consideration of absorption spectra. In a crystal these vibrations form the high-frequency part of the phonon spectrum, or internal phonons (IP). The coupling of the exciton to IPs, i.e. the vibronic coupling, is strong in real systems. Practically, it is almost entirely determined by the intramolecular electron–vibrational interaction, and the interaction constant can be found from the spectrum of an isolated molecule. Usually the IP dispersion can be neglected. As a result of the vibronic interaction, the electronic excitons in the absorption spectrum are followed by different types of vibronic (i.e. electron–vibrational) excitations. In doped crystals, complexes may arise which include one electronic exciton and one or several IPs bound to the impurity centre (a local vibron). Below we shall briefly consider one case of the formation of local vibrons.

The theory of vibronic excitations can be developed in two cases: firstly, when the vibronic interaction is weak (we will not deal with it below), and secondly, when the phonon frequency $\nu \gg \mathfrak{M}$, i.e. it exceeds substantially the width of the exciton band. In the latter case the exciton and the phonon can be considered as two stable particles interacting by a certain law. In other words, the problem changes from being one of field theory into a quantum-mechanical, i.e., dynamical problem. Therefore this theory, developed by Rashba (1966, 1968) is called the dynamic theory of vibronic spectra. Beyond these two limiting cases, the theory has been developing only as interpolation schemes: that of the dynamic coherent potential (Sumi 1974, 1975) and its generalization (Friesner and Silbey 1982).

The dynamic theory shall be used in what follows. Its total Hamiltonian has the form

$$H = H_{\mathrm{e}} + H_{\mathrm{ex}} + H_{\mathrm{ph}} + H_{\mathrm{v}} + H'_{\mathrm{imp}}. \tag{3.15}$$

Here H_{e} and H_{ex} are determined by eqs. (2.2) and (2.3),

$$H_{\mathrm{ph}} = \nu \sum_{n\alpha} b^{+}_{n\alpha} b_{n\alpha}, \tag{3.16}$$

where $b_{n\alpha}$ are the phonon annihilation operators,

$$\begin{aligned} H_{\mathrm{v}} = \gamma^2 \sum_{n\alpha\, m\beta} M_{n\alpha\, m\beta} a^{+}_{n\alpha} a_{m\beta} \left(b^{+}_{n\alpha} b_{m\beta} + b^{+}_{m\beta} b_{n\alpha} - b^{+}_{n\alpha} b_{n\alpha} - b^{+}_{m\beta} b_{m\beta} \right) \\ + \Delta_{\nu} \sum_{n\alpha} a^{+}_{n\alpha} a_{n\alpha} b^{+}_{n\alpha} b_{n\alpha} \end{aligned} \tag{3.17}$$

is the operator of the vibronic interaction in a perfect crystal. In (3.17) γ is the dimensionless coupling constant of the linear intramolecular electron–vibrational interaction

$$H_{\mathrm{int}} = \gamma \nu a^{+}_{n\alpha} a_{n\alpha} \left(b^{+}_{n\alpha} + b_{n\alpha} \right).$$

The resulting Franck–Condon shift of the minimum of the adiabatic potential equals $E_{FC} = -\gamma^2\nu$. The constant $\Delta_\nu = \nu^* - \nu$ is the shift in the vibrational frequency due to electronic excitation of the molecule; usually $\Delta_\nu < 0$, $|\Delta_\nu| \ll \nu$. It should be noted, by the way, that all transfer integrals $M_{\boldsymbol{n}\alpha\boldsymbol{m}\beta}$ appearing in this chapter are supposed to be renormalized ones. They are related to the values of the corresponding integrals in a rigid molecule $M^0_{\boldsymbol{n}\alpha\boldsymbol{m}\beta}$ by relationships $M_{\boldsymbol{n}\alpha\boldsymbol{m}\beta} = M^0_{\boldsymbol{n}\alpha\boldsymbol{m}\beta}\exp(-\gamma^2)$. It is essential that $\gamma \neq 0$ only for totally symmetrical (TS) vibrations. On the other hand, $\Delta_\nu \neq 0$ always, and as a result of $\nu \gg \mathfrak{M}$ the shift Δ_ν can compete with $\mathfrak{M}$ (i.e. $|\Delta_\nu| \gtrless \mathfrak{M}$) even at $|\Delta_\nu| \ll \nu$.

The four terms in eq. (3.15) listed above determine one-phonon vibronic spectra of perfect crystals. In a system of two particles (exciton + phonon) there always exist dissociated (two-particle) states and bound (one-particle) states can exist. The latter appear when H_v is sufficiently large compared with $\mathfrak{M}$. If $\gamma = 0$, as is the case for nontotally symmetrical (NTS) phonons, the problem becomes identical to the problem of the monomer isotopic centre (section 3.1), with Δ_ν playing the role of Δ. The analysis of the intrinsic vibronic absorption in benzene, naphthalene and anthracene from the viewpoint of the dynamic theory has been presented by Broude et al. (1981).

The last term in (3.15) corresponds to the interaction of the exciton and the phonon with a monomer impurity centre located at the site $\boldsymbol{n}_0\alpha_0$:

$$H'_{\text{imp}} = H^{\Delta}_{\text{imp}} + H^{\gamma}_{\text{imp}}, \tag{3.18}$$

$$H^{\Delta}_{\text{imp}} = \Delta_{\text{ex}} a^+_{\boldsymbol{n}_0\alpha_0} a_{\boldsymbol{n}_0\alpha_0} + \Delta_{\text{ph}} b^+_{\boldsymbol{n}_0\alpha_0} b_{\boldsymbol{n}_0\alpha_0} + \Delta_{\text{ex-ph}} a^+_{\boldsymbol{n}_0\alpha_0} a_{\boldsymbol{n}_0\alpha_0} b^+_{\boldsymbol{n}_0\alpha_0} b_{\boldsymbol{n}_0\alpha_0}, \tag{3.19}$$

$$H^{\gamma}_{\text{imp}} = \gamma(\gamma' - \gamma) \sum_{\boldsymbol{n}\alpha \neq \boldsymbol{n}_0\alpha_0} M_{\boldsymbol{n}_0\alpha_0\boldsymbol{n}\alpha} a^+_{\boldsymbol{n}_0\alpha_0} a_{\boldsymbol{n}\alpha} \Big(b^+_{\boldsymbol{n}_0\alpha_0} b_{\boldsymbol{n}\alpha} + b^+_{\boldsymbol{n}\alpha} b_{\boldsymbol{n}_0\alpha_0}$$

$$- b^+_{\boldsymbol{n}_a\alpha_0} b_{\boldsymbol{n}_0\alpha_0} - b^+_{\boldsymbol{n}\alpha} b_{\boldsymbol{n}\alpha} \Big) + (\boldsymbol{n}_0\alpha_0 \rightleftarrows \boldsymbol{n}\alpha). \tag{3.20}$$

Here Δ_{ex} and Δ_{ph} are the isotopic shift of the electronic level (which coincides with Δ of section 3.1) and the vibrational frequency, respectively, $\Delta_{\text{ex-ph}}$ is the isotopic change in the shift Δ_ν of vibration frequency, γ' is the value of the constant γ for a guest molecule. Thus the exciton and the phonon interact both with a guest molecule and with each other.

Krivenko et al. (1977, 1978) have investigated the vibronic absorption of the naphthalene N-h_8 guest in the N-d_8 host in the region of vibronic transition including TS phonon $\nu = 758\ \text{cm}^{-1}$.

The experimental results are shown in fig. 13. A_1 is the band of the intrinsic vibronic absorption corresponding to participation of this phonon. It is a

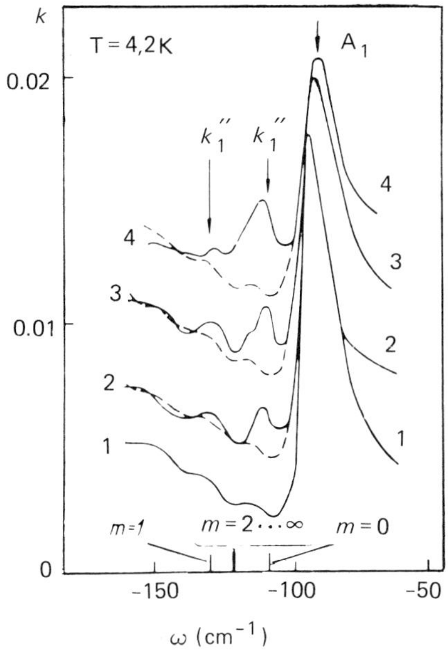

Fig. 13. Singlet vibronic absorption spectrum of isotopically doped naphthalene N-d_8. A_1 is a one-phonon vibronic band of the host with participation of a phonon $\nu = 758\ \text{cm}^{-1}$. K_1' and K_1'' are vibronic bands due to the N-h_8 guest. Four curves correspond to different concentrations of N-h_8: (1) 0%, (2) 2%, (3) 5%, (4) 10%. Light is polarized $\perp$ **b**-axis. Curves 2–4 are shifted with respect to ordinates. The dashed line shows absorption of the host crystal (Krivenko et al. 1977, 1978).

one-particle band, a direct analogue of the A_0 band in the A_0–B_0 exciton doublet. There is no corresponding one-particle B_1 band in the spectrum; it is substituted by the wide-band two-particle absorption. Arrows point to the maxima of the impurity satellites of the band, K_1' and K_1''. Indicated underneath are the calculated positions of the satellites and their serial numbers. The calculations do not involve any fitted parameters: all parameters have been previously determined from the spectra of free molecules, electron excitons and intrinsic vibronic spectra.

The level $m = 0$ corresponds to an exciton–phonon wave function whose maximum is attained when an exciton and a phonon are both located at the impurity site $\boldsymbol{n}_0\alpha_0$. Since the optical vibronic transition is an intramolecular one, such a configuration with an exciton and phonon at the same site corresponds to the maximum oscillator strength f_0. The bands with $m \geqslant 1$ correspond to the wave functions in which the exciton is located predominantly at the $\boldsymbol{n}_0\alpha_0$ site, and the maximum probability of the phonon location is attained on the host molecules, the farther from $\boldsymbol{n}_0\alpha_0$, the larger m.

Therefore f_m drop rapidly as m increases, and their sum is $\sim 0.25 f_0$. All bands with $m \geqslant 1$ merge into one band K_1'', as can be seen from fig. 13, it broadens rapidly and vanishes with growth in the impurity concentration. This is to be expected, considering that the phonon resides mainly off the impurity.

The vibronic spectra of defect centres are patterned on the same principle as the spectra of monomer centres, but they are more intricate. They have been investigated in naphthalene by Ostapenko and Shpak (1972) and Brovchenko et al. (1984).

4. *Exciton spectra of mixed crystals*

In this section we shall consider the spectra of mixed crystals in which the concentrations of both isotopic components, C_1 and C_2, are comparable with each other, $C_1 \sim C_2 \sim 0.5$. Such crystals form a continuous sequence of solid solutions; the orientation of molecules with different isotopic composition is practically the same. For simplicity's sake we assumed that the isotopic component 1 possesses an exciton spectrum of lower frequency.

At first glance, it may seem that in a system with $C_1 \sim C_2$ and $\mathfrak{M} \sim \Delta$, the violation of crystal order is so strong that the main features of the exciton absorption spectrum, i.e., well-defined, sharply polarized bands, should disappear; one would think that the spectrum would be diffuse and weakly polarized. Below it is shown that this is not so, and the spectra are more interesting and informative than could be expected.

4.1. Stability of multiplet structure in the spectra of mixed crystals

Figure 14 shows the singlet-absorption spectrum of benzene with complex isotopic composition obtained by Broude and Onoprienko (1961). For benzene $\mathfrak{M} \approx \Delta_1 \approx 30$ cm^{-1}. It is seen that the spectrum consists of relatively narrow bands whose width is comparable with the width of the A_0–C_0 bands. The bands are sharply polarized. One gains the impression that there are several exciton doublets with strongly perturbed polarization ratios. As seen from fig. 14, the multiplet structure is stable at all values of C_s $(s = 1, 2)$.

This intriguing fact shows that in a system where the long-range order is manifested in the position of lattice sites and orientation of molecules but lacking in other respects (the random discrete isotopic level shift), there exist exciton-like states. It is instructive to compare mixed crystals with amorphous monoisotopic films of kindred substances.

Lee and Gan (1977) have investigated pentacene films on a quartz substrate. The films prepared at high temperatures of the substrate, consist of

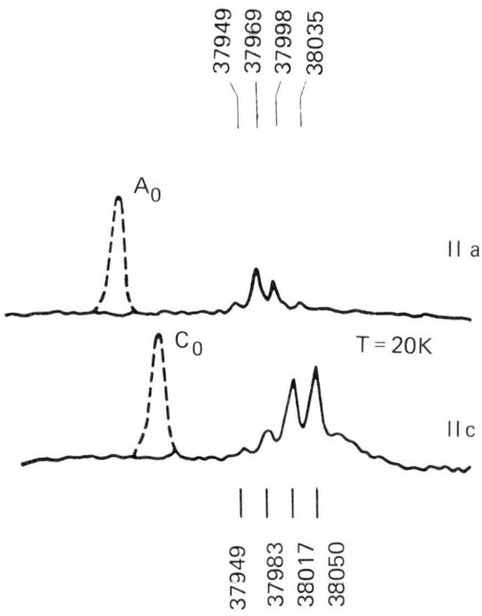

Fig. 14. The singlet absorption spectrum of mixed deuterobenzene crystals in the region of exciton transitions. Composition: B-d_6: 24%, B-hd_5: 41%, B-h_2d_4: 24%, B-h_3d_3: 5%, B-h_6: 6% (Broude and Onoprienko 1961). For comparison, shown on the left are the A_0 and C_0 components of the pure B-h_6 exciton doublet.

relatively large crystallites oriented with the *ab* plane parallel to the substrate surface. The homogeneity and the size of the crystallites are reduced as the temperature of the substrate T_s is decreased. At $T_s < -90\,°C$ the films are amorphous. Figure 15 displays the absorption spectrum changing with T_s. With the appearance of crystallites a long-wave band corresponding to electron transition begins to split into two components whose frequencies coincide at $T_s \gtrsim 100\,°C$ with the frequencies of exciton components in the spectrum of single crystals found by Prikhot'ko et al. (1969). It follows from fig. 15 that amorphization totally destroys the multiplet structure of the spectrum.

4.2. *Approximation of average amplitudes and classification of spectra*

Light absorption in a mixed crystal is described by expression (2.13) when $G^0(\omega, \boldsymbol{k}=0)$ is substituted by $G(\omega, \boldsymbol{k}=0)$, the Green function of a mixed crystal at zero momentum. Figure 14 suggests that at small $k \ll k_{Br}$ the behaviour of the Green function is similar in some respects to the behaviour of the Green function of a perfect crystal. The average-amplitude approximation

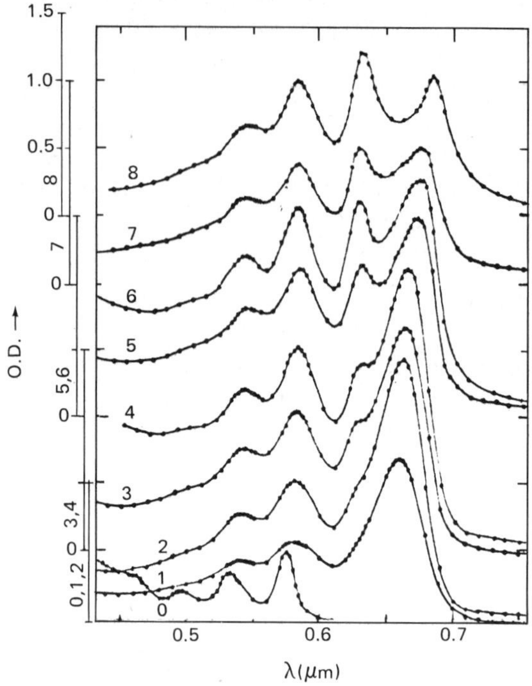

Fig. 15. Optical density spectra of pentacene films at different substrate temperatures. (0) is the pentacene spectrum in benzene. (1) $T_s = -155\,°C$, (2) $T_s = -90\,°C$, (3) $T_s = -45\,°C$, (4) $T_s = -3\,°C$, (5) $T_s = 25\,°C$, (6) $T_s = 45\,°C$, (7) $T_s = 100\,°C$, (8) $T_s = 155\,°C$ (Lee and Gan 1977).

(AAA) first proposed by Broude and Rashba (1961) * is based on the substitution:

$$G_{\boldsymbol{n}\alpha\boldsymbol{m}\beta}(\Omega) \Rightarrow \begin{cases} G^{rs}_{\alpha\beta}(\Omega, \boldsymbol{n}-\boldsymbol{m}) & \boldsymbol{n}\alpha \neq \boldsymbol{m}\beta \\ C_s G^{ss}_{\alpha\alpha}(\Omega, 0) & \boldsymbol{n}\alpha = \boldsymbol{m}\beta \end{cases} \tag{4.1}$$

which should be applied when an r-type molecule occurs at the $\boldsymbol{n}\alpha$ site, and an s-type molecule at the $\boldsymbol{m}\beta$ site; all $G^{rs}_{\alpha\alpha}(\Omega, 0) = 0$ at $r \neq s$. It can be shown that the best choice of $G^{rs}_{\alpha\beta}$ is

$$G^{rs}_{\alpha\beta}(\Omega) = \sum_{\tau} a^{\tau}_{r\alpha} a^{\tau *}_{s\beta} / (\Omega - \mathscr{E}_{\tau}). \tag{4.2}$$

* In lattice dynamics an analogous approximation was advanced by Chen et al. (1966) and is called the approximation of isodisplacements.

Here a^{τ} and $\mathscr{E}_{\tau}$ are eigenvectors and eigenvalues of an auxiliary set of equations

$$\sum_{\beta s} C_s L_{\alpha\beta} a_{s\beta} = (\mathscr{E} - E_r) a_{r\alpha}. \tag{4.3}$$

The coefficients $L_{\alpha\beta}$ are determined by eq. (2.6), and the eigenvectors adhere to the orthonormalization condition

$$\sum_{r\alpha} C_r a_{r\alpha}^{\tau *} a_{r\alpha}^{\tau'} = \delta_{\tau\tau'}. \tag{4.4}$$

Substitution (4.1) for the G-function corresponds to the selection of trial wave functions in the form of

$$\psi_{\tau k}(\boldsymbol{n}\alpha) \Rightarrow N^{-1/2} a_{r\alpha}^{\tau} \exp(\mathrm{i}\boldsymbol{k}\boldsymbol{n}_{\alpha}). \tag{4.5}$$

It is evident from eq. (4.5) that $a_{\boldsymbol{n}\alpha}^{\tau}$ have the meaning of excitation amplitudes in the crystal quantum state τ of the r-type molecules occupying the α-position. Since the r-index is absent on the left-hand side of eq. (4.3), it is convenient to introduce coefficients

$$b_{\alpha}^{\tau} = (\mathscr{E}_{\tau} - E_r) a_{r\alpha}^{\tau}. \tag{4.6}$$

They satisfy the equation

$$\sum_{\beta} L_{\alpha\beta} b_{\beta} = \epsilon b_{\alpha} \tag{4.7}$$

closely resembling eq. (2.5), while the equation for E_{τ} is

$$\frac{1}{\epsilon} = \sum_{s} \frac{C_s}{\mathscr{E} - E_s}. \tag{4.8}$$

E_s is the term of the s-type molecules in a crystal (i.e., an analogue of E_{ρ}, eq. (2.2)). For each spectrum component μ, the left-hand ϵ is equal to ϵ_{μ}, the energy of the corresponding absorption band of a perfect crystal. For each spectrum component μ, eq. (4.8) determines the roots $\mathscr{E}_{q\mu}$ where q numbers the multiplets. The number of roots equals the number of isotopic components of a mixed crystal.

From (4.8) it follows that

$$\sum_{q} \mathscr{E}_{q\mu} = \epsilon_{\mu} + \sum_{s} E_s \tag{4.9}$$

holds for each μ. It is evident from (4.9) that in AAA the sum of Davydov splittings over all multiplets is independent of C_s and equals the Davydov splitting in a pure crystal.

The conductivity tensor $\hat{\sigma}(\Omega)$ is

$$\hat{\sigma}(\Omega) = \frac{\pi}{\Omega v} \sum_{q\mu} \frac{\boldsymbol{j}_\mu \otimes \boldsymbol{j}_\mu^*}{\epsilon_\mu^2 \sum_s C_s / (\mathscr{E}_{q\mu} - E_s)^2} \delta(\Omega - \mathscr{E}_{q\mu}). \tag{4.10}$$

This expression is a generalization of (2.14).

Thus, the AAA enables one to obtain the position of all the bands (eq. (4.8)) and their intensities (eq. (4.10)) within the framework of elementary expressions. All these quantities are directly expressed through ϵ_μ and $\boldsymbol{j}_\mu$, the position of bands and their intensities in a pure crystal.

However, a high price must obviously be paid for the simplicity of the solutions thus obtained. It lies mainly in the fact that AAA fails to yield the shape for the absorption bands of a mixed crystal: all bands in AAA have a δ-function shape (eq. (4.10)). When compared with experimental data, each $\mathscr{E}_{q\mu}$ should be correlated with the centres of gravity of the respective bands. Besides, AAA does not reveal the right position of impurity bands in the limits $C_1 \to 0$ and $C_2 \to 0$; this approximation takes no account of "repulsion" (section 3.1).

The principal qualitative conclusions following from the AAA are easier to obtain from a two-component crystal when eq. (4.8) for each μ reduces to a quadratic equation. The behaviour of the roots of this equation versus C_s changes qualitatively at the pont $\epsilon_\mu = \Delta \equiv E_2 - E_1$.

Figure 16 displays two limiting cases. They permit the two basic regimes of absorption spectrum behaviour to be elucidated.

(a) $\Delta > \epsilon_\mu$. *Two-mode regime*

With C_1 increasing, the impurity band, e.g. E_1, splits into a doublet. At $C_1 = 1$ these bands coincide with the exciton doublet in a perfect crystal. At each C_1 two bands are seen in each of the spectrum components; each of those two corresponds throughout the concentration range to the prevalent excitation of one of the isotopic components. In the range of $\Delta \gg \epsilon_\mu$ the dependence of $\mathscr{E}$ energies on C_1 is linear (fig. 16a).

(b) $\Delta < \epsilon_\mu$. *One-mode regime*

In each component of the spectrum the exciton band of one pure crystal continually passes into the band of a second crystal. Therefore it cannot correspond all along to one isotopic component. There is always a second band which connects the impurity bands of the first component in the second one and those of the second component in the first one. In the range $\Delta \ll \epsilon_\mu$ the energies $\mathscr{E}$ are again linearly dependent on C_1.

The next conclusion follows from eq. (4.10): the outer bands of doublets are

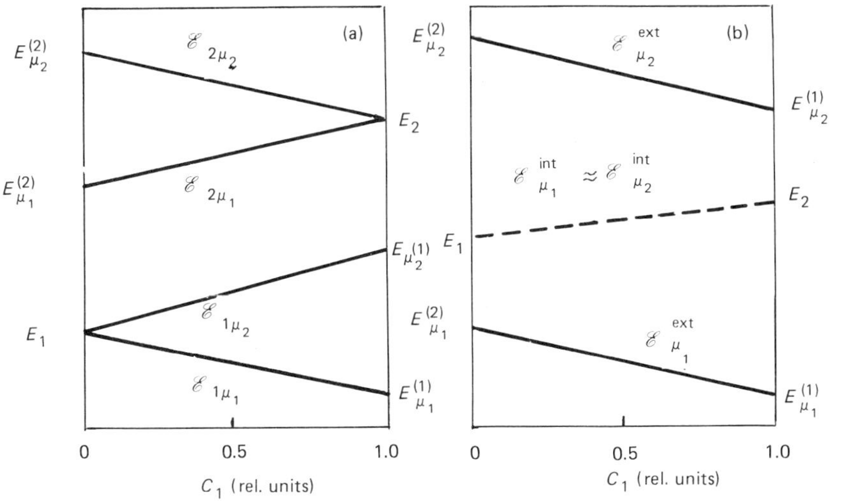

Fig. 16. Two-mode (a) and one-mode (b) regimes of concentration dependences of the positions of exciton absorption bands $\mathscr{E}_{q\mu}$ in mixed crystals. $E_\mu^{(s)} = E_s + \epsilon_\mu$ is the position of the μ-component of the exciton multiplet in a perfect crystal of the s-type.

strengthened (bands $\mathscr{E}_{2\mu_2}$ and $\mathscr{E}_{1\mu_1}$ in fig. 16a) and the inner ones ($\mathscr{E}_{2\mu_1}$ and $\mathscr{E}_{1\mu_2}$) weakened. This tendency is most pronounced in the single-mode regime; the dashed line in fig. 16b corresponds to a very weakened band. These conclusions are in excellent agreement with the general view of the spectrum in fig. 14.

The spectra with two-mode behaviour allow us to establish convincingly the genesis of the exciton multiplet, i.e. its creation out of an impurity level. This was first done by Sheka (1963b) for a solution of N-h_8 and N-d_8. The single-mode spectra offer no means of establishing the genesis of bands but permit the exciton nature of the bands to be ascertained; the continuous dependence of the band frequency on composition indicates a collective nature of the absorption.

To conclude, fig. 17 shows by way of illustration the position of the bands and their intensities as a function of concentration for isotopically mixed crystals of benzene. As seen from fig. 17b, the AAA well conveys a drastic change in intensities within the range of small C_1. In terms of the spectra of isolated impurities, this corresponds to a strong Rashba effect.

It should be emphasized that the AAA possesses an important virtue: it permits classification and description of the spectra of mixed crystals using minimum information on pure crystals (the position of the absorption bands and the isotopic shift Δ). All the other, more refined methods demand much more comprehensive data ($\rho_0(\epsilon)$, $\epsilon_\mu(\boldsymbol{k})$, etc.).

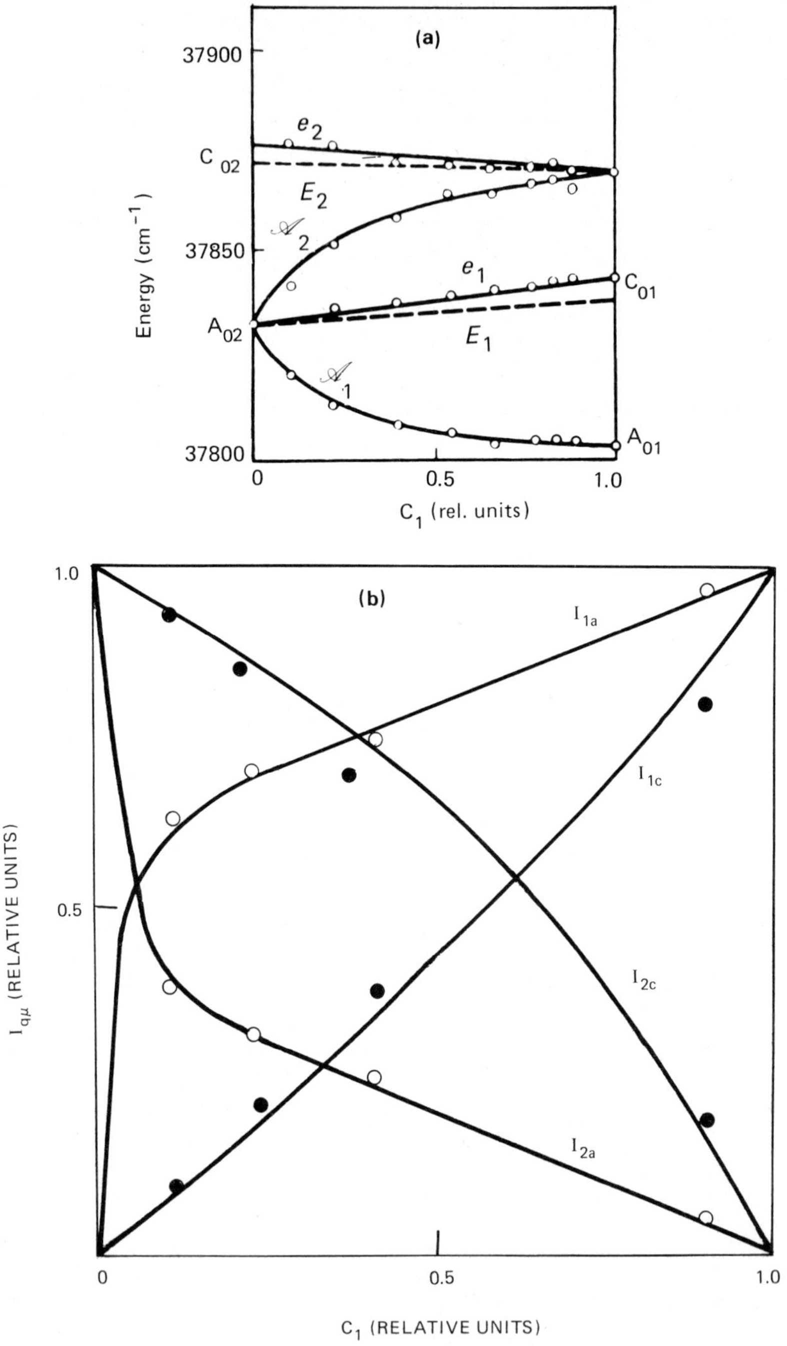

Fig. 17. Singlet spectra of mixed crystals of B-h_6 and B-d_6 benzenes. (a) (top) Position of bands versus concentration. The behaviour is a two-mode one, close to break-up: a small change in parameters would transfer the spectrum into a one-mode regime. (b) (bottom) Intensity of bands versus concentration. Solid lines: theory (AAA). Points: experiment (Broude and Kochubei 1964).

4.3. *Optical and energy spectra of mixed crystals*

Of course, the AAA can only pretend to give a rough description of a spectrum: it does not describe the spectral distribution of the intensities $\hat{\sigma}(\Omega)$ and the distribution of the density of states (DOS) $\rho(E)$. A rigorous theory can establish only general theorems on localization of the spectrum and its behaviour near singularities. Therefore some approximate methods have been developed, such as coherent potential approximation (CPA) and the method of the average T-matrix, both described in detail in the review by Elliott et al. (1974), as well as numerical methods, such as Dean's method (1972) for calculating $\rho(E)$ and the method of Napier factorization (MNF) (Mil'man 1975) for calculating $\hat{\sigma}(\Omega)$. Below we shall briefly formulate the main idea of some of these methods and the main results and statements. Proceeding from this, we shall analyze the experimental data of Broude and Leiderman (1971) on singlet excitons in naphthalene.

If energy and frequency are calculated from

$$\bar{E} = C_1 E_1 + C_2 E_2, \tag{4.11}$$

then the principal equation of CPA takes the form

$$G_0(\omega - u(\omega)) = \frac{v}{Z(2\pi)^3} \sum_{\mu} \int \frac{\mathrm{d}^3 k}{\omega - u(\omega) - \epsilon_{\mu}(\boldsymbol{k})}$$

$$= \int \frac{\rho_0(\epsilon)\, \mathrm{d}\epsilon}{\omega - u(\omega) - \epsilon}. \tag{4.12}$$

Here the functions $G_0(\omega)$ and $\epsilon_{\mu}(\boldsymbol{k})$ are the same as for a perfect crystal. The complex potential $u(\omega)$ has the meaning of a self-energy part, and the main assumption of the CPA is that u does not depend on $\boldsymbol{k}$. The Green function in the momentum representation is

$$G_{\mu}(\omega \boldsymbol{k}) = \{\omega - u(\omega) - \epsilon_{\mu}(\boldsymbol{k})\}^{-1}, \tag{4.13}$$

and the density of states is

$$\rho_{\mathrm{CP}}(\epsilon) = -\frac{1}{\pi} \mathrm{Im}\, G_0(\epsilon - u(\epsilon)). \tag{4.14}$$

It follows from (4.13) and (4.14) that in CPA all physical quantities: $\rho_{\mathrm{CP}}(\epsilon)$ and $\sigma^{\mu}_{\mathrm{CP}} \propto \mathrm{Im}\, G_{\mu}(\omega, \boldsymbol{k} = 0)$ are determined in terms of $\rho_0(\epsilon)$ and $\epsilon_{\mu}(\boldsymbol{k} = 0)$.

By way of example fig. 18 shows the spectrum of crystals N-h_8 + N-d_8 calculated in CPA using $\rho_0(\epsilon)$ found by Colson et al. (1968) and close to that

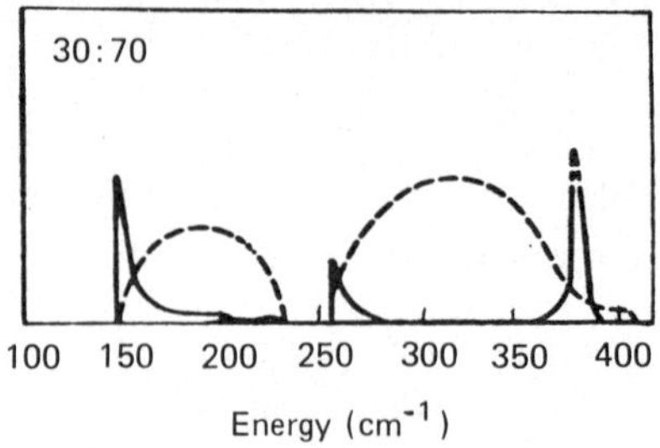

Fig. 18. Density of states (dashed lines) and absorption spectrum (solid lines) of $Nh_8 + Nd_8$ mixed crystals calculated in CPA (singlet excitons). The concentration ratio of 30:70 is shown in the figure. The intensities are normalized so that in a one-component crystal the intensities of the A_0 and B_0 absorption bands coincide (Hong and Robinson 1970).

shown in fig. 9. The spectrum ρ_{CP} consists of two separate regions. It can be shown, however (Lifshitz 1964), that a gap in the spectrum of a mixed crystal arises only if $\Delta > 2\mathfrak{M}$. Since $\Delta \approx 120$ cm^{-1} and $2\mathfrak{M} \approx 170$ cm^{-1} in naphthalene, this criterion is violated, and in the region of the CPA gap the density of states $\rho(E)$ is nonzero, however small. We shall call this region a pseudogap. The $\rho(E)$ tails also stretch beyond the left and right edges of ρ_{CP}. The exact edges of the spectrum are separated by $2\mathfrak{M} + \Delta \approx 300$ cm^{-1}. The low-frequency edge of the high density of states will be called the pseudobottom of the spectrum.

A correlation between general theorems and the CPA calculations provided a means for developing a classification of energy spectra summarized by Hoshen and Jortner (1972). To this purpose, it is convenient to introduce r_2^0, the second moment of the function $\rho_0(\epsilon)$ with respect to its centre of gravity. Usually, r_2^0 is numerically small compared with $\mathfrak{M}^2$: for model density (3.5) $r_2^0 = \mathfrak{M}^2/4$. Then the main types of behaviour of $\rho(E)$ can be classified in the following manner:

(1) Separated bands $\Delta > 2\mathfrak{M}$. A true gap exists.

(2) Quasi-separated bands $2\mathfrak{M} > \Delta \gtrsim \mathfrak{M}$. A pseudogap exists.

(3) An incipient pseudogap $\mathfrak{M} \gtrsim \Delta \gtrsim (r_2^0)^{1/2}$. A pseudogap or a weaker dip exists in $\rho(\epsilon)$.

(4) The amalgamation limit $\mathfrak{M} < (r_2^0)^{1/2}$.

Cases 1 and 2 are united by one term: the persistent case (see also Onodera and Toyozawa 1968).

Reconsider fig. 18. Two absorption curves correspond to each region of $\rho_{CP}(\epsilon)$. The curves with the low- and the high-frequency peak are polarized $\perp \boldsymbol{b}$ and $\|\boldsymbol{b}$, respectively. It is essential that the absorption band widths are considerably smaller than the widths of corresponding regions of the curve $\rho_{CP}(\epsilon)$. This important fact was first established by Onodera and Toyozawa (1968). It seems to elucidate the cause of high efficiency of AAA (see section 4.2).

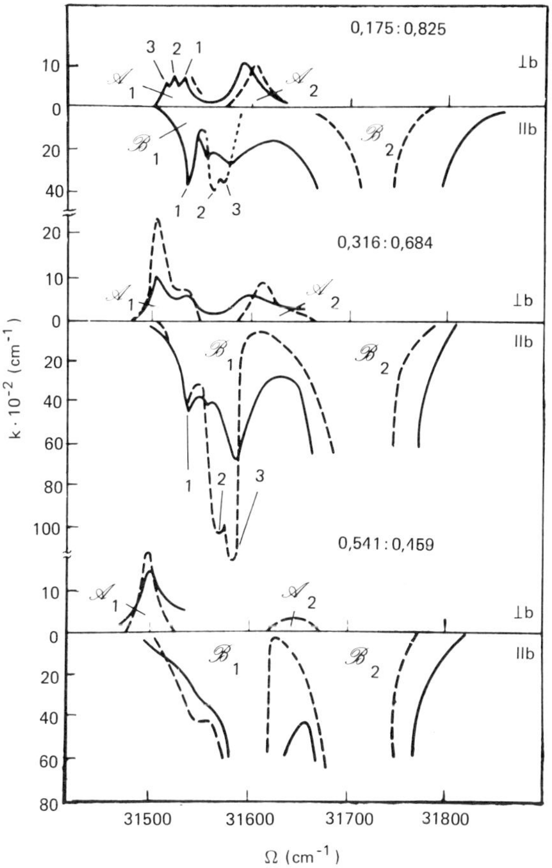

Fig. 19. The singlet absorption spectra of $Nh_8 + Nd_8$ mixed crystals in polarized light. The concentration ratio is indicated in the figure. Solid curves: experiment (Broude and Leiderman 1971); dashed lines: calculation by NFM (Krivenko et al. 1975). 1, 2, and 3 are the bands of monomers, dimers, and trimers, respectively.

The experimental data on deuteronaphthalene spectra are shown in fig. 19. When compared with fig. 18, the most striking feature is the presence of a distinct cluster structure which is sharply defined in the low-frequency part of the spectrum at the concentration of N-h_8 $C_1 = 0.175$ and does not disappear entirely even when C_1 increases to 0.32. This result seems to be unexpected. Besides, the low-frequency $\mathscr{B}$-band * has contours as well-defined as those of the low-frequency $\mathscr{A}$-band; according to fig. 18, one might have expected the

* For absorption bands of mixed crystals we use script letters $\mathscr{A}$ and $\mathscr{B}$.

former to be very diffuse. At $C_1 = 0.32$ the experiment fails to reveal any perceptible pseudogap: the maximum of the low-frequency $\mathscr{B}$-band and the high-frequency $\mathscr{A}$-band coincide practically. Apparently, the accuracy of CPA is higher in the high-frequency part of the spectrum where no manifestation of a cluster structure is observed, and also in mixed crystals with lesser values of the $\Delta/\mathfrak{M}$ ratio. So far, however, experimental data on such systems are lacking.

The necessity of a more detailed description of these experimental data has led to a numerical calculation of the absorption spectrum by NFM (Krivenko et al. 1975). Let us return to eqs. (2.13) and (2.14), substitute $G^0 \to G$ (cf. section 3.1) and assume $\boldsymbol{j}_\mu^* = \boldsymbol{j}_\mu$. Then absorption is determined by the function

$$-\operatorname{Im} G_\mu(\omega, \boldsymbol{k}=0) = \frac{\pi}{N} \sum_{\boldsymbol{n}\alpha\boldsymbol{m}\beta} B_\alpha^{\mu *} \langle \boldsymbol{n}\alpha | \delta(E-H) | \boldsymbol{m}\beta \rangle B_\beta^\mu . \tag{4.15}$$

Equation (4.15) is an explicit expression for $\operatorname{Im}\{G\}$ in $\boldsymbol{k}\mu$ representation, H is the Hamiltonian of a disordered crystal. Mil'man (1975) proposed a convenient technique for the computation of the averages such as (4.15). It is based on a well-known representation of the δ-function

$$\delta(E-H) = \lim_{\theta \to 0} \frac{1}{\theta\sqrt{\pi}} \exp\{-(E-H)^2/\theta^2\}, \tag{4.16}$$

with a subsequent application of Napier factorization

$$\delta(E-H) = \lim_{\theta \to 0} \lim_{\mathscr{M} \to \infty} \frac{1}{\theta\sqrt{\pi}} \left[1-(E-H)^2/\mathscr{M}\theta^2\right]^{\mathscr{M}}. \tag{4.17}$$

The theoretical curves shown in fig. 19 have been obtained by this method. For H the values of $M_{\boldsymbol{n}\alpha\boldsymbol{m}\beta}$ proposed by Hong and Kopelman (1971a) were used, which ensured that the shape of $\rho_0(\epsilon)$ was similar to one used in calculations by CPA (fig. 18). As can be seen, the agreement of theory with experiment improves considerably, and the remaining discrepancies can be explained by the broadening of the spectrum due to the effect of phonons and the inaccuracy of H employed.

4.4. *Vibronic spectra of mixed crystals*

It is obvious that the analysis of the vibronic spectra of mixed molecular crystals is an even more involved problem than the problem of electronic excitons in mixed crystals. One can form a notion about it, proceeding from the dynamic theory of vibronic spectra (section 3.3). In this approach the vibronic absorption is associated with simultaneous formation of two interacting particles, an electronic exciton and an internal phonon. In a disordered crystal each of these particles moves in a random field: an isotopic shift of electron energy Δ_{ex} for the exciton, and the shift of vibrational frequency Δ_{ph} for the phonon. According to eqs. (3.18)–(3.20), the presence of disorder also modifies the interaction Hamiltonian. Obviously, it is difficult to establish some kind of general regularities for these spectra, and their full description can be made only on the basis of a numerical calculation which has not yet been performed.

There exists, however, a special case when a narrow band of a vibron is liable to appear in these spectra. This vibron can be assigned almost entirely to the excitation of molecules in a single isotopic component. This possibility, which we are going to discuss as applied to an NTS phonon, exists due to the following factors:

(1) Vibronic transition occurs as an intramolecular transition, i.e. the matrix element of the transition corresponds to the creation of an exciton and an internal phonon at the same site.

(2) Dispersion in internal phonon bands is small. Therefore an NTS phonon created by light can be considered as fixed at the same site where it has been created.

(3) The energy of the created electronic exciton is equal to the difference in energy of the light quantum and the vibrational quantum of the molecule in its ground electronic state. Usually this energy gets into the continuous exciton spectrum of the mixed crystal, and the exciton can leave its "own" phonon. This provides for the wide-band absorption. An exception to this rule is observed when the exciton energy falls within (i) the region off the spectrum ($\rho(E)=0$) or the region with a low $\rho(E)$ (tails of DOS, pseudogap; section 4.3), or else (ii) the region of the continuum where the Anderson localization occurs (section 7.1). Then the local state of the exciton which results is stationary or quasi-stationary, and the absorption band is narrow.

To date, the vibronic spectra of the system N-h_8 + N-d_8 have been studied experimentally (absorption spectra: Sheka and Terenetskaya (1971); reflection spectra: Tokura et al. (1980)). These experimental data are very interesting and basically agree with each other. The peak assigned to the low-frequency vibron (N-h_8) is seen throughout the concentration range. Undoubtedly, this is due to the fact that at all concentrations the exciton energy falls outside the continuum or in its low-frequency tail. The high-frequency vibronic peak (N-d_8) changes its shape in a complicated manner at $C_{d_8} \lesssim 0.2$ and then rapidly broadens and disappears. Of the above authors, the former describe this behaviour to the exciton level getting into a pseudogap at $C_{d_8} \lesssim 0.1$–0.2, and the latter, to the Anderson transition in a lower subband at $C_{d_8} \approx 0.1$–0.15. We feel that the first interpretation agrees better with certain details of the experimental results, although the role of factor (ii) may also prove to be important. This problem calls for further investigation.

5. Reconstruction of exciton band structure. Band-to-band transitions (BTBT)

As seen from sections 3 and 4, the main problem of the spectroscopy of excitons is the determination of exciton band structure, to wit, dispersion law $\epsilon_\mu(\boldsymbol{k})$ and density of states $\rho_0(\epsilon)$. This furnishes key data for a description of a wide class of optical spectra. Some methods of determining a spectrum have already been mentioned above, such as the one based on dependence of the spectrum of monomer centers on the isotopic shift (section 3.1) and another one based on the spectra of aggregate centers (section 3.2). In both cases the spectrum is reconstructed from experimental data by means of a specific theoretical treatment. Some other methods proceeding from impurity absorption in the region of the ionization continuum or else from the shape of the spectrum of vibronic two-particle absorption have also been proposed. However, since they have not yet been applied experimentally for electron excitons, we do not dwell on them here.

Now we shall briefly describe yet another and, apparently, the most direct method of reconstructing $\rho_0(\epsilon)$, the BTBT method. It has been employed successfully as two different versions.

5.1. Vibronic band-to-band transitions (VBTBT)

The usual scheme of VBTBT is shown in fig. 20. It corresponds to direct vibronic transitions. In these transitions the quasi-momentum of a vanishing

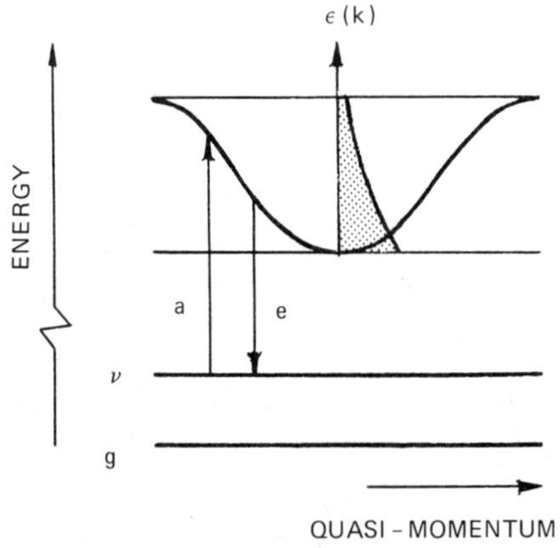

Fig. 20. Scheme of vibronic band-to-band transitions. Energy levels: (g) crystal ground state, (ν) the band of an internal phonon with frequency ν (dispersion is neglected); at the top: the exciton band with dispersion law $\epsilon(\boldsymbol{k})$. Shaded profile: Boltzmann energy distribution of excitons. Shown are the transitions corresponding to light absorption (a) and emission (e).

(created) internal phonon coincides with the quasi-momentum of a created (vanishing) exciton. The intensity of the absorption spectrum is proportional to the average Planck occupation numbers for phonons $\bar{n} = \bar{n}(\nu)$, $\nu = \nu(\boldsymbol{k})$ being the IP dispersion law. At low temperatures $T \ll \nu$ which are usually employed, $\bar{n}$ is proportional to $\exp(-\nu/T)$. Such absorption "maps" the whole exciton band, since the transitions occur into the states with all $\boldsymbol{k}$. Back transitions observed in the fluorescence spectrum practically occur only from that part of the exciton band which has the width $\sim T$; therefore the "map" of the whole band appears only if $T \sim 2\mathfrak{M}$. Since all the VBTBT are only visible at $T \neq 0$, their spectroscopy is named hot-band spectroscopy.

After Broude et al. (1963) had observed the temperature dependence of VBTBT in the fluorescence spectrum of naphthalene, Rashba (1963) showed that under the conditions commonly met in molecular crystals (with the exception of the most narrow bands of triplet excitons), in particular, $\nu(\boldsymbol{k}) =$ const, the VBTBT spectra in absorption $A(\Omega)$ and emission $E(\Omega)$ can be expressed directly in terms of the density of states $\rho_0(\epsilon)$:

$$A(\Omega) \propto \rho_0(\Omega + \nu - E_\rho), \tag{5.1}$$

$$E(\Omega) \propto \rho_0(\Omega + \nu - E_\rho) \exp\left\{-\frac{1}{T}(\Omega + \nu - E_\rho - \epsilon_{\min})\right\}. \tag{5.2}$$

Here $\epsilon_{\min}$ is the energy at the bottom of the exciton band. Thus the measurement of $A(\Omega)$ and $E(\Omega)$ yields a direct method of reconstruction of $\rho_0(\epsilon)$. It is by this method that Colson et al. (1968) have determined $\rho_0(\epsilon)$ for singlet excitons in benzene and naphthalene.

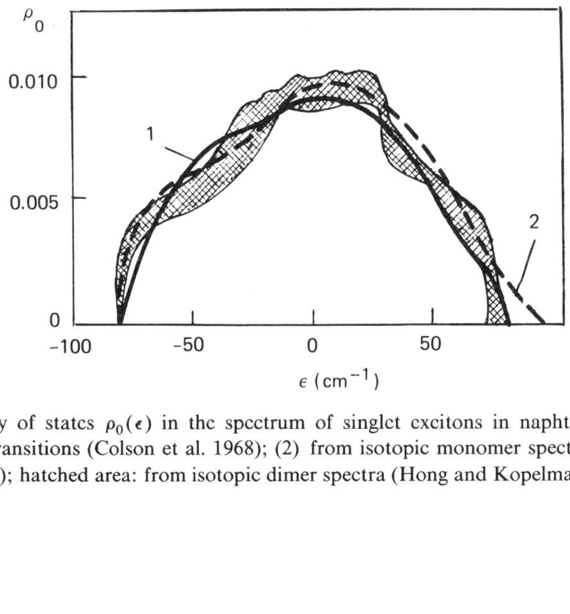

Fig. 21. Density of states $\rho_0(\epsilon)$ in the spectrum of singlet excitons in naphthalene. (1) From band-to-band transitions (Colson et al. 1968); (2) from isotopic monomer spectra (Rabin'kina et al. 1970); hatched area: from isotopic dimer spectra (Hong and Kopelman 1971a).

Figure 21 permits us to judge how close is the agreement of the $\rho_0(\epsilon)$ curves obtained by three different methods for naphthalene.

The VBTBT method determines $\rho_0(\epsilon)$ in crystals with the two- and three-dimensional spectra (2D and 3D), while in crystals with the quasi-1D spectrum it permits reconstruction of the dispersion law, since in the latter case $\epsilon(k)$ and $\rho_0(\epsilon)$ are related by a simple relation

$$\frac{\mathrm{d}\epsilon(k)}{\mathrm{d}k} = \frac{a}{2\pi}\rho_0^{-1}(\epsilon). \tag{5.3}$$

Here a is the lattice spacing properly determined. For a lattice of the DBN type (section 3.2) a represents the intermolecular distance in a stack.

While the stationary experiments on VBTBT permit the measurement of $\rho_0(\epsilon)$, the pulse measurements, where excitons with $\boldsymbol{k} \approx 0$ are created at the moment $t = 0$ and then transient spectra of band-to-band luminescence are

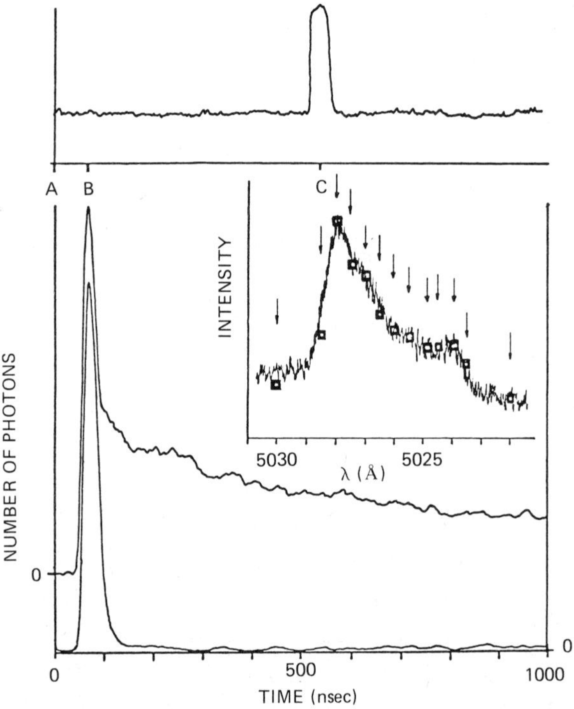

Fig. 22. Vibronic band-to-band transitions in DBN. Top trace: the output from the multiplier tube following the single-laser excitation pulse. Bottom trace: the scattered light of the laser pulse. Middle trace: the time dependence of emission on the 502.75 nm wavelength falling within the region of VBTBT. Decrease in intensity is characteristic for exciton phosphorescence. Insertion: data points and arrows show the variation with the spectrometer setting of the intensity of the slowly decaying component of the emission shown in the middle trace. The continuous line shows a steady-state VBTBT spectral profile. $T = 21$ K (Leong and Hanson 1983).

observed, make it possible to evaluate the relaxation times of band excitons. Experiments of this kind recently carried out on pure DBN by Leong and Hanson (1983) have shown that the exciton thermalization time does not exceed 50 ns at 21 K. A monotonic drop of the middle trace in fig. 22 and the data represented in the insert give conclusive evidence of this fact.

5.2. *Spin band-to-band transitions (SBTBT)*

Another scheme of band-to-band transitions (SBTBT) has been proposed and successfully used by Francis and Harris (1971) for investigation of triplet excitons. To elucidate this scheme, it is essential to discuss in detail the structure of the triplet exciton energy spectrum.

It has been mentioned above that the energy band width of triplet excitons is determined by an exchange interaction of neighbouring molecules, and usually $2\mathfrak{M} \sim 1$–$30\ \mathrm{cm}^{-1}$ in organic crystals. This magnitude is small due to the numerical smallness of overlap integrals in van der Waals crystals. Had the triplet states of molecules been triply degenerate in the absence of a magnetic field (in accordance with the three orientations of the spin $S = 1$), then the same would have been valid for the spin bands, too.

However, the triplet degeneracy of molecular levels is lifted due to the dipole–dipole and spin–orbit interactions. This splitting is small, and therefore, although the orbital momentum is quenched, the spin remains to be an acceptable quantum number. Let us assume that there are three distinct directions in a molecule: x, y, and z, as a consequence of its symmetry. For these molecules the zero-field spin Hamiltonian is generally written as

$$\begin{aligned} H_{\mathrm{mol}} &= \boldsymbol{S} \cdot \hat{\boldsymbol{T}} \cdot \boldsymbol{S} = -\left(XS_x^2 + YX_y^2 + ZS_z^2 \right) \\ &= \mathfrak{D}\left(S_z^2 - \tfrac{1}{3}\boldsymbol{S} \cdot \boldsymbol{S} \right) + \mathfrak{E}\left(S_x^2 + S_y^2 \right). \end{aligned} \tag{5.4}$$

In eq. (5.4) the energy is reckoned from E_ρ^{T}, the centre of gravity of the energy spectrum of triplet excitons. For the molecules in question, the order of magnitude characteristic for parameters $\mathfrak{D}$, $\mathfrak{E}$ is equal to $\mathfrak{D}, \mathfrak{E} \sim 3000$ MHz $\sim 10^{-1}\ \mathrm{cm}^{-1}$. Of the same scale are E_q^0, $q = x, y, z$, the three eigenvalues of the Hamiltonian H_{mol}. It is essential that $\mathfrak{D}, \mathfrak{E} \ll \mathfrak{M}$. Therefore in this approximation the energy spectrum of an exciton should consist of three identical bands (T_x, T_y, and T_z) of width $2\mathfrak{M}$ shifted with respect to each other by energies $E_q^0 \sim \mathfrak{D}, \mathfrak{E} \ll 2\mathfrak{M}$:

$$E_q(\boldsymbol{k}) = E_\rho^{\mathrm{T}} + \epsilon_0^{\mathrm{T}}(\boldsymbol{k}) + E_q^0, \qquad q = x, y, z. \tag{5.5}$$

Now explicit account must be taken of the spin–orbit interaction. Its influence is primarily manifest in the existence of phosphorescence since it is precisely the spin–orbit interaction that admixes singlet states to the triplet state and makes optical transitions allowed. The same interaction taken in the second order of perturbation theory introduces into the spectrum the terms which are dependent on both q and $\boldsymbol{k}$. Indeed, each of the q-states interacts with a definite set of singlet bands (having the same symmetry). As the energy separation of those bands and their dispersion law $\epsilon_q^{\mathrm{S}}(\boldsymbol{k})$ depend on q, the denominators in the equations of perturbation theory

$$\left(E_\rho^{\mathrm{S}} - E_\rho^{\mathrm{T}}\right) + \left(\epsilon_q^{\mathrm{S}}(\boldsymbol{k}) - \epsilon_0^{\mathrm{T}}(\boldsymbol{k})\right),$$

as well as the corresponding numerators, also depend on q. Assuming the second term in the last expression to be small, we obtain instead of eq. (5.5):

$$\begin{aligned} E_q(\boldsymbol{k}) &= E_\rho^{\mathrm{T}} + E_q^0 + \epsilon_0^{\mathrm{T}}(\boldsymbol{k}) + \epsilon_q(\boldsymbol{k}), \\ \epsilon_q(\boldsymbol{k}) &= \left(\tilde{E}_{\mathrm{SO}}/\tilde{E}_{\mathrm{ST}}\right)^2 \tilde{\epsilon}_q^{\mathrm{S}}(\boldsymbol{k}). \end{aligned} \tag{5.6}$$

Here $\tilde{E}_{\mathrm{SO}}/\tilde{E}_{\mathrm{ST}}$ is the mean value of the ratio of the spin–orbit interaction energy to the spacing between the T-band and singlet bands, and $\tilde{\epsilon}_q^{\mathrm{S}}(\boldsymbol{k})$ is the "effective" dispersion law in the singlet bands. It can be seen from eq. (5.6) that, even if the dispersion law $\epsilon_0^{\mathrm{T}}(\boldsymbol{k})$ is one dimensional, the correction $\epsilon_q(\boldsymbol{k})$ may include an appreciable 3D contribution, because it comes from singlet bands whose anisotropy is usually much less than that of triplet bands. The absolute value of this term, however, is very small due to the smallness of $\tilde{E}_{\mathrm{SO}}$ in organic molecules free from heavy atoms. If one supposes after Francis and Harris (1971) that

$$\epsilon_0^{\mathrm{T}}(\boldsymbol{k}) = -\mathfrak{M}_{\mathrm{T}} \cos(\boldsymbol{k}a), \qquad \tilde{\epsilon}^{\mathrm{S}} = -\mathfrak{M}_{\mathrm{S}} \cos(\boldsymbol{k}a), \tag{5.7}$$

$$\mathfrak{M}_{\mathrm{S}} \sim 100\ \mathrm{cm}^{-1}, \quad \mathfrak{M}_{\mathrm{T}} \sim 1\ \mathrm{cm}^{-1}, \quad \tilde{E}_{\mathrm{ST}} \sim 10^4\ \mathrm{cm}^{-1}, \quad \tilde{E}_{\mathrm{SO}} \sim 10\ \mathrm{cm}^{-1},$$

then

$$\epsilon_q(\boldsymbol{k}) \equiv \tilde{\epsilon}(\boldsymbol{k}) = f_{\mathrm{r}}\epsilon_0^{\mathrm{T}}(\boldsymbol{k}), \qquad f_{\mathrm{r}} = \left(\tilde{E}_{\mathrm{SO}}/\tilde{E}_{\mathrm{ST}}\right)^2 (\mathfrak{M}_{\mathrm{S}}/\mathfrak{M}_{\mathrm{T}}) \sim 10^{-4}. \tag{5.8}$$

Factor $f_{\mathrm{r}} \ll 1$ is called the reduction factor.

The electron spin resonance (ESR) in the zero magnetic field includes all transitions between q and q' levels induced by a microwave field. They occur with conservation of quasi-momentum at the frequencies $E_{qq'}(\boldsymbol{k}) = E_q(\boldsymbol{k}) - E_{q'}(\boldsymbol{k})$ and, obviously, are SBTBT. These transitions can be detected optically (Schmidt and van der Waals 1968, Buckley et al. 1970) owing to the q-dependence of the oscillator strength $f(q)$ for various triplet transitions. Experimental data for one of the three possible transitions in 1,2,3,4-tetrachlorobenzene

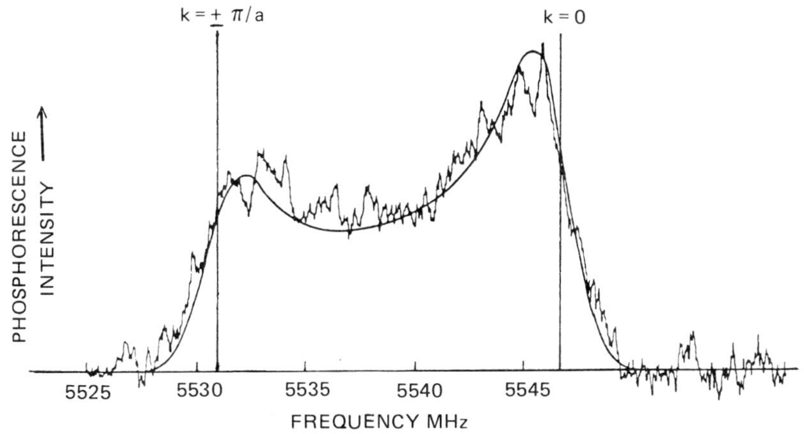

Fig. 23. Optically detected spin band-to-band transitions in TCB. $T = 3.2$ K. Solid curve: theory, including the damping of one-dimensional excitons introduced phenomenologically (Francis and Harris 1971).

(TCB) are presented in fig. 23. The general shape of the spectrum with two maxima at the edges shows that it is close to a one-dimensional one. This is consistent with the absence of the Davydov splitting, as this is a measure of the interaction between different stacks of molecules. The thermal distribution of excitons in the band is responsible for the asymmetry of the curve. The difference in the height of the maxima corresponds to $2\mathfrak{M}_T \approx 1.3\ \text{cm}^{-1}$. A stronger high-frequency maximum has been ascribed to the point $\boldsymbol{k} = 0$, i.e. it has been proposed that it is at the bottom of the exciton band. More recently Dlott and Fayer (1976) have come to the conclusion that $\boldsymbol{k} = 0$ corresponds to the top of the band. The most convincing evidence of this fact was given by van Strien et al. (1980).

The shape of the curve in fig. 23 with maximum near its edges shows unambiguously that the band is of 1D structure and that the excitons are coherent (see section 7.1). If the smearing of the edges of the spectrum is ascribed entirely to the scattering of excitons, then it is possible to obtain a lower bound for the mean free time and mean free length of excitons: $\tau_f \gtrsim 10^{-9}$ s, $l_f \gtrsim 50$–100 Å (Francis and Harris 1971). The weakness of inelastic scattering results from the fact that in a narrow band $2\mathfrak{M}_T \ll \theta_D$ (θ_D is the Debye temperature) one-phonon scattering is forbidden, and at $T \ll \theta_D$ the phase volume for the two-phonon (Compton) scattering of excitons is very small. The intensity of elastic scattering of excitons is determined by the degree of sample perfection. In general, it increases with decreasing band width.

Besides scattering, both elastic and inelastic, the rounding-off of the edges of the SBTBT curve can be brought about by even more mechanisms. First of all, it is a non-one-dimensional behaviour of functions $\epsilon_q(\boldsymbol{k})$. Although their absolute value is, according to eqs. (5.7) and (5.8), small (about $10^{-2}\%$ of ϵ_0^{T}) and so the one-dimensionality of the dispersion law remains practically unaffected, the SBTBT spectrum is determined by an integral

$$\sigma(\omega) \propto \int \frac{\mathrm{d}^3 k}{(2\pi)^3} \exp(-E_q(\boldsymbol{k})/T)$$

$$\times \delta\left[\left(E_{q'}^0 - E_q^0\right) + \left(\epsilon_{q'}(\boldsymbol{k}) - \epsilon_q(\boldsymbol{k})\right) - \omega\right]. \qquad (5.9)$$

Here ϵ_0^{T} is cancelled, and it is the functions $\epsilon_q(\boldsymbol{k})$ which enter into the argument of the δ-function. Their non-one-dimensionality leads to the smoothing of edge singularities in $\sigma(\omega)$ due to integration over transverse components of $\boldsymbol{k}$. Another mechanism of smoothing the edges of the SBTBT spectrum, the space inhomogeneity associated with interruptions in TCB stacks, has been indicated by Francis and Harris (1971). This mechanism is equivalent to a strong impurity back-scattering.

From these facts it transpires that the obtained $\tau_{\mathrm{f}} \gtrsim 10^{-9}$ s is indeed a lower bound for τ_{f} only. Reliable data on inelastic components of scattering are given in section 7.3. It should be stressed that in 1D systems the problem of the relation between the impurity and the phonon scattering is of particular importance. The answer to this question determines whether the 1D localization exists and whether the transport is of a hopping or band type (section 7.1).

6. *Two-level systems. Coherence and dephasing*

Sections 3.1 and 3.2 have dealt with the energy spectrum of local excitons. The present section concerns another characteristic of the wave function of the local exciton, to wit, its coherence. For the sake of simplicity we shall consider the impurity center as a two-level system. Let us start, for instance, from the Hamiltonian (5.4). It has three eigenstates, T_x, T_y, T_z. If the external perturbing field is in resonance with the energy difference of the two of them, for instance, $Y - Z$, we can exclude the third state, in this case T_x, from consideration. Thus we come to a two-level system. Methods for investigation of systems such as this one have been developed in the physics of magnetic resonance (see reviews by Abragam 1961, Farrar and Becker 1971). In the last decade they have been successfully applied to local triplet molecular excitons. It is of prime importance that ways have been found of optically detecting the processes in a triplet system through investigating the phosphorescence. It substantially increases the sensitivity of experimental methods. The language of developed methods is based on the techniques of Feynman et al. (1957) (below as FVH) geometrizing the two-level system dynamics in convenient terms of the theory of magnetic resonance. That is why we shall begin with a brief description of this method.

There are excellent reviews by Schmidt and van der Waals (1979) and Harris and Breiland (1978) on this field. These reviews have greatly influenced our understanding of the problems under discussion, and we refer the reader to these sources which offer a far more comprehensive treatment of the phenomena.

6.1. Dynamics of quasi-spin

In this section we consider a two-level system corresponding to T_z and T_y states of a molecule with spin $S = 1$ *. Such a subsystem is conveniently described as a system with quasi-spin $s = 1/2$. This means that all operators are written as matrices 2×2 and expanded in Pauli matrices σ_i.

Introduce the density matrix

$$\hat{\rho} = \begin{vmatrix} \rho_{zz} & \rho_{zy} \\ \rho_{yz} & \rho_{yy} \end{vmatrix}, \qquad \mathrm{Tr}\{\hat{\rho}\} = \rho_{zz} + \rho_{yy} = 1. \tag{6.1}$$

The second equation is the normalization condition. The equation of motion for $\hat{\rho}$ takes the usual form

$$\frac{\mathrm{d}\hat{\rho}}{\mathrm{d}t} = \mathrm{i}[\hat{\rho}, \hat{\boldsymbol{H}}], \qquad \hat{\boldsymbol{H}} = \begin{vmatrix} (Z-Y)/2 & V_{zy} \\ V_{yz} & (Y-Z)/2 \end{vmatrix}, \tag{6.2}$$

V is a perturbation operator.

The mean values of quasi-spin projections $s_i = \mathrm{Tr}\{\sigma_i \hat{\rho}\}$ are equal to

$$s_1 = \tfrac{1}{2}(\rho_{zy} + \rho_{yz}), \qquad s_2 = \tfrac{1}{2}\mathrm{i}(\rho_{zy} - \rho_{yz}), \qquad s_3 = \tfrac{1}{2}(\rho_{zz} - \rho_{yy}). \tag{6.3}$$

These expressions suggest the form of the vector $\boldsymbol{r} = (r_1, r_2, r_3)$ which can be substituted for the $\hat{\rho}$-matrix: $r_i = 2s_i$. Similarly, vector $\boldsymbol{\Omega}$ substituting $\hat{\boldsymbol{H}}$ is

$$\Omega_1 = V_{zy} + V_{yz}, \qquad \Omega_2 = \mathrm{i}(V_{zy} - V_{yz}), \qquad \Omega_3 = (Z - Y). \tag{6.4}$$

Ω_1 and Ω_2 yield the probabilities of transitions produced by a driving

* We select them in such a manner solely to be specific; in the experiments described below various pairs of levels were chosen.

high-frequency field, and Ω_3 the difference in energy levels. Now we use the matrix representation:

$$\hat{\rho} = \tfrac{1}{2}\hat{\boldsymbol{E}} + \tfrac{1}{2}\boldsymbol{r\sigma} = \tfrac{1}{2}\hat{\boldsymbol{E}} + \tfrac{1}{2}(r_1\sigma_x + r_2\sigma_y + r_3\sigma_z), \tag{6.5}$$

$$\hat{\boldsymbol{H}} = \tfrac{1}{2}\boldsymbol{\Omega\sigma} = \tfrac{1}{2}(\Omega_1\sigma_x + \Omega_2\sigma_y + \Omega_3\sigma_z). \tag{6.6}$$

Here $\hat{\boldsymbol{E}}$ is a unit matrix, and $(Z+Y)/2$ is chosen as the zero point of energy. One can verify directly that eq. (6.2) is equivalent to the equation

$$\frac{\mathrm{d}\boldsymbol{r}}{\mathrm{d}t} = \boldsymbol{\Omega}(t) \times \boldsymbol{r}. \tag{6.7}$$

When applying these relations to a pure state

$$\psi(t) = a(t)T_z + b(t)T_y \tag{6.8}$$

we have

$$r_1 = ab^* + a^*b, \qquad r_2 = \mathrm{i}(ab^* - a^*b), \qquad r_3 = aa^* - bb^*. \tag{6.9}$$

The transitions between T_z and T_y are induced by the magnetic field $\tilde{\boldsymbol{H}} \parallel x$; this results from the form of matrices $\hat{S}_i$ of the spin $S=1$ in the x, y, z basis. Hence, the term in the Hamiltonian describing the time-dependent driving field may be written as $V(t) = \gamma\tilde{H}\hat{S}_x\cos(\omega t)$. Then $\Omega_1 = 0$, $\Omega_2 = 2|\gamma|\tilde{H}\cos(\omega t) = 2\omega_{\mathrm{R}}\cos(\omega t)$, where $\omega_{\mathrm{R}} = |\gamma|\tilde{H}$ is the Rabi frequency. Using eq. (6.8), it is easily seen that $\langle\psi|\hat{S}_x|\psi\rangle = -r_2$, i.e. r_2 directly gives transverse magnetization. The r_3 component determines the population difference in the T_z and T_y states.

If $\tilde{H} = 0$, then $\Omega_1 = \Omega_2 = 0$ and $\boldsymbol{\Omega} = \Omega_3\boldsymbol{e}_3$; here and below $\boldsymbol{e}_i \to \boldsymbol{e}_1, \boldsymbol{e}_2, \boldsymbol{e}_3$ denotes three unit vectors in the $\boldsymbol{r}$-vector space. Under these conditions the motion of the $\boldsymbol{r}$-vector is, according to eq. (6.7), a precession around the $\boldsymbol{e}_3$ direction with the Larmor frequency $Z - Y = \omega_0$ (to be specific, take $Z > Y$). The frequency of precession is determined by zero-field splitting.

The picture is more intricate at $\tilde{H} \neq 0$. In this case it is convenient to change to the rotating reference system which is determined by three unit vectors:

$$\begin{aligned} &\tilde{\boldsymbol{e}}_1(\omega t) = \boldsymbol{e}_1\cos(\omega t) + \boldsymbol{e}_2\sin(\omega t), \\ &\tilde{\boldsymbol{e}}_2(\omega t) = \boldsymbol{e}_3 \times \tilde{\boldsymbol{e}}_1(\omega t) \qquad \text{and} \qquad \boldsymbol{e}_3. \end{aligned} \tag{6.10}$$

It is obvious that $\tilde{\boldsymbol{e}}_2(\varphi) = \mathrm{d}\tilde{\boldsymbol{e}}_1(\varphi)/\mathrm{d}\varphi$ and $(\tilde{\boldsymbol{e}}_1\tilde{\boldsymbol{e}}_2) = 0$. Bearing in mind that $\delta\boldsymbol{r}/\delta t = \mathrm{d}\boldsymbol{r}/\mathrm{d}t - \boldsymbol{\omega}\times\boldsymbol{r}$, $\boldsymbol{\omega} = \omega\boldsymbol{e}_3$ in the rotating frame, and omitting in the right-hand side of eq. (6.7) the terms oscillating with the frequency 2ω (as is usually done in considering the resonances), we obtain

$$\frac{\delta \boldsymbol{r}}{\delta t} = \boldsymbol{\Omega}_0 \times \boldsymbol{r}, \qquad \boldsymbol{\Omega}_0 = \omega_{\mathrm{R}} \tilde{\boldsymbol{e}}_2 - \Delta\omega \boldsymbol{e}_3, \qquad \Delta\omega = \omega - \omega_0. \tag{6.11}$$

Thus, in the rotating frame, the problem reduces again to precession, but this time around an axis lying in the $\tilde{\boldsymbol{e}}_2\boldsymbol{e}_3$ plane. At resonance $\omega = \omega_0$ the $\boldsymbol{r}$-vector precesses around the direction $\tilde{\boldsymbol{e}}_2$ with frequency ω_{R}. The same is valid in a strong driving field, when $\omega_{\mathrm{R}} \gg |\Delta\omega|$. In general, the precession frequency is

$$\omega_{\mathrm{eff}} = \sqrt{\omega_{\mathrm{R}}^2 + (\Delta\omega)^2}\,. \tag{6.12}$$

At $\omega = \omega_0$ the description of a system in the rotating frame is, evidently, equivalent to the transformation into the interaction representation.

Equation (6.7) is basic for the FVH method. It allows us to visualize the dynamics of the two-level system and, due to its coincidence with the equation of motion of the magnetic moment, to introduce phenomenologically relaxation terms by analogy with the Bloch equation. It should be taken into account in some cases that the number of excited atoms (quasi-spins) is not fixed, i.e. $\mathrm{Tr}\{\hat{\rho}\} \neq 1$; this may be important as applied to fluorescence. In this case feeding and decay terms are phenomenologically introduced (Harris and Breiland 1978). Here we restrict ourselves to writing an analogue to the Bloch equation:

$$\frac{\mathrm{d}\boldsymbol{r}}{\mathrm{d}t} = \boldsymbol{\Omega} \times \boldsymbol{r} - \frac{r_1\boldsymbol{e}_1 + r_2\boldsymbol{e}_2}{T_2^*} - \frac{r_3 - r_3^{\mathrm{eq}}}{T_1}\boldsymbol{e}_3, \tag{6.13}$$

as it provides the definition of the two relaxation times: the transverse one T_2^* and the longitudinal one T_1. The former governs phase relaxation and the latter, the relaxation of the population difference in the T_z and T_y states. Determination of these times is the purpose of the majority of experiments.

6.2. Optical detection of coherence

Phase information on the state of a system is to be found in nondiagonal components of $\hat{\rho}$, i.e. in-plane components of the $\boldsymbol{r}$-vector. The discovery of EPR on triplet excitons in naphthalene and anthracene (Haarer and Wolf 1970) have opened the way for employing the traditional methods of magnetic resonance. An optical modification of these methods, however, has proved to be even more effective. It is based on double resonances, or, in other words, on detecting the evolution of the spin subsystem by a fluorescence channel. It reliably detects about 10^4 spins of local excitons in a sample.

The main difficulty inherent in this method of detection is of a principal nature and can be seen from the following (Harris 1971). The probability $W(t)$ of optical transition from the $\psi(t)$ state (eq. (6.8)) to the ground state of a molecule $|0\rangle$ is

$$W(t) \propto \left(|P_y|^2 + |P_z|^2\right) - \left(|P_y|^2 - |P_z|^2\right) r_3 + \mathrm{Re}\left[\left(P_y^* P_z\right) r_+(t)\right],$$
$$r_\pm(t) = r_1(t) \pm \mathrm{i} r_2(t). \tag{6.14}$$

Here we have used (6.9) and denoted the matrix elements of the transitions from T_y and T_z into $|0\rangle$ as P_y and P_z. The coefficients $a(t)$ and $b(t)$ change with t as $\exp(-\mathrm{i}Zt)$ and $\exp(-\mathrm{i}Yt)$. Hence $r_3 = \text{const}$, and the only term in $W(t)$ which reveals quantum beats with the frequency $\omega_0 = Z - Y$ (that is, discloses the presence of coherence) is none other than the last term in eq. (6.14). It does not vanish only when $(P_y^* P_z) \neq 0$, i.e. P_y and P_z are not mutually orthogonal. This is only possible in molecules with low symmetry.

It is seen from eq. (6.14) that in general $W(t)$ reflects only the time dependence of r_3, the population difference in the y- and z-states. Thus in order to detect coherence optically it must be transferred from the $r_\pm$ plane to the r_3 axis. An adequate means involving additional or probe pulse methods was offered by Breiland et al. (1973, 1975). The trick is that at the moment when the $\boldsymbol{r}$-vector phase in the $\boldsymbol{e}_1\boldsymbol{e}_2$ plane should be determined, a $\pi/2$ pulse is applied to the system. This short pulse of a strong microwave field with the frequency $\omega \approx \omega_0$ and duration t is such that $\boldsymbol{r}$ rotates, according to eqs. (6.11) and (6.12), through an angle $\omega_{\mathrm{R}} t = \pi/2$ around the $\tilde{\boldsymbol{e}}_2$ axis. As a result, the luminescence signal after the pulse is proportional to $(\tilde{\boldsymbol{e}}_1 \boldsymbol{r})^2$. Different versions of this method allow us to trace the whole picture of the motion of the in-plane component of $\boldsymbol{r}$.

There are magnetic resonance techniques which permit optical detection without recourse to the above-mentioned trick. An example of this type is the transient nutation. If a high-frequency field is applied at the moment $t = 0$, the quasi-spin considered in the rotating frame begins precessing around the axis $\boldsymbol{\Omega}_0$ from its initial position $r_1 = r_2 = 0$, $r_3 \neq 0$ with frequency ω_{eff} (see eq. (6.12)), and

$$r_3(t) = r_3(0) \cos(\omega_{\mathrm{eff}} t). \tag{6.15}$$

In the on-resonance regime ($\Delta\omega = 0$) this motion is the nutation with angular velocity ω_{R}. According to eq. (6.14), if $|P_y|^2 \neq |P_z|^2$, the intensity of luminescence will be nonzero and will follow the law $I(t) = I_0(t) + I_1 \cos(\omega_{\mathrm{eff}} t)$. Figure 24 shows the phosphorescence of quinoline in durene obtained with this method by Schmidt et al. (1971). The saturation region of

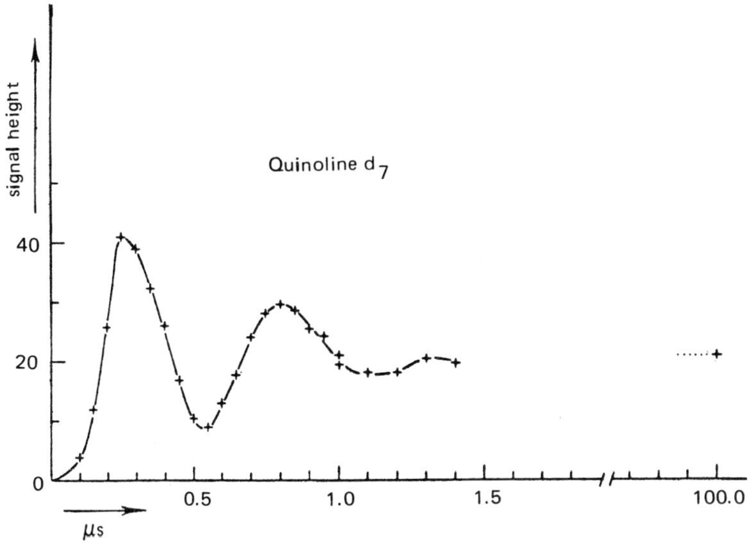

Fig. 24. Transient nutation of quinoline-d_7 in durene-d_{14}. $T = 1.25$ K (Schmidt et al. 1971).

the curve corresponds to equalization of populations in the T_y and T_z states due to dephasing with time T_2^*.

6.3. *Quenching of the influence of inhomogeneous broadening*

The rate at which the phase coherence is lost is determined by two factors. The first factor is that of irreversible phase-destructive events which are described by spin memory time T_M (or T_2) and lead to a decrease of r_1, $r_2 \rightarrow 0$ for homogeneous spin packets. The second factor is the inhomogeneous broadening, i.e. dissimilar ω_0 and ω_R for different spins. The statistical spread of ω_0 is usually characterized by a quantity $\Delta\omega_{1/2}$, the half-width at half-height. Usually the situation becomes complicated by the fact that besides the space dependence there is also the time dependence of random fields, e.g. the flip-flop motion of nuclear spins which affects electron levels and shortens T_2; these processes are usually called spectral diffusion. When studying physical mechanisms of spin relaxation, it is highly desirable to suppress external mechanisms of broadening as much as possible and to isolate T_M. A brief description of some of the methods in use follows.

One of them is the rotary echo. If under a transient nutation regime (section 6.2) at the moment $t_1 = \tau$ the phase of the field is shifted by 180°, this results, according to eq. (6.10), in $\tilde{e}_2$ changing its sign. And in the case of resonance ($\Delta\omega = 0$) it follows from eq. (6.11) that Ω_0 also changes sign. Hence the direction of nutation reverses in the rotating frame. Therefore at the moment

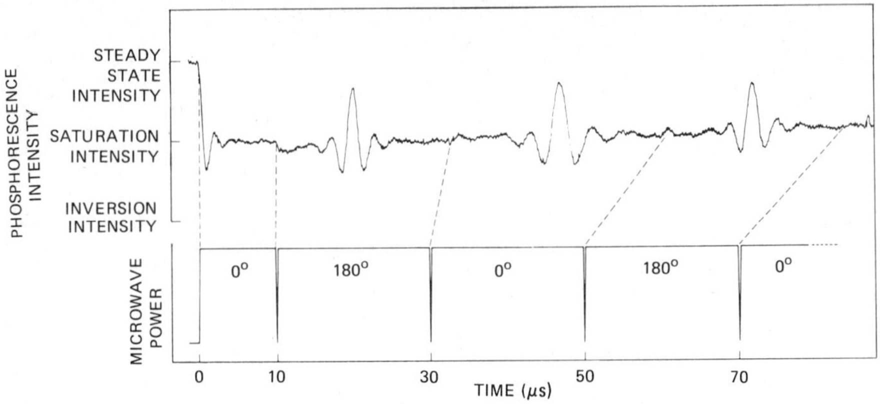

Fig. 25. Optically detected electron spin rotary echo train for Y-trap in TCB-h_2. $T = 2$ K (Harris et al. 1973).

$t_2 = 2\tau$, the spins all return to their original positions (the same which existed at the moment $t = 0$), no matter how strong the spread of ω_R. Consequently, at that moment the initial "magnetization" (i.e., the population difference in both levels) will be restored, and the fluorescence pulse will arise. That is what is called rotary echo. It was assumed above that eq. (6.11) holds true, i.e. $T_M = \infty$. At a finite T_M the magnetization at $t = 2\tau$ is less than that at $t = 0$, and their ratio permits T_M to be determined. Some cases require application of rotary echo trains, that is, the 180° phase shifts at the moments τ, 3τ, 5τ, etc. Such a train, first optically detected by Harris et al. (1973) is represented in fig. 25. $T_2 \sim 600$ μs thus obtained is far greater than $T_2 \sim 8$ μs obtained for the same system by the Hahn echo (see below). This is due to the fact that a strong microwave field $\tilde{H} \approx 1.8$ G constantly affecting the system in the rotary echo regime, substantially "decouples" the electron spin from the mechanisms of spectral diffusion.

The method of the two-pulse echo, or Hahn echo, makes it possible to exclude dephasing involved by the spread of precession frequencies ω_0. If at the initial moment $t = 0$ when $r_1 = r_2 = 0$, $r_3 = r$, the quasi-spins are all acted upon by a $\pi/2$ pulse with the same Rabi frequency, then the $\boldsymbol{r}$-vector transforms from the r_3 position to the r_1 position (i.e., saturation of the YZ transition is attained), and the free precession of quasi-spins commences in the laboratory frame. Due to the ω_0 spread, the dephasing of the quasi-spins will occur in a time $T_2^* \sim (\Delta\omega_{1/2})^{-1}$. Next, a π pulse is applied at the moment $t = \tau$. It rotates the spins around $\tilde{e}_2$ through an angle π. This means that in the rotating reference system the phase $\varphi_- = (\omega_0 - \omega)\tau$ reckoned from $\tilde{e}_1$ changes instantly and takes a value $\varphi_+ = \pi - \varphi_- = \pi - (\omega_0 - \omega)\tau$. In the course of the subsequent free precession the phase changes as $\varphi(t) = \varphi_+ + (\omega_0 - \omega)(t - \tau)$

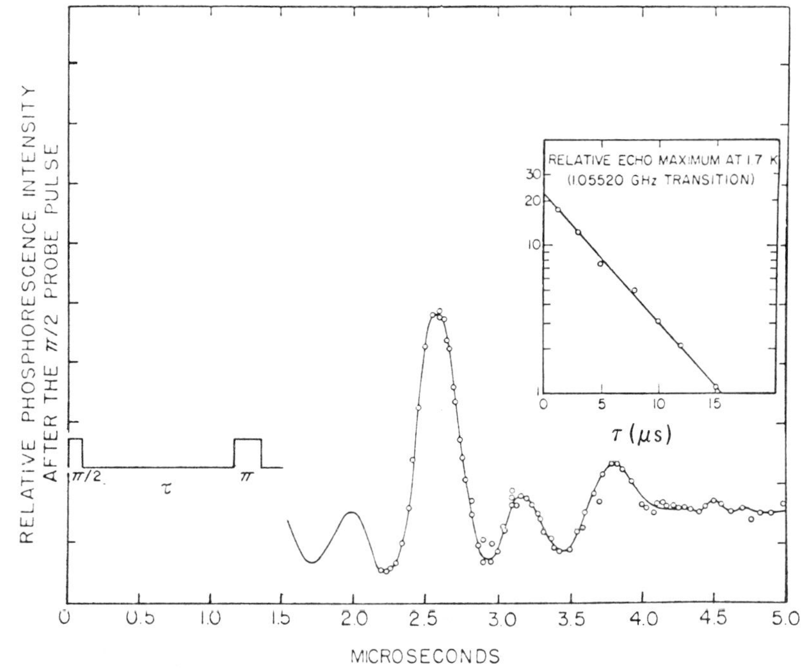

Fig. 26. Optically detected Hahn's echo for 2,3-dichloroquinoxaline in TCB. $T = 1.7$ K. The highest peak is the echo. Other maxima are due to a superfine interaction (Breiland et al. 1973).

and becomes $\varphi(2\tau) = \pi$ at the moment $t = 2\tau$. Thus the phase spread due to the inhomogeneous broadening becomes completely compensated at the moment $t = 2\tau$. Therefore the original coherent state is reconstructed when $T_M = \infty$. This phenomenon is called the Hahn echo. The intensity of the echo allows measurement of T_2.

The system emits microwaves at the moment just following the $\pi/2$ pulse and again at the moment of the echo ($t = 2\tau$), i.e., when there exists a coherent oscillating magnetic moment. It was observed by Schmidt (1972) on quinoline monomers in durene. An additional pulse (see section 6.2) allows us to transfer the coherent signal to the $\boldsymbol{e}_3$ axis and obtain the phosphorescence signal. Breiland et al. (1973) detected the Hahn echo optically for a 2,3-dichloroquinoxaline guest in TCB host (fig. 26). T_2 was determined to be 5 μs from the τ-dependence of the echo.

In the spin-locking method a $\pi/2$ pulse which tilts r_3 in the r_1 position is instantly followed by the 90° phase shift in the microwave field $\tilde{H}$, after which the field acts upon the system during a time τ. When the field phase shifts 90°, the $\tilde{e}_1$ and $\tilde{e}_2$ directions interchange, while $\boldsymbol{r}$ is left unaltered. As a result, $\boldsymbol{r}$ is located in the direction of a new vector $\tilde{e}_2$, i.e. $\boldsymbol{r}$ becomes fixed in

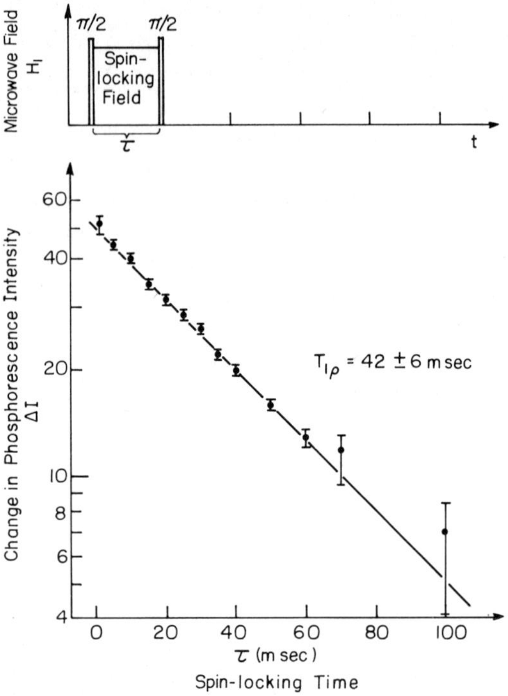

Fig. 27. Fluorescence relaxation time $T_{1\rho}$ (close to T_2) for TCB-d_2 in durene-d_{14}. $T = 2$ K. Shown at the top is the sequence of microwave pulses (according to Harris and Breiland 1978).

the rotating reference system. If the field $\tilde{H}$ is sufficiently strong, coherence should be very long-lived in the spin-locked regime due to the averaging of the effects of spectral diffusion. The slow decay of the spin-locked state can be optically observed by means of short additional $\pi/2$ pulses. These pulses should be applied in a corresponding phase, e.g. after an additional 90° phase shift restoring $\tilde{e}_1$ and $\tilde{e}_2$ to their initial positions in the rotating frame. Such an experiment was carried out by Harris et al. (1973); the results are shown in fig. 27. The large value of coherence time obtained (42 ms) is only slightly less than the average lifetime (63 ms) of two spin sublevels coupled by the spin-locking field.

6.4. *Dimers or "mini-excitons"*

Dimers are objects intermediate between monomers and excitons. They approximate monomers in simplicity and can be described by the techniques developed for two-level systems. On the other hand, dimers are the easiest systems for which the one question that is of the utmost importance for excitons may be raised: is the electron excitation transfer between the molecules coherent or does the hopping mechanism operate here? For this reason, the dimer is sometimes considered as a "mini-exciton".

Schwoerer and Wolf (1967) pioneered the study of a convenient model system based on naphthalene, to wit, the N-h_8 dimer in N-d_8. Below we are going to discuss only triplet excitons of that system. It is known from spectroscopic data that for N_1N_2 pairs (fig. 1) the magnitude of the exchange splitting of levels $\delta E \approx 2M_{12} \approx 2.5$ cm^{-1}, M_{12} being an exchange integral; so δE is similar to the Davydov splitting. The lower excited level apparently corresponds to the antisymmetric wave function of the dimer Φ_-. The general structure of the spectrum is the same as described in section 5.2 for excitons. The leading term of the Hamiltonian is δE, the term next in magnitude is the zero-field splitting (almost the same as in a monomer), and the difference $\Delta = \omega_0^+ - \omega_0^- \equiv \omega_0^+ - \omega_0$ in the zero-field splitting of the level corresponding to the Φ_- and Φ_+ states is an analogue of $\epsilon_q(\boldsymbol{k})$.

Bottler et al. (1976) observed for a N_1N_2 pair a two-pulse echo in an external magnetic field and determined the temperature dependence of T_{M} as well as of the T_yT_z transition frequency in the zero field ($T_M(T)$ and $\omega_0(T)$, respectively). The dephasing mechanism is associated with $\Delta \neq 0$. Due to the thermal excitation of the system resulting in its promotion from the Φ_- into the Φ_+ state, the spin precesses for a certain time fraction at a frequency $\omega_0 + \Delta$. Since thermal excitation is a stochastic process, it leads to dephasing. The same mechanism, creating a certain effective interaction between levels, involves a temperature shift of frequency of the T_yT_z transition in the ground state: $\omega_0(T) - \omega_0(T=0) \equiv \epsilon(T)$. These quantities are described by equations (van 't Hof and Schmidt 1975)

$$T_{\mathrm{M}}^{-1} = \frac{1}{\tau}\frac{\Delta^2\tau^2}{1+\Delta^2\tau^2}\exp(-\delta E/T), \tag{6.16}$$

$$\epsilon(T) = \frac{\Delta}{1+\Delta^2\tau^2}\exp(-\delta E/T). \tag{6.17}$$

Here τ is the time of de-excitation of the Φ_- state:

$$W = \tau^{-1}\exp(-\delta E/T). \tag{6.18}$$

Experimental data are presented in figs. 28 and 29. The exponential dependence of $\epsilon(T)$ and T_{M}^{-1} (in the low-temperature region) determines $\delta E = 2.5$ cm^{-1} in agreement with spectroscopic data. The remaining parameters are $\Delta = -14.5$ MHz and $\tau = 5 \times 10^{-10}$ s. Using these data and eq. (6.18) one finds the temperature at which $W\Delta \approx 1$; it is equal to $T_\Delta = 1.7$ K. This explains the maximum in the curve in fig. 28, as frequent transitions between the Φ_-

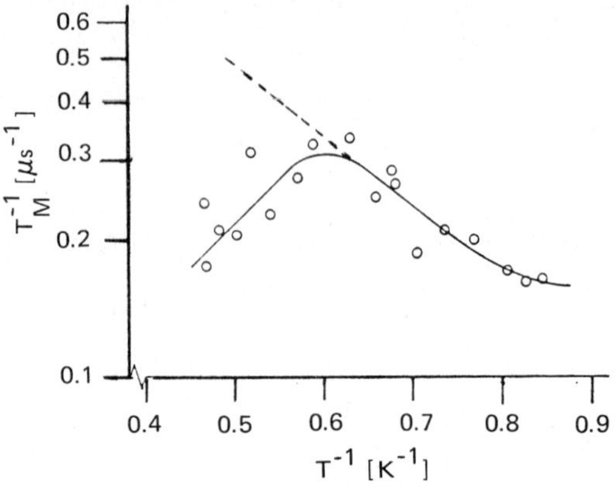

Fig. 28. Temperature dependence of T_M^{-1} for N_1N_2 pairs in naphthalene, determined from the high-field ESR (Bottler et al. 1976).

and Φ_+ states lead to suppression of dephasing by a mechanism analogous with motional narrowing.

From these data it is possible to determine the degree of the Φ_- state coherence: $W/\delta E \approx 10^{-3}$ at 1.2 K and $\approx 10^{-2}$ at 4.2 K, so the dimer N-h_2 exhibits coherent behaviour in the whole temperature range involved. Although this conclusion could also be drawn qualitatively from the distinct spectroscopic resolution of the $\Phi_-\Phi_+$ doublet, the quantitative measure of coherence called for a special investigation. An important achievement is the measurement of τ which otherwise can be found only from complicated and not very reliable kinetic measurements.

Bottler et al. (1977) have succeeded in carrying out similar measurements for the N_1N_1 dimer which can be considered as a "mini-exciton" model for excitons in linear chains. It has also been possible to measure the Hahn echo

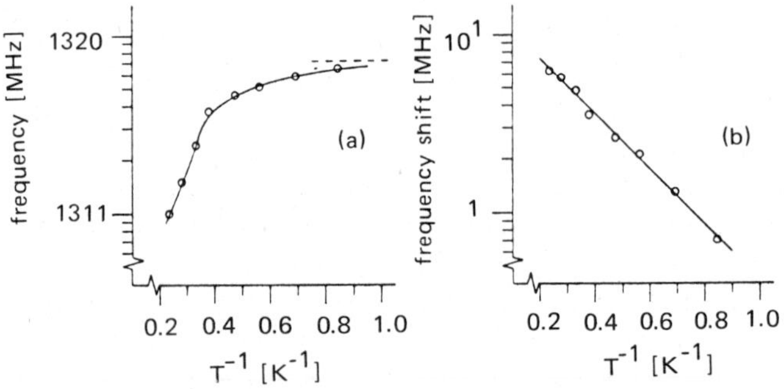

Fig. 29. (a) Temperature dependence of the resonance frequency $\omega_0(T)$ of the YZ zero-field transition for N_1N_2 pairs in naphthalene. (b) Frequency shift $\epsilon(T)$ for this resonance as a function of inverse temperature (Bottler et al. 1976).

in the zero field. The time τ has been found to be two orders of magnitude longer than that for the N_1N_2 pair.

6.5. Trapping and inter-impurity transitions

There are examples of the dephasing of impurity centres that are even more closely connected with the problem of exciton transport than the example given in section 6.4. They are being considered in the present section.

The spin-locking technique allows so efficient a suppression of the influence of various dephasing mechanisms that only those listed below can still be active in monomeric centres under favourable conditions: (1) spin relaxation along the spin-locking field as well as (2) the disappearance of the local exciton due to radiative or nonradiative decay or (3) due to its transition to the exciton band or other impurity centres.

The main idea of the experiment is as follows (Fayer and Harris 1974, Brenner et al. 1975). An impurity-doped crystal is investigated under the condition of stationary excitation. Then a constant flux of excitons flows through impurity centres owing to dynamic equilibrium arising in the presence both of feeding and decay. If at the moment $t = 0$ the system is switched over to the spin-locking regime, all the quasi-spins which relate to the population existing at the moment $t = 0$ will be locked in the $\tilde{e}_2$ position and immobile in the rotating frame (see section 6.3). If the field $\tilde{H}$ is sufficiently strong, the number of locked carriers will change only as a result of decay, promotion into the exciton band or transition to other centres. If in the other centre the exciton level possesses higher energy, then the rate of carrier losses from the locked state can be written as

$$W = W_D + W_{Pr}, \qquad W_{Pr} = \tau^{-1} \exp(-\Delta E/T). \tag{6.19}$$

Here W_D is the contribution made by the first mechanism, ΔE is the energy of exciton promotion in the most effective of the last two mechanisms, τ is the time of back transition (which may depend on ΔE and T). All the excitons which will be trapped by impurity centres after the moment $t = 0$, will undergo nutation (section 6.2) in the $\tilde{e}_1 e_3$ plane. If the duration of the spin-locking regime is long enough, the angular distribution of coming spins is uniform in the $\tilde{e}_1 e_3$ plane. Under these conditions the transition is saturated, and for such an ensemble $r_3(t) = 0$. An additional $\pi/2$ pulse rotates both this ensemble and the population of the spin-locked position around the $\tilde{e}_1$ direction. The former is lying in the $\tilde{e}_1 \tilde{e}_2$ plane now. Consequently, $r_3 = 0$ as before, and its contribution to phosphorescence remains the same. The second ensemble, however, arrives in a position along e_3, and its contribution to phosphorescence gives a jump. Therefore a jump arises also in the time

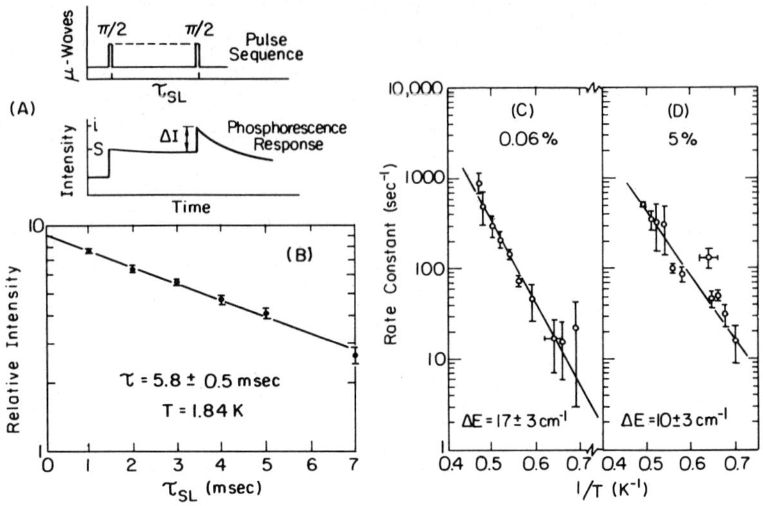

Fig. 30. Determination of local exciton promotion probability by the optically detected spin-locking method. (A) A schematic for the $\tilde{H}(t)$ dependence and the phosphorescence intensity $I(t)$. (B) A typical decay of a spin-locked spin ensemble for a crystal with a TCB-h_2 concentration of 0.06% at 1.84 K. (C) and (D) show the dependence of ln W_{Pr} on $1/T$ for the 0.06% and 5% crystals, respectively (Lewellyn et al. 1975).

dependence of total phosphorescence. Its dependence on time τ of the spin-locking determines W (see fig. 30A).

In their experiments Lewellyn et al. (1975) have studied the following system. In the TCB-d_2 host there are shallow traps TCB-*hd* to the extent of $\approx 5\%$ and deep traps TCB-h_2 to the extent of $\approx 5\%$ or $\approx 0.06\%$. Their depth is 11 cm^{-1} and 22 cm^{-1}, respectively. W_D was measured independently by the traps TCB-h_2 in durene where $\Delta E = 1460$ cm^{-1} and therefore thermal promotion was completely excluded. The measurement of $W_{Pr}(T)$ (see fig. 30B) permitted ΔE to be determined for both crystals TCB-d_2. ΔE corresponds to promotion of the local exciton from the TCB-h_2 guest into the exciton band of the host in the 0.06% crystal, and to promotion onto the shallow TCB-*hd* traps in the 5% crystal (fig. 30C, D). The length of time that the exciton stays at deep traps calculated in that study are all-important in understanding the mechanism of exciton transport in disordered crystals. In the same paper, these authors investigated the temperature dependence of the intensity of phosphorescence from deep and shallow traps in both the crystals. This question, however, involving the rate of exciton migration over traps will be discussed in section 7.

7. Band states: transport, localization, coherence

It can be seen from section 6 that the problem of coherence is sufficiently clear when applied to discrete levels of local excitons. This concerns both its theoretical aspects and the experimental methods of investigating the phase memory. The literature on molecular excitons reveals quite another situation for exciton states belonging to the energy range where their density $\rho(E)$ is nonzero and a continuous function of E. This is accounted for by real objective factors: the considerably greater intricacy of the problem and the multiplicity of situations arising. As a result, even terminology applied by different authors differs drastically. Therefore we shall now define the basic concepts in order to avoid ambiguous formulations and statements in the remainder of the section.

7.1. General definitions

In an ideal crystal the basic concept is that of band structure, that is, the existence of the quasi-particle dispersion law and the density of states $\rho_0(E)$ related to it (section 2.1). The existence of group velocity $\boldsymbol{v} = \text{grad}_k E(\boldsymbol{k})$ provides for the possibility of particle motion.

A weak scattering by phonons leads to a normal transport mechanism which we shall call a band mechanism and which is described by the Boltzmann kinetic equation. The criteria of band transport are: $\ell_{\text{ph}} \gg a$ or $\tau_{\text{ph}} \mathfrak{M} \gg 1$, where ℓ_{ph} and τ_{ph} are the length and the time of the exciton free path, respectively, as related to scattering by phonons, a is the lattice spacing. The weak exciton–phonon scattering hardly affects $\rho(E)$, but however, smooths van Hove singularities. If the scattering by phonons is so strong that $\ell_{\text{ph}} \sim a$, then the transport is of the diffusive type with diffusion coefficient $\mathscr{D} \sim a^2/\tau$, where $\tau \sim \mathfrak{M}^{-1}$ is the mean time of exciton transfer between neighbouring sites. In this case the band description and the notion of exciton dispersion law are not applicable. Moreover, the exciton has no energy spectrum and no stationary states. Only the density of states $\rho(E)$ retains its meaning, but it is entirely different from the corresponding function in a rigid lattice. The only trait they can have in common is the characteristic scale $\mathfrak{M}^{-1}$, and this is only when the characteristic frequency of phonons responsible for scattering meets the condition $\nu_{\text{eff}} \precsim \mathfrak{M}$. Such transport will be termed incoherent. In a pure crystal it can occur only at sufficiently high temperatures.

Usually, the introduction of small amounts of impurities also leads to the band transport: ℓ_{ph} is simply substituted by ℓ_i, the free path length with respect to impurity scattering. It should be noted, however, that there is a qualitative difference between these quantities which is critical for a number of important situations, namely, when the impurity scattering is elastic and the phonon scattering is inelastic (the extent of its inelasticity is determined by parameter values of the exciton and phonon spectra).

In the presence of impurities the band transport is violated in those regions of the energy range where the behaviour of wave functions changes qualitatively, i.e. a new type of states comes into being. Such are the localized states. Their distinguishing feature is the decrease of their wave functions at infinity. If we exceed the limits of the approximation employed when considering isolated guest centers (section 3), it turns out that on both sides of the initial energy band there occur tails of $\rho(E)$ corresponding to localized states. As the guest concentration C increases, the localized states also penetrate deeper and deeper into the band itself. There are critical values $E_{de}(C)$ on the energy scale which are called mobility edges; the term "diffusion edges" is more convenient, however, for our purposes. We have localized states on one side of each diffusion edge and nonlocalized or extended ones, on the other side (Mott 1967). Needless to say, these nonlocalized states do not coincide with plane waves $|\mu\boldsymbol{k}\rangle$, describing excitons in a perfect crystal. Still, like the functions $|\mu\boldsymbol{k}\rangle$, they do not decrease at infinity and they describe the states with finite particle fluxes. In the region between the edges of diffusion, the states are all extended. If C and the perturbing potential of the guest are sufficiently high, the upper and lower diffusion edges close up, and the extended states totally disappear. This is the Anderson transition (1958).

The exciton retains spectrum and stationary states throughout the concentration range. Only the band structure of the spectrum would disappear.

The general picture of transport changes qualitatively near the diffusion region. Only the activation transport is possible in the region of localized states: on absorbing a phonon, the exciton transfers either into another localized state (hopping transport) or beyond the diffusion edge where it can move freely. Since the localization length $\ell_\ell(E)$ grows infinitely as the diffusion edge is approached, the hopping length also increases. In the region of extended states the transport follows the band scheme, because the scattering by phonons limits the transport. In this region the exciton diffusion coefficient $\mathscr{D}(E)$ exhibits critical behaviour near $E = E_{de}$: either it rises abruptly from zero to a finite value or increases continuously but rapidly by some scaling law $\mathscr{D}(E) \propto |E - E_{de}|^{\alpha}$, $\alpha > 0$. The latter type of behaviour now seem to be more probable.

The picture described above refers to the systems with a 3D spectrum. In the systems with a 1D spectrum a lot of things look different. According to Mott and Twose (1961), the states in the latter systems are all localized at an arbitrary degree of disorder, with localization length $\ell_\ell \sim \ell_i$, i.e. of the order of a usual free path length with respect to impurity scattering. If the scattering is weak, the energy spectrum of the system is only slightly affected by localization: running exciton waves are transformed into spectral-narrow packets of standing waves. The type of transport is determined not so much by the absolute value of ℓ_i as by its ratio to the free path length relative to phonon scattering ℓ_{ph}. If $\ell_i \ll \ell_{ph}$, the transport will be of the hopping type throughout the band. If, however, $\ell_i \gg \ell_{ph}$, the localization is destroyed, and the transport will be of the usual band type. In this simplest formulation, the effect of phonons is reduced to the suppression of guest localization. Nevertheless, under certain conditions the phonons may strengthen the guest localization or create the momentary, or dynamical, localization in a perfect crystal (Rashba et al. 1977, Cohen 1977).

The systems with a 2D spectrum occupy an intermediate position between the 1D and 3D systems. There exists "weak localization" in them (Abrahams et al. 1979, Gor'kov et al. 1979).

We now turn to the discussion of the coherence of exciton states in imperfect crystals. Let the interaction with phonons be entirely lacking at first. Since the impurities create a static potential, localized states as well as extended ones are all coherent. It means simply that their wave functions vary with time as $\psi(\boldsymbol{r}, t) \propto \exp(-\mathrm{i}Et)$, where E is the energy of the respective state. For instance, such a phenomenon as the appearance of localized states inside the exciton band is purely an effect of interference which is possible only in the case of a strict coherence of states. This interference nature of localization is particularly pronounced in terms of Berezinskii's diagram technique (1973) permitting ℓ_ℓ to be calculated. The same mechanism shows up in 3D systems under light scattering near exciton resonances. Taking interference in the processes of the multiple elastic scattering of excitons into account results in the doubling of intensity of the backward resonance scattering of light (Ivchenko et al. 1977).

The above considerations apply to stationary states. As to the restricted space–time packets, however, the impurities also make a contribution to dephasing due to overrunning of phase shifts during the motion of excitons against the background of a random guest potential. The methods of excluding this type of dephasing were considered in section 6.3.

In order to violate the phase coherence under the condition of space homogeneity, it is necessary that dynamic interactions modulating E and resulting in energy relaxation * should be introduced into the consideration. In so doing, however, the coherence time τ_φ may not coincide with the time of energy relaxation. Let us consider as a simple example a system with the quasi-elastic scattering where the average energy transfer δE in single collisions occurring with frequency $\tau^{-1} \equiv \tau_{ph}^{-1}$, fulfills the condition $\delta E \ll \epsilon$ (ϵ varies within the limits of the exciton band). If ΔE_n is the change in energy after n collisions, that $|\Delta E_n| \sim \delta E(t/\tau)^{1/2}$, $t = n\tau$. The time τ_φ is determined from the condition of self-consistency:

$$\tau_\varphi \sim |\Delta E_n|^{-1} \sim \frac{1}{\delta E}\sqrt{\frac{\tau}{\tau_\varphi}},$$

* A trivial corollary is the shortening of T_1 owing to nonradiative transitions.

whence (Altshuler et al. 1982)

$$\tau_\varphi \sim \left(\frac{\tau}{(\delta E)^2} \right)^{1/3}. \tag{7.1}$$

This quantity can substantially differ from the energy relaxation time

$$\tau_\epsilon \sim \tau \left(\frac{\epsilon}{\delta E} \right)^2. \tag{7.2}$$

The term “coherent transport” is frequently used in the literature on excitons. One sees that, taken literally, this notion is totally inapplicable to 1D systems, since the coherent states cannot provide transport, i.e. particle flux, because of localization. So, this term can be only used in a somewhat wider sense. Namely, the transport might be considered as coherent if there exist in the system long-lived stationary states whose uncertainty in energy $\delta\epsilon \sim \tau^{-1} \ll \epsilon$. In such an interpretation of the term both the band and the hopping transport are coherent in the case when the scattering is quasi-elastic. The collisions with average time τ limit the transport in the first case ($\mathscr{D} \sim v^2\tau$, v is the mean velocity), and open a transport channel ($\mathscr{D} \sim \ell_\ell^2/\tau$) in the second case.

7.2. Critical phenomena in exciton luminescence

Kopelman et al. (1975a, b) have discovered the critical behaviour of the exciton transport in mixed crystals. They investigated the luminescence of mixed crystals $Nh_8 + Nd_8$ containing small amounts ($C \sim 10^{-3}$) of betamethylnaphthalene guest (BMN). The concentration of N-h_8 (C_g) varied over a wide range. BMN furnishes a deep trap (“supertrap”) in N-d_8, and N-h_8, a relatively shallow one (“guest”). Thus, in the region of singlet transition their depths are $\approx 400\ \mathrm{cm}^{-1}$ and $49\ \mathrm{cm}^{-1}$, respectively. In experiments carried out at $T = 2$–4 K, the luminescence spectrum consists of the luminescence of N-h_8 and BMN. The phenomenon under discussion in this section was discovered by Kopelman et al. (1975a, b) and consisted in the drastic redistribution of intensity near a certain concentration of N-h_8. The phenomenon was observed both in the fluorescence spectrum and the phosphorescence spectrum, but critical concentrations in the spectra proved to be different: $C_g \approx 0.5$–0.6 and $C_g \approx 0.08$, respectively. It turned out afterwards that those values and the shape of the curve

$$P(C_g) = I_s/(I_s + I_g) \tag{7.3}$$

depend on temperature and C_{BMN}. In eq. (7.3), I_s and I_g are the luminescence

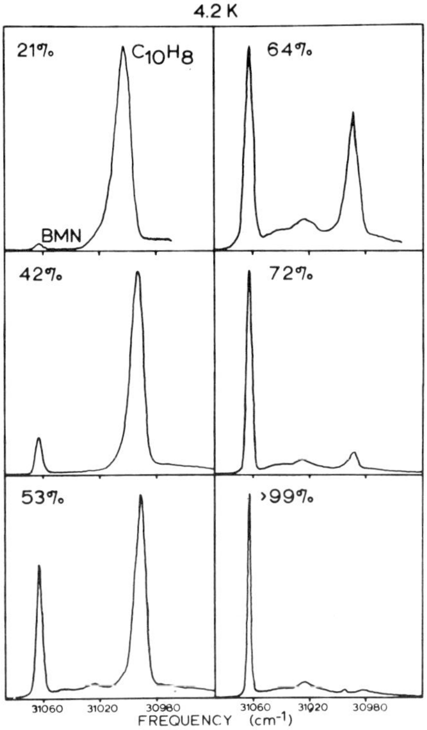

Fig. 31. Supertrap and guest fluorescence for a ternary crystal system, $Nd_8 + Nh_8 + \text{BMN}$. The guest concentration is shown in the figure. The concentration ratio is constant: $C_s/C_g = 10^{-3}$. $T = 4.2$ K. The BMN band corresponds to the 0–0 transition, the N-h_8 band, to a vibronic transition with a phonon $\nu = 509\ \text{cm}^{-1}$ (Gentry and Kopelman 1982).

intensities of the supertraps and the guest molecules, respectively. A detailed description of the experimental material and various approaches to its interpretation are presented in a review by Kopelman (1983). Since, from our viewpoint, a consecutive interpretation of these very interesting data leaves much to be elucidated as yet, we shall only dwell on those points which seem the most essential.

Figure 31 shows the latest experimental data. According to the classification of section 4.3, a mixed crystal $Nh_8 + Nd_8$ falls into category 2, a spectrum with pseudogap. As C_g increases, the lower subband broadens, and the fluorescence from the states positioned near its pseudobottom shifts permanently into the low-frequency part of the spectrum. Originally, the contribution of BMN to the total emission is very small. But it increases drastically within a narrow interval of concentrations; $P = 1/2$ at $C_g \approx 0.6$. A strong percolation of excitons from the guest band to supertraps begins in this

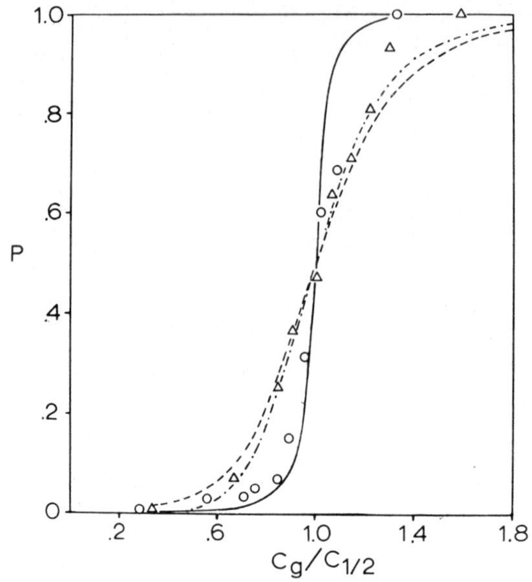

Fig. 32. The supertrap fraction P in the total fluorescence flux in a ternary system $Nd_8 + Nh_8 +$ BMN as a function of reduced concentration ($C_{1/2}$ is the N-h_8 concentration for which $P = 1/2$). Experimental data (circles) 1.8 K, (triangles) 4.2 K. Solid curve: a classical percolation calculation for a square lattice. The other curves correspond to models which are not considered here, therefore they may be regarded as a guide to the eye only (Gentry and Kopelman 1982).

concentration range. The steepness of the increase in $P(C_g)$ is illustrated in fig. 32. It diminishes with increasing temperature. Kopelman's review contains an important statement, namely, triplet excitons feature universality. It means that the data obtained at different T and C_s fall on the same curve when they are plotted in the variables used in fig. 32.

The data analogous to figs. 31 and 32 have been obtained for benzenes $Bh_6 + Bd_6$ in which pyrasene serves as a supertrap (Colson et al. 1977) and for some other systems as well. Thus this phenomenon is of a general nature.

The original interpretation proposed by Kopelman et al. (1977) is based on the classical scheme of percolation theory. For its application, the clusters comprising exclusively the N-h_8 molecules are introduced. The construction of a cluster imposes the restriction that among the nearest neighbours of each molecule of a cluster, there should be at least one molecule belonging to the same cluster. In the naphthalene lattice it is the neighbouring translationally nonequivalent molecules of N_1 and N_2 type that are the nearest ones (fig. 1). Such a construction corresponds to selecting on a plane net a way of connecting the nearest neighbours. It is assumed that clusters are not connected with each other and, if an exciton gets in a cluster containing at least one supertrap, then in a lifetime τ it will reach that supertrap and be trapped.

Otherwise the luminescence of excitons is observed in the guest band. In such a model, the critical percolation concentration C_p and the shape of the curve $P = P(C_g)$ are amenable to computerizing. The fact that so simple a scheme encompasses some important feature of the phenomena, is seen from fig. 32. The computation gives a wholly satisfactory value of the critical concentration $C_{cr} \approx C_p$ for singlet excitons. It fails to explain, however, a substantially smaller value for triplets. Subsequent improvement introduced into the theory (Kopelman 1983) (taking into account interactions with next-nearest neighbours, clusters with leaking, etc.) have not affected the principal results.

At the same time, a straightforward application of this classical percolation scheme disregarding quantum effects involves intrinsic difficulties. The percolation onto supertraps with probability $P \approx 1/2$ means that the average number of molecules in a cluster is about $\approx C_g/C_s \approx 10^3$. In such large clusters the states would be localized, at least partially. At any rate, this is true for the states in the lower part of the spectrum which forms the spectrum of guest luminescence. Quantum effects would result in localization owing to different molecules of the cluster being nonequivalent. This nonequivalence is associated with the presence of ramification and the interaction of individual molecules with the molecules of the neighbouring clusters. This interaction is not small, as can be readily seen from the perturbation theory. For instance, the interaction of two guests separated by one host molecule is equal to $\sim M^2/\Delta$ in the second order perturbation theory. Here $M \approx \mathfrak{M}/Z$, Z is the number of nearest neighbours, Δ is the isotopic shift. For singlets in naphthalene $M \approx M_{12} \approx 20\ \text{cm}^{-1}$, $\Delta \approx 120\ \text{cm}^{-1}$, M/Δ is not small, being about 0.2 (section 3), and the number of points where two clusters almost come in contact with each other is very large at $C_g \approx 0.5$. Therefore the cluster model does not eliminate difficulties, but to a great extent, shifts them on to the consideration of the transport inside the cluster with leaking to the other clusters taken into account.

At the same time, a number of other approaches (Klafter and Jortner 1977, Broude et al. 1981) have been proposed for treating experimental data making allowance for Anderson localization. It seems to us that in treating so intricate a problem involving kinetic aspects (the transport to supertraps should occur during the lifetime τ) the two approaches, those of percolation and localization, must be considered as complementing each other rather than as strictly alternative (Ahlgren and Kopelman 1980, Klafter and Jortner 1980). Presented below is argumentation similar to that we have proposed in the book by Broude et al. (1981, section 4.6).

In a guest system, localization must arise as a consequence of nondiagonal disorder due to random distribution of impurities at the lattice sites. This system is essentially a lattice analogue of the Lifshitz model. According to Lifshitz (1964) all states are localized at $C_g \ll 1$. On the contrary, the states are all delocalized at $C_g = 1$. Let the radius of guest centers be close to the lattice constant a. This is valid at $\Delta \approx 120\ \mathrm{cm}^{-1}$ (section 3.1). Since the problem involves no length-dimensional parameter other than a under these conditions, it is obvious that the concentration C_ℓ at which localization starts being destroyed in the middle of the band, must be of the order of unity. If we apply the Mott criterion for localization in the guest band (Mott and Davis 1971, section 2.7.1) to the problem in hand, then $C_\ell \approx 1/16$. At $C_g < C_\ell$ the guest states (i.e. the states whose energy is below the pseudogap) are all localized. The above considerations hold true if the result, proven for a regular atomic arrangement and being to the effect that in an entirely nondiagonal disorder the Anderson transition is absent and the extended states always persist in the middle of the spectrum (Economou and Antoniou 1977, Weaire and Srivastava 1977), is inapplicable to the discrete Lifshitz model. We believe this assumption to be highly probable. But even if this result is valid for the Lifshitz model, it will little affect our subsequent reasonings, since at a large $\mathfrak{M}$ the peripheral part of the spectrum where localization exists for a certainty is the most important, while at a small $\mathfrak{M} \sim 1\ \mathrm{cm}^{-1}$ localization can easily occur on account of the weak diagonal disorder always present in disordered systems.

Let us begin with the case of singlets. The energy band is wide for them, therefore for guest excitons there exist deep localized states in relatively small but compact clusters (or similar "bulges" on the branches of large clusters). They may turn out to be serious rivals of supertraps in capturing excitons. In such a system the distance between the diffusion edges and the positions of the majority of the guest levels active in exciton capture and fluorescence, can prove to be well above T. Therefore the activated promotion to the edge of diffusion for the exciton singly captured by one of these levels cannot have a chance to occur in the relatively short radiative lifetime of a singlet exciton ($\tau > 10^{-7}$ s). The hopping diffusion would also be slow due to the substantial spatial separation of relatively deep guest levels. As a result, the exciton flux to the supertraps will be suppressed.

The clue to understanding the problem is apparently to be found in the fact that only when approaching $C_g \approx 0.5$ does a distinct pseudobottom of the spectrum occur, and the fluorescence band is close to it. This can be understood, e.g., from fig. 19. It shows that the low-frequency structure stemming from small clusters only starts being smeared from $C_g \approx 0.3$. Approximately the same conclusion can be drawn from histograms of $\rho(\epsilon)$ (Hong and Kopelman 1971b). The same follows from fig. 31 where the guest fluorescence band narrows as C_g increases up to the value ≈ 0.5. It is natural to suppose that the diffusion edge falls within the limits of this half-width which in its turn is comparable with $T \approx 2$–4 K. Therefore, the percolation to the traps through the extended and the strongly overlapped localized states in the lifetime τ indeed becomes possible. At the same time, it is quite natural to suppose the existence of correlation between the smoothing of the fine structure near the pseudobottom and the classical percolation: the appearance of a ramified net of "passages" should integrate single centers into a united system and thus promote formation of a distinct pseudobottom. This assumption agrees with Allen's (1980) approach; he showed that a number of results of Anderson localization theory can be reproduced in terms of classical diffusion, limited by the Heisenberg uncertainty principle. However, this apparently exhausts the connection of the classical percolation with the quantum-mechanical transport of excitation to supertraps.

At present the experimental data on the dependence of the parameter P on the intensity of excitation are lacking. Meanwhile, as the intensity increases, an ever increasing number of low-positioned impurity levels are filled with excitons . This makes it possible to fill all localized states and reach the level of percolation. Such an experiment successfully carried out by Ovsyankin and Fedorov (1982) on $SrYb_2F_2$ (a disordered system based on the lattice of tysonite LaF_3) would be also very useful for interpreting the data on molecular excitons.

The migration mechanism for triplets seems to be essentially different. Here the matrix element of excitation transfer is small: $M_{12} \approx 1\ cm^{-1}$ (Hanson 1970), and already at $T \approx 2$ K there is no energy limitation for the transport. In this connection, it is interesting to compare the estimate of $C_\ell \approx 0.06$ as obtained above with the experimental value of the critical concentration $C_{cr} \approx 0.08$. The proximity of these values suggests that at $C_g \approx 0.06$–0.08 the Anderson transition, which can be interpreted in terms of the Lifshitz model, occurs in the triplet band.

A delayed fluorescence resulting from the fusion of two triplet excitons into a singlet one (Avakian and Merrifield 1968) also features an unusual behaviour in mixed crystals. Instead of a normal bimolecular law, one observes a more rapid increase (up to the twentieth power) of the intensity of emission as a function of the intensity of excitation (Kopelman 1983). Evidently, this is due to either the Poisson distribution for the simultaneous presence of several triplets in one small cluster or the above-mentioned nonlinear effect associated with the filling of levels. An interpretation of the data on triplet–triplet annihilation in naphthalene has been recently proposed by Kopelman et al. (1984) in terms of reaction kinetics on fractals, i.e. self-similar structures (Mandelbrot 1977, Rammal and Toulouse 1983). This interpretation is based on the statement that the percolating cluster is self-similar.

Vankan and Veeman (1982) observed a very interesting effect of doping – the enhancement of intrinsic phosphorescence (i.e. phosphorescence of the host). No exciton phosphorescence was found in the nominally "pure" DBN, since the excitons were captured by traps. The introduction of a DBN-d_6 guest (15%) into a DBN-h_6 host created a strong intrinsic phosphorescence. The mechanism of its appearance was very simple. DBN-d_6 was breaking the stacks of DBN-h_6 into separate "cages" where excitons were locked and radiated if the "cage" was free from a trap. Similar phenomena were also observed (Patel and Hanson 1982) in single crystals of 4,4′-dimethylbenzophenone as a result of a photochemical reaction, because the appearance of radiation defects brought about the intensification of exciton phosphorescence.

7.3. Characteristic times

It has been shown in section 5.2 that the SBTBT method permits the lower bound for the exciton relaxation time to be found. A number of efforts has been recently performed to determine the relaxation time from different phenomena and correlate the results. Not dwelling on contradictions arising here, we shall briefly state the results of van Strien et al. (1980) who apparently were the first to reliably measure inelastic relaxation times.

The experiments were carried out on TDB at $T \geqslant 1.12$ K. Using a laser pulse with a duration of 15 ns and a spectral width of 0.2 cm^{-1}, excitons were created at the top of the triplet exciton band (the band width is 1.36 cm^{-1}). The two-pulse spin echo was employed for detecting the excitons (section 6.3). The microwave field frequency was varying within the range of the whole width of the SBTBT spectrum, which permitted the separate detection of excitons with different $\boldsymbol{k}$. Microwave pulses with a width ~ 100 ns were capable of shifting with respect to each other and to the laser pulse. This provided a means for measuring the times of increase and decrease in population and phase memory time T_M. The basic data are presented in fig. 33.

It is seen from the figure that the time scale of inelastic scattering is of an order of microseconds. For excitons with $\boldsymbol{k} \approx 0$ at $T = 1.13$ K the time of signal decay $T_s = 2.7$ μs and $T_M = 1.5$ μs. The time T_s decreases rapidly with

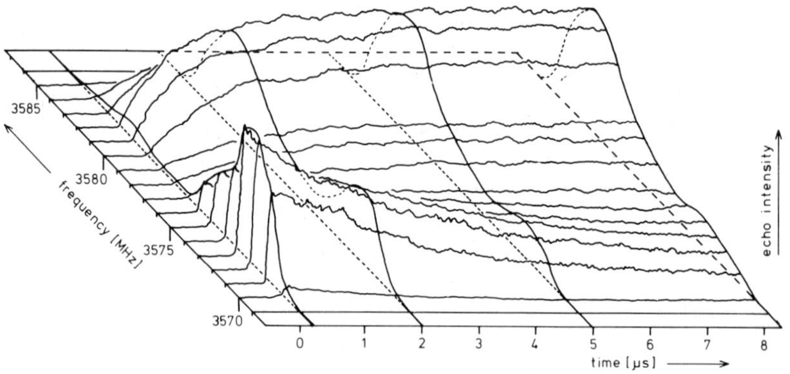

Fig. 33. A three-dimensional representation of time evolution of the spin-echo lineshape for triplet excitons in TCB. Time is reckoned from the laser flash creating excitons at the top of the exciton band. $T = 1.13$ K (Van Strien et al. 1980).

increasing temperature; $T_s = 0.8$ μs at $T = 2.5$ K. The light emission from triplet states and the spin–lattice relaxation ($T_1 = 135$ μs) may be neglected on this scale. The authors ascribe the difference between T_s and T_M to impurity scattering. Although the detailed mechanism of relaxation, apparently consisting of several successive two-phonon (Compton) events of exciton scattering by acoustic phonons, is not clear as yet, there is no doubt that the inelastic scattering time exceeds by almost three orders the lower bound for τ_f obtained from SBTBT and adduced in section 5.2.

Now consider possible mechanisms of the low-temperature relaxation of excitons by phonons in TCB. It is essential that the one-phonon scattering of excitons in a perfect crystal is forbidden by conservation laws because of $2\mathfrak{M}$ being small compared to the Debye frequency. In imperfect crystals it can occur on account of transferring the momentum to the defect (guest). This mechanism is equivalent to that of the inelastic impurity scattering. In perfect crystals the Compton (two-phonon) process of inelastic scattering of the exciton by phonons is possible. The main contribution is made by phonons with frequency ν exceeding T several times over, since the intensity of scattering grows with the phonon energy (due to an increase in the phase volume as well as in the matrix element for the coupling via the deformation potential), and this growth is only restricted by a decrease in Planck occupation numbers $\bar{n}(\nu/T)$. If the effective frequencies ν are large, $\nu \gg \mathfrak{M}$, then we can write down a Hamiltonian of the type of eq. (3.17) for the interaction of corresponding phonons with the exciton. At a small $M \approx \mathfrak{M}/Z$ the first term containing the linear interaction constant may be less than the second term. In this case the probability of the Compton process is entirely determined by nonlinear exciton–phonon interactions and therefore its rate cannot be associated with the probabilities of one-phonon processes at all.

Another approach to the experimental investigation of exciton scattering is based on measuring the shape of exciton absorption bands and their temperature dependence. Thus, Burland et al. (1977a, b) measured and discussed the absorption spectra of TCB and BMN, respectively.

In this connection we would like to consider still further this problem which was already dwelt upon in section 2.2. Omitting the common factor, the expression for conductivity at a frequency ω can be written as

$$-\operatorname{Im} G(\omega, \boldsymbol{k}=0) = -\operatorname{Im}\frac{1}{\omega - \epsilon_0(\boldsymbol{k}=0) - \Sigma(\omega, \boldsymbol{k}=0)}. \tag{7.4}$$

Upon separating the real and the imaginary part of Σ:

$$\Sigma(\omega, \boldsymbol{k}=0) = \Sigma_R(\omega) - \mathrm{i}\pi\Gamma(\omega) \tag{7.5}$$

eq. (7.4) is transformed to

$$-\operatorname{Im} G(\omega, \boldsymbol{k}=0) = \frac{\pi\Gamma(\omega)}{(\omega - \epsilon(0) - \Sigma_{\mathrm{R}}(\omega))^2 + (\pi\Gamma(\omega))^2}. \tag{7.6}$$

Σ_{R} and Γ are related to each other by the dispersion relation

$$\Sigma_{\mathrm{R}}(\omega) = \fint \mathrm{d}\omega' \frac{\Gamma(\omega')}{\omega - \omega'}. \tag{7.7}$$

The bar means that the integral is taken in the sense of the Gauchy principal value.

For the time being, let us restrict ourselves to the simplest case of only one exciton band and only one phonon band so that all indices in eqs. (2.16) and (2.17) can be omitted. Denote $g(\boldsymbol{k}=0, \boldsymbol{q}) \equiv q(\boldsymbol{q})$. Then at $T=0$ in the second order of perturbation theory for exciton–phonon interactions the following expression for Γ holds:

$$\Gamma(\omega) = \int \frac{\mathrm{d}^3 q}{(2\pi)^3} g^2(\boldsymbol{q}) \delta(\omega - \nu(\boldsymbol{q}) - \epsilon(\boldsymbol{q})), \tag{7.8}$$

where $\nu(\boldsymbol{q})$ and $\epsilon(\boldsymbol{q})$ are dispersion laws in the phonon and exciton bands. According to the conservation law appearing under the integral in eq. (7.8), it is seen that $\Gamma(\omega)$ is not the decay probability of any real state. The only exception is the point $\omega = \epsilon(0)$, when an intermediate virtual state through which the optical transition is effected becomes a real one and so the conservation laws are fulfilled: $\epsilon(\boldsymbol{k}=0) = \nu(\boldsymbol{q}) + \epsilon(\boldsymbol{q})$. At this point only $\Gamma(\omega)$ acquires the same sense of the level width that it has in the Weisskopf and Wigner theory (1930). Clear physical grounds exist for this fact: there is no true absorption (dissipation) in the case of an isolated atom, and the "absorption line" is the line of resonance scattering. For the excitons, on the contrary, the absorption is a true one, since in this case there exists a mechanism of dissipation: the phonon emission.

Equations (7.6) and (7.7) are inconvenient for gaining any information whatsoever on $\Gamma(\omega)$ from $\operatorname{Im} G(\omega, 0)$. It is much more convenient to resolve eq. (2.18) with respect to Σ, using the well-known dispersion relation for the retarded time–temperature Green function G (see e.g., Abrikosov et al. 1962);

$$G'(\omega) = \frac{1}{\pi} \fint \frac{G''(\omega')}{\omega' - \omega} \mathrm{d}\omega', \qquad G \equiv G' + \mathrm{i}G''. \tag{7.9}$$

Since $G'(\omega) \approx \omega^{-1}$ at $|\omega| \to \infty$,

$$-\int G''(\omega')\, \mathrm{d}\omega' = \pi, \tag{7.10}$$

with integration in eqs. (7.9) and (7.10) extended to the whole of the absorption spectrum. It follows from eqs. (2.18), (7.5) and (7.9) that

$$\Gamma(\omega) = -\pi G''(\omega) \Big/ \left[\left(\fint \frac{G''(\omega')}{\omega - \omega'} \mathrm{d}\omega' \right)^2 + (\pi G''(\omega))^2 \right]. \tag{7.11}$$

The G'' function is directly derived from the experimentally measured $\sigma(\omega) \propto G''(\omega)$ through the use of the normalization condition (7.10). Such a reversion of eq. (2.18) for reconstructing the parameters of the initial Hamiltonian from measured functions has been proposed by Rabin'kina et al. (1970) for the case of vibronic spectra and has then been applied in determining $\Gamma(\omega)$ for singlet excitons in naphthalene (Sheka 1975) and in reconstructing the density of states in the IP energy spectrum (see the review in Belousov 1982). Also note that eq. (7.11) is valid for an arbitrary mechanism of exciton scattering (impurities, phonons), yet the spectrum $\sigma(\omega)$ is supposed to be free from inhomogeneous broadening.

It is seen from eq. (7.8) that even in the simplest case knowledge of $\Gamma(\omega)$ fails to supply direct information on the magnitude of relaxation times. It only determines a "weighted DOS", i.e., the product of the density of states in the spectrum of a fictitious quasi-particle with dispersion law $\epsilon_{\text{eff}}(\boldsymbol{q}) = \epsilon(\boldsymbol{q}) + \nu(\boldsymbol{q})$, and the mean square of the interaction constant for the energy ϵ_{eff}. Although this is indirect information, the constants of interaction can be evaluated from the function $\Gamma(\omega)$ far better than directly from the observed spectrum (at any rate, in the case of a weak exciton–phonon coupling when eq. (7.8) is valid). It is from these coupling constants and the spectra of the exciton and phonons that the relaxation times controlled by one-phonon scattering should be derived.

As applied to the crystals with a narrow exciton band, the results discussed above in this section show that $\Gamma(\omega)$ can be determined by one-phonon processes, while the free-path time can be determined by two-phonon processes with a perfectly independent coupling constant. Therefore there is no reason for direct correlation of Γ^{-1} and τ_{f}.

Taking into account all the aforesaid, the inferences about the changes in the exciton transport mechanism in TCB (Burland et al. 1977a) based on the fact that there are small temperature changes in the shape of the exciton absorption band at a minor general increase of its width, are not convincing.

In order to illustrate on a model how little information the general shape of the curve $\sigma(\omega)$ gives of the intensity of scattering, we present here expressions for a particular example: a 1D exciton with a narrow band ($\mathfrak{M} \ll \nu_{\text{D}}$, ν_{D} is the Debye frequency) interacts with 1D acoustic phonons $\nu(q) = sq$ via the deformation potential ($g^2(q) \propto q$); s is the velocity of sound. Omitting $\mathfrak{M}$ and taking into account the renormalization shift of the exciton level, we obtain from eqs. (7.8) and (7.7):

$$\Gamma(\omega) = \gamma\omega, \qquad 0 \leqslant \omega \leqslant \nu_D,$$
$$\Sigma_R(\omega) = -\gamma\omega \exp\left(\frac{\nu_D - \omega}{\omega}\right), \tag{7.12}$$

$\gamma = \text{const}$ is the dimensionless coupling constant. It follows directly from eqs. (7.12) and (7.6) that

$$\sigma(\omega) \propto -G''(\omega, k=0) = \frac{\pi\gamma/\omega}{[1 + \gamma \ln((\nu_D - \omega)/\omega)]^2 + (\pi\gamma)^2}. \tag{7.13}$$

The singularity at $\omega \to 0$ is cut off by impurity scattering or by taking into account higher orders of perturbation theory. At $\gamma \ll 1$ the behaviour of eq. (7.13) is determined by the law $G'' \propto \omega^{-1}$ almost throughout the region $0 < \omega < \nu_D$, and only in the close vicinity of ν_D is this trend abruptly cut off by the logarithm in the denominator. Clearly the half-width of such a curve is determined by the cutoff of eq. (7.13) at $\omega = 0$ and by the magnitude of ν_D; it is practically independent of the coupling constant γ.

We hope that this discussion will to a certain degree elucidate the basic cause of the contradictions involved in the magnitude of characteristic times determined from the absorption spectra and using some other, more direct methods.

8. *Vibrational excitons*

In all the foregoing we have discussed electron excitons. However, in molecular crystals there is a large number of spectrum branches which can be considered with good reason as vibrational excitons. They are the branches of a phonon spectrum originating from intramolecular vibrations. They have been figured above as IPs. Since the intramolecular interactions prevail over the intermolecular ones, the IP frequencies are appreciably higher and dispersion of the frequencies appreciably lower than the respective values for LP. Such phonon branches can be calculated and experimentally studied by methods of crystal lattice dynamics. The lower part of the IP spectrum is by now amenable for investigation by the neutron scattering technique. The data on the phonon spectrum of a naphthalene crystal were obtained by Natkaniec et al. (1980).

At the same time, these phonons exhibit properties which are typical for molecular excitons, e.g., Davydov splitting. Therefore it is natural to consider them as vibrational excitons and to study them also with the methods the spectroscopy of electron excitons has in store. The main hindrance is the small width of vibrational exciton bands which rarely exceeds 10 cm^{-1}. But the scale of the width of triplet exciton bands is the same, and yet we saw above (section 3.1) that it was already possible in doped crystals to observe in triplet excitons the phenomena which had been formerly studied in singlet excitons only. Therefore, we can hope for success in studying vibrational excitons by methods of electron exciton spectroscopy, especially when starting with the widest bands of IP.

At the present time, the methods of exciton spectroscopy have been applied with much success to the investigation of vibrational excitons in a number of inorganic crystals: CO_2, $NaNO_3$, NH_4Cl. The band widths in them are of the order of 50 cm^{-1}, and it is on these materials that Belousov, Pogarev and collaborators have succeeded in carrying out a practically complete "exciton" program. Compared to the electron excitons, the vibrational excitons prove to be in some respect even more rewarding objects, since the disguising effect of LP is weaker (the corresponding anharmonism is smaller than the exciton–phonon coupling). We do not dwell here on these studies in detail, as the investigated compounds are beyond the scope of the present paper, and the results have been described in Belousov's review (1982).

Aside from the problem of stationary states, the vibrational excitons, like the electron ones, bring forth all the problems associated with their decay and scattering. Recently the method of coherent anti-Stokes Raman scattering (CARS) has been used to advantage for measuring the relaxation rate of vibrational excitons. The method is described in detail by Laubereau and Kaiser (1978) and we will only briefly formulate its basic idea. Coherent phonons are created in a crystal with the help of the first pulse generated by one laser or, usually, several lasers operating simultaneously. Next, by means of a series of trial laser pulses the intensity of anti-Stokes scattering is measured, and the rate of phonon relaxation is judged by its decay. Since the intensity of Raman scattering depends not only on the number of phonons, but also on fulfillment of phase-matching conditions (i.e., $\boldsymbol{k}$-matching), the relaxation comprises not only the damping of phonons (their fission) but also a change in their momentum distribution. The measurements performed on pure crystals have shown that at low temperatures the signal damping is exponential with time $T_1 \sim 100$ ps. Such results have been obtained for some excitons in a naphthalene crystal (Decola et al. 1980, Hesp and Wiersma 1980, Dlott et al. 1982) and for the 991 cm^{-1} exciton in a benzene crystal (Ho et al. 1981). In all cases relaxation is assigned to phonon fission. According to calculations of Rigini et al. (1983) the 991 cm^{-1} exciton in benzene decays into an 860 cm^{-1} exciton and two lattice phonons.

The CARS method can also be employed in studying mixed crystals. Figure 34 shows the dependence of the dephasing time on composition. At a 15% concentration of B-d_6 the dephasing time decreases by nearly one half. The authors ascribe this fall to the strengthening of the exciton decay in mixed

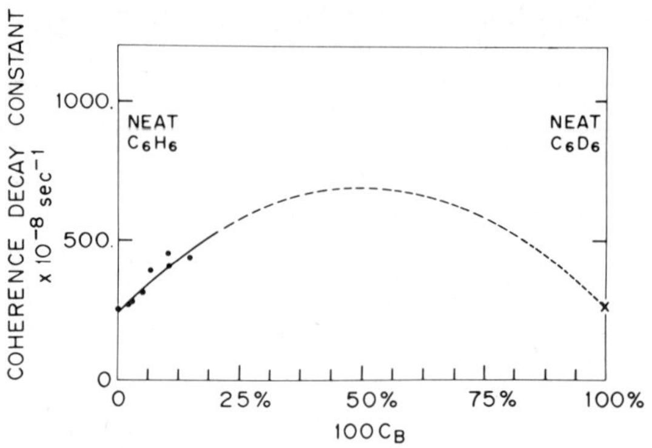

Fig. 34. Coherence decay rate in mixed crystals of $Bd_6 + Bh_6$ deuterobenzenes as a function of the B-d_6 concentration (Ho et al. 1983).

crystals. The parabolic dependence on concentration which follows from their semiphenomenological theory is shown in fig. 34 by a dashed line. Unfortunately, the choice of the 991 cm^{-1} vibration prevents the CARS peculiarities specific for excitons from being revealed experimentally, since at a decay time $T_1 \approx 40$ ps the uncertainty in energy $\hbar/T_1$ practically coincides with the exciton band width $\approx 1\ cm^{-1}$ (Ho et al. 1981).

We are unaware of other experimental data on the measurement of CARS in mixed crystals.

9. Conclusions

Organic polymers and molecular crystals belong to the substances in which the narrow bands of exciton absorption had been detected and identified at the dawn of the physics of excitons. But, during the ensuing two decades, the development of the spectroscopy of molecular excitons was largely impeded by the lack of information on the structure of exciton bands, i.e. on the exciton dispersion law $\epsilon(\boldsymbol{k})$. The situation which has arisen has been partly elucidated in section 1. The methods of calculating exciton bands based on molecular wave functions proved to be ineffective, while the methods of reconstructing the exciton bands from experimental data, developed in other fields of solid state physics, were absolutely inapplicable to molecular excitons. For them, an entirely new system of techniques had to be created. As is seen from the bulk of this chapter, all methods which so far can possibly be applied in reconstructing the energy spectrum of molecular excitons from experimental data, are based on the information obtained from the spectra of doped and mixed crystals. The only exceptions are the values of Davydov splitting and the band-to-band transition method. Thus, the study of imperfect molecular crystals has become a method for determining the parameters of perfect crystals.

This specific role that the spectroscopy of imperfect crystals plays in the spectroscopy of molecular excitons in general has lead to a concentration of efforts on research in the field of doped crystals and crystals having lattice defects. It has resulted in the discovery of the phenomena and the development of the methods which go beyond the limits of the spectroscopy of organic crystals and have acquired the importance in the physics of non-metallic crystals in general: lattice dynamics, spectroscopy of magnetics and semiconductors, etc. When evaluating the effectiveness of these techniques by the accuracy they provide, the results are as follows: in crystals with the quasi-1D spectrum the function $\epsilon(k)$ is reconstructed with high precision. In crystals with a 3D spectrum, the density of states $\rho_0(\epsilon)$ is reconstructed reliably enough, while the accuracy of determining $\epsilon(\boldsymbol{k})$ is lower. Nevertheless, the obtained accuracy proves to be sufficient, for instance, for development of the quantitative theory of vibronic spectra of pure and doped crystals. All ideas up to now suggested, introducing essential alterations in the methods of determination of $\epsilon(\boldsymbol{k})$, are based upon experiments on the scattering of Röntgen quanta, electrons, neutrons, etc. The difficulty lies in the necessity of a compromise between a large transfer of energy ($\sim$ some eV) and momentum ($\sim$ Brillouin momentum) and a high resolution of $\sim 10^{-4}$ eV. Time will show the prospects of these methods as applied to molecular excitons.

In the last 10–15 years the potentialities of the experimental methods for investigation of molecular excitons have undergone a substantial transformation. It is connected with (1) the introduction of laser techniques, combining a high spectral and time resolution with a high beam intensity and (2) the use of the magnetic resonances in combination with optical detection of them (providing for the record sensitivity). Owing to these new methods, it has been possible to observe polariton effects in a very weak transition in naphthalene crystals, to make impressive progress in the study of triplet excitons, to determine their band structure (for a band width as small as $\sim 1.5\ \mathrm{cm}^{-1}$!), to develop the spectroscopy of local triplet excitons, and to advance far in the measurements of the rate of different processes and in the grasp of the mechanisms of decay and dephasing of quasi-stationary states. Here technical difficulties brought about by the narrowness of the energy spectrum are more than compensated by unique potentialities of the magnetic resonance technique.

The discovery of critical phenomena in the low-temperature luminescence of imperfect crystals has drawn closer together the spectroscopy of molecular excitons and a vast area of research into Anderson and Mott localization, conducted on various materials. The overwhelming majority of studies is performed on electrons for which the effects of disorder and Coulomb interaction are hard to discern. The case is somewhat different from molecular excitons, and so the investigations on electrons and molecular excitons may complement each other. Yet another difference between the electrons and the molecular excitons lies in their statistics: the Pauli statistics for excitons forbids the presence of two excitons at the same site but not in the same quantum state. Therefore of some interest is the change in the spectrum of local states having considerable spatial extension as they become filled by more than one exciton. The control of the dephasing rate by magnetic resonance can be very effective for the diagnostics of localization of the triplet excitons. In the crystals where the dipole transition to the bottom of the exciton band is allowed (and its oscillator strength is not too small), interesting peculiarities may manifest themselves if it is possible, due to the existence of a "bottle neck", to accumulate a major portion of excitons in the energy range where the role of polariton effects (i.e., retardation) is rather small. Then a logarithmic suppression of Anderson localization should arise, while in anisotropic crystals, with their nonanalytic energy spectra, the four- or five-dimensional behaviour of excitons would be observed (in contrast to electrons for which only lower dimensionalities are possible). Finally, it is essential for the excitons that the process of percolation to luminescence centers is a kinetic one and must occur during the exciton lifetime τ. The excitons are not unique, however, in this respect; the kinetic aspect is also peculiar to photoconductivity of disordered semiconductors. It should be noted with regret that, for all the beauty of the observed phenomena and the promise they offer for application to biological objects, their detailed mechanism is still not quite elucidated. Also, as yet neither 3D nor 1D localization has been reliably observed on molecular excitons.

However, the present-day level of understanding the problem and more refined experimental techniques, together with the rapid growth in the number of studies permit us to look ahead optimistically.

This article was completed July 1984.

List of abbreviations

DOS – density of states
BTBT – band-to-band transitions
VBTBT – vibronic BTBT
SBTBT – spin BTBT
ESR – electron spin resonance
LP – lattice phonon
IP – internal phonon
TS phonon – totally symmetric phonon
NTS phonon – nontotally symmetric phonon

AAA – average-amplitude approximation
CPA – coherent-potential approximation
NFM – Napier factorization method
1D, 2D, 3D – one-, two-, and three-dimensional
FVH – Feynman, Vernon and Hellworth
BMN – betamethylnaphthalene
TCB – 1,2,3,4-tetrachlorobenzene
DBN – 1,4-dibromonaphthalene
N – naphthalene
B – benzene

References

Abragam, A., 1961, The Principles of Nuclear Magnetism (Oxford University Press, Oxford).
Abrahams, E., P.W. Anderson, D.C. Licciardello and T.W. Ramakrishnan, 1979, Phys. Rev. Lett. **42**, 673.
Abrikosov, A.A., L.P. Gor'kov and I.E. Dzyaloshinskii, 1962, Methods of Quantum Field Theory in Statistical Physics (Fizmatgiz, Moscow) (English Translation, 1963) (Prentice-Hall, Englewood Cliffs, New York).
Agranovich, V.M., 1959, Zh. Eksp. & Teor. Fiz. **37**, 430 [1960, Sov. Phys.-JETP **10**, 307]
Agranovich, V.M., and V.L. Ginzburg, 1979, Spatial Dispersion in Crystal Optics and the Theory of Excitons (Nauka, Moscow) (English Translation Springer, Berlin, 1984).
Agranovich, V.M., and R.M. Hochstrasser, eds, 1983, Spectroscopy and Excitation Dynamics of Condensed Molecular Systems (North-Holland, Amsterdam).
Ahlgren, D.C., and R. Kopelman, 1980, J. Chem. Phys. **73**, 1005.
Allen, P.B., 1980, J. Phys. C: Solid State Phys. **13**, L-667.
Altshuler, B.L., A.G. Aronov and D.E. Khmelnitsky, 1982, J. Phys. C: Solid State Phys. **15**, 7367.
Anderson, P.W., 1958, Phys. Rev. **109**, 1492.
Avakian, P., and R.E. Merrifield, 1968, Mol. Cryst. **5**, 37.
Belousov, M.V., 1982, in: Excitons, eds E.I. Rashba and M.D. Sturge (North-Holland, Amsterdam) p. 771.
Benderskii, V.A., V.Kh. Brikenstein, N.A. Vidmont, V.V. Korshunov and Ye.Ya. Misochko, 1980, Mol. Cryst. Liq. Cryst. **57**, 47.
Benk, H., H. Haken and H. Sixi, 1982, J. Chem. Phys. **77**, 5730.
Berezinskii, V.L., 1973, Zh. Eksp. & Teor. Fiz. **65**, 1251 [1974, Sov. Phys.-JETP **38**, 620].
Blackman, M., 1933, Z. Phys. **86**, 421.
Bottler, B.J., C.J. Nonhof, J. Schmidt and J.H. van der Waals, 1976, Chem. Phys. Lett. **43**, 210.
Bottler, B.J., A.J. van Strien and J. Schmidt, 1977, Chem. Phys. Lett. **49**, 1977.
Breiland, W.G., C.B. Harris and A. Pines, 1973, Phys. Rev. Lett. **30**, 158.
Breiland, W.G., H.C. Brenner and C.B. Harris, 1975, J. Chem. Phys. **62**, 3458.
Brenner, H.C., J.C. Brock, M.D. Fayer and C.B. Harris, 1975, Chem. Phys. Lett. **33**, 471.
Broude, V.L., 1980, Mol. Cryst. Liq. Cryst. **57**, 5.
Broude, V.L., and S.M. Kochubei, 1964, Fiz. Tverd. Tela **6**, 354 [1964, Sov. Phys.-Solid State **6**, 285].
Broude, V.L., and A.V. Leiderman, 1971, Pis'ma v Zh. Eksp. & Teor. Fiz. **13**, 426 [JETP Lett. **13**, 302].
Broude, V.L., and M.I. Onoprienko, 1961, Opt. & Spektrosk. **10**, 634 [Opt. & Spectrosc. **10**, 334].
Broude, V.L., and E.I. Rashba, 1961, Fiz. Tverd. Tela **3**, 1941 [Sov. Phys.-Solid State **3**, 1415].

Broude, V.L., A.F. Prikhot'ko and E.I. Rashba, 1959, Usp. Fiz. Nauk **67**, 99 [Sov. Phys.-Usp. **67**(2), 38].
Broude, V.L., E.I. Rashba and E.F. Sheka, 1961, Dokl. Akad. Nauk SSSR **139**, 1085 [1962, Sov. Phys.-Dokl. **6**, 718].
Broude, V.L., E.F. Sheka, M.T. Shpak and L.G. Shpakovskaja, 1963, Opt. & Spektrosk. Collected Papers 1, Lumin. p. 98.
Broude, V.L., A.I. Vlasenko, E.I. Rashba and E.F. Sheka, 1965, Fiz. Tverd. Tela **7**, 2094 [Sov. Phys.-Solid State **7**, 1686].
Broude, V.L., A.V. Leiderman and T.G. Tratas, 1971, Fiz. Tverd. Tela **13**, 3624 [Sov. Phys.-Solid State **13**, 3058].
Broude, V.L., N.A. Wildmont, V.V. Korshunov, A.A. Maksimov and I.I. Tartakovskii, 1977, Pis'ma v Zh. Eksp. & Teor. Fiz. **26**, 436 [JETP Lett. **26**, 311].
Broude, V.L., E.I. Rashba and E.F. Sheka, 1981, Spectroscopy of Molecular Excitons (Energoizdat, Moscow) (English Translation Springer, Berlin, 1985).
Brovchenko, I.V., L.L. Valkunas, G.B. Vektaris, N.I. Ostapenko and M.T. Shpak, 1984, Pis'ma v Zh. Eksp. & Teor. Fiz. **39**, 105.
Buckley, M.J., C.B. Harris and A.H. Maki, 1970, Chem. Phys. Lett. **4**, 591.
Burland, D.M., D.E. Cooper, M.D. Fayer and C.R. Cochanour, 1977a, Chem. Phys. Lett. **52**, 279.
Burland, D.M., U. Konzelmann and R.M. Macfarlane, 1977b, J. Chem. Phys. **67**, 1926.
Chen, Y.S., W. Shockley and G.L. Pearson, 1966, Phys. Rev. **151**, 648.
Cohen, M.H., 1977, in: Organic Conductors and Semiconductors, Lecture Notes in Physics, Vol. 65 (Springer, Berlin) p. 225.
Colson, S.D., D.M. Hanson, R. Kopelman and G.W. Robinson, 1968, J. Chem. Phys. **48**, 2215.
Colson, S.D., S. George, T. Keyes and V. Vaida, 1977, J. Chem. Phys. **67**, 4941.
Davydov, A.S., 1948, Zh. Eksp. & Teor. Fiz. **18**, 210.
Davydov, A.S., 1968, Theory of Molecular Excitons (Nauka, Moscow) (English Translation, 1971) (Plenum Press, New York).
Davydov, A.S., and E.I. Rashba, 1957, Ukrain. Phys. Zh. **2**, 226.
Dean, P., 1972, Rev. Mod. Phys. **44**, 127.
Decola, R.L., R.M. Hochstrasser and H.P. Trommsdorff, 1980, Chem. Phys. Lett. **72**, 1.
Dlott, D., C.L. Schosser and E.L. Chronister, 1982, Chem. Phys. Lett. **90**, 386.
Dlott, D.D., and M.D. Fayer, 1976, Chem. Phys. Lett. **41**, 305.
Doberer, U., H. Port and H. Bank, 1982, Chem. Phys. Lett. **85**, 253.
Dolganov, V.K., and E.F. Sheka, 1971, Zh. Eksp. & Teor. Fiz. **60**, 2230 [Sov. Phys.-JETP **33**, 1198].
Economou, E.N., and P.D. Antoniou, 1977, Solid State Commun. **21**, 285.
Elliott, R.J., J.A. Krumhansl and P.L. Leath, 1974, Rev. Mod. Phys. **46**, 465.
Farrar, T.G., and E.D. Becker, 1971, Pulse and Fourier Transform NMR (Academic Press, New York).
Fayer, M.D., and C.B. Harris, 1974, Chem. Phys. Lett. **25**, 149.
Feynman, R.P., F.L. Vernon and R.W. Hellworth, 1957, J. Appl. Phys. **28**, 49.
Francis, A.H., and C.B. Harris, 1971, Chem. Phys. Lett. **9**, 181, 188.
Franck, J., and E. Teller, 1938, J. Chem. Phys. **6**, 861.
Frenkel, J., 1931, Phys. Rev. **37**, 17, 1276.
Friesner, R., and R. Silbey, 1982, Chem. Phys. Lett. **93**, 107.
Gentry, S.T., and R. Kopelman, 1982, Chem. Phys. Lett. **93**, 264.
Ginzburg, V.L., 1958, Zh. Eksp. & Teor. Fiz. **34**, 1593 [Sov. Phys.-JETP **7**, 813].
Glockner, E., and H.C. Wolf, 1969, Z. Naturforsch. A **24**, 934.
Gor'kov, L.P., A.I. Larkin and D.E. Khmel'nitskii, 1979, Pis'ma v Zh. Eksp. & Teor. Fiz. **30**, 248 [JETP Lett. **30**, 228].
Gribnikov, Z.S., and E.I. Rashba, 1958, Zh. Tekhn. Fiz. **28**, 1948 [Sov. Phys.-Techn. Phys. **3**, 1790].

Haarer, D.P., and H.C. Wolf, 1970, Mol. Cryst. Liq. Cryst. **10**, 359.
Hanson, D.M., 1970, J. Chem. Phys. **52**, 3409.
Hanson, D.M., 1983a, in: Molecular Electronic Devices, ed. F.L. Carter (Marcel Dekker, New York) p. 89.
Hanson, D.M., 1983b, in: Spectroscopy and Excitation Dynamics of Condensed Molecular Systems, eds V.M. Agranovich and R.M. Hochstrasser (North-Holland, Amsterdam) p. 621.
Harris, C.B., 1971, J. Chem. Phys. **54**, 972.
Harris, C.B., and W.G. Breiland, 1978, in: Laser and Coherence Spectroscopy, ed. J.I. Steinfeld (Plenum Press, New York) p. 373.
Harris, C.B., R.L. Schlupp and H. Schuch, 1973, Phys. Rev. Lett. **30**, 1019.
Hesp, B.H., and D.A. Wiersma, 1980, Chem. Phys. Lett. **75**, 423.
Ho, F., W.-S. Tsay, J. Trout and R.M. Hochstrasser, 1981, Chem. Phys. Lett. **83**, 5.
Ho, F., W.-S. Tsay, J. Trout, S. Velsko and R.H. Hochstrasser, 1983, Chem. Phys. Lett. **97**, 141.
Hochstrasser, R.M., and G.R. Meredith, 1977, J. Chem. Phys. **67**, 1273.
Hochstrasser, R.M., and G.J. Small, 1966, J. Chem. Phys. **45**, 2270.
Hochstrasser, R.M., and J.D. Whiteman, 1972, J. Chem. Phys. **56**, 5945.
Hong, H.-K., and R. Kopelman, 1970, Phys. Rev. Lett. **25**, 1030.
Hong, H.-K., and R. Kopelman, 1971a, J. Chem. Phys. **55**, 724.
Hong, H.-K., and R. Kopelman, 1971b, J. Chem. Phys. **55**, 5380.
Hong, H.-K., and G.W. Robinson, 1970, J. Chem. Phys. **52**, 825.
Hopfield, J.J., 1958, Phys. Rev. **112**, 1555.
Hoshen, J., and J. Jortner, 1972, J. Chem. Phys. **56**, 933.
Huang, K., 1951, Proc. R. Soc. A **208**, 352.
Ivchenko, E.L., G.E. Pikus, B.S. Razbirin and A.I. Starukhin, 1977, Zh. Eksp. & Teor. Fiz. **72**, 2230 [Sov. Phys.-JETP **45**, 1172].
Klafter, J., and J. Jortner, 1977, Chem. Phys. Lett. **49**, 410.
Klafter, J., and J. Jortner, 1980, J. Chem. Phys. **73**, 1004.
Kopelman, R., 1983, in: Spectroscopy and Excitation Dynamics of Condensed Molecular Systems, eds V.M. Agranovich and R.M. Hochstrasser (North-Holland, Amsterdam) p. 139.
Kopelman, R., E.M. Monberg, F.W. Ochs and P.N. Prasad, 1975a, Phys. Rev. Lett. **34**, 1506.
Kopelman, R., E.M. Monberg, F.W. Ochs and P.N. Prasad, 1975b, J. Chem. Phys. **62**, 292.
Kopelman, R., E.M. Monberg and F.W. Ochs, 1977, Chem. Phys. **21**, 373.
Kopelman, R., P.W. Klymko, J.S. Newhouse and L. Anacker, 1984, Phys. Rev. B **29**, 3747.
Koster, G.F., and J.C. Slater, 1954, Phys. Rev. **95**, 1167.
Krivenko, T.A., A.V. Leiderman and E.I. Rashba, 1975, Fiz. Tverd. Tela **17**, 137 [Sov. Phys.-Solid State **17**, 78].
Krivenko, T.A., A.V. Leiderman, E.I. Rashba and E.F. Sheka, 1977, Pis'ma v Zh. Eksp. & Teor. Fiz. **25**, 538 [JETP Lett. **25**, 503].
Krivenko, T.A., E.I. Rashba and E.F. Sheka, 1978, Mol. Cryst. Liq. Cryst. **47**, 119.
Kronenberger, A., 1930, Z. Phys. **63**, 494.
Kronenberger, A., and P. Pringsheim, 1926, Z. Phys. **40**, 75.
Laubereau, A., and W. Kaiser, 1978, Rev. Mod. Phys. **50**, 607.
Lee, K.O., and T.T. Gan, 1977, Chem. Phys. Lett. **51**, 120.
Leong, B., and D. Hanson, 1983, J. Chem. Phys. **78**, 1613.
Lewellyn, M.T., A.H. Zewail and C.B. Harris, 1975, J. Chem. Phys. **63**, 3687.
Lifshitz, I.M., 1947, Zh. Eksp. & Teor. Fiz. **17**, 1017, 1076.
Lifshitz, I.M., 1964, Usp. Fiz. Nauk **83**, 617 [1965, Sov. Phys.-Usp. **7**, 549].
Lysenko, E.E., 1936, Zh. Eksp. & Teor. Fiz. **6**, 787.
Mandelbrot, B.B., 1977, Fractals: Form, Chance and Dimension (Freeman, San Francisco).
Meletov, K.P., E.I. Rashba and E.F. Sheka, 1979, Pis'ma v Zh. Eksp. & Teor. Fiz. **29**, 184 [JETP Lett. **29**, 165].

Meletov, K.P., E.I. Rashba and E.F. Sheka, 1980, Mol. Cryst. Liq. Cryst. **57**, 65.
Mil'man, P.D., 1975, cited in Krivenko et al. (1975).
Mott, N.F., 1967, Adv. Phys. **16**, 49.
Mott, N.F., and E.A. Davis, 1971, Electronic Processes in Non-Crystalline Materials (Clarendon Press, Oxford).
Mott, N.F., and W.D. Twose, 1961, Adv. Phys. **10**, 107.
Natkaniec, I., E.I. Bokhenkov, D. Dorner, J. Kalus, G. McKenzie, G.S. Pawley, U. Schmelzer and E.F. Sheka, 1980, J. Phys. C **13**, 4265.
Obreimov, I.B., 1927, Zh. Russk. Fiz. Khim. Obshch. **59**, 549 (in Russian).
Obreimov, I.B., and W.J. de Haas, 1928, Comm. from Leiden, No. 191; 1929, ibid. No. 204 c.
Ochs, F.W., P.N. Prasad and R. Kopelman, 1974a, Chem. Phys. **6**, 253.
Ochs, F.W., P.N. Prasad and R. Kopelman, 1974b, Chem. Phys. Lett. **29**, 290.
Onodera, Y., and Y. Toyozawa, 1968, J. Phys. Soc. Jpn. **24**, 341.
Osad'ko, I.S., and A.R. Chigirev, 1981, Fiz. Tverd. Tela **23**, 538 [Sov. Phys.-Solid State **23**, 302].
Ostapenko, N.I., and M.T. Shpak, 1972, Pis'ma v Zh. Eksp. & Teor. Fiz. **16**, 513 [JETP Lett. **16**, 365].
Ostapenko, N.I., V.I. Sugakov and M.T. Shpak, 1973, in: Excitons in Molecular Crystals (Naukova Dumka Publ., Kiev) p. 92 (in Russian).
Ostapenko, N.I., G.Yu. Khotyaintseva, M.P. Chernomorets and M.T. Shpak, 1978, Pis'ma v Zh. Eksp. & Teor. Fiz. **27**, 452 [JETP Lett. **27**, 423].
Ostapenko, N.I., G.Yu. Khotyaintseva, M.P. Chernomorets and M.T. Shpak, 1979, Phys. Status Solidi (b) **93**, 493.
Ovsyankin, V.V., and A.A. Fedorov, 1982, Pis'ma v Zh. Eksp. & Teor. Fiz. **35**, 199 [JETP Lett. **35**, 245].
Patel, J.S., and D.M. Hanson, 1982, Chem. Phys. **72**, 35.
Pekar, S.I., 1957, Zh. Eksp. & Teor. Fiz. **33**, 1022 [1958, Sov. Phys.-JETP **6**, 785].
Pekar, S.I., 1959, Zh. Eksp. & Teor. Fiz. **36**, 451 [Sov. Phys.-JETP **9**, 314].
Pekar, S.I., 1982, Crystal Optics and Additional Light Waves (Naukova Dumka Publ., Kiev) (English Translation W.A. Benjamin, Reading, MA, 1983).
Philpott, M.R., 1968, J. Chem. Phys. **49**, 4537.
Port, H., D. Vogel and H.C. Wolf, 1975, Chem. Phys. Lett. **34**, 23.
Port, H., D. Rund and H.C. Wolf, 1981, Chem. Phys. **60**, 81.
Prikhot'ko, A.F., 1944, J. Phys. USSR **8**, 257.
Prikhot'ko, A.F., 1949, Zh. Eksp. & Teor. Fiz. **19**, 383.
Prikhot'ko, A.F., A.F. Skorobogat'ko and L.I. Sikora, 1969, Opt. & Spektrosk. **26**, 966.
Rabin'kina, N.V., E.I. Rashba and E.F. Sheka, 1970, Fiz. Tverd. Tela **12**, 3569 [Sov. Phys.-Solid State **12**, 2898].
Rammal, R., and G. Toulouse, 1983, J. Phys. Lett. **44**, L-13.
Rashba, E.I., 1957, Opt. Spektrosk. **2**, 568.
Rashba, E.I., 1962, Fiz. Tverd. Tela **4**, 3301 [Sov. Phys.-Solid State **4**, 2417].
Rashba, E.I., 1963, Fiz. Tverd. Tela **5**, 1040 [Sov. Phys.-Solid State **5**, 757].
Rashba, E.I., 1966, Zh. Eksp. & Teor. Fiz. **50**, 1064 [Sov. Phys.-JETP **23**, 708].
Rashba, E.I., 1968, Zh. Eksp. & Teor. Fiz. **54**, 542 [Sov. Phys.-JETP **27**, 292].
Rashba, E.I., A.A. Gogolin and V.I. Mel'nikov, 1977, in: Organic Conductors and Semiconductors, Lecture Notes in Physics, Vol. 65 (Springer, Berlin) p. 265.
Rigini, R., P.F. Fracassi and R.G. Della Valle, 1983, Chem. Phys. Lett. **97**, 308.
Robinette, S.L., and G.J. Small, 1976, J. Chem. Phys. **65**, 837.
Scheibe, G., 1941, Z. Elektrochem. **47**, 73.
Schmidt, J., 1972, Chem. Phys. Lett. **14**, 411.
Schmidt, J., and J.H. van der Waals, 1968, Chem. Phys. Lett. **2**, 640.
Schmidt, J., and J.H. van der Waals, 1979, in: Time Domain Electron Spin Resonance, eds L. Kevan and R.N. Schwarts (Wiley, New York) p. 343.

Schmidt, J., W.G. van Dorp and J.H. van der Waals, 1971, Chem. Phys. Lett. **8**, 345.
Schwoerer, M., and H.C. Wolf, 1967, Mol. Cryst. **3**, 177.
Sheka, E.F., 1963a, Fiz. Tverd. Tela **5**, 2316 [Sov. Phys.-Solid State **5**, 1718].
Sheka, E.F., 1963b, Izv. Akad. Nauk USSR Ser. Fiz. **27**, 503 [Bull. Acad. Sci. USSR, Phys. Ser. **27**, 501].
Sheka, E.F., 1970, Fiz. Tverd. Tela **12**, 1167 [Sov. Phys.-Solid State **12**, 911].
Sheka, E.F., 1972, in: Physics of Impurity Centres in Crystals, ed. G.S. Zavt (Acad. Sci. Eston. SSR Publ., Tallinn) p. 431.
Sheka, E.F., 1975, Mol. Cryst. Liq. Cryst. **29**, 323.
Sheka, E.F., and I.P. Terenetskaya, 1971, Fiz. Tverd. Tela **13**, 1071 [Sov. Phys.-Solid State **13**, 889].
Stevenson, S.H., and G.J. Small, 1983, Chem. Phys. Lett. **95**, 18.
Stevenson, S.H., C.K. Johnson and G.J. Small, 1981, J. Phys. Chem. **85**, 2709.
Sugakov, V.I., 1982, in: Excitons, eds E.I. Rashba and M.D. Sturge (North-Holland, Amsterdam) p. 709.
Sumi, H., 1974, J. Phys. Soc. Jpn. **36**, 770.
Sumi, H., 1975, J. Phys. Soc. Jpn. **38**, 825.
Talapatra, G.B., and T.N. Misra, 1980, Chem. Phys. Lett. **76**, 42.
Tokura, Y., T. Koda and I. Nakada, 1980, J. Phys. Soc. Jpn. **49**, Supp. A: Proc. 15th Conf. Physics of Semiconductors, Kyoto p. 417.
Tolpygo, K.B., 1950, Zh. Eksp. & Teor. Fiz. **20**, 497.
Toyozawa, Y., 1958, Prog. Theor. Phys. **20**, 53.
Van Strien, A.J., J.F.C. van Kooten and J. Schmidt, 1980, Chem. Phys. Lett. **76**, 7.
Van 't Hof, C.A., and J. Schmidt, 1975, Chem. Phys. Lett. **36**, 460.
Vankan, J.M.J., and W.S. Veeman, 1982, Chem. Phys. Lett. **92**, 519.
Weaire, D., and V. Srivastava, 1977, Solid State Commun. **23**, 863.
Weisskopf, V., and E. Wigner, 1930, Z. Phys. **63**, 54.

CHAPTER 6

Phonon Multimode Spectra: Biphonons and Triphonons in Crystals with Defects

V.M. AGRANOVICH and O.A. DUBOVSKY

Institute of Spectroscopy
USSR Academy of Sciences
Troitsk, Moskovskaya Oblast, 142092
USSR

In memory of Nicholas M. Weinstein

Optical Properties of Mixed Crystals
Edited by
R.J. Elliott and I.P. Ipatova

Contents

1. Introduction 300
2. Biphonons in perfect crystals in the overtone spectrum region 302
   2.1. Qualitative discussion. Model Hamiltonian 302
   2.2. Schrödinger equation and dispersion relations for biphonons 305
   2.3. Width of the biphonon zone 309
      2.3.1. Linear crystals 309
      2.3.2. Three-dimensional crystals 312
   2.4. Biphonons in crystals with several molecules per unit cell 316
3. Local biphonons in crystals with isotopic substitution impurities 318
   3.1. Where does the formation of local states begin in a spectrum of optical vibrations? Effects of strong anharmonicity 318
      3.1.1. General remarks 318
      3.1.2. Local biphonon in an $^{14}NH_4Br$ crystal containing the isotopic substitution impurity ^{15}N 320
   3.2. Two-particle states in a crystal having defects: basic equations and biphonon localization at an anharmonicity defect 321
   3.3. Secular equation for a biphonon in a crystal having an isolated isotopic substitution impurity 325
      3.3.1. Two-particle states at $A = 0$ 326
      3.3.2. General form of the secular equation ($A \neq 0$) 328
   3.4. Certain limiting cases 330
      3.4.1. Limiting case of strong anharmonicity 330
      3.4.2. Limiting case of high values of isotopic shift (vacancy defect) 331
      3.4.3. Relative arrangement of the levels of two-particle states in a crystal having an isotopic impurity 332
   3.5. Local biphonons in crystals having two molecules per unit cell 335
   3.6. Biphonon scattering and decay processes upon interaction with an isolated isotopic impurity 340
      3.6.1. Elastic scattering process for a biphonon 344
      3.6.2. Stripping and pickup processes 346
   3.7. Local states spectrum with Fermi resonance 348
      3.7.1. General statement of the problem 348
      3.7.2. Case of strong anharmonicity ($A \gg T$) 352
      3.7.3. Results of a numerical calculation for the one-dimensional model 354

4. Two-particle states in isotopically disordered crystals 358
   4.1. Coherent potential approximation (CPA) for two-particle states 358
      4.1.1. Introduction 358
      4.1.2. The CPA equation 359
      4.1.3. Two limiting cases of disorder and behavior in the limit $x \to 0$ 362
   4.2. CPA calculations of the biphonon spectrum in disordered one-dimensional crystals 367
   4.3. CPA calculations of the biphonon spectrum in disordered three-dimensional crystals 370
   4.4. Quasi-biphonons in impurity crystals of NH_4Cl. A comparison of the theoretical and experimental data 374
   4.5. Two-particle vibrations of disordered one-dimensional crystals: a numerical solution of the secular equation 378
   4.6. Another limiting case of an isotopically disordered crystal: the van Kranendonk model 381
5. Bound three-phonon complexes (triphonons) 383
   5.1. The perfect crystal 383
   5.2. Triphonon localization at an isotopic defect 390
   5.3. A new type of three-phonon complexes 391
References 396

1. Introduction

The interaction of atoms and molecules in crystals causes collective excitations, leading to states of diverse types: phonons, excitons, magnons, plasmons, etc. The characteristics of these elementary quasi-particles, such as their spectra, kinetic coefficients, etc., have been the object of intensive research over many decades. In recent years, concurrently with this research, many more experimental and theoretical investigations have been conducted on systems in which, under well-defined conditions, the interaction of one-quantum (elementary) collective excitations can and does lead to the formation of bound complexes of these excitations. These intensive investigations are stimulated both by advances in experimental techniques, leading to improved sensitivity of the recording devices, and by general progress in condensed matter theory.

This chapter deals mainly with the simplest of the above-mentioned complexes, i.e. biphonons. The special features of biphonons in ideal crystals have already been discussed in Volume 4 of this series (Agranovich 1983). In accordance with the general subject matter of the present volume, we shall consider biphonon spectra in crystals with defects.

It is well-known that the investigation of elementary excitations in imperfect crystals, i.e. in crystals containing violations of the crystalline structure, forms one of the largest specializations of solid-state physics. Theoretical investigations in this area were begun by Lifshitz (1947, 1956) and by Montroll and Potts (1955, 1956) in the harmonic approximation. Their papers were the first to show that the presence of defects, in general, leads to the formation of excited states (local phonons), localized in the vicinity of the defects.

As a result of neutron scattering experiments and optical investigations it was subsequently established that the above-mentioned states appear in spectra of various types in the form of separate lines and bands located outside the continuum of zone states.

If, however, the perturbation produced by the defect is insufficient to split off a local level, then so-called quasi-local states may be formed (Brout and Visscher 1962, Kagan and Iosilevsky 1962); for a review of these investigations and their history, see, for instance, Maradudin (1966).

The simple picture described above is valid only when the defect concentration is low, so that the interaction between defects can be neglected and the problem can be approached assuming only a single isolated defect. For crystals with a large number of defects it is necessary to take defect interaction into account, paying attention, among other matters, to the possibility of the formation of binary, ternary or even larger clusters. In this case special calculation methods are required (see, for example, Lifshitz et al. 1982). One of the oldest methods of taking defects into consideration is that of additive refraction (Agranovich 1974, Agranovich and Galanin 1982). This method was devised in order to calculate the dielectric constant of molecular mixtures and, within its scope, the presence of defect clusters is not taken into account. Similar results are given by the isoshift model of Chang and Mitra (1971), the average T-matrix model and other calculation methods that are relatively simple and related in concept.

More complex methods of calculation were also required for a more adequate description of crystal systems with defects. Such methods were developed by Soven (1967), Taylor (1967), Yonezawa and Matsubara (1966a, b), Velicky et al. (1968), and Onodera and Toyozawa (1968). These investigators formulated the coherent potential approximation (CPA), that subsequently found wide application in the theory of disordered matter. The fundamentals of the modern theory of disordered crystal spectra can be found, for instance, in the review by Elliott et al. (1974) and in the book by Lifshitz et al. (1982) mentioned above.

In connection with the aforesaid we point out that all the investigations cited above dealt with the influence of defects on the spectra of one-particle elementary excitations only (phonons, excitons, etc.). Many qualitatively new features appear in the spectra of many-particle excitations in crystals with defects. An appropriate theory, requiring in this case that anharmonicity be taken into account, becomes substantially more complex. This will be demonstrated in the following by using as an example a crystal in which there are only two optical phonons. Experimental investigations of many-particle local states by, for example, Belousov et al. (1982) have also been started recently. These investigations yielded experimental confirmation of the possibility, predicted by Agranovich (1970), of the formation of a local biphonon in isotopically substitutional crystals in which no local state is formed in the region of one-particle excitations. The role of anharmonicity turns out to be the decisive factor in this case: local states are first formed in the overtone region and only when anharmonicity is taken into consideration.

In this chapter, when discussing the spectra of two-particle states in disordered crystals with a high impurity concentration we shall resort to a method which is an analog of the CPA method from the theory of one-particle states. This type of method is interpolational and, for instance, for one-particle states at low defect concentrations, should lead to accurate relations determining the levels of local one-particle states. It is clear then that this method, when applied to the theory of two-particle states under the same conditions (i.e., at low impurity concentrations), should properly describe two-particle states localized on an isolated defect (local biphonons). Hence the development of a theory dealing with such local biphonons is of importance, not only in analyzing the spectra of crystals with a low impurity concentration, but also in devising various kinds of interpolation theories intended for the investigation of spectra of the two-particle states of crystals with a high impurity concentration.

Section 2 of this chapter is a brief account of aspects of the theory of two-particle states in a perfect crystal (general equations, the Green function and dispersion relations) that will be required subsequently for constructing a theory of biphonon states in crystals with defects. Discussed in section 3 are the special features of two-particle spectra (local and quasi-local biphonons, and the conditions required for their formation) in crystals of various structures containing isolated defects: isotopic substitution of impurities. The fourth section treats special features and problems of the coherent potential approximation when applied to two-particle states in disordered crystals. Some results of numerical calculations for the spectra of biphonon states in one- and three-dimensional crystals are given, and available data of experimental investigations are discussed.

Section 5 deals with the special features of three-particle spectra (free and local triphonons).

2. *Biphonons in perfect crystals in the overtone spectrum region*

2.1. Qualitative discussion. Model Hamiltonian

It is possible, even from purely qualitative reasoning, to come to the conclusion that the effect of anharmonicity of optical vibrations in crystals can be very large indeed for the region of the overtone spectrum of intramolecular vibrations. It is a fact that in isolated molecules the anharmonicity energy A is usually of the order of 1 to 3% of the energy of a quantum of the fundamental vibrations (here and henceforth, the anharmonicity energy A is defined as half the shift of the overtone energy E_2 with respect to twice the energy of the fundamental tone: $E_2 = 2\hbar\Omega - 2A$, where Ω is the frequency of the fundamental tone). For example, at $\Omega = 1000\ \text{cm}^{-1}$, the anharmonicity energy A ranges from 10 to 30 cm^{-1}. At the same time, the energy of intermolecular interaction, which determines the width T_1 of the phonon energy band for the above-mentioned region of frequencies Ω, as follows, for instance, from measurements of Raman spectra of many crystals, also turns out to be of the order of 10 cm^{-1}. Therefore, cases in which $T_1/A \approx 1$ are possible. Here, the spectrum of optical vibrations in the region of overtones or of combination tones may have an extremely complex nature.

Let us consider, for example, a molecular crystal with σ molecules per unit cell. Assume that in a molecule there is a nondegenerate dipole vibration with frequency Ω. With anharmonicity of the third and fourth orders taken into account, the energy of the lowest excited states of the oscillator can be represented by the expression (Landau and Lifshitz 1959)

$$E_p = p\hbar\Omega - A(p^2 + p), \tag{1}$$

where $p = 0, 1, 2$ is the quantum number of the oscillator and A is a certain positive factor, the anharmonicity constant, which we shall assume to be known. If we introduce the Bose creation and annihilation operators B_n^+ and B_n for a quantum of oscillation in molecule n, then the energy operator of the molecule, neglecting its interaction with its surroundings, can be represented in the form

$$\hat{H}_n = \hbar\Omega B_n^+ B_n - A(B_n^+ B_n B_n^+ B_n + B_n^+ B_n).$$

The subindex n used above is a complex one: $n = \boldsymbol{n}\alpha$, where $\boldsymbol{n}$ is an integral vector and $\alpha = 1, 2, \dots, \sigma$ is the number of the molecule in the unit cell.

Since, by virtue of commutation relations for operators B_n and B_n^+, we have the operator identity $(B_n^+ B_n)^2 = B_n^{+2} B_n^2 + B_n^+ B_n$, we can write

$$\hat{H}_n = \hbar\omega B_n^+ B_n - A B_n^{+2} B_n^2, \tag{2}$$

where $\hbar\omega = \hbar\Omega - 2A$.

Next we take into account the motion of the excitation through the crystal. Note, to begin with, that if there are two quanta of molecular vibrations in the crystal and they are localized on different molecules, the energy of the crystal, neglecting the intermolecular interaction, is equal to $E = 2\hbar\omega$. But if both quanta are localized on a single molecule, then owing to intramolecular anharmonicity, $E = 2\hbar\omega - 2A$. Thus, resorting to the language of quasi-particles, we can assert that intramolecular anharmonicity in the case under discussion leads to a reduction in the energy of the crystal as quasi-particles approach one another, and thereby facilitates their attraction. However the localization of quasi-particles (intramolecular phonons) on a single molecule leads to an increase in the kinetic energy of their relative motion. Since, in order of magnitude, this energy is equal to the width of the phonon energy zone, states of phonons bound to one another are certainly established if the anharmonicity energy is large compared to the width of the phonon energy zone. In the overtone frequency region, in this case, along with the energy zone of two-particle states corresponding to the independent motion of two phonons, states of biphonons which are located somewhat lower on the frequency scale (at $A > 0$), are also established. In the limiting case of weak intermolecular interaction, biphonons are transformed into states of isolated molecules, excited to the second vibrational level. If, on the contrary, the anharmonicity is weak ($|A| < T_1$), then, in general, in three-dimensional crystals (one-dimensional ones are discussed below), no states of biphonons are formed, and the spectrum has only a zone (or zones) of two-particle states.

The aforesaid qualitative considerations (see also Agranovich 1983) are confirmed by an analysis which requires that intermolecular interactions be systematically taken into account. The operator of this interaction, neglecting the effects of anharmonicity, is obviously of the form

$$\hat{H}_{\text{int}} = \sum_{n,m}{}' V_{nm} B_n^+ B_m, \tag{3}$$

where V_{nm} is a matrix element of the energy operator for the interaction of molecules n and m, corresponding to the transition of one quantum of vibrations from molecule n to molecule m. If the anharmonicity of vibrations in molecules n and m is taken into consideration then, along with operator (3), terms of the order of BBB^+, BBB^+B^+, etc. appear in the operator of intermolecular interaction. Henceforth we shall assume that the main (governing) factor is the intramolecular anharmonicity taken into account in eq. (2), whereas the anharmonicity due to intermolecular interaction will be ignored. Such an approximation is evidently good enough for many molecular crystals, but it is frequently applied to semiconductors of the silicon and germanium types, and also to the crystals of inert gases (see, for instance, Jindal and Pathak 1977, Pathak and Jindal 1974).

In this approximation the complete Hamiltonian is

$$\hat{H} = \sum_n \hat{H}_n + \hat{H}_{\text{int}}$$

or

$$\hat{H} = \sum_n \hbar\omega B_n^+ B_n + \sum_{nm}{}' V_{nm} B_n^+ B_m - A \sum_n B_n^{+2} B_n^2. \tag{4}$$

If we transform from site representation to a Fourier representation then, for crystals with a single molecule per unit cell

$$B_{\boldsymbol{n}} = \frac{1}{\sqrt{N}} \sum_{\boldsymbol{k}} B_{\boldsymbol{k}} \exp\{\mathrm{i}\boldsymbol{k}\boldsymbol{n}\}, \tag{5}$$

where N is the number of molecules in the crystal, so that operator (4) assumes the form

$$\hat{H} = \sum_{\boldsymbol{k}} \varepsilon(\boldsymbol{k}) B_{\boldsymbol{k}}^{+} B_{\boldsymbol{k}} - AN^{-1} \sum_{\boldsymbol{k}_1 \boldsymbol{k}_2 \boldsymbol{k}_3} B_{\boldsymbol{k}_1}^{+} B_{\boldsymbol{k}_2}^{+} B_{\boldsymbol{k}_3} B_{\boldsymbol{k}_1 + \boldsymbol{k}_2 - \boldsymbol{k}_3}, \tag{6}$$

where

$$\varepsilon(\boldsymbol{k}) = \hbar\omega + \sum_{m}{}' V_{nm} \exp\{\mathrm{i}\boldsymbol{k}(\boldsymbol{m} - \boldsymbol{n})\}.$$

This is precisely the expression used for $\hat{H}$ in a number of investigations, for example by Ruvalds and Zawadowski (1970). It follows from eq. (6) that the approximation being discussed, i.e., an approximation in which only intramolecular anharmonicity is taken into account, actually corresponds to neglecting the dependence of constant A on the total wavevector $\boldsymbol{k}_1 + \boldsymbol{k}_2$ of the interacting particles.

Appearing first of all in crystals containing isotopic substitution impurities (as previously mentioned, this is precisely the type of disordered crystals that we shall discuss in sections 3 and 4) is diagonal disorder. It follows from the form of Hamiltonian $\hat{H}_n$ [see eq. (2)], that in taking intramolecular anharmonicity into account, the presence of such disorder indicates that not only is the energy $\hbar\omega$ a random function of n, but so is the anharmonicity constant A. After giving due attention to the aforesaid we come to the conclusion that in an isotopically disordered crystal, Hamiltonian $\hat{H}$ has the form

$$\hat{H} = \sum_{n} \hbar\omega_n B_n^{+} B_n + \sum_{nm}{}' V_{nm} B_n^{+} B_n - \sum_{n} A_n B_n^{+2} B_n^{2}. \tag{7}$$

If quantity V_{nm} is assumed to be random in the same way as quantities ω_n and A_n, then, in eq. (7), nondiagonal disorder is also taken into account.

2.2. Schrödinger equation and dispersion relations for biphonons

Operator $\hat{H}$ [see eq. (4)] commutes with the operator of the total number of vibrational quanta. Hence the number of these quanta is specified in steady states of the crystal.

There are no such quanta in the ground state. We shall denote by $|0\rangle$ the corresponding wave function. The anharmonicity term makes no contribution to the energy of states having a single quantum, i.e. the states $|1\rangle$. Hence the energy spectrum of one-particle states is determined by Hamiltonian (4) with $A = 0$.

The wave function of such one-phonon states with energy ε can be written in the form of the expansion

$$|1\rangle = \sum_n \Psi_n B_n^+ |0\rangle,$$

in which the coefficients Ψ_n satisfy the system of equations

$$(\varepsilon - \hbar\omega)\Psi_n - \sum_m{}' V_{nm}\Psi_m = 0.$$

For the optical phonon of the μth branch (where $\mu = 1, 2, \ldots, \sigma$, and σ is the number of molecules per unit cell)

$$\Psi_n(\boldsymbol{k}) = \frac{u_{\alpha\mu}(\boldsymbol{k})}{\sqrt{N}} \exp\{\mathrm{i}\boldsymbol{kn}\},$$

where $\boldsymbol{k}$ is the phonon wavevector. The quantities $u_{\alpha\mu}(\boldsymbol{k})$ satisfy the system of equations

$$\left[\varepsilon_\mu(\boldsymbol{k}) - \hbar\omega\right] u_{\alpha\mu}(\boldsymbol{k}) = \sum_\beta V^{\alpha\beta}(\boldsymbol{k}) u_{\beta\mu}(\boldsymbol{k}),$$

where

$$V^{\alpha\beta}(\boldsymbol{k}) = \sum_m V_{\boldsymbol{nm}}^{\alpha\beta} \exp\{\mathrm{i}\boldsymbol{k}(\boldsymbol{n} - \boldsymbol{m})\},$$

whereas $\varepsilon_\mu(\boldsymbol{k})$ (where $\mu = 1, 2, \ldots, \sigma$) are the roots of the equations

$$\det\left|(\varepsilon - \hbar\omega)\delta_{\alpha\beta} - V^{\alpha\beta}(\boldsymbol{k})\right| = 0.$$

For the orthonormalized states $|1\rangle$, the coefficients $u_{\alpha\mu}(\boldsymbol{k})$ are assumed to satisfy the system of relations

$$\sum_\alpha u_{\alpha\mu}^*(\boldsymbol{k}) u_{\alpha\mu'}(\boldsymbol{k}) = \delta_{\mu\mu'}, \tag{8}$$

$$\sum_\mu u_{\alpha\mu}^*(\boldsymbol{k}) u_{\beta\mu}(\boldsymbol{k}) = \delta_{\alpha\beta}. \tag{9}$$

In a crystal with one molecule per unit cell, the coefficient $u_\alpha = 1$, whereas the phonon energy is $\varepsilon(\boldsymbol{k}) = \hbar\omega + V(\boldsymbol{k})$.

Of interest to us in the following are states $|2\rangle$ with two vibrational quanta. In a similar manner these states can be represented in the form

$$|2\rangle = \sum_{nm} \Psi_{nm} B_n^+ B_m^+ |0\rangle, \qquad \Psi_{nm} = \Psi_{mn}, \tag{10}$$

where the quantity Ψ_{nm} denotes the wave function of states of a crystal with two vibrational excitations in the coordinate representation. Of interest to us, however, are states $|2\rangle$ with a definite energy E. Therefore the quantity Ψ_{nm} should satisfy the Schrödinger equation

$$\hat{H}|2\rangle = E|2\rangle.$$

After substituting relations (10) into this equation and making use of the commutation relation for the Bose operators B_n and B_m^+ $(B_n B_m^+ - B_m^+ B_n = \delta_{nm})$ we find that the quantities Ψ_{nm} satisfy the system of equations

$$(E - 2\hbar\omega + 2A\delta_{nm})\Psi_{nm} = \sum_l (V_{nl}\Psi_{lm} + V_{ml}\Psi_{ln}), \tag{11}$$

where $\delta_{nm} = \delta_{\alpha\beta}\delta_{\boldsymbol{nm}}$ (recall that $n \equiv \boldsymbol{n}, \alpha$).

It proves convenient, in the following, to transform to the Fourier expansion

$$\Psi_{nm} = \Psi_{\boldsymbol{nm}}^{\alpha\beta} = \sum_{\boldsymbol{k}_1\boldsymbol{k}_2} \tilde{\Psi}_{\boldsymbol{k}_1\boldsymbol{k}_2}^{\alpha\beta} \exp\{\mathrm{i}\boldsymbol{k}_1\boldsymbol{n} + \mathrm{i}\boldsymbol{k}_2\boldsymbol{m}\} \tag{12}$$

or

$$\Psi_{nm} = \sum_{\boldsymbol{kK}} \tilde{\Psi}_{(\boldsymbol{K}+\boldsymbol{k})/2,(\boldsymbol{K}-\boldsymbol{k})/2}^{\alpha\beta} \exp\{\mathrm{i}(\boldsymbol{K}/2)(\boldsymbol{n}+\boldsymbol{m}) + \mathrm{i}(\boldsymbol{k}/2)(\boldsymbol{n}-\boldsymbol{m})\}, \tag{13}$$

where $\boldsymbol{K} = \boldsymbol{k}_1 + \boldsymbol{k}_2$ and $\boldsymbol{k} = \boldsymbol{k}_1 - \boldsymbol{k}_2$. The solutions of eq. (11) can be selected so that they transform according to one of the irreducible representations of the translation group. For these states (with total wavevector $\boldsymbol{K}$)

$$\Psi_{nm}(\boldsymbol{K}) = \exp\{\mathrm{i}(\boldsymbol{K}/2)(\boldsymbol{n}+\boldsymbol{m})\} \times \sum_{\boldsymbol{K}} \tilde{\Psi}_{(\boldsymbol{K}+\boldsymbol{k})/2,(\boldsymbol{K}-\boldsymbol{k})/2}^{\alpha\beta} \exp\{\mathrm{i}(\boldsymbol{k}/2)(\boldsymbol{n}-\boldsymbol{m})\}. \tag{14}$$

It is clear that the function

$$\varphi_{\alpha\beta}(\boldsymbol{K}, \boldsymbol{n}-\boldsymbol{m}) = \sum_{\boldsymbol{k}} \tilde{\Psi}_{(\boldsymbol{K}+\boldsymbol{k})/2,(\boldsymbol{K}-\boldsymbol{k})/2}^{\alpha\beta} \exp\{\mathrm{i}(\boldsymbol{k}/2)(\boldsymbol{n}-\boldsymbol{m})\} \tag{15}$$

is the wave function of relative motion of two phonons, and that for bound states (i.e., for biphonons) it should decrease with an increase in $|\boldsymbol{n}-\boldsymbol{m}|$.

Let us derive the dispersion equation that determines the dependence of the biphonon vibrational energy $E(\boldsymbol{K})$ on the total wavevector, and, wherever possible, determines this relation.

We consider, firstly, crystals with one molecule per unit cell. We substitute expansion (12) into eq. (11) and make use of the representation for the Kronecker delta function $\delta_{nm}=N^{-1}\Sigma_q \exp \mathrm{i}\boldsymbol{q}(\boldsymbol{n}-\boldsymbol{m})\}$, where N is the total number of unit cells. Then the system of equations (11) is converted to the form

$$\sum_{k_1 k_2} [E-\varepsilon(\boldsymbol{k}_1)-\varepsilon(\boldsymbol{k}_2)]\,\tilde{\Psi}_{k_1 k_2}\exp\{\mathrm{i}\boldsymbol{k}_1\boldsymbol{n}+\mathrm{i}\boldsymbol{k}_2\boldsymbol{m}\}$$

$$=-2AN^{-1}\sum_{k_1' k_2' q}\tilde{\Psi}_{k_1' k_2'}\exp\{\mathrm{i}(\boldsymbol{k}_1'+\boldsymbol{q})\boldsymbol{n}+\mathrm{i}(\boldsymbol{k}_2'-\boldsymbol{q})\boldsymbol{m}\},$$

where $\varepsilon(\boldsymbol{k})=\hbar\omega+V(\boldsymbol{k})$ is the energy of a phonon with wavevector $\boldsymbol{k}$. The transformation on the right-hand side of this equation from summation over $\boldsymbol{k}_1'$, $\boldsymbol{k}_2'$ and $\boldsymbol{q}$ to summation over $\boldsymbol{k}_1=\boldsymbol{k}_1'+\boldsymbol{q}$, $\boldsymbol{k}_2=\boldsymbol{k}_2'-\boldsymbol{q}$ and $\boldsymbol{q}$ reduces the equation to the form

$$\sum_{k_1 k_2}\left\{[E-\varepsilon(\boldsymbol{k}_1)-\varepsilon(\boldsymbol{k}_2)]\,\tilde{\Psi}_{k_1 k_2}+2AN^{-1}\sum_{q}\tilde{\Psi}_{k_1-q,k_2+q}\right\}$$

$$\times\exp\{\mathrm{i}\boldsymbol{k}_1\boldsymbol{n}+\mathrm{i}\boldsymbol{k}_2\boldsymbol{m}\}=0.$$

Thus the system of secular equations (11) for functions Ψ_{nm} complies with the following system of secular equations for the functions $\tilde{\Psi}_{k_1 k_2}$:

$$\tilde{\Psi}_{k_1 k_2}+2AN^{-1}G_{k_1 k_2}(E)\sum_{k_1'+k_2'=k_1+k_2}\tilde{\Psi}_{k_1' k_2'}=0, \tag{16}$$

where $G_{k_1 k_2}(E)$ is the two-particle Green function with anharmonicity neglected (i.e. at $A=0$)

$$G_{k_1 k_2}(E)=[E-\varepsilon(\boldsymbol{k}_1)-\varepsilon(\boldsymbol{k}_2)]^{-1}. \tag{17}$$

From eq. (16) we can readily obtain the dispersion equation that determines the spectral dependence of $E(\boldsymbol{K})$ for biphonons.

As a matter of fact, since, according to eq. (16),

$$\tilde{\Psi}_{k_1 k_2}=-2AG_{k_1 k_2}(E)Q(\boldsymbol{K}), \tag{18}$$

where $\boldsymbol{K} = \boldsymbol{k}_1 + \boldsymbol{k}_2$, the quantity

$$Q(\boldsymbol{K}) \equiv N^{-1} \sum_{\boldsymbol{k}_1' + \boldsymbol{k}_2' = \boldsymbol{K}} \tilde{\Psi}_{\boldsymbol{k}_1' \boldsymbol{k}_2'} \tag{19}$$

satisfies the equation

$$Q(\boldsymbol{K}) \left[1 + 2AN^{-1} \sum_{\boldsymbol{k}_1 + \boldsymbol{k}_2 = \boldsymbol{K}} G_{\boldsymbol{k}_1 \boldsymbol{k}_2}(E) \right] = 0$$

which, like $\tilde{\Psi}_{\boldsymbol{k}_1 \boldsymbol{k}_2}$ is nonzero only for the values of E and $\boldsymbol{K}$ that are linked by the dispersion relation

$$F(E, \boldsymbol{K}) \equiv 1 + 2AI(E, \boldsymbol{K}) = 0, \tag{20}$$

where

$$I(E, \boldsymbol{K}) = \frac{1}{N} \sum_{\boldsymbol{k}_1} \frac{1}{E - \varepsilon(\boldsymbol{k}_1) - \varepsilon(\boldsymbol{K} - \boldsymbol{k}_1)}. \tag{21}$$

If, at a given value of $\boldsymbol{K}$, eq. (20) is satisfied for a certain real value of the energy $E = E(\boldsymbol{K})$, lying outside the zone $\varepsilon(\boldsymbol{k}_1) + \varepsilon(\boldsymbol{K} - \boldsymbol{k}_1)$ of two-particle states, then, according to eq. (18), the function $\tilde{\Psi}_{\boldsymbol{k}_1 \boldsymbol{k}_2}(\boldsymbol{k}_1 + \boldsymbol{k}_2 = \boldsymbol{K})$ for this value of E has no singularities (poles) even for real values of $\boldsymbol{k}_1$ and $\boldsymbol{k}_2$. This means that the wave function of relative motion of two phonons, i.e. function $\varphi(\boldsymbol{K}, \boldsymbol{n} - \boldsymbol{m})$ [see eq. (15)] decreases with an increase in $|\boldsymbol{n} - \boldsymbol{m}|$ and, in this way, corresponds to the bound state of the phonons (i.e. a biphonon). Before proceeding with the derivation of an analogous dispersion equation for biphonons in crystals having several molecules per unit cell, we shall discuss the width of the biphonon zone, an important problem for local biphonon theory.

2.3. *Width of the biphonon zone*

2.3.1. *Linear crystals*

We begin by considering features of the dispersion law $E(\boldsymbol{K})$ determined by relationship (20). For this purpose we shall first discuss the simplest model of a one-dimensional crystal, with one molecule per unit cell as previously. Moreover, we shall only take into account the interaction between closest neighbors ($V_{n,n-1} = V_{n,n+1} = V$). In this approximation the phonon energy is

$$\varepsilon(k) = \hbar\omega + 2V \cos(ka),$$

where a is a lattice constant. After substituting this relation into eq. (21) and converting by ordinary means from summation over k_1 to integration within the limits of the first Brillouin zone, we obtain, as the result of simple transformations of the integrand, the dispersion equation

$$1+2A(a/2\pi)\int_{-\pi/a}^{\pi/a}\{E-2\hbar\omega-4V\cos(Ka/2)\cos[(k-K/2)a]\}^{-1}\,\mathrm{d}k$$
$$=0. \tag{22}$$

The integral in eq. (22) can be readily calculated. As a result, for values of E outside the zone of dissociated states (i.e., for E values for which $|E-2\hbar\omega|\geqslant 4V$) we have

$$1-2A\{(E-2\hbar\omega)^2-[4V\cos(Ka/2)]^2\}^{-1/2}=0.$$

It follows from this equation that the biphonon dispersion law is determined by the relation

$$E(K)=2\hbar\omega-\sqrt{(2A)^2+[4V\cos(Ka/2)]^2}. \tag{23}$$

The equation obtained above for the energy of a biphonon in a perfect linear chain was derived by making use of the general dispersion relation (20), which is valid for crystals of arbitrary dimensionality. At the same time, in the given special case of a perfect one-dimensional crystal, there exists, as can readily be seen by a direct substitution, an accurate analytical solution of the basic equations (11), which does not make use of integral representations (Agranovich 1984). This solution for the wave function Ψ_{nm}, corresponding to a biphonon with wavevector $\boldsymbol{K}$, is of the form

$$\Psi_{nm}=\exp\{\mathrm{i}K[(n+m)/2]-\kappa|n-m|\}.$$

Here the constant $\kappa(K)$ of attenuation of the function of relative biphonon motion and the biphonon energy $E(K)$ are determined as the corresponding solutions of a system of two equations that directly follow from the diagonal $(n=m)$ and nondiagonal $(n\neq m)$ equations (11):

$$E-2\hbar\omega+2A=4V\,\mathrm{e}^{-\kappa a}\cos(Ka/2),$$

$$E-2\hbar\omega=4V\cosh(\kappa a)\cos(Ka/2).$$

Simultaneous solution of this system of equations yields for $E(K)$, as can

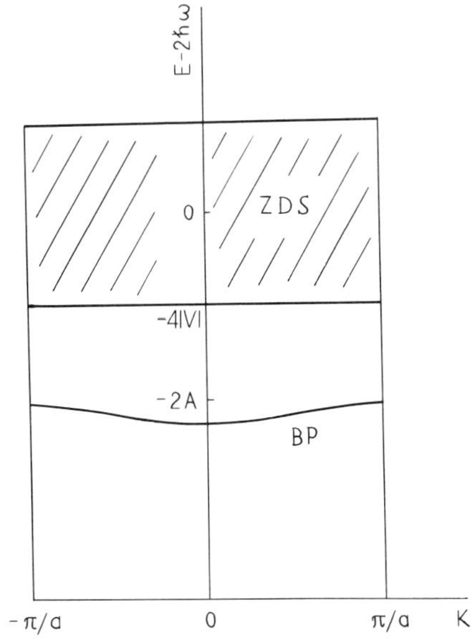

Fig. 1. Spectrum of biphonon vibrations in a perfect crystalline chain ($A > 2|V|$): BP is the biphonon zone and ZDS is the zone of dissociated two-particle states. The BP zone is completely split from the ZDS.

readily be seen, eq. (23), whereas the quantity $\kappa(K)$ is correspondingly determined, for example, from the second equation of this system as

$$\kappa(K) = a^{-1} \operatorname{arc\,cosh}\{[E(K) - 2\hbar\omega]/4V \cos(Ka/2)\}.$$

The solution for Ψ_{nm} given above can, naturally, also be obtained from the general relation (14) by the necessary integration in the complex plane k, taking into account the poles $\tilde{\Psi}_{k,K-k}$ [see eq. (16)] and making use of residue theory.

Shown in figs. 1 and 2 are curves of the dependence $E(K)$, plotted from eq. (23), for two relationships between A and $|V|$. The hatched area corresponds to the energy zone $E(k_1, k_2) = 2\hbar\omega + 2V[\cos(k_1 a) + \cos(k_2 a)]$ of two-particle free states, so that its minimum corresponds to the energy $E_{\min} = 2\hbar\omega - 4|V|$. Figure 1 is for the situation in which $A > 0$ and, moreover, $A > 2|V|$. In this case the zone of bound states (biphonon zone) is split off from the zone of dissociated states over the whole region of variation of K. If, on the contrary, $2|V| > A$, then biphonons exist only in a certain restricted region $|K| \leqslant K_0$. Of importance is the fact (and this could be expected from general considerations) that biphonons are formed in a one-dimensional crystal at any, even the smallest, value of A. However, the value of A determines the size of

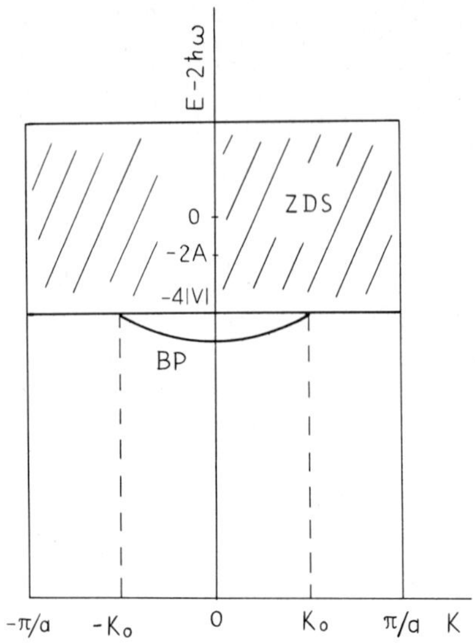

Fig. 2. Spectrum of biphonon vibrations in a perfect crystalline chain ($0 < A < 2|V|$). The BP zone is partially split off from the ZDS.

the wavevector region within which the biphonon exists. Note still another fact: in the given model, regardless of the sign of the effective mass of optical phonons, at $K \approx 0$ biphonon states always have a positive effective mass.

It is necessary, in addition, to point out a fact that is very important for further discussion. The width of the biphonon zone in a given crystal depends strongly on the value of A (see figs. 1 and 2). In particular, at $A > 2|V|$ this width is

$$T_2 = \sqrt{(2A)^2 + (4V)^2} - 2A.$$

Of interest is an estimate of the biphonon zone width at strong anharmonicity, i.e., at $A \gg |V|$. Using the relation given above we have

$$T_2 = (2V)^2/A. \tag{24}$$

Thus, with an increase of the anharmonicity constant, the biphonon zone width decreases as the ratio T^2/A, where T is the halfwidth of the zone of one-particle states.

2.3.2. Three-dimensional crystals

At large values of A, such a dependence of the biphonon zone width on T and A (i.e. $T_2 \approx T^2/A$), is also valid for crystals of arbitrary dimensionality

(Agranovich 1970). Indeed, let us consider again the general dispersion equation [eq. (20)]. For sufficiently high values of the anharmonicity constant A, the levels of biphonon states are split off by the amount $-2A$ with small corrections determined by the magnitude of intermolecular interaction.

Assuming that

$$E(\boldsymbol{K}) = 2\hbar\omega - 2A + \delta E(\boldsymbol{K}),$$

where $\delta E(\boldsymbol{K})$ is a small addition, we expand in the sum of eq. (20) in terms of the small value $[V(\boldsymbol{k}_1) + V(\boldsymbol{k}_2) - \delta E(\boldsymbol{K})]/2A$ up to quadratic terms. After carrying out some simple transformations we obtain the following expression for $\delta E(\boldsymbol{K})$:

$$\delta E(\boldsymbol{K}) = \frac{1}{N} \sum_{\substack{\boldsymbol{k}_2\boldsymbol{k}_2 \\ (\boldsymbol{k}_1+\boldsymbol{k}_2=\boldsymbol{K})}} [V(\boldsymbol{k}_1) + V(\boldsymbol{k}_2)] - \frac{1}{2A}\frac{1}{N} \sum_{\substack{\boldsymbol{k}_1\boldsymbol{k}_2 \\ (\boldsymbol{k}_1+\boldsymbol{k}_2=\boldsymbol{K})}} [V(\boldsymbol{k}_1) + V(\boldsymbol{k}_2)]^2. \tag{25}$$

Since, by definition, matrix element V_{nm} has no diagonal elements, the summation

$$\sum_{\boldsymbol{k}} V(\boldsymbol{k}) = 0.$$

Hence the first summation on the right-hand side of eq. (25) is also equal to zero, so that

$$\delta E(\boldsymbol{K}) = -\frac{1}{2A}\frac{1}{N} \sum_{\substack{\boldsymbol{k}_1,\boldsymbol{k}_2 \\ (\boldsymbol{k}_1+\boldsymbol{k}_2=\boldsymbol{K})}} [V(\boldsymbol{k}_1) + V(\boldsymbol{k}_2)]^2.$$

Thus, with an accuracy to a numerical factor of the order of unity, we obtain, as in eq. (24), the following estimate for the zone width of biphonon states:

$$T_2 \approx T^2/2A. \tag{26}$$

Let us continue the analysis of biphonon spectra for three-dimensional crystals. Certain typical features of this spectrum, i.e. features of the dependence $E(\boldsymbol{K})$ can also be examined if, in calculating the integral of eq. (21), we make use of an approximation for the dependence of the energy $\varepsilon(\boldsymbol{k}_1) + \varepsilon(\boldsymbol{k}_2)$ on the sum $\boldsymbol{K} = \boldsymbol{k}_1 + \boldsymbol{k}_2$ and the difference $\boldsymbol{k} = (\boldsymbol{k}_1 - \boldsymbol{k}_2)/2$ wavevectors. This approximation (see, for instance, Drchal and Velicky 1976) becomes the exact representation for the one-dimensional crystal discussed above. In this approximation, the quantity $\varepsilon(\boldsymbol{k}_1)+\varepsilon(\boldsymbol{k}_2)$, appearing in eq. (21) is to be replaced as follows:

$$\begin{aligned}\varepsilon(\boldsymbol{k}_1)+\varepsilon(\boldsymbol{k}_2) &\equiv \varepsilon(\boldsymbol{K}/2+\boldsymbol{k})+\varepsilon(\boldsymbol{K}/2-\boldsymbol{k}) \\ &= 2\hbar\omega + 2Tl(\boldsymbol{K}/2)l(\boldsymbol{k}),\end{aligned} \tag{27}$$

where $l(\boldsymbol{k})$ is a certain dimensionless function determining the dispersion $\varepsilon(\boldsymbol{k})=\hbar\omega+Tl(\boldsymbol{k})$ of one-particle states. Making use of eq. (27) and going over in eq. (21) from summation to the integral taken over the frequency spectrum of one-particle states, we obtain

$$\begin{aligned}&\frac{1}{N}\sum_{\substack{\boldsymbol{k}_1\boldsymbol{k}_2\\(\boldsymbol{k}_1+\boldsymbol{k}_2=\boldsymbol{K})}}\frac{1}{E-\varepsilon(\boldsymbol{k}_1)-\varepsilon(\boldsymbol{k}_2)} \\ &\quad=\frac{1}{2l(K/2)}\int\frac{\rho(x)\,\mathrm{d}x}{(E-2\hbar\omega)/2l(\boldsymbol{K}/2)-x},\end{aligned} \tag{28}$$

where the function $\rho(x)$ is the density of one-particle states:

$$\rho(x)=\frac{1}{N}\sum_{\boldsymbol{k}}\delta[x-\varepsilon(\boldsymbol{k})+\hbar\omega]. \tag{29}$$

After using, in addition, the "elliptical" model for $\rho(x)$ (Onodera and Toyozawa 1968) we have

$$\rho(x)=\frac{2}{\pi T^2}\sqrt{T^2-x^2}; \qquad |x|\leqslant T, \tag{30}$$

thereby taking the square-root features of the spectrum properly into account for a three-dimensional crystal near the zone boundaries. Then the integral in eq. (28) is readily calculated at $|E-2\hbar\omega|>2Tl(K/2)$ (Gradshtein and Ryzhik 1971). As a result, i.e., after using approximations (27) and (30), dispersion equation (20) assumes the form

$$-\frac{1}{2A}=\frac{2}{[2Tl(\boldsymbol{K}/2)]^2}\left\{E-2\hbar\omega+\sqrt{(E-2\hbar\omega)^2-[2Tl(\boldsymbol{K}/2)]^2}\right\}. \tag{31}$$

Before analyzing this relationship we point out that by definition $[\varepsilon(\boldsymbol{k})=\hbar\omega+Tl(\boldsymbol{k})]$, function $l(\boldsymbol{k})$ varies in the range from -1 to 1, and that the lower edge of the zone of dissociated two-particle states corresponds to the energy $E_{\min} = 2\hbar\omega - 2T$. When $A > 0$, for example, biphonon states can be formed only when $E < E_{\min}$. In this case, the condition for the existence of a biphonon with the wavevector $\boldsymbol{K}$ can be written in the form

$$E(\boldsymbol{K}) < E_{\min} = 2\hbar\omega - 2T.$$

Next, after solving eq. (31) for E, we find that the solution outside the zone of two-particle states exists only when $A > T/2$. In this case the solution is of the form

$$E(\boldsymbol{K}) = 2\hbar\omega - 2A - T^2\left[l(\boldsymbol{K}/2)\right]^2/2A. \tag{32}$$

Consequently, the condition for the existence of a biphonon can be written in the form

$$2A + \frac{T^2}{2A} l^2\left(\frac{\boldsymbol{K}}{2}\right) \geqslant 2T, \quad \text{when} \quad A > \frac{T}{2} \tag{33}$$

or in the form

$$l^2\left(\frac{\boldsymbol{K}}{2}\right) \geqslant 4\left(\frac{A}{T}\right)\left(1 - \frac{A}{T}\right).$$

These relations are valid for any $\boldsymbol{K}$ values if $A > T$. In this case both

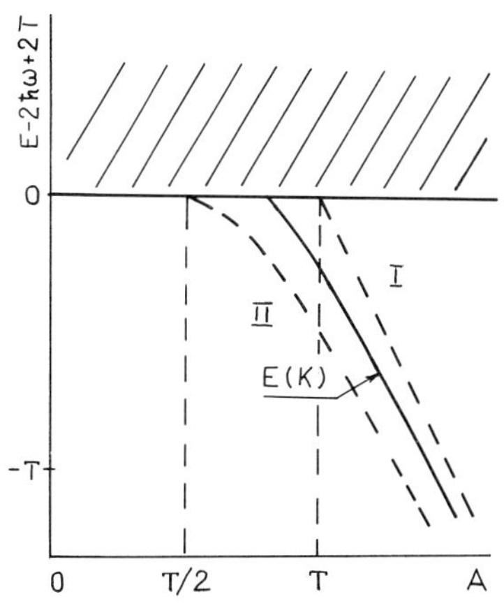

Fig. 3. Dependence of the energy of a biphonon with wavevector $\boldsymbol{K}$ on the anharmonicity constant A in a three-dimensional crystal. Dashed line I ($E = 2\hbar\omega - 2A$) corresponds to the upper boundary of the biphonon zone; dashed curve II ($E = 2\hbar\omega - 2A - T^2/2A$), to its lower boundary. At $A < T$ the upper boundary of the biphonon zone is determined by the lower boundary ($E_{\min} = 2\hbar\omega - 2T$) of the zone of dissociated states. This zone corresponds to the hatched area in the figure.

boundaries of the biphonon zone: $E_{max} = 2\hbar\omega - 2A$ (upper boundary) and $E_{min} = 2\hbar\omega - 2A - T^2/2A$ (lower boundary) lie below the value $2\hbar\omega - 2T$, i.e., below the lower edge of the zone of dissociated two-particle states. Hence, there are states with all possible values of $\boldsymbol{K}$ in the split-off biphonon zone. If, however, $E_{min} < 2\hbar\omega - 2T$, while $E_{max} > 2\hbar\omega - 2T$ (as is the case when $T/2 < A < T$), the only states formed in the biphonon zone are ones with $\boldsymbol{K}$ values that comply with the inequality $1 > l^2(\boldsymbol{K}/2) > l^2(\boldsymbol{K}_0/2)$, where $l^2(\boldsymbol{K}_0/2) = (4A/T)(1 - A/T)$. At $A = T/2$ the value of E_{min} is $2\hbar\omega - 2T$ so the biphonon is no longer formed (see fig. 3). Note also that when $A > T$, and the whole biphonon zone is split off, its width, $E_{max} - E_{min}$, is $T^2/2A$, which confirms the estimate made previously.

Even though a number of approximations were made in the aforesaid, the results obtained provide, nevertheless, a correct idea of the role played by various factors that determine the conditions for the formation of the biphonon zone, as well as its structure.

2.4. Biphonons in crystals with several molecules per unit cell

In a crystal having several molecules per unit cell, the coefficients $\tilde{\Psi}^{\alpha\beta}_{\boldsymbol{k}_1\boldsymbol{k}_2}$ [(see eq. (12)] can be conveniently presented in the form of the expansion

$$\tilde{\Psi}^{\alpha\beta}_{\boldsymbol{k}_1\boldsymbol{k}_2} = N^{-1} \sum_{\mu_1\mu_2} \Psi^{\mu_1\mu_2}_{\boldsymbol{k}_1\boldsymbol{k}_2} u_{\alpha\mu_1}(\boldsymbol{k}_1) u_{\beta\mu_2}(\boldsymbol{k}_2); \tag{34}$$

the coefficients u are determined in section 2.2. If, in using eq. (34), we substitute relation (12) into secular equation (11) and make use of the conditions of orthonormality and completeness [see eqs. (8) and (9)], for the quantities $\Psi^{\mu_1\mu_2}_{\boldsymbol{k}_1\boldsymbol{k}_2}$ we obtain the system of secular equations

$$\Psi^{\mu_1\mu_2}_{\boldsymbol{k}_1\boldsymbol{k}_2} = -2AG^{\mu_1\mu_2}_{\boldsymbol{k}_1\boldsymbol{k}_2}(E) \sum_{\nu} u^*_{\nu\mu_1}(\boldsymbol{k}_1) u^*_{\nu\mu_2}(\boldsymbol{k}_2) Q_\nu(\boldsymbol{K}), \tag{35}$$

where $\boldsymbol{K} = \boldsymbol{k}_1 + \boldsymbol{k}_2$

$$Q_\nu(\boldsymbol{K}) = N^{-1} \sum_{\substack{\mu'\mu''\boldsymbol{k}'\boldsymbol{k}'' \\ (\boldsymbol{k}'+\boldsymbol{k}''=\boldsymbol{K})}} u_{\nu\mu'}(\boldsymbol{k}') u_{\nu\mu''}(\boldsymbol{k}'') \Psi^{\mu'\mu''}_{\boldsymbol{k}'\boldsymbol{k}''}, \tag{36}$$

and

$$G^{\mu_1\mu_2}_{\boldsymbol{k}_1\boldsymbol{k}_2}(E) = \left[E - \varepsilon_{\mu_1}(\boldsymbol{k}_1) - \varepsilon_{\mu_2}(\boldsymbol{k}_2)\right]^{-1}. \tag{37}$$

On the basis of eqs. (35) and (36), it is readily evident that the function $Q_\alpha(\boldsymbol{K})$, where $\alpha = 1, 2, \ldots, \sigma$ satisfies the following system of equations:

$$Q_\alpha(\boldsymbol{K}) + 2A\sum_\beta I_{\alpha\beta}(\boldsymbol{K})Q_\beta(\boldsymbol{K}) = 0, \tag{38}$$

in which the matrix

$$I_{\alpha\beta}(E, \boldsymbol{K}) = N^{-1} \sum_{\substack{\mu_1\mu_2\boldsymbol{k}_1\boldsymbol{k}_2 \\ (\boldsymbol{k}_1+\boldsymbol{k}_2=\boldsymbol{K})}} u_{\alpha\mu_1}(\boldsymbol{k}_1)u_{\alpha\mu_2}(\boldsymbol{k}_2)u^*_{\beta\mu_1}(\boldsymbol{k}_1) \cdot u^*_{\beta\mu_2}(\boldsymbol{k}_2)G^{\mu_1\mu_2}_{\boldsymbol{k}_1\boldsymbol{k}_2}(E). \tag{39}$$

Thus, the biphonon energy $E(\boldsymbol{K})$ is determined from the condition that the following determinant becomes zero:

$$|\delta_{\alpha\beta} + 2AI_{\alpha\beta}(E, \boldsymbol{K})| = 0, \tag{40}$$

which is a generalization of condition (20).

In particular, in crystals with two molecules per unit cell, eq. (40) is of the form

$$1 + 2A(I_{11} + I_{22}) + 4A^2(I_{11}I_{22} - I_{12}I_{21}) = 0.$$

It follows from this equation that in the crystals being discussed it is possible to have, in general, biphonons of two types. The dispersion of these biphonons is determined by solving one of the equations

$$1 + 2AI^{(\pm)}(E, \boldsymbol{K}) = 0, \tag{41}$$

where

$$I^{(\pm)}(E, \boldsymbol{K}) = \tfrac{1}{2}\left[I_{11} + I_{22} \pm \sqrt{(I_{11} - I_{22})^2 + 4I_{12}I_{21}}\right].$$

Equation (41) is analogous in form to that of eq. (20), but, since the functions $I^{(+)}(E, \boldsymbol{K})$ and $I^{(-)}(E, \boldsymbol{K})$ are, in general, different, there may be various situations for one and the same value of the anharmonicity constant A. These situations include: (a) both eqs. (41) have roots $E(\boldsymbol{K})$ outside the zone of two-particle dissociated states, in which case two biphonon branches are formed; (b) only one of eqs. (41) has a root $E(\boldsymbol{K})$ inside the indicated region of the spectrum, in which case only one biphonon branch is formed; and (c) neither of eqs. (41) has a solution. Situation (c) never occurs for one-dimensional crystals, but it may occur in three-dimensional crystals when $A < A_0^{(+)}(\boldsymbol{K})$, $A_0^{(-)}(\boldsymbol{K})$, where $A_0^{(\pm)}(\boldsymbol{K})$ are certain limiting values of the anharmonicity constant, determining the region in which solutions of eqs. (41) exist.

It follows, in this manner, from the above that the number of biphonon branches is not, in general, equal to the number of molecules per unit cell (Agranovich 1970).

In section 4 we shall discuss the question of the extent to which the above-mentioned features of biphonon formation are subject to qualitative or quantitative changes in crystals having isotopic substitution impurities.

3. *Local biphonons in crystals with isotopic substitution impurities*

3.1. *Where does the formation of local states begin in a spectrum of optical vibrations? Effects of strong anharmonicity*

3.1.1. *General remarks*

We begin, in this section, a discussion of features of the biphonon spectrum in imperfect crystals. For the sake of simplicity we shall assume that the crystal contains only the simplest point defects: isotopic substitution impurities. Before going over to a general statement of the theory of local biphonons, we shall make several qualitative remarks concerning the effect of anharmonicity on the spectrum of local optical vibrations in a crystal having an isotopic impurity (here Δ denotes the isotopic shift). We point out, first of all, that in the harmonic approximation, for instance in a crystal with one molecule per unit cell, the frequency ω_ℓ of local vibrations split off the zone $\varepsilon(k)$ of nondegenerate optical vibrations satisfies the equation

$$1 = \frac{\Delta}{N} \sum_{\boldsymbol{k}} \frac{1}{\hbar\omega_\ell - \varepsilon(\boldsymbol{k})}. \tag{42}$$

It is a well-known fact (see, for example, Lifshitz 1956) that in three-dimensional crystals this equation has a solution for ω_ℓ, outside the frequency zone $\varepsilon(\boldsymbol{k})/\hbar$, only at sufficiently large values of $|\Delta|$, i.e. $|\Delta| \geqslant \Delta_c$, where $\Delta_c \approx T_1$ and T_1 is the width of the optical phonon zone in a perfect crystal. If the preceding inequality is complied with, local states appear, not only in the region $\omega_\ell \approx \Omega$, but also in the region of the second and higher overtones, i.e. at the frequencies $2\omega_\ell$, $3\omega_\ell$, etc. The aforesaid is, of course, obvious when one considers that all of these states correspond to different quantum numbers $(n = 1, 2, \ldots)$ of a harmonic oscillator, i.e. the normal local vibrations of a crystal having an isotopic impurity. If, however, inequality $|\Delta| > \Delta_c$ is not complied with, i.e. if $|\Delta| < \Delta_c$, then eq. (42) has no solutions for ω_ℓ that lie outside the zone $\varepsilon(\boldsymbol{k})$. In this latter case, the presence of an isotopic defect does not lead to the formation of local states. It is clear that in the harmonic approximation such states do not appear either in the fundamental tone region (i.e. at $\omega_\ell \approx \Omega$) or in the overtone region.

Taking anharmonicity into account can qualitatively change the pattern of the local states spectrum. Specifically, it can lead to the formation of such states even in cases ($|\Delta| < \Delta_c$) when in the harmonic approximation there are no states whatsoever localized in the region of the defect (Agranovich 1970).

Let us assume that the anharmonicity constant A is large compared to the width $T_1 = 2T$ (where T is the halfwidth) of the phonon zone. In this case, as has been shown in section 2.3, the width of the biphonon zone is of the order of T_1^2/A, i.e. small compared to the width T_1 of the optical phonon zone. Under these same conditions (i.e. at sufficiently strong anharmonicity, when the biphonon energy is $E \approx 2\hbar\omega - 2A$), the zonal additions (which are of the order of the phonon zone width T_1) can be neglected in expression (17) for the Green function. In this approximation, the Green function $G_{\boldsymbol{k}_1\boldsymbol{k}_2}(E)$ practically ceases to depend on $\boldsymbol{k}_1$ and $\boldsymbol{k}_2$, and the quantity $\tilde{\Psi}_{\boldsymbol{k}_1\boldsymbol{k}_2}$ [see eq. (18)] is found to be only a function of the total wavevector $\boldsymbol{K} = \boldsymbol{k}_1 + \boldsymbol{k}_2$, while function $\varphi_{\alpha\beta}(\boldsymbol{K}, \boldsymbol{n} - \boldsymbol{m})$ [see eq. (15)] is proportional to $\delta_{\boldsymbol{nm}}$. In this limiting case of strong anharmonicity, the biphonon state $|2\rangle$ [see eq. (10)] is the superposition of the states of twofold excited molecules. It is clear then that an elementary generalization of an equation of the type of (42) can be used to find their local states. Specifically, the equation for the frequency of a local biphonon $\omega_\ell^{(2)}$, i.e. the localized state split off the biphonon zone, can be written (see also section 3.3) as follows:

$$1 = \frac{2\tilde{\Delta}}{N} \sum_{\boldsymbol{K}} \frac{1}{\hbar\omega_\ell^{(2)} - E(\boldsymbol{K})}, \tag{43}$$

where $E(\boldsymbol{K})$ is the energy of a biphonon with wavevector $\boldsymbol{K}$ in a perfect crystal, $2\tilde{\Delta}$ is the isotopic shift of the twofold excited state of an impurity molecule:

$$\tilde{\Delta} = \Delta - (A' - A), \tag{44}$$

and A' is its anharmonicity constant. Since the states of the biphonon, like those of the phonon, are characterized by specifying only a single value of the wavevector, an analysis of eq. (43) is analogous to that of eq. (42). On the basis of the results of such an analysis, which we have already used for phonons, it can be contended that the level of a local biphonon is formed if

$$2|\tilde{\Delta}| > \tilde{\Delta}_c, \tag{45}$$

where $\tilde{\Delta}_c \approx T_2$ is the width of the biphonon zone ($T_2 \sim T_1^2/A$) or, for example, at $A' = A$:

$$2|\Delta| > \tilde{\Delta}_c. \tag{46}$$

It is clear that inequality (45) or (46) can be complied with even in the case when

$$|\tilde{\Delta}| < \Delta_c,$$

i.e. when the following inequalities are simultaneously complied with (for example, at $A' = A$):

$$T_1^2/2A \ll |\Delta| \ll T_1. \tag{47}$$

Complying with the right-hand inequality does not lead to the formation of local states in the region of the fundamental tone. However, if we comply with the second (left-hand) inequality, splitting off the level of a local biphonon is provided for. This demonstrates the importance of the role of anharmonicity in forming the spectra of local states. In the limiting situation under discussion the spectrum of local states begins at $E = 2\hbar\Omega$ rather than at $E \approx \hbar\Omega$. Clearly, such an effect can also occur for three-phonon vibrations, etc. [the theory of bound three-phonon complexes was developed by Dubovsky (1985), see section 5].

We now point out yet another property of eq. (43); from this equation it follows that when $|\Delta| > \Delta_c$, but when the absolute value of the quantity $2\tilde{\Delta}$ is less than the biphonon zone width (when $A' - A \neq 0$, i.e. an anharmonicity defect), a local state in the region $2\hbar\Omega$ is not implemented, though one exists in the region $\hbar\Omega$. Thus, in this way, anharmonicity can lead not only to the stabilization of local states in the overtone region, but also, in general, to their inhibition.

The reasoning given above is based on an analysis of eq. (43), which was derived in a strong anharmonicity approximation. A more general theory of local biphonons is discussed in subsection 3.2.

3.1.2. *Local biphonon in an $^{14}NH_4Br$ crystal containing the isotopic substitution impurity ^{15}N*

Before beginning this discussion we cite here the results obtained in experimental research conducted by Belousov et al. (1982). In these investigations they studied the spectrum of local states in the frequency region $\omega \approx 2800$ cm^{-1} in a $^{14}NH_4Br$ crystal (a cubic crystal of group T_d^1) having as an impurity the isotope ^{15}N. It was first shown experimentally in these investigations that, in accordance with predictions (Agranovich 1970), anharmonicity actually can lead to the formation of a local biphonon under conditions in which no local phonons exist.

In the Raman spectrum of a pure crystal of $^{14}NH_4Br$, two narrow lines (of a width not larger than 0.2 cm^{-1}) are observed in the region $\omega_4(\boldsymbol{k} \approx 0)$ of one-phonon transitions. The lines correspond to TO (1397.5 cm^{-1}) and LO (1413.5 cm^{-1}) phonons. Upon introducing the isotope ^{15}N, only negligible broadening and shift of these lines occur in crystals $^{15}N_x{}^{14}N_{1-x}H_4Br$ (where $x = 0.05$), and no local one-phonon vibration is set up. Its absence and the typically one-mode nature of the one-phonon spectrum are explained by the fact that the isotopic shift $\Delta = -6$ cm^{-1} (Price et al. 1960) is small compared to the width of the one-phonon zone ($2T = 36$ cm^{-1}). Such a value of the one-phonon zone width follows from measurements of the Raman spectrum for the region of $\omega_4(\boldsymbol{k}) + \omega_4(-\boldsymbol{k})$ band. The picture of a spectrum in this region for a natural isotope ^{15}N content ($x = 0.0037$) is shown in fig. 4a. The band 2795–2867 cm^{-1} corresponds to two-particle transitions $\omega_4(\boldsymbol{k}) + \omega_4(-\boldsymbol{k})$. Its low-frequency boundary coincides with twice the frequency $\omega_4(\boldsymbol{k} \approx 0)$ of a TO phonon and its width (according to the selection rules) is equal to twice the width of the phonon zone $\omega_4(\boldsymbol{k})$. The narrow line at 2792.5 cm^{-1} (see fig. 4a), located below the zone of dissociated two-particle states, corresponds, obviously, to the excitation of a biphonon. Its occurrence in a second-order spectrum is thereby an indication of quite strong anharmonicity with the characteristic constant $A \approx 30$ to 35 cm^{-1}.

An additional line at $\hbar\omega_1' = 2788$ cm^{-1} was found even in a crystal with a natural content of the isotope ^{15}N by Belousev et al. (1982). When the concentration is increased to $x = 0.05$ and $x = 0.3$, the integrated intensity of this line increases (see fig. 4b). It is exactly this fact that enables us to regard this line as being correspondingly coupled to the impurity of the two-phonon state (local biphonon).

Experimental investigations of two-phonon spectra of crystals with defects are continuing. This is what makes the further theoretical analysis of local multiple-particle states in crystals a timely object of research.

3.2. *Two-particle states in a crystal having defects: basic equations and biphonon localization at an anharmonicity defect*

Considering the general theory, we point out that the Hamiltonian of a crystal having impurities can be written in the following form:

$$\hat{H} = \sum_{n\alpha} \hbar\omega_{n\alpha} B_{n\alpha}^{+} B_{n\alpha} + \sum_{n\alpha \neq m\beta} V_{nm}^{\alpha\beta} B_{n\alpha}^{+} B_{m\beta} - \sum_{n\alpha} A_{n\alpha} B_{n\alpha}^{+2} B_{n\alpha}^{2}, \tag{48}$$

where $\hbar\omega_{n\alpha} = \hbar\omega$ if there is a molecule of the base substance at the site $\boldsymbol{n}\alpha$, and $\hbar\omega_{n\alpha} = \hbar\omega + \Delta$ if it is an impurity molecule. Analogously, $A_{n\alpha} = A$ is the

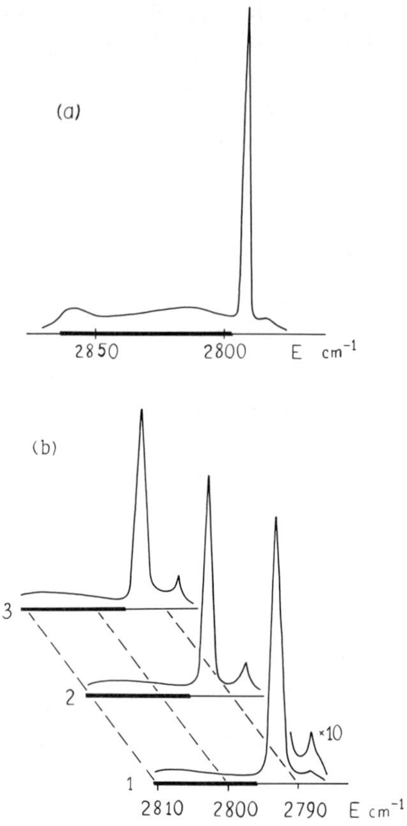

Fig. 4. Raman scattering spectra for $^{15}N_x{}^{14}N_{1-x}H_4Br$ crystals in the region of $\omega_4(\boldsymbol{k})+\omega_4(-\boldsymbol{k})$ transitions. A heavy line distinguishes the region of two-particle transitions of unbound phonons. The peak at 2788 cm^{-1} corresponds to the excitation of local biphonons. Local phonons do not exist. Concentrations are (a) $x=0.0037$; (b): (1) $x=0.0037$, (2) $x=0.05$ and (3) $x=0.3$.

intramolecular anharmonicity constant for molecules of the base substance, whereas $A_{n\alpha}=A'$ for impurity molecules. The procedure for deriving a system of equations for the wave functions $\Psi_{nm}^{\alpha\beta}$ of two-particle states is entirely similar to that given in section 2. This system is of the form

$$\left[E-\hbar\omega_{n\alpha}-\hbar\omega_{m\beta}+2A_{n\alpha}\delta_{\alpha\beta}\delta_{nm}\right]\Psi_{nm}^{\alpha\beta}$$

$$=\sum_{p\gamma}\left[V_{np}^{\alpha\gamma}\Psi_{pm}^{\gamma\beta}+\Psi_{np}^{\alpha\gamma}V_{pm}^{\gamma\beta}\right]. \tag{49}$$

The condition that the determinant $D_{nm,pq}^{\alpha\beta\gamma\nu}(E)$ of this infinite $(N\to\infty)$ system of equations is equal to zero determines the spectrum of two-particle states of a crystal with impurities.

Assuming that an isolated substitution impurity occupies the site $\boldsymbol{n}=0$ ($\alpha=1$ for crystals having several molecules per unit cell), we shall discuss several general properties of the determinant $\hat{D}(E)$. It follows from the general form of the system of equations (49), for example, that for a defect with only a changed anharmonicity constant ($A' \neq A$, $\Delta=0$), the determinant $\hat{D}(E)$ is linear with respect to A'. This latter is associated with the fact that quantity A' is present only in the matrix element $D_{00,00}$ of the determinant, and is absent in the other elements (for the sake of simplicity, we consider, in the following, crystals having a single molecule per unit cell). If, however, the defect has a term shift $\Delta \neq 0$, then the determinant of the system of equations (49) contains the quantity Δ in all the diagonal elements $D_{0\boldsymbol{n},0\boldsymbol{n}}$, where $\boldsymbol{n}$ is an arbitrary site. Thus, in contrast to the determinant for one-particle states, which is linear with respect to Δ [this leads to eq. (42), linear with respect to Δ in the region of two-particle states], the determinant $D(E)$ is, in general, a polynomial of degree of N in Δ. This predetermines, generally speaking, the complexity of an exact dispersion equation, which governs the spectrum of local biphonons.

Of interest, without doubt, in this connection is an exact solution, corresponding to one or another of the limiting situations. One such solution (for the model of a substitutional impurity with $A' \neq A$, but $\Delta=0$) will be considered at the end of this section.

Since in this case $A_{\boldsymbol{n}} = A + (A' - A)\delta_{\boldsymbol{n}0}$, the system of equations (49) can be rewritten as follows:

$$D^0_{\boldsymbol{nm},\boldsymbol{pq}}\Psi_{\boldsymbol{pq}} = -2(A'-A)\delta_{\boldsymbol{n}0}\delta_{\boldsymbol{m}0}\Psi_{00}, \tag{50}$$

where the matrix

$$D^0_{\boldsymbol{nm},\boldsymbol{pq}} = (E - 2\hbar\omega + 2A\delta_{\boldsymbol{nm}})\delta_{\boldsymbol{np}}\delta_{\boldsymbol{mq}} - V_{\boldsymbol{np}}\delta_{\boldsymbol{mq}} - V_{\boldsymbol{mq}}\delta_{\boldsymbol{np}}. \tag{51}$$

Next we introduce the Green function

$$G_{\boldsymbol{nm},\boldsymbol{pq}}(E) = \left[D^0(E)\right]^{-1}_{\boldsymbol{nm},\boldsymbol{pq}}. \tag{52}$$

With its help we obtain

$$\Psi_{\boldsymbol{nm}} = -2(A'-A)G_{\boldsymbol{nm},00}\Psi_{00}, \tag{53}$$

and putting $\boldsymbol{n}=\boldsymbol{m}=0$, we find that for states with $\Psi_{00} \neq 0$ the energy of a two-particle state when a defect is present ($\Delta=0$, $A' \neq A$) satisfies the equation

$$1 = -2(A'-A)G_{00,00}(E). \tag{54}$$

The Green function $G_{\boldsymbol{nm},\boldsymbol{pq}}$ is determined, according to eq. (52), by the equation

$$D^0_{\boldsymbol{nm},\boldsymbol{pq}} G_{\boldsymbol{pq},\boldsymbol{n}'\boldsymbol{m}'} = \delta_{\boldsymbol{nn}'}\delta_{\boldsymbol{mm}'}. \tag{55}$$

It is readily evident, for instance, that by direct substitution into eq. (55) [see also eqs. (17) and (20)], the following relationship is valid for this function:

$$\begin{aligned} G_{\boldsymbol{nm},\boldsymbol{pq}}(E) = {} & \frac{1}{N^2} \sum_{\boldsymbol{K},\boldsymbol{k}} G_{\boldsymbol{K}/2+\boldsymbol{k},\boldsymbol{K}/2-\boldsymbol{k}}(E) \\ & \times \exp\{\mathrm{i}(\boldsymbol{K}/2)(\boldsymbol{n}+\boldsymbol{m}-\boldsymbol{p}-\boldsymbol{q}) + \mathrm{i}\boldsymbol{k}(\boldsymbol{n}-\boldsymbol{m}-\boldsymbol{p}+\boldsymbol{q})\} \\ & \times \Bigg[1 - \frac{2A}{F(E,\boldsymbol{K})}\frac{1}{N}\sum_{\boldsymbol{k}'} G_{\boldsymbol{K}/2+\boldsymbol{k}',\boldsymbol{K}/2-\boldsymbol{k}'}(E) \\ & \times \exp\{\mathrm{i}(\boldsymbol{k}-\boldsymbol{k}')(\boldsymbol{p}-\boldsymbol{q})\}\Bigg]. \end{aligned} \tag{56}$$

Consequently, the value of $G_{00,00}(E)$ appearing in eq. (54), after taking eqs. (20) and (21) into account, equals

$$G_{00,00} = \frac{1}{N}\sum_{\boldsymbol{K}} \frac{I(E,\boldsymbol{K})}{F(E,\boldsymbol{K})}, \tag{57}$$

so that eq. (54) can be written in the form

$$1 = -2(A'-A)\frac{1}{N}\sum_{\boldsymbol{K}} \frac{I(E,\boldsymbol{K})}{F(E,\boldsymbol{K})}. \tag{58}$$

It is interesting to compare this equation with eq. (43) if in the latter (i.e. in the expression for $\tilde{\Delta}$) we assume the isotopic shift to be absent, i.e., we assume that $\Delta = 0$. It is quite clear that eq. (58) is a more general equation than eq. (43) with $\Delta = 0$, because the value of the anharmonicity constant is arbitrary in eq. (58), whereas in eq. (43) this anharmonicity is assumed to be strong. Let us show in what way eq. (58) is transformed into eq. (43) (with $\Delta = 0$) at high values of A. Recall, for this purpose, that the zeros of the function $F(E, \boldsymbol{K})$, according to eq. (20), determine the value of the biphonon energy $E(\boldsymbol{K})$.

Hence at E values close to $E(\boldsymbol{K})$, the above-mentioned function can be represented in the form

$$F(E, \boldsymbol{K}) = \left(\frac{\partial F}{\partial E}\right)_{E=E(\boldsymbol{K})} [E - E(\boldsymbol{K})]. \tag{59}$$

At the same time, according to eqs. (20) and (21), for the same values of E, the quantity $I(E, \boldsymbol{K}) = -1/2A$, whereas

$$\left(\frac{\partial F}{\partial E}\right)_{E=E(\boldsymbol{K})} = -\frac{2A}{N}\sum_{\boldsymbol{k}}[E - \varepsilon(\boldsymbol{K}/2 + \boldsymbol{k}) - \varepsilon(\boldsymbol{K}/2 - \boldsymbol{k})]^{-2} \approx -\frac{1}{2A}. \tag{60}$$

On the basis of the aforesaid, we actually do come to the conclusion that at sufficiently large A values eq. (58) is transformed into eq. (43) (with $\Delta = 0$). We point out yet another special case: assume that $\Delta = 0$ and $A = 0$, i.e. that only an impurity molecule has anharmonicity ($A' \neq 0$). Since, in this case $A = 0$, there are no biphonons in the perfect lattice of the base substance. But, in general, a local state can be formed in the region of the overtone $E \approx 2\hbar\Omega$. Its energy is determined by eq. (58) at $A = 0$, i.e. from the equation

$$1 = -\frac{2A'}{N}\sum_{\boldsymbol{K}} I(E, \boldsymbol{K}) = -2A' \int \frac{\rho_2(E')}{E - E'} \mathrm{d}E',$$

where $\rho_2(E')$ is the density of two-particle states with energy E'. In its form this equation is analogous to eq. (42) for local one-particle states.

It is self-evident that the assumption that the isotopic shift $\Delta = 0$ is quite artificial. When both anharmonicity of the matrix and isotopic shift are simultaneously taken into account, the equation determining the energy of the local biphonons turns out to be considerably more complicated.

3.3. *Secular equation for a biphonon in a crystal having an isolated isotopic substitution impurity*

As has already been emphasized, it is necessary, in deriving the secular equation for the general case (i.e., when $\Delta \neq 0$ and $A \neq 0$), to find the value of the determinant of an infinite system of equations (49). This procedure is extremely cumbersome. In order to simplify its discussion, we point out that the condition under which this determinant equals zero, i.e. the secular equation for E, can be written as

$$Z_1(E, \Delta) = AZ_2(E, A, \Delta). \tag{61}$$

We have introduced two functions, Z_1 and Z_2, into this relationship. As indicated, Z_1 depends only on E and Δ, and is equal to the determinant of the system of equations (49) at $A = 0$. Its zeros determine the spectrum of two-particle states in a harmonic crystal having an isolated isotopic impurity. Function Z_2 depends on both Δ and A, remains finite when $A \to 0$ and, by itself, has no physical meaning. Nevertheless, it is convenient to make use of presentation (61) because it enables the form of function $Z_1(E, \Delta)$, which is independent of A, to be determined on the basis of harmonic crystal theory (i.e. at $A = 0$).

3.3.1. Two-particle states at $A = 0$

Assuming as before that we are dealing with a crystal with one molecule per unit cell, let us make use of expansion (12) (omitting the indices α and β). Then, from the system of equations (49), for the function $\tilde{\Psi}_{\boldsymbol{k}_1\boldsymbol{k}_2}$ we obtain a system of equations:

$$\begin{aligned}&[E - 2\hbar\omega - V(\boldsymbol{k}_1) - V(\boldsymbol{k}_2)]\,\tilde{\Psi}_{\boldsymbol{k}_1\boldsymbol{k}_2}\\ &\quad = -2AQ(\boldsymbol{k}_1 + \boldsymbol{k}_2) + \Delta[\varphi(\boldsymbol{k}_1) + \varphi(\boldsymbol{k}_2)],\end{aligned} \tag{62}$$

where for $Q(\boldsymbol{k}_1 + \boldsymbol{k}_2)$ the notation of eq. (19) is used, whereas

$$\varphi(\boldsymbol{k}) = \frac{1}{N}\sum_{\boldsymbol{k}'} \tilde{\Psi}_{\boldsymbol{k}'\boldsymbol{k}} \equiv \frac{1}{N}\sum_{\boldsymbol{k}'} \tilde{\Psi}_{\boldsymbol{k},\boldsymbol{k}'}. \tag{63}$$

From eq. (62) at $A = 0$ we obtain

$$\tilde{\Psi}_{\boldsymbol{k}_1\boldsymbol{k}_2} = \Delta G_{\boldsymbol{k}_1\boldsymbol{k}_2}(E)[\varphi(\boldsymbol{k}_1) + \varphi(\boldsymbol{k}_2)].$$

Taking into account eq. (63), we find that

$$\varphi(\boldsymbol{k}_2) = \Delta N^{-1} \sum_{\boldsymbol{k}_1} G_{\boldsymbol{k}_1\boldsymbol{k}_2}\varphi(\boldsymbol{k}_1) + \varphi(\boldsymbol{k}_2)\Delta N^{-1}\sum_{\boldsymbol{k}_1} G_{\boldsymbol{k}_1\boldsymbol{k}_2}(E),$$

i.e. that

$$\varphi(\boldsymbol{k}_2) = S(E, \boldsymbol{k}_2)\Delta N^{-1}\sum_{\boldsymbol{k}_1} G_{\boldsymbol{k}_1\boldsymbol{k}_2}(E)\varphi(\boldsymbol{k}_1), \tag{64}$$

where by definition

$$S(E, \mathbf{k}) = \left[1 - \Delta N^{-1} \sum_{\mathbf{k}'} G_{\mathbf{k}'\mathbf{k}}(E)\right]^{-1}. \tag{65}$$

The determinant of the system of equations (64), i.e., the quantity $Z_1(E, \Delta)$:

$$Z_1(E, \Delta) = \left| \delta_{\mathbf{k}_1\mathbf{k}_2} - \Delta S(E, \mathbf{k}_1) N^{-1} G_{\mathbf{k}_1\mathbf{k}_2}(E) \right| \tag{66}$$

can have zeros corresponding to various types of two-particle states. Specifically, function $Z_1(E, \Delta)$ should have roots at a value of E equal to the total energy of two local phonons, if any are formed. As previously mentioned, the energy $\hbar\omega_\ell$ of a local phonon satisfies eq. (42), i.e. in the notation of eq. (17), the equation

$$1 = 2\Delta N^{-1} \sum_{\mathbf{k}} G_{\mathbf{k}\mathbf{k}}(E), \tag{67}$$

where $E = 2\hbar\omega_\ell$. Moreover, function $Z_1(E, \Delta)$ vanishes at energy values E equal to the sum of the energy $\hbar\omega_\ell$ of a local phonon and the energy $\varepsilon(k)$ of a free (unlocalized) phonon, or to the sum of the energies $\varepsilon(k) + \varepsilon(k')$ of two free phonons. Evaluating the determinant (66), we find that function $Z_1(E, \Delta)$ is a polynomial of degree N in Δ and, when $N \to \infty$, can be written in the form

$$Z_1(E, \Delta) = 1 + \Delta a_1 + \Delta^2 a_2 + \ldots, \tag{68}$$

where

$$a_1 = t_1; \quad a_2 = \frac{t_1^2}{2!} + t_2; \quad a_3 = \frac{t_1^3}{3!} + t_1 t_2 + t_3, \quad \ldots, \quad \text{etc.}$$

For an arbitrary m value, the quantity a_m is equal to the sum of all kinds of terms of the type $t_i^p \ldots t_j^{p'}/p! \ldots p'!$, where $pi + \ldots + p'j = m$. Thus, for instance,

$$a_4 = \frac{t_1^4}{4!} + \frac{t_1^2}{2!} t_2 + t_1 t_3 + \frac{t_2^2}{2!} + t_4;$$

$$a_5 = \frac{t_1^5}{5!} + \frac{t_1^3}{3!} t_2 + \frac{t_1^2}{2!} t_3 + \frac{t_2^2}{2!} t_1 + t_2 t_3 + t_1 t_4 + t_5, \quad \text{etc.}$$

In these relationships the quantities t_1, $t_2, \ldots$ are determined as follows:

$$t_1 = -\frac{1}{N}\sum_{\boldsymbol{k}} \tilde{G}_{\boldsymbol{kk}}(E),$$

$$t_2 = -\frac{1}{2N^2}\sum_{\boldsymbol{k}_1\boldsymbol{k}_2} \tilde{G}_{\boldsymbol{k}_1\boldsymbol{k}_2}(E)\tilde{G}_{\boldsymbol{k}_2\boldsymbol{k}_1}(E),$$

$$\ldots$$

$$t_n = -\frac{1}{nN^n}\sum_{\boldsymbol{k}_1\boldsymbol{k}_2\ldots\boldsymbol{k}_n} \tilde{G}_{\boldsymbol{k}_1\boldsymbol{k}_2}(E)\tilde{G}_{\boldsymbol{k}_2\boldsymbol{k}_3}(E)\ldots\tilde{G}_{\boldsymbol{k}_n\boldsymbol{k}_1}(E), \quad \text{where} \quad n > 1$$

and

$$\tilde{G}_{\boldsymbol{kk}'}(E) = \sqrt{S(E,\boldsymbol{k})S(E,\boldsymbol{k}')}\,G_{\boldsymbol{kk}'}(E).$$

It is of importance that series (68) can also be identically written in the form of the product

$$Z_1(E,\Delta) = \left[1 - 2\frac{\Delta}{N}\sum_{\boldsymbol{k}} G_{\boldsymbol{kk}}(E)\right] \times \left[1 - \Delta a_1 + \Delta^2 a_2 - \ldots + (-1)^n \Delta^n a_n + \ldots\right], \tag{69}$$

and the proof of the validity of this expression for $Z_1(E,\Delta)$ can be obtained, for example, by correlating series (68) and (69), after first multiplying them by $\prod_{\boldsymbol{k}} S^{-1}(E,\boldsymbol{k})$.

In view of the foregoing concerning the roots of equation $Z_1(E,\Delta) = 0$, the occurrence of the first factor in eq. (69) [see also eq. (67)] is quite natural. Unfortunately, further factorization of eq. (69) does not lead to appreciable simplifications. Hence, in the subsequent discussion we shall use eq. (69).

The expression within the second pair of brackets of eq. (69) vanishes at E-values equal to the sum of the energies of the delocalized and localized phonons, or of two different ($\boldsymbol{k}_1 \neq \boldsymbol{k}_2$) delocalized phonons. At energy values $E = 2\varepsilon(\boldsymbol{k})$ (i.e. for the states when $\boldsymbol{k}_1 = \boldsymbol{k}_2$), the expression within the first set of brackets in eq. (69) vanishes.

3.3.2. General form of the secular equation ($A \neq 0$)

In order to obtain the right-hand side of eq. (61), we should return to the discussion of the initial system of equations (62). It follows from this system of equations that at $A \neq 0$

$$\tilde{\Psi}_{k_1k_2} = \Delta G_{k_1k_2}(E)[\varphi(k_1) + \varphi(k_2)] - 2AG_{k_1k_2}(E)Q(k_1 + k_2). \qquad (70)$$

After making use of this relation and the notation of eq. (21) we find that

$$Q(E, K) = \Delta[1 + 2AI(E, K)]^{-1} \times \frac{1}{N} \sum_{\substack{k'k'' \\ (k'+k''=K)}} G_{k'k''}(E)[\varphi(k') + \varphi(k'')]. \qquad (71)$$

Taking eq. (70) into consideration, as well as the definition (63), we have the final equation for the quantities $\varphi(k)$:

$$\varphi(k_1) = \Delta S(E, k_1)\left\{\frac{1}{N}\sum_{k_2} G_{k_1k_2}(E)\varphi(k_2) - 2A\frac{1}{N^2}\sum_{K} G_{k_1,K-k_1}(E)\frac{1}{1 + 2AI(E, K)} \times \sum_{\substack{k'k'' \\ (k'+k''=K)}} G_{k'k''}(E)[\varphi(k') + \varphi(k'')]\right\}. \qquad (72)$$

It follows from this equation that the condition that its determinant equals zero yields eq. (61), where the function $Z_1(E, \Delta)$ is determined by relationship (69), whereas

$$Z_2(E, A, \Delta) = \Delta N^{-1}\sum_{K} \frac{\sum_{n=0}(-\Delta)^n b_{n+1}(E, K)}{1 + 2AI(E, K)}, \qquad (73)$$

where

$$b_1 = T_1(K); \qquad b_2 = N^{-1}\sum_{K'}{}' [T_1(K)\tilde{T}_1(K, K') + T_2(K, K')], \text{ etc.} \qquad (74)$$

We have not written out here the expressions for the other coefficients b_n

(where $n = 3, 4, \ldots$) because they will not be required for the following (see also, in this connection, Agranovich and Dubovsky 1981). In eqs. (74)

$$T_1(E, \boldsymbol{K}) = -N^{-1}\sum_{\boldsymbol{k}} g_{\boldsymbol{kk}}(\boldsymbol{K});$$

$$\tilde{T}_1(\boldsymbol{K}, \boldsymbol{K}') = -N^{-1}\sum_{\boldsymbol{k}} \tilde{g}_{\boldsymbol{kk}}(\boldsymbol{K}, \boldsymbol{K}');$$

$$T_2(\boldsymbol{K}, \boldsymbol{K}') = -N^{-2}\sum_{\boldsymbol{k}_1\boldsymbol{k}_2} g_{\boldsymbol{k}_1\boldsymbol{k}_2}(\boldsymbol{K})\tilde{g}_{\boldsymbol{k}_2\boldsymbol{k}_1}(\boldsymbol{K}, \boldsymbol{K}'),$$

where

$$g_{\boldsymbol{k}_1\boldsymbol{k}_2}(\boldsymbol{K}) = 4S(E, \boldsymbol{k}_1)G_{\boldsymbol{k}_1,\boldsymbol{K}-\boldsymbol{k}_1}(E)G_{\boldsymbol{k}_2,\boldsymbol{K}-\boldsymbol{k}_2}(E),$$

$$\tilde{g}_{\boldsymbol{k}_1\boldsymbol{k}_2}(\boldsymbol{K}, \boldsymbol{K}') = \tfrac{1}{2}g_{\boldsymbol{k}_1\boldsymbol{k}_2}(\boldsymbol{K}')[I(E, \boldsymbol{K}') - I(E, \boldsymbol{K})]^{-1} - S(E, \boldsymbol{k}_1)G_{\boldsymbol{k}_1\boldsymbol{k}_2}(E). \quad (75)$$

The prime on the summation sign over the wavevector $\boldsymbol{K}'$ in eq. (74) indicates that, in the summation, values of $\boldsymbol{K}'$ must be omitted for which $I(E, \boldsymbol{K}') = I(E, \boldsymbol{K})$. This means that in going from summation to the integral over $\boldsymbol{K}'$, the required integral is to be understood in the sense of the principal value.

3.4. Certain limiting cases

3.4.1. Limiting case of strong anharmonicity

We point out, first of all, that as $\Delta \to 0$, eq. (61) [see also eqs. (69) and (73)] has solutions $E = E(\boldsymbol{K})$, determined from the condition

$$1 + 2AI(E, \boldsymbol{K}) = 0.$$

As was to be expected, this condition coincides with the dispersion equation (20) for a biphonon in a perfect crystal.

Assume now that $\Delta \neq 0$, but that a case with strong anharmonicity occurs, i.e., $A \gg T$, where T is the halfwidth of the phonon zone. In this case, biphonons are certain to be formed in a perfect crystal (see section 2). In the presence of an isotopic impurity and when $|\Delta| < 2A$, the spectrum of local two-particle states turns out to differ essentially from that found in a harmonic crystal. As a matter of fact, by virtue of the smallness of the ratios $|\Delta/2A|$ and $|2T/A|$, for energies E close to the biphonon zone energy

$$G_{k_1k_2}(E) = [E - \varepsilon(\boldsymbol{k}_1) - \varepsilon(\boldsymbol{k}_2)]^{-1} \approx -\frac{1}{2A},$$

$$S(E, \boldsymbol{K}) \approx 1,$$

$\Delta^n a_n \propto (\Delta/A)^n$ (where $n \geqslant 1$); $\Delta^{n+2} b_{n+1} \propto (\Delta/A)^{n+2}$. Hence, omitting small terms, eq. (61) can be written as follows:

$$1 = -\frac{\Delta}{A} \sum_{\boldsymbol{K}} \frac{1}{F(E, \boldsymbol{K})}. \tag{76}$$

If we now take eqs. (59) and (60) into account, the preceding relation, as is readily evident, reduces to eq. (43) at $A' = A$. As has already been pointed out, this equation leads to a local two-particle state even at sufficiently small values of $|\Delta| < T$, when no local states appear in the one-particle spectrum.

3.4.2. Limiting case of high values of isotopic shift (vacancy defect)

Of special interest, along with the limiting situations discussed above, is the problem of the spectrum of local states in the case when the maximum characteristic quantity is the amount of isotopic shift Δ (i.e., when $|\Delta| \gg A$ and $|\Delta| \gg T$). At such Δ values, local states occur in one-particle, two-particle and other multiple-particle states even in the harmonic approximation. The energy, for instance, of a two-particle state of this type is equal to $E = 2\hbar\omega + 2\Delta$. It is very strongly displaced both from the biphonon zone and from the zone of two-phonon dissociated states. In this connection, the natural question to ask is: can, in the situation being discussed, "shallow" local states also be formed in the vicinity of the phonon or biphonon zone?

At large values of $|\Delta|$, resonance of the impurity and matrix is excluded. Such an impurity behaves like a vacancy and the question formulated above, is, in fact, reduced to the question: can the vacancy (for $|\Delta| \to \infty$) lead to the formation of localized states? For a one-particle spectrum [see eq. (42)] there are no such states as $|\Delta| \to \infty$ (where $\sigma = 1$) (Agranovich 1968). For the region of the spectrum of two-particle states, it is necessary to return to eq. (61) and to consider its form as $|\Delta| \to \infty$. Let us first examine the left-hand side of this equation. On the basis of relations (66) and (65), we come to the conclusion that the left-hand side of eq. (61), as $|\Delta| \to \infty$ and with a finite E, tends to a finite value. This means that within the indicated limit of large $|\Delta|$ values, it is sufficient, in order to find the left-hand side of eq. (61), to find the term proportional to $1/\Delta$ in the expression $(1 - \Delta a_1 + \Delta^2 a_2 - \cdots)$ [see eq. (69)]. In a similar manner, it can readily be shown that the right-hand side of eq. (61) remains finite as $|\Delta| \to \infty$. As a result of passing to this limit, eq. (61) can be written as follows:

$$2N^{-1}\sum_{k} G_{kk}(E) \sum_{n=1} (-1)^n \tilde{a}_n = AN^{-1}\sum_{K} \frac{\sum_{n=0}(-1)^n \tilde{b}_n}{1+2AI(E,\,\boldsymbol{K})}, \tag{77}$$

where

$$\tilde{a}_1 = N^{-1}\sum_{k} G^{\mathrm{II}}_{kk}(E);$$

$$\tilde{a}_2 = \tilde{a}_1 N^{-1}\sum_{k} \tilde{G}^{\mathrm{I}}_{kk}(E) - N^{-2}\sum_{k_1 k_2} G^{\mathrm{I}}_{k_1 k_2}(E) G^{\mathrm{II}}_{k_2 k_1}(E); \ldots$$

Functions G^{I} and G^{II} can be determined by suitable calculations, the quantities b_n are readily obtained from the expressions for b_n when it is considered that $\lim_{|\Delta|\to\infty} \Delta S(E,\,\boldsymbol{k}) = -[(1/N)\Sigma_{k'} G_{kk'}(E)]^{-1}$ *. Here we restrict ourselves to the most interesting case of strong anharmonicity, in which $[T/A] \ll 1$. In this limit for the region of energies E close to the biphonon zone, terms containing $\tilde{a}_n$ and $\tilde{b}_n$ can be omitted at $n \geqslant 2$ in eq. (77), so that eq. (77) assumes the form

$$N^{-1}\sum_{K} \frac{1}{F(E,\,\boldsymbol{K})} = 0. \tag{78}$$

As has been repeatedly pointed out, the zero values of the function $F(E,\,\boldsymbol{K})$ determine the position of the biphonon zone in a perfect crystal. Even if we neglect biphonon dispersion [in this approximation $F(E) = 1 + 2A/(E - 2\hbar\omega)$] eq. (78) cannot be satisfied at finite E values. It follows, therefore, from the aforesaid that biphonons localized near a vacancy, along with local phonons, are not formed in crystals having one molecule per unit cell discussed here. In the following we shall show that the situation is changed in a crystal of more complex structure.

3.4.3. Relative arrangement of the levels of two-particle states in a crystal having an isotopic impurity

Before beginning our discussion on these levels, we have two remarks to make.

* These results were obtained together with A.V. Orlov.

First of all let us consider the general arrangement of the energy levels of two-particle states in a crystal having one molecule per unit cell and an isotopic impurity. This arrangement is shown schematically in fig. 5. Shown here is the dependence of the position of the levels on the isotopic shift, assuming anharmonicity to be strong and that the biphonon zone is separated from the zone of dissociated two-particle states by a finite gap. In the diagram the horizontal band F + F of states corresponds to the zone of the energies of two free phonons; the horizontal band BP corresponds to the zone of energies of a biphonon in a perfect crystal; the inclined band F + L of states corresponds to the energies of two phonons, one of which (L) is localized at the isotopic defect, whereas the other (F) is free. The line LBP describes the dependence of the level of the local biphonon on the amount of isotopic shift Δ. In the region of large $|\Delta|$ values, the F + L band is displaced a distance proportional to Δ, whereas the line LBP is displaced a distance proportional to 2Δ. Since we assume the crystal to be three dimensional, states L and likewise states LBP are formed only at $|\Delta|$ values larger than certain critical values.

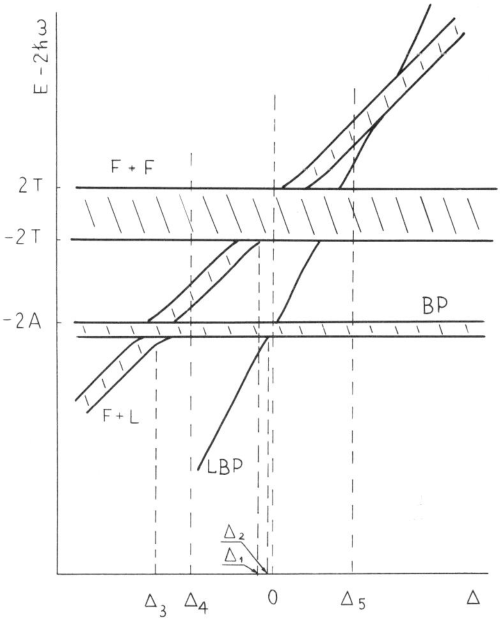

Fig. 5. Dependence of the spectrum of two-particle vibrations on the isotopic shift (one molecule per unit cell). Shown are the term LBP of local biphonon vibrations, the band BP of free biphonons, the zone F+L of states of the free phonon + the local phonon and the zone F+F of dissociated states.

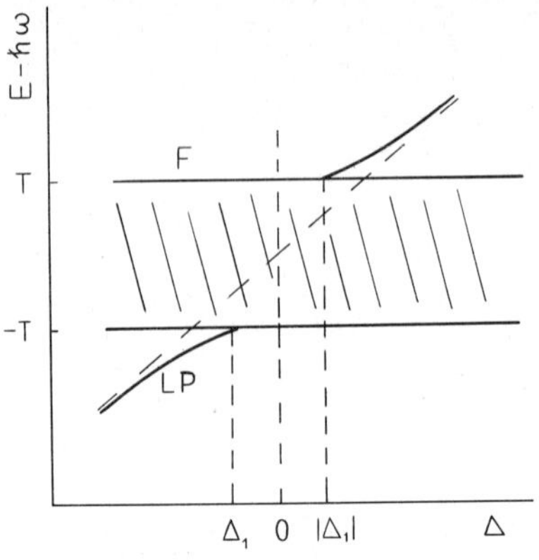

Fig. 6. Dependence of the energy of local phonons on the isotopic shift.

For local states L and for states LBP, these critical values Δ_1 and Δ_2 differ, as is shown in fig. 5 ($\Delta_2 > \Delta_1$). For the purpose of comparison, a schematic diagram of one-particle states as a function of Δ is shown in fig. 6.

Moreover, we point out that zone F + L, as is also evident in fig. 5, intersects the biphonon zone BP at certain values of Δ. Consequently, processes prove to be feasible at these Δ values for transforming free biphonons into states of the F + L type. Here, one of the phonons is localized at the impurity and the other is transformed into a free phonon. This process is analogous to a direct nuclear stripping reaction with a deuteron (Butler 1957, Austern 1970). Note the absence of symmetry in the relative arrangement of the levels with respect to the sign of Δ, which is in no way surprising inasmuch as the sign of the anharmonicity constant A in fig. 5 is assumed to be positive. As will be shown in section 4 this fact also manifests itself in the formation of the density of states in disordered crystals having a large number of impurities.

Our second remark concerns the region of existence of the local biphonon level at various values of A and Δ. It has been emphasized above that at large A values, levels of local biphonons can exist even at sufficiently small $|\Delta|$ values [see eq. (47)]. This poses the question: is the formation of a local biphonon level feasible at small A and small $|\Delta|$ values, when there is no local phonon level in the (one-particle) spectrum and, at the same time, when there is no zone of biphonon levels in the spectrum of the perfect crystal. In order to clear up the aforesaid and to answer the question, let us direct our attention to fig. 7, in which A_c is the critical value of the anharmonicity constant (the biphonon level exists in a perfect crystal only at $A \geqslant A_c$), and where Δ_c is the critical value of the isotopic shift (the local phonon level is formed only at $|\Delta| \geqslant \Delta_c$). The heavy curve in this figure, plotted on the basis of a numerical solution of the general equation (61) for a three-dimensional

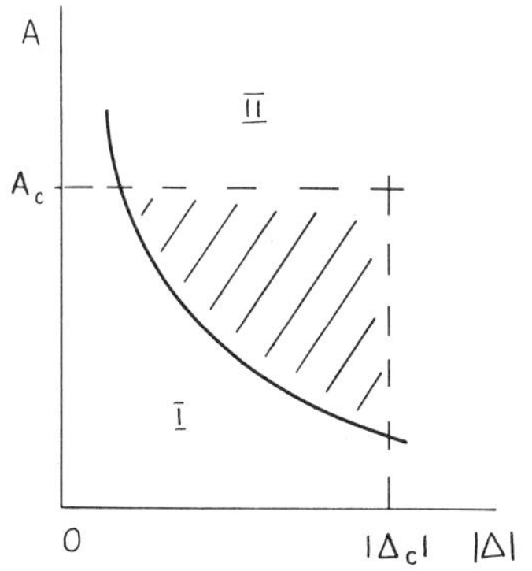

Fig. 7. Regions of A and $|\Delta|$ values at which the formation of local biphonons is possible. The region of $A < A_c$ and $\Delta < |\Delta_c|$, in which the formation of local biphonons is possible, is the hatched area.

crystal, divides plane $(A, |\Delta|)$ into region I, in which no local biphonon state is formed, and region II, in which such a state is formed. Of interest is the fact (evident in fig. 7) that in the region of existence of local biphonons (region II) there actually is a (hatched) region of A and Δ values, such that $A < A_c$ and $|\Delta| < \Delta_c$. Thus, we can give an affirmative answer to the question posed above: even weak anharmonicity ($A < A_c$) at low $|\Delta|$ values (where $|\Delta| < \Delta_c$) facilitates the formation of a local biphonon.

The results of the solution of eq. (61), used to find the boundaries of regions I and II, were obtained for a model of semi-elliptical density of one-particle states. Further refinement of these results is of interest.

3.5. Local biphonons in crystals having two molecules per unit cell

The timeliness of a discussion of the special features of local biphonon spectra in this type of crystals stems from the fact that in recent years quite a number of new effects have been discovered in the region of the two-particle states spectrum, precisely in impurity crystals having two molecules per unit cell, such as $CaCO_3$, $NaNO_3$, etc. In the present section we shall deal only with problems of local biphonon theory, endeavoring mainly to clear up and emphasize the qualitatively new results obtained in going over from crystals with one molecule per unit cell to crystals of more complex structure. However, we shall begin our discussion with the spectrum of one-phonon local states.

In this type of crystals the spectrum of one-phonon states for nondegenerate oscillators contains σ branches $\varepsilon_\mu(\boldsymbol{k})$, where $\mu = 1, 2, \ldots, \sigma$, and σ is the number of molecules per unit cell. If there is an isotopic substitution impurity in the crystal, for example, at site $n = 0$, α, then, in the harmonic approximation, the energies of one-particle states (see, for instance, Agranovich 1968) are determined by the equation

$$1 = \frac{\Delta}{N} \sum_{\mu k} \frac{|u_{\alpha\mu}(\boldsymbol{k})|^2}{E - \varepsilon_\mu(\boldsymbol{k})}, \tag{79}$$

where the amplitude $u_{\alpha\mu}(\boldsymbol{k})$ is defined in section 2.2. It follows from an analysis of this equation that the levels of local or quasi-local states can be formed, at sufficiently high $|\Delta|$ values, both outside the region of zones of one-particle states, as well as between the zones. In particular, in a crystal with two molecules per unit cell and under conditions in which zones $\varepsilon_\mu(\boldsymbol{k})$, where $\mu = 1, 2$, do not overlap, the dependence of the energies of local states on the quantity Δ is shown schematically in fig. 8. The energy of the lower local state $E_1(\Delta)$ decreases at large $|\Delta|$ values (where $\Delta < 0$) according to a linear law ($E_1 \approx \hbar\omega + \Delta$). At the same time, the upper (gap) local state tends, at large $|\Delta|$ values, to a certain finite limit E_v. It follows from eq. (79) that the limiting values (as $|\Delta| \to \infty$) of the energy of the upper branch is determined by the equation

$$0 = \frac{1}{N} \sum_{\mu k} \frac{|u_{\alpha\mu}(\boldsymbol{k})|^2}{E_v - \varepsilon_\mu(\boldsymbol{k})}. \tag{80}$$

Since the quantity E_v corresponds to the limit $|\Delta| \to \infty$, it is actually equal to

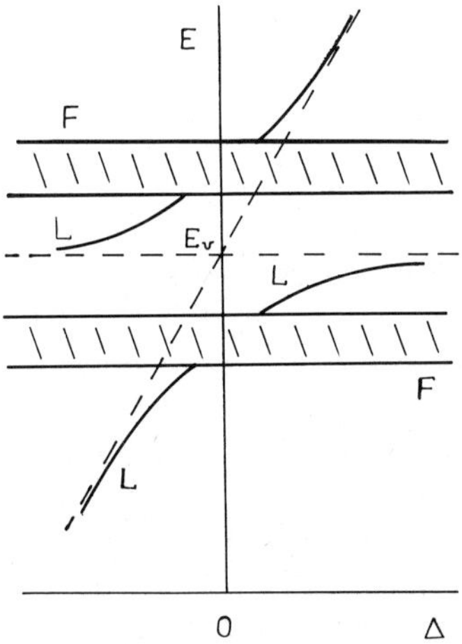

Fig. 8. Dependence on the isotopic shift of the energy of local phonons in a crystal with two molecules per unit cell.

the energy of a phonon localized near a vacancy (see section 3.4). It is clear that such a state turned out to be feasible only due to the presence of two zones $\varepsilon_\mu(\boldsymbol{k})$, where $\mu = 1, 2$; each of whose contributions to the right-hand side of eq. (80), by virtue of the inequalities $\varepsilon_1(\boldsymbol{k}) < E_v < \varepsilon_2(\boldsymbol{k})$, have opposite signs and compensate each other.

The special features mentioned above of the spectrum of local one-particle states are also significantly manifested in the spectra of local biphonons. Owing to the cumbersome derivations we shall first discuss the qualitative pattern of the dependence on Δ of the levels of these states. For this purpose we shall make use of results of numerical calculations of the biphonon spectra for a closed crystalline chain with two molecules per unit cell and with one isotopic defect (Agranovich et al. (1983). An electronic computer was used to determine the values of the energies E that make the determinant of the system of equations (49) vanish.

Results obtained in calculating the spectra as functions of the anharmonicity constant A and the amounts of isotopic shift are given in figs. 9 and 10. Only the interaction of the nearest neighbors in the chain were taken into account in these calculations.

Plotted in fig. 9 is the dependence of the spectrum of biphonons in an ideal chain on the anharmonicity constant. As is evident from fig 9, at sufficiently high values of the anharmonicity constant ($A > A_c$), a doublet of bands, I and II, is formed in the biphonon spectrum (in an ideal chain with $V^{12} > V^{11}$). As

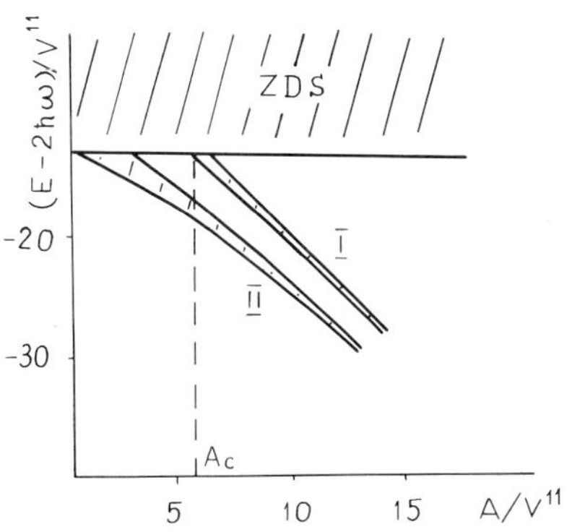

Fig. 9. Dependence of the spectrum of biphonon vibrations of a perfect crystalline chain on the anharmonicity constant A (with $V^{12} = 5.0\ V^{11}$ and $\Delta = 0.0$).

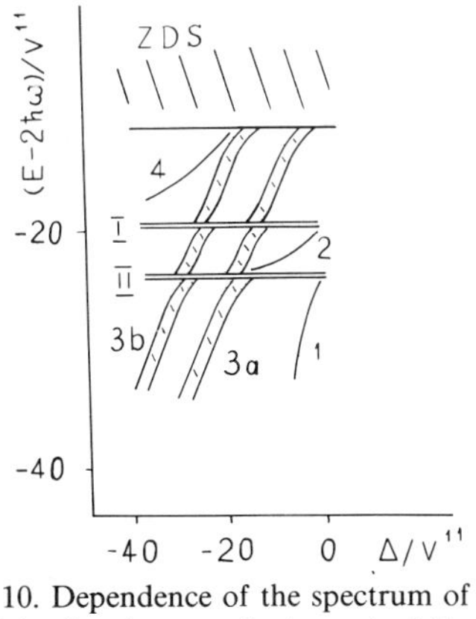

Fig. 10. Dependence of the spectrum of two-particle vibrations on the isotopic shift (with two molecules per unit cell). Indicated are the doublets of the terms of local biphonon vibrations 1, 2 and 4; the zones of free biphonons I and II; states of the free phonon plus the local phonon 3a and 3b; and the zone of dissociated states ZDS ($A = 10\ V^{11}$ and $V^{12} = 5V^{11}$).

A increases, the energy of the bands decreases, the splitting between bands I and II decreases in proportion to $(V^{12})^2/A$ and the bands become narrower [the bandwidth is $(V^{11})^2/A$]. When the anharmonicity constant is reduced in the region $A < A_c$, the high-frequency component of the doublet (band I) merges with the zone of unbound phonons (i.e. with the zone ZDS of dissociated states). Thus, when $A < A_c$, only the low-frequency band II remains in the biphonon spectrum. As $A \to 0$ this band also merges with the ZDS.

The development of the terms of local biphonons as a function of the isotopic shift Δ is shown in fig. 10. The horizontal bands I and II correspond to the doublet of free biphonon zones, not bound to an impurity. Bands 3a and 3b correspond to the two-particle state, in which one of the excitations "sits" on the defect (i.e. corresponds to a local phonon), whereas the other is a free phonon of type a or b, respectively. States of this type have already been mentioned in the preceding section with respect to crystals having one molecule per unit cell. The energies $E_3^{a,b}(\Delta)$ of these states as $|\Delta| \to \infty$ behave as $E_3^{a,b}(\Delta) \approx \Delta$. Zones 3a and 3b, as is evident in fig. 10, intersect, at certain Δ_i values (where $i = 1, 2, 3, 4$), the biphonon zones I and II. Consequently, at these Δ_i values, processes turn out to be possible for transforming free biphonons I or II into states 3a or 3b. Here one of the phonons is localized at the impurity, whereas the other is transformed into a free a- or b-phonon. Figure 10 shows how term 1 of the local biphonon is split off the low-frequency biphonon zone II when $|\Delta|$ is increased. At high $|\Delta|$ values, this level genetically corresponds to the twofold excited state of an impurity with energy of approximately $2(\hbar\omega + \Delta)$. The behavior of this term as Δ varies coincides basically with the behavior of the analogous term in crystals having one molecule per unit cell.

At the same time (see fig. 10) term 2 of the local biphonon is split off the lower edge of the high-frequency zone of free biphonons (off band I). The dependence of the energy of this level on Δ differs substantially from the dependence on Δ of the high-frequency component 2 of the doublet of local phonons (see fig. 8). As $|\Delta|$ increases this term of the local biphonon does not tend asymptotically to a limit like term 2 in fig. 8. Instead it approaches the intersection of bands 3a and II, ending in this intersection at a certain value of $\Delta = \Delta_2$.

Also shown in fig. 10 is how term 4 of the local biphonon is split off the region of intersection of band 3b and the zone of two-particle dissociated states at sufficiently high values of $|\Delta| > \Delta_5$. With an increase of $|\Delta|$, the energy of this term decreases, approaching the high-frequency edge of zone I of free biphonons. At high $|\Delta|$ values, this term is at a distance of approximately $(V^{12})^2/A$ from the high-frequency edge of biphonon zone I and tends, as $|\Delta| \to \infty$, to a certain finite limit E_{4v}; it corresponds to a biphonon localized in the vicinity of a vacancy .

In crystals with several molecules per unit cell and an isotopic impurity at the site $n=0$, $\alpha=1$, the system of equations for the coefficients is of the form

$$(E-2\hbar\omega+2A\delta_{\alpha\beta}\delta_{nm})\Psi^{\alpha\beta}_{nm}-\sum_{\nu p}\left[V^{\alpha\nu}_{np}\Psi^{\nu\beta}_{pm}+\Psi^{\alpha\nu}_{np}V^{\nu\beta}_{pm}\right]$$

$$=\Delta\left[\delta_{\alpha 1}\delta_{n0}+\delta_{\beta 1}\delta_{m0}\right]\Psi^{\alpha\beta}_{nm}. \tag{81}$$

It can be shown, in a way analogous to that employed in section 3.3, that at arbitrary values of A and Δ, and with $\sigma=2$ (two molecules per unit cell), the equation determining the energy of a local biphonon can be written as follows:

$$\left[1-2\Delta N^{-1}\sum_{\mu\boldsymbol{k}}|u_{1\mu}(\boldsymbol{k})|^2G^{\mu\mu}_{\boldsymbol{kk}}(E)\right]\left[1+\sum_{n=1}(-\Delta)^n a_n(E)\right]$$

$$=A\Delta N^{-1}\sum_{\mu\boldsymbol{K}}\left[F_\mu(E,\boldsymbol{K})\right]^{-1}\sum_{n=0}(-\Delta)^n b^{\mu}_{n+1}(E,\boldsymbol{K}). \tag{82}$$

This equation is a generalization of relation (61). Quantities $G^{\mu\mu}_{\boldsymbol{kk}}(E)$ in eq. (82) are determined by eq. (37); the corresponding relations between the quantities a_n and t_n are also valid, but quantity t_n is redefined:

$$t_1=-N^{-1}\sum_{\mu\boldsymbol{k}}\tilde{G}^{\mu\mu}_{\boldsymbol{kk}}(E),$$

$$t_2=-(1/2N^2)\sum_{\mu_1\mu_2\boldsymbol{k}_1\boldsymbol{k}_2}\tilde{G}^{\mu_1\mu_2}_{\boldsymbol{k}_1\boldsymbol{k}_2}(E)\tilde{G}^{\mu_2\mu_1}_{\boldsymbol{k}_2\boldsymbol{k}_1}(E),\ldots,$$

where

$$\tilde{G}^{\mu_1\mu_2}_{\boldsymbol{k}_1\boldsymbol{k}_2}(E)=u_{1\mu_1}(\boldsymbol{k}_1)u^*_{1\mu_2}(\boldsymbol{k}_2)\sqrt{S_{\mu_1}(E,\boldsymbol{k}_1)S_{\mu_2}(E,\boldsymbol{k}_2)}\,G^{\mu_1\mu_2}_{\boldsymbol{k}_1\boldsymbol{k}_2}(E),$$

$$S_\mu(E,\boldsymbol{k})=\left[1-\Delta N^{-1}\sum_{\mu_1\boldsymbol{k}_1}\left|u_{1\mu_1}(\boldsymbol{k}_1)\right|^2G^{\mu\mu_1}_{\boldsymbol{kk}_1}(E)\right]^{-1}.$$

Quantities $b^{\mu}_n(E,\boldsymbol{K})$, in a way analogous to eq. (74), are expressed in terms of quantities $T^{\mu}_n(K)$.

We shall not give the corresponding relations here because they are too cumbersome and will not be used in the following. The general equation (82) derived above may only be applied by investigators who are ready to conduct numerical calculations on the energies of local biphonons (making use of one or another model of the density of one-particle states). But a quantitative analysis of eq. (82) indicates that the most important features in the evolution of the spectrum of local two-particle states (features that were noted in the discussion on the numerical calculations with an electronic computer for the simplest one-dimensional model), are of a sufficiently general nature and should be kept in mind when analyzing experimental results.

3.6. *Biphonon scattering and decay processes upon interaction with an isolated isotopic impurity*

Because the biphonon is a compound quasi-particle, processes become possible during its interaction with a defect (for example, an isotopic impurity) that have no analog in our dealings with phonons. To sum up their essential feature, it should be recalled again that the existence of an isolated defect may be accompanied by the formation of local states of phonons (local phonons), and also leads to processes of free phonon scattering by the defect. Such effects also occur, of course, for the biphonon and, in particular, the local states of the biphonon (local biphonons) have already been dealt with in the preceding subsections. In this subsection we shall consider processes for the elastic scattering of a free biphonon by an isolated defect. Besides, we shall determine the cross section of a process that has no analog when dealing with a phonon; namely, the stripping process that we have previously mentioned in discussing the results of computer calculations of the spectra of two-particle vibrations of a linear crystal having an isotopic defect. As a result of this process, a phonon localized at the defect is formed in the crystal at the position of the initial zone biphonon. Also formed is a free phonon leaving the defect (the reverse process—"pickup"— is obviously also feasible). To calculate the cross sections of the above-mentioned processes of scattering and decay (assuming, for the sake of simplicity, one molecule per unit cell of the crystal), we shall resort to the exact system of equations (72). The energy E in eq. (72) should, for this purpose, be assumed equal to the energy of a biphonon in a perfect crystal, corresponding to a certain value of its wavevector $\boldsymbol{K}_0$ [i.e., $E = E(\boldsymbol{K}_0)$, where $E(\boldsymbol{K})$ is the biphonon zone in a perfect crystal]. The aforesaid signifies that energy E satisfies eq. (20). Hence, by virtue of eqs. (70) and (71), a known uncertainty appears in finding the wave functions $\Psi_{nm}(\boldsymbol{K}_0, E)$ of a biphonon in the coordinate representation [see eq. (12)], requiring summation over $\boldsymbol{K}$. This uncertainty is associated with the calculations of the pole contribution, and is due to the possibility at $E = E(\boldsymbol{K}_0)$ of adding the solution $\tilde{\Psi}^{(0)}_{\boldsymbol{k}_1\boldsymbol{k}_2}$ [i.e. the solution of eq. (70) with $\Delta = 0$ and $\boldsymbol{k}_1 + \boldsymbol{k}_2 = \boldsymbol{K}_0$, corresponding to a biphonon wave incident on the defect] to the solution of eq. (70). Consequently, eq. (16) is valid for its wave function. With an accuracy to an amplitude factor that is of no importance here

$$\tilde{\Psi}^{(0)}_{\boldsymbol{k}_1\boldsymbol{k}_2}(\boldsymbol{K}_0, E) = NG_{\boldsymbol{k}_1\boldsymbol{k}_2}(E), \quad \text{with} \quad \boldsymbol{k}_1 + \boldsymbol{k}_2 = \boldsymbol{K}_0. \tag{83}$$

Hence, on the basis of definition (63),

$$\varphi^{(0)}(\boldsymbol{k}, \boldsymbol{K}_0) \equiv N^{-1}\sum_{\boldsymbol{k}'} \tilde{\Psi}^{(0)}_{\boldsymbol{k}'\boldsymbol{k}}(\boldsymbol{K}_0, E) = G_{\boldsymbol{k},\boldsymbol{K}_0-\boldsymbol{k}}(E). \tag{84}$$

With such a choice of solution of the equation given above with $\Delta = 0$, the detour of the above-mentioned pole should be such that terms of the order of Δ yield only diverging waves.

Note that in the coordinate representation the wave function of an incident biphonon is

$$\Psi^{(0)}_{\boldsymbol{nm}}(\boldsymbol{K}_0, E) = \exp\{\tfrac{1}{2}\mathrm{i}\boldsymbol{K}_0(\boldsymbol{n}+\boldsymbol{m})\}\kappa(\boldsymbol{K}_0, \boldsymbol{n}-\boldsymbol{m}),$$

where the wave function of relative motion is

$$\kappa(\boldsymbol{K}_0, \boldsymbol{n}-\boldsymbol{m}) = \sum_{\substack{\boldsymbol{k}_1,\boldsymbol{k}_2 \\ (\boldsymbol{k}_1+\boldsymbol{k}_2=\boldsymbol{K}_0)}} G_{\boldsymbol{k}_1\boldsymbol{k}_2}(E)\exp\{\tfrac{1}{2}\mathrm{i}(\boldsymbol{k}_1-\boldsymbol{k}_2)(\boldsymbol{n}-\boldsymbol{m})\}. \tag{85}$$

Taking into account the aforesaid, eq. (70) for $E = E(\boldsymbol{K}_0)$ can be written as follows [see also eq. (71)]:

$$\begin{aligned}\tilde{\Psi}_{\boldsymbol{k}_1\boldsymbol{k}_2}(\boldsymbol{K}_0, E) = {}& \tilde{\Psi}^{(0)}_{\boldsymbol{k}_1\boldsymbol{k}_2}(\boldsymbol{K}_0, E) + \Delta G_{\boldsymbol{k}_1\boldsymbol{k}_2}(E)[\varphi(\boldsymbol{k}_1)+\varphi(\boldsymbol{k}_2)] \\ & -2AG_{\boldsymbol{k}_1\boldsymbol{k}_2}(E)Q(E+\mathrm{i}\gamma, \boldsymbol{K}_0),\end{aligned} \tag{86}$$

where $\gamma \to +0$, so that instead of eq. (72) we now obtain for $\varphi(\boldsymbol{k})$ a system of inhomogeneous linear equations:

$$\sum_{\boldsymbol{k}_2}(L^{-1})_{\boldsymbol{k}_1\boldsymbol{k}_2}\varphi(\boldsymbol{k}_2) = \varphi^{(0)}(\boldsymbol{k}_1, \boldsymbol{K}_0). \tag{87}$$

In eqs. (87) the matrix

$$(L^{-1})_{\boldsymbol{k}_1\boldsymbol{k}_2} = S^{-1}(E, \boldsymbol{k}_1)\delta_{\boldsymbol{k}_1\boldsymbol{k}_2} + \Delta R_{\boldsymbol{k}_1\boldsymbol{k}_2}, \tag{88}$$

in which function $S(E, \boldsymbol{k})$ is determined by eq. (65), whereas the matrix

$$R_{\boldsymbol{k}_1\boldsymbol{k}_2} = -\frac{1}{N}G_{\boldsymbol{k}_1\boldsymbol{k}_2}(E) + \frac{4A}{N^2}\sum_{\boldsymbol{k}_2'}\frac{G_{\boldsymbol{k}_1\boldsymbol{k}_2'}(E)G_{\boldsymbol{k}_1+\boldsymbol{k}_2'-\boldsymbol{k}_2,\boldsymbol{k}_2}(E)}{1+2AI(\boldsymbol{k}_1+\boldsymbol{k}_2', E+\mathrm{i}\gamma)}. \tag{89}$$

Next we solve eq. (87) for $\varphi(\boldsymbol{k})$:

$$\varphi(\boldsymbol{k}_1) = \sum_{\boldsymbol{k}_2} L_{\boldsymbol{k}_1\boldsymbol{k}_2}\varphi^{(0)}(\boldsymbol{k}_2, \boldsymbol{K}_0). \tag{90}$$

This, or its equivalent relationship (86) determines the total wave function, equal to the sum of the incident biphonon wave $(E, \boldsymbol{K}_0)$ and various kinds of diverging waves. The diverging biphonon waves are formed as a result of elastic scattering of the biphonon by an impurity.

Diverging phonon waves can also be formed in relation (86), but only under the condition that for some values of $\boldsymbol{K}_0$ the energy conservation law can be complied with, i.e.

$$E(\boldsymbol{K}_0) = \varepsilon_\ell + \varepsilon(\boldsymbol{k}), \tag{91}$$

where $\varepsilon_\ell = \hbar\omega_\ell$ is the energy of a local phonon, and $\varepsilon(\boldsymbol{k})$ is the energy of a free phonon with wavevector $\boldsymbol{k}$. The energy conservation law written out above corresponds exactly to the stripping reaction. It is clear that this law [i.e. eq. (91)] can be complied with only with definite relations between the quantities A, Δ and T.

Turning now to the determination in eq. (86) of the amplitude of scattered waves and the cross sections of the processes, we point out that their values are found by the pole contributions of function $Q(E + \mathrm{i}\gamma)$ [see eq. (86)] that occur in transforming to the coordinate representation and that require summation over $\boldsymbol{k}_1$ and $\boldsymbol{k}_2$.

We restrict ourselves in the following to an analysis only under conditions when the anharmonicity is strong, i.e. when the anharmonicity constant A is large compared to the halfwidth of the phonon zone. In this limiting case

$$G_{\boldsymbol{k}'\boldsymbol{k}''}(E) \approx -1/2A, \tag{92}$$

so that the expression $R_{\boldsymbol{k}'\boldsymbol{k}''}$ for the matrix [see eq. (89)] is simplified as follows:

$$\begin{aligned} R_{\boldsymbol{k}'\boldsymbol{k}''} &= \frac{1}{N}\left\{\frac{1}{2A} + \frac{1}{AN}\sum_{\boldsymbol{k}}[1 + 2AI(E + \mathrm{i}\gamma, \boldsymbol{k})]^{-1}\right\} \\ &\equiv \frac{1}{N}R(E), \end{aligned} \tag{93}$$

i.e., in the same way as function $G_{\boldsymbol{k}'\boldsymbol{k}''}(E)$, it ceases to depend upon the value of $\boldsymbol{k}'$ and $\boldsymbol{k}''$. In this approximation, eq. (87) assumes the form

$$\Delta R(E)\sum_{\boldsymbol{k}}\varphi(\boldsymbol{k}) + S^{-1}(E, \boldsymbol{k}_1)\varphi(\boldsymbol{k}_1) = \varphi^{(0)}(\boldsymbol{k}_1, \boldsymbol{K}_0).$$

The solution of this equation is simple and we give here only the final result:

$$\varphi(\boldsymbol{k}) = S(E, \boldsymbol{k})\varphi^{(0)}(\boldsymbol{k}, \boldsymbol{K}_0) - \frac{\Delta R(E) N^{-1} \sum_{\boldsymbol{k}'} S(E, \boldsymbol{k}')\varphi^{(0)}(\boldsymbol{k}', \boldsymbol{K}_0)}{1 + \Delta R(E) N^{-1} \sum_{\boldsymbol{k}'} S(E, \boldsymbol{k}')}. \tag{94}$$

If we now recall eqs. (84) and (92), we find in the approximation being discussed that

$$\varphi(\boldsymbol{k}) = \frac{S(E + \mathrm{i}\gamma, \boldsymbol{k})}{1 + \Delta R(E) N^{-1} \sum_{\boldsymbol{k}'} S(E + \mathrm{i}\gamma, \boldsymbol{k}')} \varphi^{(0)}(\boldsymbol{k}, \boldsymbol{K}_0). \tag{95}$$

After substituting this relation into eq. (86) and transforming to a coordinate representation, we have the following expression for the two-particle wave function:

$$\begin{aligned} \Psi_{\boldsymbol{nm}}(E, \boldsymbol{K}_0) &= \Psi^{(0)}_{\boldsymbol{nm}}(E, \boldsymbol{K}_0) + \frac{\Delta}{(2A)^2 D_0(E + \mathrm{i}\gamma)} \frac{1}{N^2} \sum_{\boldsymbol{k}'\boldsymbol{k}''} \exp\{\mathrm{i}[\boldsymbol{k}'\boldsymbol{n} + \boldsymbol{k}''\boldsymbol{m}]\} \\ &\times \Bigg\{ S(E + \mathrm{i}\gamma, \boldsymbol{k}') + S(E + \mathrm{i}\gamma, \boldsymbol{k}'') + F^{-1}(\boldsymbol{k}' + \boldsymbol{k}'', E + \mathrm{i}\gamma) \\ &\times \frac{1}{N} \sum_{\substack{\boldsymbol{k}_1'\boldsymbol{k}_2' \\ (\boldsymbol{k}_1' + \boldsymbol{k}_2' = \boldsymbol{k}' + \boldsymbol{k}'')}} [S(E + \mathrm{i}\gamma, \boldsymbol{k}_1') + S(E + \mathrm{i}\gamma, \boldsymbol{k}_2')] \Bigg\}. \end{aligned} \tag{96}$$

In this equation we have used the notation of eq. (20), as well as the approximate relation (92) and eq. (84). We have also introduced the notation

$$D_0(E) = 1 + \Delta R(E) N^{-1} \sum_{\boldsymbol{k}'} S(E, \boldsymbol{k}'). \tag{97}$$

For the energy region $E \approx E(\boldsymbol{K}_0)$ and when $A \gg T$, the quantity $D_0(E)$ coincides, with an accuracy to small terms of approximately $(T/A) \ll 1$, with the determinant $D(E)$ of the system of equations (72). This conclusion requires no further substantiation if we take into account that $D_0(E)$ is a determinant of the same system of equations, but for the case when approximation (92) is used.

3.6.1. Elastic scattering process for a biphonon

Since the condition $D(E)=0$ determines the energy of local biphonons, and in eq. (96) the energy E is equal to the free biphonon energy $E(\boldsymbol{K}_0)$, the role played by the factor $D_0(E)$ in eq. (96) becomes far more important in the case when the level of the local biphonon is close to the free biphonon zone [here $|D_0(E)|$ is of low value]. It is clear that such a case may occur only at sufficiently small values of $|\Delta| \ll A$ (see fig. 5). Here $S(E, \boldsymbol{k}) \to 1$ and

$$D_0(E) \approx 1 + \frac{\Delta}{A}\frac{1}{N}\sum_{\boldsymbol{K}}\frac{1}{F(E, \boldsymbol{K})}. \tag{98}$$

The condition $D_0(E)=0$ coincides exactly with eq. (76). In this case eq. (96) is also simplified to

$$\Psi_{\boldsymbol{nm}}(E, \boldsymbol{K}_0) = \Psi_{\boldsymbol{nm}}^{(0)}(E, \boldsymbol{K}_0) + \frac{2\Delta}{(2A)^2 D_0(E)}\frac{1}{N^2}\sum_{\boldsymbol{k}'\boldsymbol{k}''}\frac{\exp\{\mathrm{i}(\boldsymbol{k}'\boldsymbol{n}+\boldsymbol{k}''\boldsymbol{m})\}}{F(E+\mathrm{i}\gamma, \boldsymbol{k}'+\boldsymbol{k}'')}. \tag{99}$$

For the values of E being discussed we could use by virtue of the presumed inequality $A \gg T$, eqs. (59) and (60). Therefore, eq. (99) can also be rewritten in the form

$$\Psi_{\boldsymbol{nm}}(E, \boldsymbol{K}_0) = \Psi_{\boldsymbol{nm}}^{(0)}(E, \boldsymbol{K}_0) - \frac{2\Delta}{2AD_0(E)}\frac{1}{N^2}\sum_{\boldsymbol{k}'\boldsymbol{k}''}\frac{\exp \mathrm{i}(\boldsymbol{k}'\boldsymbol{n}+\boldsymbol{k}''\boldsymbol{m})}{E(\boldsymbol{K}_0)-E(\boldsymbol{k}'+\boldsymbol{k}'')+\mathrm{i}\gamma}.$$

With an accuracy to small terms of the order of $(T/A)^2$, the function

$$\Psi_{\boldsymbol{nm}}^{(0)}(E, \boldsymbol{K}_0) \approx -\frac{1}{2A}\delta_{\boldsymbol{nm}} \exp\{\tfrac{1}{2}\mathrm{i}\boldsymbol{K}_0(\boldsymbol{n}+\boldsymbol{m})\}.$$

Next, in the second term of eq. (99) summation over $\boldsymbol{K}=\boldsymbol{k}'+\boldsymbol{k}''$ and $\boldsymbol{q}=\boldsymbol{k}'-\boldsymbol{k}''$ gives

$$\Psi_{\boldsymbol{nm}}(E, \boldsymbol{K}_0) = -\delta_{\boldsymbol{nm}}\frac{1}{2A}\Psi_{\boldsymbol{n}}(E, \boldsymbol{K}_0),$$

where

$$\Psi_{\boldsymbol{n}}(E, \boldsymbol{K}_0) = \exp\{\mathrm{i}\boldsymbol{K}_0\boldsymbol{n}\} - \frac{2\Delta}{D_0(E)}\frac{1}{N}\sum_{\boldsymbol{K}}\frac{\exp\{\mathrm{i}\boldsymbol{K}\boldsymbol{n}\}}{E(\boldsymbol{K}_0) - E(\boldsymbol{K}) + \mathrm{i}\gamma}. \tag{100}$$

Thus, as was to be expected, in the case when $A \gg T$ and for values of $\Delta < 0$, such that $|\Delta| < |\Delta_3|$ (see fig. 5), the problem of biphonon scattering by an isotopic defect becomes formally analogous to the problem of phonon scattering. This latter problem (in the approximation of isotropic effective mass) was solved by Dubovsky and Konobeev (1965) and we shall apply its result here. In the aforementioned approximation, the biphonon energy is

$$E(\boldsymbol{K}) = E(0) + M(|\boldsymbol{K}|/K_m)^2,$$

where M is the width of the biphonon zone and K_m is the maximum value of $|\boldsymbol{K}|$. Integration in eq. (100) at large values of $|\boldsymbol{n}|$ yields the expression

$$\Psi_{\boldsymbol{n}} = \exp\{\mathrm{i}\boldsymbol{K}_0\boldsymbol{n}\} + \Psi\frac{\exp\{\mathrm{i}|\boldsymbol{K}_0|\cdot|\boldsymbol{n}|\}}{|\boldsymbol{n}|}, \tag{101}$$

where

$$\Psi = \frac{\pi}{2K_m}\frac{\xi}{1 - \xi(1 - x^2 + \frac{1}{2}\mathrm{i}\pi x)},$$

$$\xi = 6|\Delta|/M; \qquad x = |\boldsymbol{K}_0|/K_m,$$

so that the cross section of the process of scattering a biphonon by an impurity is equal to

$$\sigma_{\mathrm{e}} = 4\pi|\Psi|^2 = \frac{\pi^3}{K_m^2}\frac{\xi^2}{(1 - \xi + \xi x^2)^2 + \frac{1}{4}\pi^2\xi^2x^2}. \tag{102}$$

It follows from eq. (43) that in the approximation of the effective mass used here (at $A' = A$) the local biphonon state exists only when $|\Delta| > \Delta_1 = M/6$. As $|\Delta| \to |\Delta_2|$ (i.e. as $\xi \to 1$), the local biphonon level approaches the bottom of the biphonon zone. Therefore, at low $|\boldsymbol{K}_0|$ values (i.e. when $x \ll 1$), the scattering cross section [see eq. 102] drastically increases and can substantially exceed the geometric scattering cross section $\sigma_0 \approx 1/K_m^2$. Such an increase in the cross section is due to the fact that the local biphonon level is close to the bottom of the zone, and is an effect that is well known in scattering theory (see, for example, Landau and Lifshitz 1959).

3.6.2. Stripping and pickup processes

Let us turn next to an analysis of eq. (96) at the Δ values for which, along with the above-discussed biphonon scattering process, the stripping process is also possible. To clear up the mathematic features of this case, we shall, for the sake of clarity, assume that for a biphonon with wavevector $\boldsymbol{K}_0$ the conservation law (91) is valid for a phonon with wavevector $\boldsymbol{k}$. This means [see eq. (91)] that the difference $E(\boldsymbol{K}_0) - \varepsilon(\boldsymbol{k}) = \varepsilon_\ell$, i.e. it is equal to the energy $\varepsilon_\ell = \hbar\omega_\ell$ of a local phonon and, consequently, according to eq. (42), the following identity is valid:

$$f(E, \boldsymbol{k}) \equiv 1 - \frac{\Delta}{N} \sum_{\boldsymbol{k}''} \frac{1}{[E(\boldsymbol{K}_0) - \varepsilon(\boldsymbol{k})] - \varepsilon(\boldsymbol{k}'')} = 0. \tag{103}$$

As is evident from eq. (65), at fixed values of $E(\boldsymbol{K}_0)$ and $\boldsymbol{k}' = \boldsymbol{k}$, the quantity $S(E, \boldsymbol{k}')$ has a pole. In the vicinity of this pole

$$f(E, \boldsymbol{k}') = \left(\frac{\partial f}{\partial E}\right)_0 [E(\boldsymbol{K}_0) - \varepsilon(\boldsymbol{k}') - \varepsilon_\ell].$$

From eq. (103) we have

$$\left(\frac{\partial f}{\partial E}\right)_{E-\varepsilon(\boldsymbol{k})=\varepsilon_\ell} = \frac{\Delta}{N} \sum_{\boldsymbol{k}'} \frac{1}{[\varepsilon_\ell - \varepsilon(\boldsymbol{k}')]^2}.$$

Since, in the case being considered (see fig. 5), $|\Delta| \approx 2A$ and, consequently, $|\Delta| \gg T$ and $|\varepsilon_\ell - \varepsilon(\boldsymbol{k}')| \approx |\Delta|$, we obtain

$$\left(\frac{\partial f}{\partial E}\right)_0 \approx \frac{1}{\Delta}.$$

In this way, in a region in which a stripping reaction is possible

$$S(E + \mathrm{i}\gamma, \boldsymbol{k}') = \frac{1}{E(\boldsymbol{K}_0) - \varepsilon(\boldsymbol{k}') - \varepsilon_\ell + \mathrm{i}\gamma}. \tag{104}$$

The term in eq. (96), proportional to the function F^{-1}, leads to the previously discussed process of elastic biphonon scattering. Taking relation (104) into account for this process only renormalizes the corresponding amplitude and it will not be discussed here. For this reason we shall direct our attention to the second and third terms of eq. (96). It can be readily seen, after taking eq. (104) into consideration, that the sum of these terms is equal to the sum $\alpha_{nm}+\alpha_{mn}$, where

$$\alpha_{nm}=\frac{\Delta^2}{(2A)^2D_0(E)}\delta_{m0}\frac{1}{N}\sum_{k_1}\frac{\exp\{\mathrm{i}k_1n\}}{E(K_0)-\varepsilon(k_1)-\varepsilon_\ell+\mathrm{i}\gamma}. \tag{105}$$

The appearance here of the function δ_{m0} corresponds to the localization of one of the phonons. At the same time, the summation over k_1 in eq. (105) describes the wave function of the free phonon departing from the defect. In the approximation of the isotropic effective mass of the phonon $[\varepsilon(k_1)=\varepsilon(0)+2T(|k_1|/K_m)^2]$, the calculation of the sum mentioned above is analogous to the calculation of the sum appearing in eq. (100). At high $|n|$ values

$$\alpha_{nm}=\tilde{\Psi}\delta_{m0}\frac{\exp\{\mathrm{i}|k|\cdot|n|\}}{|n|},$$

where

$$\tilde{\Psi}=\frac{\Delta\frac{1}{2}\pi K_m^{-1}\tilde{\xi}}{2(2A)^2D_0(E)};\quad \text{where}\quad \tilde{\xi}=\frac{3|\Delta|}{T}. \tag{106}$$

Since under the conditions for which stripping is feasible, the quantity $\Delta\approx-2A$ (recall that in all cases we are dealing here with crystals for which $A\gg T$); the value of function $D_0(E)$ at $E=E(K_0)$, according to eq. (97) [if eqs. (104), (59) and (60) are also taken into account] is equal to

$$D_0(E)\approx-2\Delta^2\left\{N^{-1}\sum_K[E-E(K)+\mathrm{i}\gamma]^{-1}\right\}$$
$$\times\left\{N^{-1}\sum_{k'}[E-\varepsilon(k')-\varepsilon_\ell+\mathrm{i}\gamma]^{-1}\right\}.$$

There are no values of the biphonon energy $E=E(K_0)$ at which this function vanishes. In the approximation of the isotropic effective masses, eq. (91), expressing the energy conservation law, can be written in the form

$$M\left(\frac{K_0}{K_m}\right)^2+C(A,\Delta)=2T\left(\frac{|k|}{K_m}\right)^2, \tag{107}$$

where $C(A,\Delta)=E(0)-\varepsilon(0)-\varepsilon_\ell$. In this same approximation

$$D_0(E)=-\tfrac{1}{2}\xi\tilde{\xi}(1-x^2+\tfrac{1}{2}\mathrm{i}\pi x)(1-y^2+\tfrac{1}{2}\mathrm{i}\pi y),$$

where [see also the notation of eqs. (101) and (106)]

$$y = |\boldsymbol{k}|/K_m.$$

The stripping reaction cross section is

$$\sigma_S = 4\pi |2A\tilde{\Psi}|^2 v_{ph}/v_{bph},$$

where $v_{ph} = [d\varepsilon(\boldsymbol{k})/dk]$ and $v_{bph} = [dE(\boldsymbol{K}_0)/dK_0]$ are the group velocities of the phonon and biphonon, respectively. In the isotropic effective mass approximation we shall determine the dependence of the cross section σ_S on $|\boldsymbol{K}_0| = K_0$ for the case when $K_0 \ll K_m$ into the explicit form. Since

$$\frac{v_{ph}}{v_{bph}} = \frac{2Tk}{MK_0} \approx \frac{2T}{M}\left(\frac{K_m}{K_0}\right),$$

cross section σ_S, with an accuracy to a numerical factor of the order of unity, is found to equal

$$\sigma_S \approx \frac{\pi^3}{K_m^2}\left(\frac{T}{A}\right)^3\left(\frac{K_m}{K_0}\right).$$

It is natural to compare this cross section with the elastic scattering cross section [see eq. (102)] for the biphonon. Since, in the case being discussed, $|\Delta| \approx 2A \gg T$ (with $\xi \gg 1$), cross section (102) is found to be of the order of π^3/K_m^2. Therefore, the stripping cross section σ_S can be larger than the biphonon elastic scattering cross section σ_e only for biphonons with sufficiently small values of $K_0 \ll K_m$. These, precisely, are the conditions under which it proves worthwhile to take stripping processes into account in considering the mechanisms for broadening biphonon lines in a crystal with defects.

We shall not discuss the pickup reaction here; it is the reverse of the stripping process considered above. Instead, we point out that along with the elastic processes we have discussed, defects can also lead to inelastic processes, which proceed, however, with the participation of quasi-particles of another kind (for instance, acoustic phonons).

3.7. Local states spectrum with Fermi resonance

3.7.1. General statement of the problem

It is well known that Fermi resonance (FR) is set up in isolated molecules in the case when the frequency ω_2 of some fundamental dipole-active oscillation turns out to be close to the frequency of an overtone of another oscillation ω_1, i.e. in the case when $\omega_2 \approx n\omega_1$, where $n = 2, 3, \dots$. The excitation of the overtone by light is usually of low intensity. If there is a level of intensive dipole-active oscillation near to the overtone, the oscillations are "mixed" as a result of intramolecular anharmonicity, and the intensity of absorption is "transferred" from the fundamental oscillation to the overtone (Herzberg 1945).

The vibration spectrum in crystals at FR has a much more complex structure than that of the spectrum in isolated molecules. This complication at FR occurs even with two-particle states (biphonons, etc.). Since, in this case, all the characteristic parameters of the problem (widths of the phonon zones, intramolecular anharmonicity constants, also including the ones that correspond to FR) turn out to be quantities of the same order of magnitude, a theoretical analysis also proves to be quite cumbersome.

Research on this problem in perfect crystals was conducted by Ruvalds and Zawadowski (1970) and by Agranovich and Lalov (1971, 1985) (see also Agranovich 1983). In crystals with defects, however, the pattern of the spectrum at FR acquires a number of new special features in connection with the possibility of forming local biphonons. These features are of particular interest owing to the development of experimental investigations (see, for example, Lisitsa et al. 1978). We shall discuss them in the following within the framework of the simplest model of a molecular crystal having one molecule per unit cell and having an isolated defect: an isotopic substitution impurity. No fundamental difficulties are encountered in further generalizing the obtained results to crystals with several molecules per unit cell.

We shall therefore take into account in the crystal two normal oscillatory modes: B and C phonons. We shall assume that the frequency $\omega_2(\boldsymbol{K})$ of a C phonon is close to twice that of a B phonon ($\omega_2 \approx 2\omega_1$). If we have an isotopic defect at site $\boldsymbol{n} = 0$, the model Hamiltonian of the crystal can be written as follows:

$$\hat{H} = \sum_{\boldsymbol{n}} \hbar\omega_{1\boldsymbol{n}} B_{\boldsymbol{n}}^{+} B_{\boldsymbol{n}} + \sum_{\boldsymbol{n}\neq\boldsymbol{m}} V_{\boldsymbol{nm}}^{(\mathrm{B})} B_{\boldsymbol{n}}^{+} B_{\boldsymbol{m}} - A \sum_{\boldsymbol{n}} B_{\boldsymbol{n}}^{+2} B_{\boldsymbol{n}}^{2}$$
$$+ \sum_{\boldsymbol{n}} \hbar\omega_{2\boldsymbol{n}} C_{\boldsymbol{n}}^{+} C_{\boldsymbol{n}} + \sum_{\boldsymbol{n}\neq\boldsymbol{m}} V_{\boldsymbol{nm}}^{(\mathrm{C})} C_{\boldsymbol{n}}^{+} C_{\boldsymbol{m}} + (\Gamma/\sqrt{2}) \sum_{\boldsymbol{n}} \left[B_{\boldsymbol{n}}^{+2} C_{\boldsymbol{n}} + C_{\boldsymbol{n}}^{+} B_{\boldsymbol{n}}^{2} \right]. \tag{108}$$

In this expression $B_{\boldsymbol{n}}^{+}$, $B_{\boldsymbol{n}}$ and $C_{\boldsymbol{n}}^{+}$, $C_{\boldsymbol{n}}$ are Bose operators for the creation and annihilation of B and C quanta of vibrations in molecule $\boldsymbol{n}$, having frequencies $\omega_{1\boldsymbol{n}} = \omega_1$ and $\omega_{2\boldsymbol{n}} = \omega_2$ at $\boldsymbol{n} \neq 0$, respectively. But $\hbar\omega_{10} = \hbar\omega_1 + \Delta_{\mathrm{B}}$ and $\hbar\omega_{20} = \hbar\omega_2 + \Delta_{\mathrm{C}}$, where Δ_{B} and Δ_{C} are isotopic shifts. In eq. (108), $V_{\boldsymbol{nm}}^{(\mathrm{B})}$ and $V_{\boldsymbol{nm}}^{(\mathrm{C})}$ are matrix elements that determine the transfer of the B or C quantum from molecule $\boldsymbol{n}$ to molecule $\boldsymbol{m}$, and the energy A determines the part of the anharmonicity of the B phonons that does not lead to the mixing of the normal B and C oscillatory modes. Also taken into consideration in eq. (108), for this reason, is the intramolecular anharmonicity with the constant Γ. This anharmonicity leads to the mixing of the excited states $B_{\boldsymbol{n}}^{+2}|0\rangle$ and $C_{\boldsymbol{n}}^{+}|0\rangle$, where $|0\rangle$ is the ground state of the crystal.

It is natural to look for the wave function of the excited states of the crystal in the form of the superposition

$$\Psi = \sum_{\boldsymbol{nm}} \Psi_{\boldsymbol{nm}} B_{\boldsymbol{n}}^{+} B_{\boldsymbol{m}}^{+} |0\rangle + \sum_{\boldsymbol{n}} \varphi_{\boldsymbol{n}} C_{\boldsymbol{n}}^{+} |0\rangle,$$

where $\Psi_{\boldsymbol{nm}}$ and $\varphi_{\boldsymbol{n}}$ are functions determined from the solution of the Schrödinger equation

$$\hat{H}\Psi = E\Psi.$$

In view of the orthogonality of states $B_{\boldsymbol{n}}^{+} B_{\boldsymbol{m}}^{+} |0\rangle$ and $C_{\boldsymbol{n}}^{+} |0\rangle$, the quantities $\Psi_{\boldsymbol{nm}}$ and $\varphi_{\boldsymbol{n}}$ should satisfy the following system of equations

$$\left(E - \hbar\omega_{1\boldsymbol{n}} - \hbar\omega_{1\boldsymbol{m}} + 2A\delta_{\boldsymbol{nm}}\right)\Psi_{\boldsymbol{nm}} - \left(\Gamma/\sqrt{2}\right)\varphi_{\boldsymbol{n}}\delta_{\boldsymbol{nm}}$$

$$= \sum_{\boldsymbol{l}} \left[V_{\boldsymbol{nl}}^{(\mathrm{B})} \Psi_{\boldsymbol{lm}} + \Psi_{\boldsymbol{nl}} V_{\boldsymbol{lm}}^{(\mathrm{B})} \right],$$

$$\left(E - \hbar\omega_{2\boldsymbol{n}}\right)\varphi_{\boldsymbol{n}} = \sqrt{2}\,\Gamma\Psi_{\boldsymbol{nn}} + \sum_{\boldsymbol{l}} V_{\boldsymbol{nl}}^{(\mathrm{C})}\varphi_{\boldsymbol{l}}. \tag{109}$$

If we put $\Gamma = 0$ in the system of equations (109), the system breaks down into two independent systems of equations: one, of $\Psi_{\boldsymbol{nm}}$, is of the type discussed in the preceding sections, whereas the other is a system of equations for the function $\varphi_{\boldsymbol{n}}$, from which the spectrum of C phonons is determined. To investigate the general case, in which $\Gamma \neq 0$, we go over to the Fourier expansion of the functions $\Psi_{\boldsymbol{nm}}$ and $\varphi_{\boldsymbol{n}}$:

$$\Psi_{\boldsymbol{nm}} = \sum_{\boldsymbol{k}_1\boldsymbol{k}_2} \tilde{\Psi}_{\boldsymbol{k}_1\boldsymbol{k}_2} \exp\{\mathrm{i}(\boldsymbol{k}_1\boldsymbol{n} + \boldsymbol{k}_2\boldsymbol{m})\},$$

$$\varphi_{\boldsymbol{n}} = \sum_{\boldsymbol{k}} \varphi_{\boldsymbol{k}} \exp\{\mathrm{i}\boldsymbol{k}\boldsymbol{n}\}. \tag{110}$$

The substitution of eq. (110) into (109) yields the system of equations

$$\left[E-\varepsilon^{(\mathrm{B})}(\boldsymbol{k}_1)-\varepsilon^{(\mathrm{B})}(\boldsymbol{k}_2)\right]\Psi_{\boldsymbol{k}_1\boldsymbol{k}_2}+2AN^{-1}\sum_{\boldsymbol{k}_1'+\boldsymbol{k}_2'=\boldsymbol{k}_1+\boldsymbol{k}_2}\tilde{\Psi}_{\boldsymbol{k}_1'\boldsymbol{k}_2'}-(\Gamma/\sqrt{2})\varphi_{\boldsymbol{k}_1+\boldsymbol{k}_2}$$

$$=(\Delta_{\mathrm{B}}/N)\left(\sum_{\boldsymbol{k}_2'}\tilde{\Psi}_{\boldsymbol{k}_1\boldsymbol{k}_2'}+\sum_{\boldsymbol{k}_1'}\tilde{\Psi}_{\boldsymbol{k}_1'\boldsymbol{k}_2}\right),$$

$$\left[E-\varepsilon^{(\mathrm{C})}(\boldsymbol{k})\varphi_{\boldsymbol{k}}-\sqrt{2}\,\Gamma N^{-1}\sum_{\boldsymbol{k}_1'+\boldsymbol{k}_2'=\boldsymbol{k}}\tilde{\Psi}_{\boldsymbol{k}_1'\boldsymbol{k}_2'}\right]=\Delta_{\mathrm{C}}\frac{1}{N}\sum_{\boldsymbol{k}_1}\varphi_{\boldsymbol{k}_1},$$

where $\varepsilon^{(\mathrm{B})}(\boldsymbol{k}_1)$ and $\varepsilon^{(\mathrm{C})}(\boldsymbol{k})$ are the energies of the B and C phonons. After eliminating functions $\varphi_{\boldsymbol{k}}$ from this system of equations, we have the following system of equations for functions $\tilde{\Psi}_{\boldsymbol{k}_1\boldsymbol{k}_2}$:

$$\left[E-\varepsilon^{(\mathrm{B})}(\boldsymbol{k}_1)-\varepsilon^{(\mathrm{B})}(\boldsymbol{k}_2)\right]\tilde{\Psi}_{\boldsymbol{k}_1\boldsymbol{k}_2}+2A_{\boldsymbol{K}}(E)N^{-1}\sum_{\boldsymbol{k}_1'+\boldsymbol{k}_2'=\boldsymbol{K}}\tilde{\Psi}_{\boldsymbol{k}_1'\boldsymbol{k}_2'}$$

$$-\tilde{\Delta}_{\mathrm{C}}(E)\Gamma^2 G_{\boldsymbol{K}}^{(\mathrm{C})}N^{-1}\sum_{\boldsymbol{k}_2''}G_{\boldsymbol{k}_2''}^{(\mathrm{C})}N^{-1}\sum_{\boldsymbol{k}_1'+\boldsymbol{k}_2'=\boldsymbol{k}_2''}\tilde{\Psi}_{\boldsymbol{k}_1'\boldsymbol{k}_2'}$$

$$=\Delta_{\mathrm{B}}N^{-1}\left[\sum_{\boldsymbol{k}_1'}\tilde{\Psi}_{\boldsymbol{k}_1'\boldsymbol{k}_2}+\sum_{\boldsymbol{k}_2'}\tilde{\Psi}_{\boldsymbol{k}_1\boldsymbol{k}_2'}\right], \tag{111}$$

where

$$G_{\boldsymbol{K}}^{(\mathrm{C})}(E)=\left[E-\varepsilon^{(\mathrm{C})}(\boldsymbol{K})\right]^{-1};$$

$$\tilde{\Delta}_{\mathrm{C}}(E)=\Delta_{\mathrm{C}}\left[1-\Delta_{\mathrm{C}}N^{-1}\sum_{\boldsymbol{K}}G_{\boldsymbol{K}}^{(\mathrm{C})}(E)\right]^{-1};$$

$$A_{\boldsymbol{K}}(E)=A-\tfrac{1}{2}\Gamma^2 G_{\boldsymbol{K}}^{(\mathrm{C})}(E);\qquad \boldsymbol{K}=\boldsymbol{k}_1+\boldsymbol{k}_2. \tag{112}$$

It follows from this equation, among other things, that FR renormalizes the anharmonicity constant A, making it dependent on the energy.

When $\Delta_{\mathrm{B}}=\Delta_{\mathrm{C}}=0$ (perfect crystal), eq. (111) leads in the usual way (see section 2.2) to the following dispersion equation

$$1+2A_{\boldsymbol{K}}(E)N^{-1}\sum_{\boldsymbol{k}_1+\boldsymbol{k}_2=\boldsymbol{K}}\left[E-\varepsilon^{(\mathrm{B})}(\boldsymbol{k}_1)-\varepsilon^{(\mathrm{B})}(\boldsymbol{k}_2)\right]^{-1}=0.$$

This, specifically, is the equation that has already been discussed in the cited references to the theory of FR in perfect crystals.

Let us consider, to begin with, a case in which the quantity Δ_C can be assumed sufficiently small, and the term proportional to $\tilde{\Delta}_C(E)$ can be omitted in eq. (111). In this special case eq. (111) becomes analogous to eq. (62), which determines the spectrum and wave function of two-particle states in a molecular crystal having an isotopic defect. The presence of FR in eq. (111) is manifested only in the replacement of constant A by the already mentioned new "effective" energy $A_{\boldsymbol{K}}(E)$ of anharmonicity. Hence, in using eq. (111) with $\Delta_C = 0$, the dispersion equation for local biphonons, derived on the basis of eq. (62), remains formally valid. It is only necessary in this equation to replace quantity A by $A_{\boldsymbol{K}}(E)$ and to insert it after the sign of summation over $\boldsymbol{K}$. Thus, for example, the right-hand side of eq. (73) assumes the form

$$\Delta_B \frac{1}{N} \sum_{\boldsymbol{K}} \frac{A_{\boldsymbol{K}}(E) \sum_{n=0} (-\Delta)^n b_{n+1}(E, \boldsymbol{K})}{1 + 2A_{\boldsymbol{K}}(E) I(E, \boldsymbol{K})},$$

etc. Naturally, the transformation in eq. (73) to $A_{\boldsymbol{K}}(E)$, i.e. taking FR into account, makes an analysis of the special features of the spectra of local states in the overtone region even more complicated. Calculation results obtained by an electronic computer on the spectra of local states will be dealt with further on. For the time being we shall discuss a case of strong harmonicity, when quantity A is much larger than the width of the B-phonon zone.

3.7.2. Case of strong anharmonicity ($A \gg T$)

This limiting case at $\Gamma = 0$ (i.e. in the absence of FR) has already been discussed in section 3.4. According to what has been previously mentioned, we can take FR into account (i.e. at $\Gamma \neq 0$) by making use of eq. (76), given in that section, if we make the substitutions $A \to A_{\boldsymbol{K}}(E)$ and $\Delta \to \Delta_B$. As has been pointed out, the equation obtained in this manner determines the spectrum of local states, not only when $A \gg T$ (where $2T$ is the width of the B-phonon zone), but also at $\Delta_C = 0$. If, according to eq. (20), we take into consideration the definition of $F(E, \boldsymbol{K})$, then, in the substitution $A \to A_{\boldsymbol{K}}(E)$, eq. (76) is transformed into the following:

$$1 = -\Delta_B \frac{1}{N} \sum_{\boldsymbol{K}} \frac{1}{A_{\boldsymbol{K}}(E)[1 + 2A_{\boldsymbol{K}}(E) I(E, \boldsymbol{K})]}. \tag{113}$$

Let us consider this equation, assuming that the energy $\varepsilon^{(C)}(\boldsymbol{K})$ of a C phonon is close to the energy $E(\boldsymbol{K})$ of a biphonon in a perfect crystal [otherwise, $A_{\boldsymbol{K}}(E) \to A$ and eq. (113) reduces to eq. (76)]. Since when $A \gg T$ [see eqs. (59) and (60)]

$$1 + 2AI(E, \boldsymbol{K}) \approx -\frac{1}{2A}[E - E(\boldsymbol{K})],$$

we find that in the region of the spectrum being considered, where the quantities $|E - E(\boldsymbol{K})|$ and $|E - \varepsilon^{(C)}(\boldsymbol{K})|$ are considered to be small, it is approximately true that

$$1 + 2A_{\boldsymbol{K}}(E)I(E, \boldsymbol{K}) \approx -\frac{1}{2A}\left\{[E - E(\boldsymbol{K})] - \frac{2\Gamma^2}{E - \varepsilon^{(C)}(\boldsymbol{K})}\right\},$$

so that eq. (113), determining the spectrum of local states, is transformed into

$$1 = 2\Delta_B \frac{1}{N}\sum_{\boldsymbol{K}} \frac{[E - \varepsilon^{(C)}(\boldsymbol{K})]^2}{\{[E - E(\boldsymbol{K})][E - \varepsilon^{(C)}(\boldsymbol{K})] - 2\Gamma^2\}\{E - \varepsilon^{(C)}(\boldsymbol{K}) - \Gamma^2/A\}}.$$

If, moreover, $\Gamma \ll A$, this equation reduces to that derived by Lalov (1974):

$$1 = 2\Delta_B \frac{1}{N}\sum_{\boldsymbol{K}} \frac{E - \varepsilon^{(C)}(\boldsymbol{K})}{[E - E(\boldsymbol{K})][E - \varepsilon^{(C)}(\boldsymbol{K})] - 2\Gamma^2}.$$

We shall not discuss the nature of the solutions of the approximate dispersion equations derived above. To illustrate the effect of FR on the

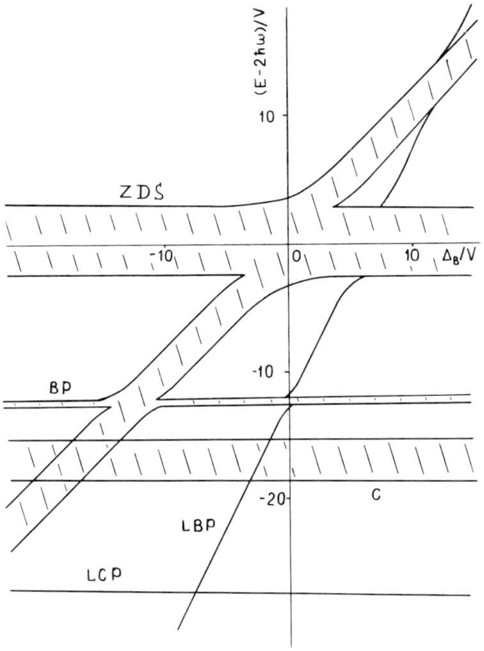

Fig. 11. Dependence of the spectrum of two-particle states on the isotopic shift Δ_B in the absence of Fermi resonance with one-phonon vibrations ($\Gamma/V = 0.0$ and $\Delta_C/V = -10$).

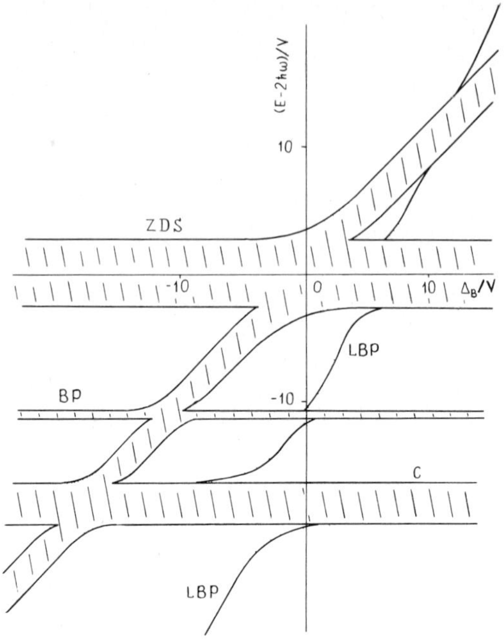

Fig. 12. Dependence of the vibrational spectrum on the isotopic shift Δ_B with Fermi resonance of interacting two-particle and one-particle states; weak interaction ($\Gamma/V = 2.0$ and $\Delta_C = 0.0$).

spectrum of local states we shall consider the results of numerical calculations. In these calculations the system of equations (109) for a one-dimensional crystal (among other cases, for $\Delta_c \neq 0$ as well) was solved by means of an electronic computer for various values of the dimensionless quantities Δ_B/V, Δ_C/V, A/V and Γ/V, where $V = V_{n,n+1}$ is the energy of resonance interaction of the nearest molecules. It was assumed, in order to reduce the number of independent parameters, that this energy is the same for both B and C phonons (i.e. the B- and C-phonon zones were taken to be of equal width). Notwithstanding the known limitations of the one-dimensional model, the results that were obtained enable a general qualitative pattern to be set up for the evolution of spectra of local excitations with FR on the basis of the main parameters of the problem.

3.7.3. Results of a numerical calculation for the one-dimensional model

To make the results of the calculations more illustrative, we consider first the spectrum of local and bulk oscillations against the parameter Δ_B/V at $\Gamma = 0$ as shown in fig. 11. This implies conditions under which there is actually no mixing of the states of the B and C phonons (for the sake of clarity we assume that the energy of the C phonon is less than that of the B biphonon:

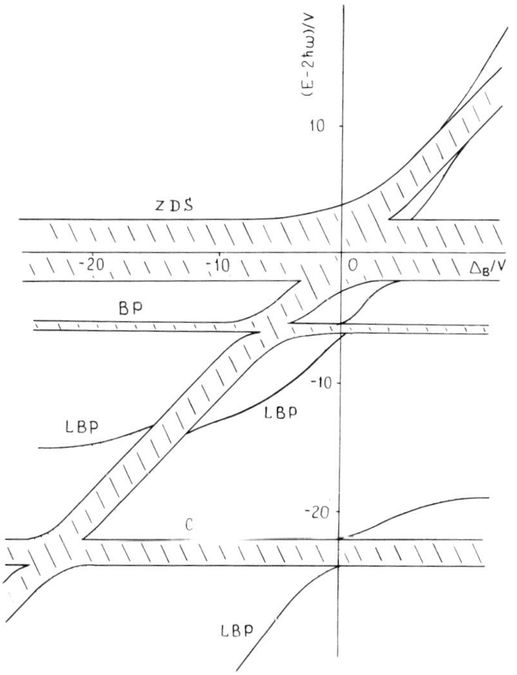

Fig. 13. Dependence of the vibrational spectrum on the isotopic shift Δ_B with Fermi resonance of interacting two-particle and one-particle states, intermediate interaction ($\Gamma/V = 6.0$ and $\Delta_C/V = 0.0$).

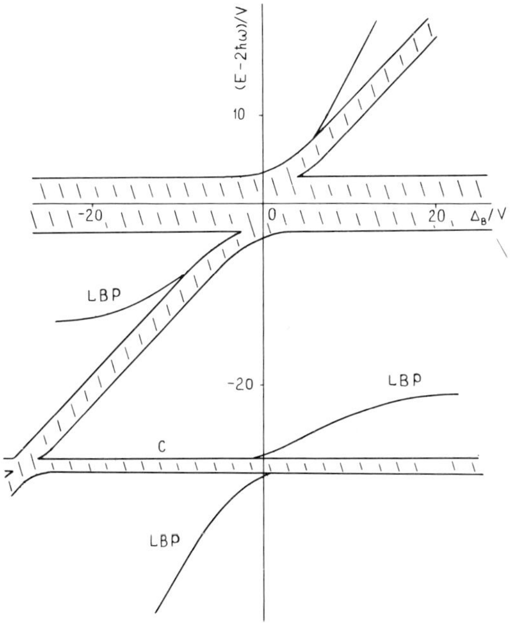

Fig. 14. Dependence of the vibrational spectrum on the isotopic shift Δ_B with Fermi resonance of interacting two-particle and one-particle states; strong interaction ($\Gamma/V = 10.0$ and $\Delta_C/V = 0.0$).

$\hbar\omega_2 - 2\hbar\omega_1 = -17V$; $A = 6V$; $\omega = \omega_1$). In this case we have simple superposition of the B- and C-phonon spectra; the line of the local B biphonon (LBP), i.e. the dependence of its energy on Δ_B/V, as well as the zone F + L of states for B phonons, "smoothly" intersect (i.e. without perturbation) both the zones of the free C phonon and the line of the local C phonon (LCP).

Shown in figs. 12, 13 and 14 are spectra of states for a series of increasing values of Γ, but at $\Delta_C = 0$. In particular, fig. 12 corresponds to the situation in which the value of Γ is still relatively low. In this case, the general pattern of the spectrum of B states is retained except that: (a) the binding energy of the B biphonon is reduced somewhat together with the effective anharmonicity constant $A_{\boldsymbol{K}}(E) = A - \frac{1}{2}\Gamma^2[E - \varepsilon^{(C)}(\boldsymbol{K})]$ and, moreover; (b) splitting of the LBP line of the local B biphonon occurs as a result of FR in the region of the C-phonon zone.

Naturally, with an increase in Γ (see fig. 13), the binding energy of free biphonons (BP) continues to decrease. But of more interest under these conditions is the effect of FR on the position of the terms of the local biphonon: the splitting of the LBP line is found to be so large that the formation of local states becomes possible even for high $|\Delta_B|$ values ($|\Delta_B| \to \infty$, the "vacancy" limit). These states on the LBP line are a superposition

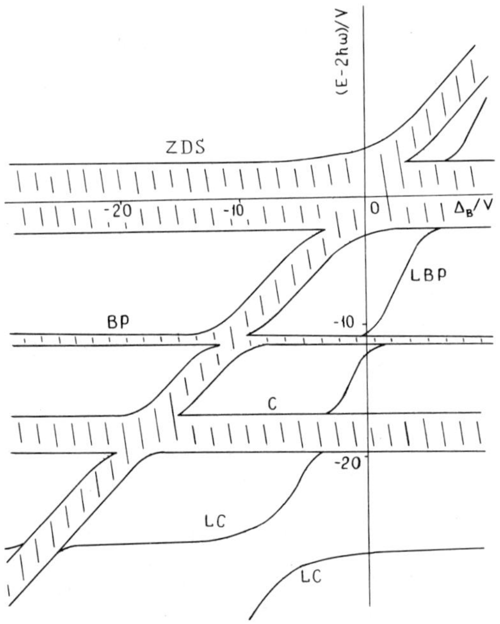

Fig. 15. Dependence of the vibrational spectrum on the isotopic shift Δ_B with Fermi resonance of interacting two-particle and one-particle states. The isotopic shift of C vibrations is nonzero ($\Gamma/V = 2.0$ and $\Delta_C/V = -10.0$).

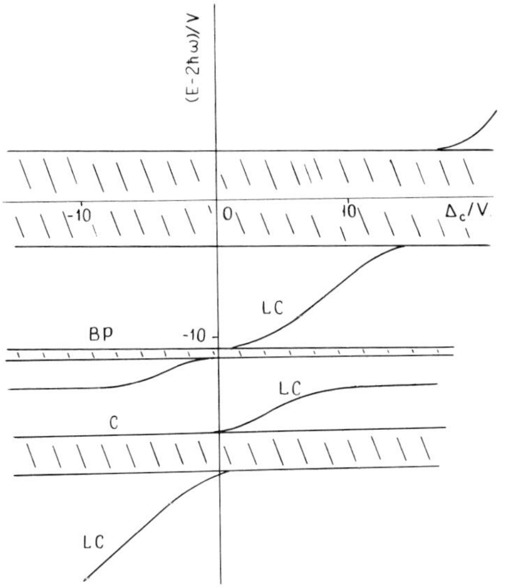

Fig. 16. Dependence of the vibrational spectrum on the isotopic shift Δ_C of C vibrations with Fermi resonance of interacting two-particle and one-particle states; weak interaction ($\Gamma/V = 2.0$ and $\Delta_B/V = 0.0$).

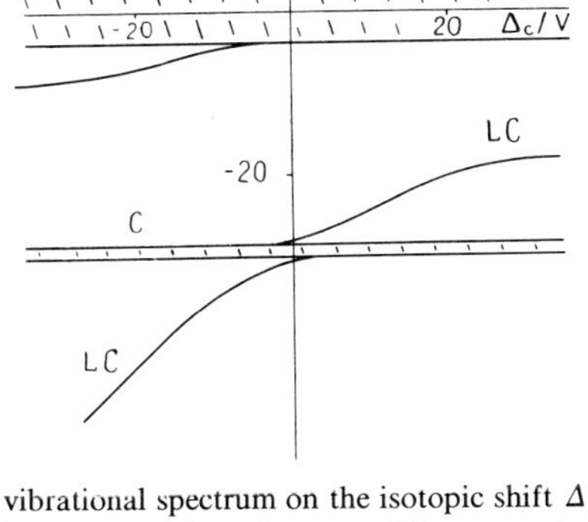

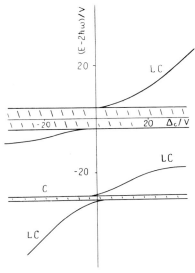

Fig. 17. Dependence of the vibrational spectrum on the isotopic shift Δ_C of C vibrations with the Fermi resonance of interacting two-particle and one-particle states; strong interaction ($\Gamma/V = 6.0$ and $\Delta_B/V = 0.0$).

of the states of the B and C phonons, so that the designation (LBP) only indicates their genesis.

Also of interest is the intersection, at certain Δ_B/V values, of the LBP with the F + L zone for B phonons. In the region of this intersection the process of LBP decay becomes possible. As a result, one of the B phonons remains localized and the second becomes free (a process opposite to this one is, of course, also possible).

Figure 14 illustrates a limiting situation, corresponding to sufficiently high Γ values [and, consequently, to low effective anharmonicity constants, see eq. (112)]. No zone of free biphonons is formed in this case. Nevertheless, local states (LBP line) are set up by the effect of FR.

Also shown in fig. 15 for the sake of completeness is a situation in which the isotopic shift of the C phonon is also nonzero ($\Delta_B \neq 0$ and $\Delta_C \neq 0$). Figures 16 and 17, on the other hand, show the features in the evolution of spectra in relation to Δ_C for two values of Γ in the case when the isotopic shift for the B phonon equals zero (i.e. $\Delta_B = 0$).

4. *Two-particle states in isotopically disordered crystals*

4.1. *Coherent Potential Approximation (CPA) for two-particle states*

4.1.1. *Introduction*

In the present section we deal with two-particle spectra of optical vibrations of mixed binary crystals of A_xB_{1-x} containing an appreciable and, in general, arbitrary concentration x of an isotopic substitution impurity. Such an investigation requires an analysis of two-particle Green functions, which, as is known (Elliott et al. 1974) are also needed for determining a great many kinetic parameters of the crystal (its thermal conductivity, electrical conductivity, etc.). Of cardinal importance, however, in the situation being considered is to take the strong anharmonicity into account. This leads to a number of special features in the dependence of the spectra of two-particle states on the concentration x of the isotopic impurity. To study these features we shall make use of the coherent potential approximation (CPA). This method has already been applied by Drchal and Velicky (1976) to solve a problem, methodically close to ours, that is posed in the theory of electron–electron interaction in disordered paramagnetic binary alloys. In the cited paper the so-called energy of contact interaction was regarded as the random parameter; in application to anharmonic phonons within the scope of Hamiltonian (48), it would correspond to random values of the anharmonicity constant $A_{n\alpha}$. Of considerably more importance, however, in isotopic mixed crystals is the random nature of the energy $\hbar\omega_{n\alpha}$. Therefore, in formulating the basic equations derived in using the coherent potential approximation, we shall regard as random both the value of the energy $\hbar\omega_{n\alpha}$ and the value of the anharmonicity constant $A_{n\alpha}$. Hence the results obtained in the above-mentioned paper of Drchal and Velicky (1976) will correspond to a certain limiting case.

We shall correlate the theoretical results discussed below with experimental data on the spectra of two-particle states obtained by Belousov and his co-workers (Belousov and Pogarev 1978). They were investigating second-order Raman scattering spectra in isotopically mixed cubic crystals of $^{15}N_x{}^{14}N_{1-x}H_4Cl$ in the spectral region $\omega \approx 2\omega_0$, where $\omega_0 = 1400$ cm^{-1}, that corresponds to an overtone of the non-degenerate dipole active oscillation of the NH_4^+ ion. The high-frequency internal vibrations of this ion interact relatively weakly with other modes of lattice vibration, so that in the indicated region of the spectrum, the crystal behaves like a molecular crystal with one molecule (ion) per unit cell. In connection with the aforesaid, we shall also assume in the following that the molecular vibration is nondegenerate and that the isotopically mixed molecular crystal being considered has one molecule per unit cell [in which case the indices, α, β and γ can be omitted in eqs. (48) and (49)].

4.1.2. The CPA equation

In accordance with eq. (49) we introduce the Green function $G_{nm,pq}$. Thus

$$G_{nm,pq}(E) = \left[D^{-1}(E)\right]_{nm,pq}, \tag{114}$$

in which the matrix

$$D_{nm,pq}(E) = [E - \hbar\omega_n - \hbar\omega_m + 2A\delta_{nm}]\delta_{np}\delta_{mq} - V_{np}\delta_{mq} - V_{mq}\delta_{np}. \tag{115}$$

It proves convenient for the following, making use of an idea advanced by Elliott et al. (1974), to introduce the so-called Green locator function $g_{nm,pq}$, coinciding with $G_{nm,pq}$ at $V_{nm} = 0$:

$$g_{nm,pq}(E) = G_{nm,pq}(E)\,|_{V=0}. \tag{116}$$

Next we expand the Green function $G_{nm,pq}$ into a perturbation theory series in terms of V_{nm}. In operator notation this expansion is of the form

$$\hat{G} = \hat{g} + \hat{g}(\hat{V}_{\rm I} + \hat{V}_{\rm II})\hat{g} + \hat{g}(\hat{V}_{\rm I} + \hat{V}_{\rm II})\hat{g}(\hat{V}_{\rm I} + \hat{V}_{\rm II})\hat{g} + \dots, \tag{117}$$

in which the matrices

$$(\hat{V}_{\rm I})_{nm,pq} = V_{np}\delta_{mq} \quad \text{and} \quad (\hat{V}_{\rm II})_{nm,pq} = V_{mq}\delta_{np}. \tag{118}$$

It follows from expansion (117) that Green function $\hat{G}$ is the solution of an equation analogous to the Dyson equation, i.e. to an equation of the form

$$\hat{G} = \hat{g} + \hat{g}(\hat{V}_{\mathrm{I}} + \hat{V}_{\mathrm{II}})\hat{G}. \tag{119}$$

In this equation, obviously, only locator $\hat{g}$ is a random quantity that depends upon the distribution of impurities in the crystal. However, in order to find the optical constants that are of interest to us it is necessary to know the average (over all possible configurations of impurity molecule arrangements) value of Green function $\overline{G}(E)$ (in which the bar indicates configuration averaging). Within the framework of one of the feasible approximation, namely coherent potential approximation, it is postulated that the sought-for averaged Green function $\overline{G}(E)$ can be found from an equation of the type (119), but having a determinate locator $\hat{\sigma}$. Thus

$$\overline{\hat{G}} = \hat{\sigma} + \hat{\sigma}(\hat{V}_{\mathrm{I}} + \hat{V}_{\mathrm{II}})\overline{\hat{G}}, \tag{120}$$

where matrix $\hat{\sigma}$ is determined from a certain condition of self-consistency. The following explains how this condition is found (Leath 1973).

It follows from eqs. (119) and (120) that

$$\hat{g}^{-1} = \hat{G}^{-1} + (\hat{V}_{\mathrm{I}} + \hat{V}_{\mathrm{II}}),$$

$$\hat{\sigma}^{-1} = (\overline{\hat{G}})^{-1} + (\hat{V}_{\mathrm{I}} + \hat{V}_{\mathrm{II}}).$$

After eliminating the matrix $\hat{V}_{\mathrm{I}} + \hat{V}_{\mathrm{II}}$ from these equations we find

$$(\overline{\hat{G}})^{-1} = \hat{G}^{-1} + (\hat{\sigma}^{-1} - \hat{g}^{-1})$$

or

$$\hat{G} = \left[1 - \overline{\hat{G}}(\hat{\sigma}^{-1} - \hat{g}^{-1})\right]^{-1}\overline{\hat{G}}.$$

This relation can also be rewritten as

$$\hat{G} = \overline{\hat{G}} + \overline{\hat{G}}\hat{T}\overline{\hat{G}}, \tag{121}$$

where the so-called T-matrix

$$\hat{T} = \left[(\hat{\sigma}^{-1} - \hat{g}^{-1})^{-1} - \overline{\hat{G}}\right]^{-1}. \tag{122}$$

If we now recall the definition of the quantity $\overline{\hat{G}}$, i.e., $\overline{\hat{G}} = \langle \hat{G} \rangle_{av}$, it follows from eq. (121) that the average value of the T-matrix should be equal to zero:

$$\overline{\hat{T}} = 0. \tag{123}$$

Equation (123), serving to determine matrix $\hat{\sigma}$ that appears in eq. (120), is exactly what is required for the condition of self-consistency.

In the theory of one-particle states, the CPA (123), in the one-site approximation ($\sigma_{nm}^{(1)} = \sigma(E)\delta_{nm}$), is of the form

$$\overline{\left\{\left[\sigma^{-1}(E) - g_{nn}^{-1}(E)\right]^{-1} - \overline{G_{nn}(E)}\right\}^{-1}} = 0, \tag{124}$$

and actually, according to eq. (120), is an equation for $\sigma(E)$. The reader can find an analysis of the conditions required for the applicability of the one-site approximation in the already cited review by Elliott et al. (1974). Hence, without going into any details of this analysis here, we shall only point out the most essential advantage of CPA. Even in its simplest one-site version, the coherent potential approximation is found to be accurate in both limiting cases in which the concentration of one or another substance is equal to zero. Consequently, the results obtained in this way can serve as interpolation equations [see the review by Elliott et al. (1974), p. 488]. Recall that in the indicated limiting cases ($x \to 0$ or $x \to 1$), eq. (124) in application, for example, to lattice vibrations, determines a function $\sigma(E)$ for which the poles of Green function $\overline{G}(E)$ coincide with the energies of the local phonons. This statement had been previously established, but only for one-particle states. This poses the question: to what extent are the advantages of CPA retained for the region of two-particle states as well, where, for instance, the spectrum of local vibrations upon strong anharmonicity is enriched by the formation of states of local biphonons? In the following we shall show that CPA is sufficiently fruitful for this purpose as well.

For the region of two-particle states, the most natural and simplest generalization of eq. (124) is a condition of the type

$$\overline{\left[\left(\sigma^{-1} - g^{-1}\right)_{nn,nn}^{-1} - \overline{G}_{nn,nn}(E)\right]^{-1}} = 0. \tag{125}$$

Henceforth we shall assume that matrix $\hat{\sigma}$ is diagonal. Thus

$$\sigma_{nm,pq}^{-1}(E) = \left[E - \Sigma(E) + 2\tilde{A}(E)\delta_{nm}\right]\delta_{np}\delta_{mq}. \tag{126}$$

We have introduced here, along with the effective anharmonicity constant $\tilde{A}(E)$, the complex quantity $\Sigma(E) = \Sigma'(E) + i\Sigma''(E)$, the self-energy part (see Elliott et al. 1974).

According to eq. (115),

$$g^{-1}_{nn,nn}(E) = E - 2E_n + 2A_n, \tag{127}$$

$E_n = E_{1,2}$ (where $E_1 = \hbar\omega_1$ and $E_2 = \hbar\omega_2$) and $A_n = A_{1,2}$ (if site $\boldsymbol{n}$ is occupied, respectively, by a molecule of the first or second kind), and making use of eq. (125), after explicitly writing in the average value, as well as eqs. (126) and (127), we obtain the equation

$$\frac{x}{\left[2E_1 - 2A_1 + 2\tilde{A}(E) - \Sigma(E)\right]^{-1} - \bar{G}_{nn,nn}(E)} + \frac{1-x}{\left[2E_2 - 2A_2 + 2\tilde{A}(E) - \Sigma(E)\right]^{-1} - \bar{G}_{nn,nn}(E)} = 0,$$

where x is the relative concentration of molecules of the first kind.

After making simple transformations, this equation assumes the form

$$\frac{x}{2E_2 - 2A_2 + 2\tilde{A}(E,x) - \Sigma(E,x)} + \frac{1-x}{2E_1 - 2A_1 + 2\tilde{A}(E,x) - \Sigma(E,x)} = \bar{G}_{nn,nn}(E), \tag{128}$$

which indicates explicitly the dependence of quantities Σ and $\tilde{A}$, not only on the energy, but on the concentration x as well.

4.1.3. Two limiting cases of disorder and behavior in the limit $x \to 0$

According to eq. (120), function $\bar{G}_{nn,nn}(E)$ is determined by the locator $\hat{\sigma}(E)$ and, in this manner, depends upon two unknown functions, $\Sigma(E)$ and $\tilde{A}(E)$, of the energy. In this general case, eq. (128) is insufficient for determining these functions, and it proves necessary to go beyond the scope of the one-site approximation. But this approximation can be applied in two limiting cases. The first and simpler one is discussed in the paper by Drchal and Velicky (1976) cited above. This case corresponds to the situation in which only the anharmonicity constant is a random quantity, so that

$$E_1 = E_2 = E_0 = \Sigma(E)/2 \qquad \text{and}$$

$$\sigma^{-1}_{nm,pq}(E) = \left[E - 2E_0 + 2\tilde{A}(E)\delta_{nm}\right]\delta_{np}\delta_{mq}.$$

Hence Green function $\bar{G}(E)$, according to eq. (120), corresponds to a perfect effective crystal with the anharmonicity $\tilde{A}(E)$. As follows from eqs. (56) and (57) [see also eq. (21)], the quantity $\bar{G}_{nn,nn}(E)$ is determined, in this case, by the relation

$$\bar{G}_{nn,nn}(E, x) = N^{-1}\sum_{K} I(E, \boldsymbol{K})[1 + 2A(E, x)I(E, \boldsymbol{K})]^{-1}, \tag{129}$$

so that eq. (128) assumes the following form

$$\frac{x}{2[\tilde{A}(E, x) - A_2]} + \frac{1-x}{2[\tilde{A}(E, x) - A_1]} = \frac{1}{N}\sum_{\boldsymbol{K}} \frac{I(E, \boldsymbol{K})}{1 + 2\tilde{A}(E, x)I(E, \boldsymbol{K})}. \tag{130}$$

Equation (130) contains only one unknown function, $\tilde{A}(E, x)$, and can therefore be used to find this function.

We intend to show that in the limit $x \to 0$ (the case $x \to 1$ can be considered in a similar way), even in the simplest one-site version of the CPA, Green function $\bar{G}(E)$ contains an additional pole among its poles, and this pole corresponds exactly to a local biphonon. We have shown above that in a lattice containing an impurity with a changed anharmonicity constant, the energy of the local biphonon is determined by eq. (58). Exactly the same kind of pole appears at low x values for Green function $\bar{G}(E, x)$ in the first order of its expansion in terms of concentration. Indeed, in the limit $x \to 0$ in eq. (130) (here, obviously, $A(E) \to A_2$), we find that

$$\left.\frac{\mathrm{d}\tilde{A}(E, x)}{\mathrm{d}x}\right|_0 = \frac{1}{2}\left[\frac{1}{2(A_1 - A_2)} + \frac{1}{N}\sum_{\boldsymbol{K}} \frac{I(E, \boldsymbol{K})}{1 + 2A_2 I(E, \boldsymbol{K})}\right]^{-1}. \tag{131}$$

At the same time at low x values

$$\bar{G}_{nn,nn}(E, x) = \bar{G}_{nn,nn}(E, 0) + x\left(\frac{\partial \bar{G}_{nn,nn}}{\partial x}\right)_{x=0},$$

or, making use of eq. (129),

$$\bar{G}_{nn,nn}(E, x) = \bar{G}_{nn,nn}(E, 0) - x\left\{\frac{1}{N}\sum_{\boldsymbol{K}} \frac{2I^2(E, \boldsymbol{K})}{[1 + 2A_2 I(E, \boldsymbol{K})]^2}\right\}\left\{\frac{\mathrm{d}\tilde{A}(E, x)}{\mathrm{d}x}\right\}_{x=0}, \tag{132}$$

which proves the statement made above, if eq. (131) is taken into account, as well as the fact that at low values of $x \to 0$, $A' = A_1$ for impurities, whereas $A = A_2$ for the matrix [see eq. (58)].

For the region of arbitrary x values, the function $\tilde{A}(E, x)$ that satisfies eq. (130) can only be found by applying numerical methods (Drchal and Velicky 1976).

Let us next discuss the second limiting case, which was investigated in CPA by Agranovich et al. (1979) and Agranovich and Dubovsky (1980). This case corresponds to the situation in which the anharmonicity constant is the same for all molecules of the crystal, i.e. $A_1 = A_2 = A = \tilde{A}(E)$, whereas the energy $\hbar\omega_n$ of a quantum is a random quantity. Therefore, according to eq. (126),

$$\sigma^{-1}_{nm,pq}(E) = [E - \Sigma(E) + 2A\delta_{nm}]\,\delta_{np}\delta_{mq}$$

and for the matrix element $\bar{G}_{nn,nn}(E)$ of the Green function we can, as before make use of expression (57):

$$\bar{G}_{nn,nn}(E) = \frac{1}{N}\sum_{K}\frac{\tilde{I}(\Sigma, E, K)}{1 + 2A\tilde{I}(\Sigma, E, K)}, \tag{133}$$

where, however,

$$\tilde{I}(\Sigma, E, K) = N^{-1}\sum_{k_1}[E - \tilde{\varepsilon}(E, k_1) - \tilde{\varepsilon}(E, K - k_1)]^{-1}, \tag{134}$$

and $\tilde{\varepsilon}(E, k)$ is the energy of an optical phonon with the wavevector k in an effective crystal. Thus

$$\tilde{\varepsilon}(E, k) = \tfrac{1}{2}\Sigma(E) + \sum_{m} V_{nm} \exp\{ik(m - n)\}. \tag{135}$$

It proves more convenient to write eq. (134) in the form

$$\tilde{I}(\Sigma, E, K) = N^{-1}\sum_{k_1}[E - \Sigma(E) - V(k_1) - V(K - k_1)]^{-1},$$

where

$$V(k) = \sum_{m} V_{nm} \exp\{ik(m - n)\}.$$

By virtue of the aforesaid, in the case being discussed eq. (128) takes the form

$$\frac{x}{2E_2-\Sigma(E,x)}+\frac{1-x}{2E_1-\Sigma(E,x)}=\bar{G}_{\boldsymbol{nn},\boldsymbol{nn}}(E,x), \tag{136}$$

$$\bar{G}_{\boldsymbol{nn},\boldsymbol{nn}}(E,x)=\frac{1}{N}\sum_{\boldsymbol{K}}\frac{\tilde{I}(\Sigma,E,\boldsymbol{K})}{1+2A\tilde{I}(\Sigma,E,\boldsymbol{K})}. \tag{137}$$

The most important difference between eq. (136), which the sought-for function $\Sigma(E, x)$ must satisfy, and eq. (130), which determine function $\tilde{A}(E, x)$, stems from the fact that in eq. (136) function $\Sigma(E, x)$ is under the sign of a two-fold integral [and not a simple integral as is $A(E, x)$ in eq. (130)]. This circumstance substantially complicates the determination of function $\Sigma(E, x)$ (see below).

Before discussing the approximations we used to solve eq. (136), let us consider the dependence $\Sigma(E, x)$ for low values of $x \to 0$ [here, obviously, $\Sigma(E, x) \to 2E_2$]. In first order with respect to x

$$\Sigma(E,x)=2E_2+x\left(\frac{\partial\Sigma}{\partial x}\right)_0,$$

where, as follows from eq. (136),

$$\left(\frac{\partial\Sigma(E,x)}{\partial x}\right)_{x=0}=\left[\frac{1}{2(E_1-E_2)}-\frac{1}{N}\sum_{\boldsymbol{K}}\frac{\tilde{I}(2E_2,E,\boldsymbol{K})}{1+2A\tilde{I}(2E_2,E,\boldsymbol{K})}\right]^{-1}. \tag{138}$$

Thus, in the approximation linear with respect to x,

$$\bar{G}_{\boldsymbol{nn},\boldsymbol{nn}}(E,x)=\bar{G}_{\boldsymbol{nn},\boldsymbol{nn}}(E,0)+x\left[\frac{\partial\bar{G}_{\boldsymbol{nn},\boldsymbol{nn}}(E,x)}{\partial\Sigma}\right]_{\Sigma=2E_2}\left[\frac{\partial\Sigma(E,x)}{\partial x}\right]_{x=0}. \tag{139}$$

The first term in this equation corresponds to a perfect crystal, in which $\Sigma = 2E_2$. The second term in eq. (139) leads to the appearance in $\bar{G}_{\boldsymbol{nn},\boldsymbol{nn}}(E, x)$ of singularities at frequencies which, according to eq. (138), are determined by the equation

$$1-2\Delta\frac{1}{N}\sum_{\boldsymbol{K}}\frac{I(E,\boldsymbol{K})}{1+2AI(E,\boldsymbol{K})}=0. \tag{140}$$

It should be pointed out, first of all, that this comparatively simple equation does not coincide with the exact and complex one [eq. (61)] that determines the whole set of two-particle states that appear when an isotopic defect is introduced. Nevertheless, eq. (140), containing two independent parameters Δ and A of the problem, correctly conveys the nature of the solution of the exact equation (61) for all limiting values of Δ and A. Let us consider, for instance, eq. (140) at low A values ($A \to 0$). In this case this equation is of the form

$$1 - 2\Delta \frac{1}{N^2} \sum_{k_1 k_2} \frac{1}{E - \varepsilon(\boldsymbol{k}_1) - \varepsilon(\boldsymbol{k}_2)} = 0, \tag{141}$$

or

$$1 - 2\Delta \int \frac{\rho_2(E')\,\mathrm{d}E'}{E - E'} = 0, \tag{142}$$

where $\rho_2(E')$ is the density of two-particle states. At high $|\Delta|$ values ($|\Delta| \gg T$), eq. (141) yields the correct value for the energy $E = 2E_\ell \approx 2\hbar\omega_2 + 2\Delta$. Besides, at low $|\Delta|$ values, i.e. under conditions in which CPA, as is known (see Elliott et al. 1974), is an exact method, eq. (141) also yields an accurate result $E \to \varepsilon(\boldsymbol{k}_1) + \varepsilon(\boldsymbol{k}_2)$. For arbitrary values of Δ, the energy E of a two-particle localized state at $A = 0$, i.e. the energy of $E = 2E_\ell$, as follows from eq. (42), satisfies the equation

$$1 - 2\Delta \frac{1}{N} \sum_{k} \frac{1}{E - 2\varepsilon(\boldsymbol{k})} = 0, \tag{143}$$

or

$$1 - 2\Delta \int \frac{\frac{1}{2}\rho_1(E'/2)\,\mathrm{d}E'}{E - E'} = 0,$$

where $\rho_1(E'/2)$ is the density of one-particle states. As can be seen, eq. (143) differs from eq. (141). But when $\Delta \to 0$, the former, in contrast to eq. (141), yields $E \to 2\varepsilon(\boldsymbol{k})$, and therefore does not permit all the two-particle states with energy $\varepsilon(\boldsymbol{k}_1) + \varepsilon(\boldsymbol{k}_2)$ to be taken into consideration.

Since, in general, $\rho_2(E') \neq \rho_1(E'/2)/2$, the solution $E = 2E_\ell$ of eq. (141) differs somewhat from the corresponding solution of eq. (143) only at $\Delta \approx \Delta_c$, when the energy $E_\ell(\Delta)$ is close to the bottom of the phonon zone. But this difference is found to be very small, however, in the approximation, for instance, of "elliptic" density ρ_1 (see fig. 18). The critical values for splitting off the local state are found to be close to each other $\{|\Delta_c^{(1)} - \Delta_c^{(2)}|/2T \approx 0.2\}$, whereas the asymptotic forms of dependence $E_\ell(\Delta)$ coincide.

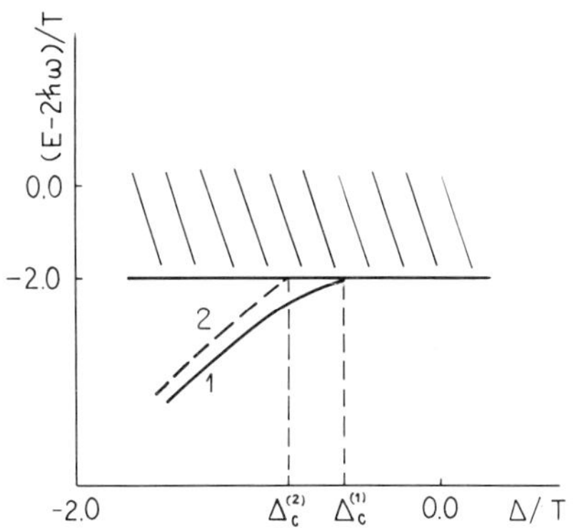

Fig. 18. Limiting dependences on the isotopic shift Δ of the energy of local two-phonon vibrations (at $A=0$). Curve 1 is the solution of eq. (143); curve 2 is the solution of eq. (141).

At high values of the constants $A \gg T$, eq. (140) is transformed into an equation for the energy of a local biphonon [see eq. (43)] (we shall not discuss the details of this transformation here, because it is similar to that carried out in section 3.4). If, however, A is arbitrary, but $|\Delta|$ is small ($\Delta \to 0$), eq. (140) is transformed into the exact equation for biphonons in a perfect crystal.

Hence it follows from the aforesaid that for all limiting values of parameters A, Δ and x of eq. (136), the CPA is found to be sufficiently accurate and can therefore serve as the basis for constructing interpolation in the region of intermediate values of these parameters. In subsequent subsections we shall apply CPA to solve certain specific problems. In addition we shall also make use of computer calculations to assess the accuracy of CPA, in particular, to find the density $\rho(E)$ of two-particle states. This density, it is known, is proportional to the imaginary part of the Green function [in our case $\rho(E) \propto \mathrm{Im}\{\bar{G}_{nn,nn}(E, x)\}$, see eq. (137)]. To assume the theoretical and experimental data we shall also require the light absorption factor $\kappa(E)$ and the Raman scattering cross section $r(E)$. These quantities are proportional to the same density of states having a low value of $\boldsymbol{K} \approx 0$. When eq. (137) is taken into account, we come to the conclusion that $\kappa(E) \sim P(E, x)$ and $r(E) \propto P(E, x)$, where, by definition,

$$P(E, x) = \mathrm{Im}\{I(\Sigma, E, 0)[1 + 2AI(\Sigma, E, 0)]^{-1}\}. \qquad (144)$$

4.2. *CPA calculations of the biphonon spectrum in disordered one-dimensional crystals*

First let us consider the simplest model of a one-dimensional crystal in the approximation of interaction of nearest neighbors. For this model, the dependence of the integral $I(\Sigma, E, \boldsymbol{K})$ on Σ can be derived analytically. Moreover, the results obtained here will be used in correlating CPA with the results of a straightforward numerical solution of the basic system of equations (49).

Biphonon states in a perfect one-dimensional crystal have been discussed in section 2.3.1. Making use of the notation of this subsection and a very simple transformation, we can represent function $I(\Sigma, E, K)$ [see eq. (134)] in the form

$$I(\Sigma, E, K) = N^{-1} \sum_{k} [E - \Sigma(E) - 4V \cos(Ka/2) \cos(ka)]^{-1}, \quad (145)$$

where, as has been previously indicated, $\Sigma(E) = \Sigma'(E) + \mathrm{i}\Sigma''(E)$.

After transforming in eq. (145) from the summation over k to the integral, we can calculate $I(\Sigma, E, K)$ analytically (Gradshtein and Ryzhik 1971). It can be shown that the real and imaginary parts of the function $I = I' + \mathrm{i}I''$ are related to $\Sigma'(E)$ and $\Sigma''(E)$ as follows:

$$I'(\Sigma, E, K) = (1/\sqrt{2}) R(\Sigma, E, K) \times \{[E - \Sigma' - 4V \cos(Ka/2)] M_{(-)}(\Sigma, E, K) + [E - \Sigma' + 4V \cos(Ka/2)] M_{(+)}(\Sigma, E, K)\},$$

$$I''(\Sigma, E, K) = (1/\sqrt{2}) R(\Sigma, E, K) \times [M_{(+)}(\Sigma, E, K) + M_{(-)}(\Sigma, E, K)] \Sigma'',$$

where the functions $R(\Sigma, E, K)$ and $M_{(\pm)}(\Sigma, E, K)$ are of the form

$$R(\Sigma, E, K) = \{[M_{(+)}(\Sigma, E, K) M_{(-)}(\Sigma, E, K)]^{-1} + (E - \Sigma')^2 + \Sigma''^2 - [4V \cos(Ka/2)]^2\}^{-1/2},$$

$$M_{(\pm)}(\Sigma, E, K) = \{[E - \Sigma' \pm 4V \cos(Ka/2)]^2 + \Sigma''^2\}^{-1/2}.$$

Evaluation of the integral in eq. (136) with respect to K and the subsequent simultaneous solution of the system of two transcendental equations (136) for $\Sigma'(E)$ and $\Sigma''(E)$ were carried out numerically for various values of the parameters $\delta = (E_1 - E_2)/2V$, A/V and x (with $E_1 > E_2$).

The results obtained in calculating Σ' (dashed line) and Σ'' (continuous line) for the value $\delta = 1.33$ are shown in fig. 19. With this choice of parameter

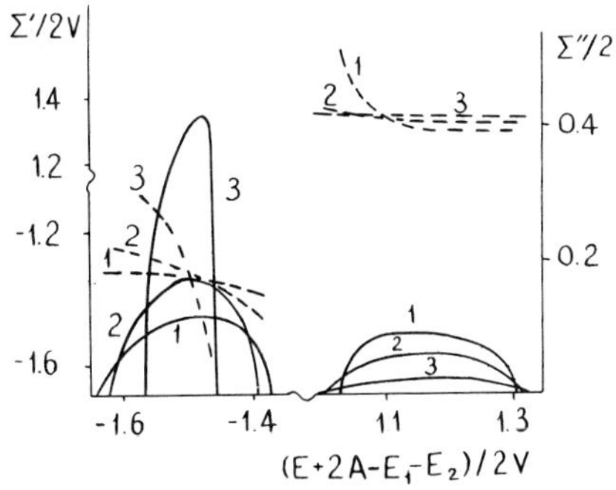

Fig. 19. Dependence of the self-energy part $\Sigma(E) = \Sigma'(E) + i\Sigma''(E)$ on the energy for a linear crystal at various impurity concentrations (continuous lines refer to Σ''; dashed lines refer to Σ'. Here $A/V = 6.0$ and $E_1/V = -E_2/V = 1.33$). Concentrations: (1) $x = 0.5$, (2) $x = 0.7$ and (3) $x = 0.9$.

δ, the zones of one-phonon vibrations overlap in perfect crystals of types 1 and 2. Naturally, the bands of unbound two-particle states also overlap. For the anharmonicity constant $A = 6V$ (case of strong anharmonicity), the biphonon zones both in crystal 1 and in crystal 2 are wholly split off in the energy region lying below the zone of unbound phonons in both perfect crystals.

Plotted in fig. 20 on the basis of the results obtained in calculating $\Sigma'(E)$ and $\Sigma''(E)$, and in applying eqs. (145) and (137), are the dependences of the density of states $\rho(E)$ on the concentration x.

Plotted in fig. 21 is the dependence $\rho(E)$ for the same value of A, but for a lower value of parameter $\delta = 0.08$. In this case a reduction of parameter δ leads to the transition from the two-mode to the one-mode behavior of the spectrum (this occurs in a situation in which the impurity and basic biphonon

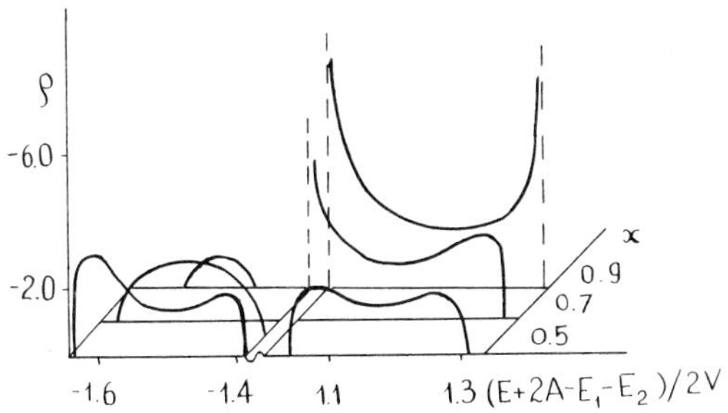

Fig. 20. Density of states of a linear crystal plotted against the concentration. Biphonon zones are separated (strong isotopic shift) (here $A/V = 6.0$ and $E_1/V = 1.33$).

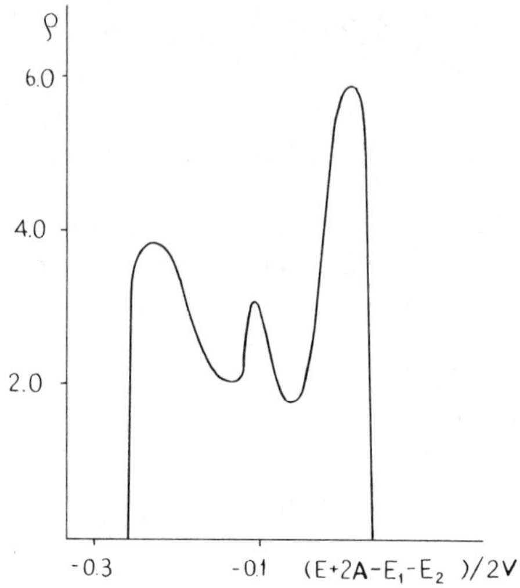

Fig. 21. Density of biphonon vibrational states of a linear crystal (weak isotropic shift) ($A/V = 6.0$ and $E_1/V = 0.08$).

zones overlap). At this a density of states peak value appears in the center of the united zone of biphonons.

4.3. *CPA calculations of the biphonon spectrum in disordered three-dimensional crystals*

If the approximation of nearest neighbors is applied for a three-dimensional crystal, the zonal contribution to the energy of one-particle excitation [$\varepsilon(\boldsymbol{k}) = \hbar\omega + V(\boldsymbol{k})$] is of the form

$$V(\boldsymbol{k}) = 2V\left[\cos(k_1 a) + \cos(k_2 a) + \cos(k_3 a)\right],$$

where k_1, k_2 and k_3 are the Cartesian components of wavevector $\boldsymbol{k}$. In this case the function $Z(\boldsymbol{q}) = V(\boldsymbol{K}/2 + \boldsymbol{q}) + V(\boldsymbol{K}/2 - \boldsymbol{q})$, appearing in the definition $I(\Sigma, E, K)$, is determined by the relation

$$Z(\boldsymbol{q}) = 4V\left[\cos(K_1 a/2)\cos(q_1 a) + \cos(K_2 a/2)\cos(q_2 a) + \cos(K_3 a/2)\cos(q_3 a)\right], \tag{146}$$

where

$$\boldsymbol{q} = (q_1, q_2, q_3), \quad |K_i a| \leqslant \pi, \quad i = 1, 2, 3.$$

To evaluate the integral in eq. (136) by means of the relation (146), in a way analogous to that employed by Drchal and Velicky (1976) we shall apply the following approximate procedure. In the expression for $I(\Sigma, E, \boldsymbol{K})$ we go from summation over $\boldsymbol{q}$ to integration, using the following model density of the values of quantity $Z(\boldsymbol{q})$:

$$\mathrm{d}\boldsymbol{q} = \frac{4\pi^2}{a^3 V^2(\boldsymbol{K}/2)} \left\{ [2V(\boldsymbol{K}/2)]^2 - Z^2(\boldsymbol{q}) \right\}^{1/2} \mathrm{d}Z. \tag{147}$$

The limiting maximum and minimum values of quantity $Z(\boldsymbol{q})$, as follows from eq. (146), are equal to $\pm 2V(\boldsymbol{K}/2)$, and the behavior of the model density (147) near these boundaries has the correct square root dependence. In the above-mentioned approximation the function $I(\Sigma, E, \boldsymbol{K})$ can be represented as follows:

$$I(\Sigma, E, \boldsymbol{K}) = N^{-1} \sum_{\boldsymbol{q}} [E - \Sigma - V(\boldsymbol{K}/2 - \boldsymbol{q}) - V(\boldsymbol{K}/2 + \boldsymbol{q})]^{-1}$$

$$= \frac{1}{2\pi V^2(\boldsymbol{K}/2)} \int_{-2|V(\boldsymbol{K}/2)|}^{2|V(\boldsymbol{K}/2)|} \frac{[4V^2(\boldsymbol{K}/2) - Z^2]^{1/2}\, \mathrm{d}Z}{E - \Sigma - Z}.$$

This integral can undergo an identity transformation. As a result the following expression can be obtained for $I(\Sigma, E, \boldsymbol{K})$:

$$I(\Sigma, E, \boldsymbol{K}) = \frac{1}{2V^2(\boldsymbol{K}/2)} \times \left\{ E - \Sigma - \left[(E - \Sigma)^2 - 4V^2(\boldsymbol{K}/2) \right] J(\Sigma, E, \boldsymbol{K}) \right\}, \tag{148}$$

where

$$J(\Sigma, E, \boldsymbol{K}) = \frac{1}{\pi} \int_{-2|V(\boldsymbol{K}/2)|}^{2|V(\boldsymbol{K}/2)|} (E - \Sigma - Z)^{-1} [4V^2(\boldsymbol{K}/2) - Z^2]^{-1/2}\, \mathrm{d}Z. \tag{149}$$

Integral (149) is analogous to the one discussed previously for a linear crystal, and it can be evaluated. As a result we obtain (with $J = J' + iJ''$):

$$I'(\Sigma, E, \boldsymbol{K}) = \frac{1}{2V^2(\boldsymbol{K}/2)} \Big\{ E - \Sigma' - \big[(E - \Sigma')^2 - (\Sigma'')^2 - 4V^2(\boldsymbol{K}/2) \big] \times J'(\Sigma, E, \boldsymbol{K}) - 2\Sigma''(E - \Sigma') J''(\Sigma, E, \boldsymbol{K}) \Big\},$$

$$I''(\Sigma, E, \boldsymbol{K}) = \frac{1}{2V^2(\boldsymbol{K}/2)} \Big\{ -\Sigma'' - \big[(E - \Sigma')^2 - (\Sigma'')^2 - 4V^2(\boldsymbol{K}/2) \big] \times J''(\Sigma, E, \boldsymbol{K}) + 2\Sigma''(E - \Sigma') J'(\Sigma, E, \boldsymbol{K}) \Big\}, \tag{150}$$

where

$$J'(\Sigma, E, \boldsymbol{K}) = (2\varphi)^{-1/2} \big\{ [E - \Sigma' - 2V(\boldsymbol{K}/2)]\, m_{(+)}^{-1} + [E - \Sigma' + 2V(\boldsymbol{K}/2)]\, m_{(-)}^{-1} \big\},$$

$$J''(\Sigma, E, \boldsymbol{K}) = (2\varphi)^{-12} \big[m_{(+)}^{-1} + m_{(-)}^{-1} \big] \Sigma''.$$

In this relationship, functions $m_{(\pm)}(\Sigma, E, \boldsymbol{K})$ and $\varphi(\Sigma, E, \boldsymbol{K})$ are determined as follows:

$$m_{(\pm)}(\Sigma, E, \boldsymbol{K}) = \big\{ [E - \Sigma' \pm 2V(\boldsymbol{K}/2)]^2 + (\Sigma'')^2 \big\}^{1/2},$$

$$\varphi(\Sigma, E, \boldsymbol{K}) = m_{(+)}(\Sigma, E, \boldsymbol{K}) m_{(-)}(\Sigma, E, \boldsymbol{K}) + (E - \Sigma')^2 + (\Sigma'')^2 - 4V^2(\boldsymbol{K}/2).$$

The relations between the quantities $I(\Sigma, E, \boldsymbol{K})$ and $\Sigma(E)$ given above [see eq. (150)] enable us to solve eq. (136) and to determine from it the quantity $\Sigma(E, x) = \Sigma'(E, x) + i\Sigma''(E, x)$.

In the same way as the previously discussed case of a linear crystal, subsequent operations (integration with respect to $\boldsymbol{K}$ and the solution of a system of two nonlinear equations for Σ' and Σ'') were carried out numerically. The integration with respect to $\boldsymbol{K}$ involved the application of the model "elliptical" density of one-phonon states.

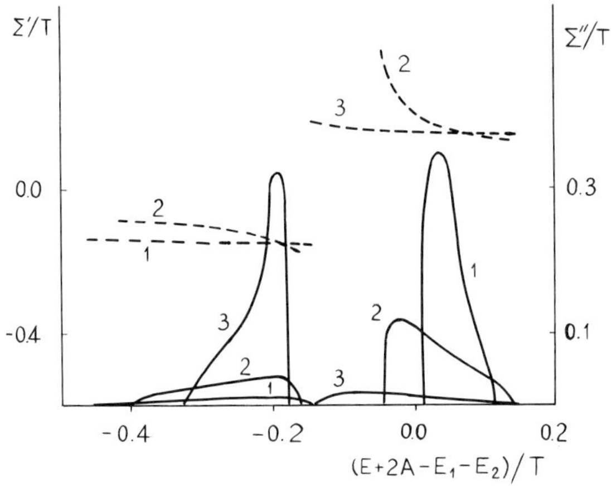

Fig. 22. Dependence of the self-energy part $\Sigma = \Sigma' + i\Sigma''$ on the energy for a three-dimensional crystal at various impurity concentrations (continuous lines refer to $\Sigma''(E)$; dashed lines refer to $\Sigma'(E)$]. Here $2A/T = 3.0$ and $2E_1/T = 0.15$. Concentrations: (1) $x = 0.1$; (2) $x = 0.5$ and (3) $x = 0.9$.

Plotted in fig. 22 are the results obtained in calculating the functions Σ' and Σ'' for various concentrations x and with $\delta = 0.15$ and $A = 3V$. There is, at these values, a considerable overlapping of the one-phonon zones, as well as the zones of two-phonon free states, for any composition of the solution (i.e. at any $x \leqslant 1$). In perfect crystals of the first or second kind the biphonon zones

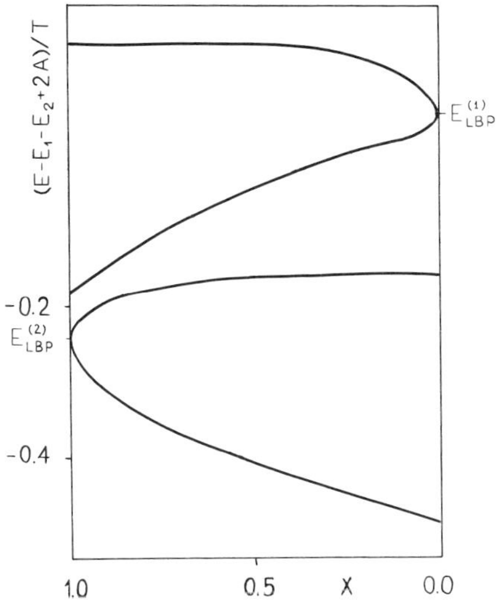

Fig. 23. Concentration dependence of the biphonon zone boundaries in a three-dimensional crystal ($2A/T = 3.0$ and $2E_1/T = 0.15$).

are also close to one another (they even have a negligible overlap). Nevertheless, in a crystalline mixture at the concentration $0.1 \leqslant x \leqslant 0.9$ separated biphonon zones already exist. The evolution of the biphonon zone boundaries when the concentration x is varied, is illustrated in fig. 23 for the case under consideration. At $x \approx 1$, the presence of an isotopic impurity of the second kind leads to the formation of the level $E_{\mathrm{LBP}}^{(2)}$ of the local biphonon. This level is located below the level of biphonon 1 and, upon a reduction of x, gradually develops into the zone of biphonon 2. At the same time, the zone of biphonon 1 is contracted into the level $E_{\mathrm{LBP}}^{(1)}$ of the local biphonon. This level now corresponds to an isolated impurity of the first kind in a crystal of the second kind. In the region of intermediate concentrations, both zones of the biphonon, as has been pointed out, are found to be well separated.

4.4. *Quasi-biphonons in impurity crystals of NH_4Cl. A comparison of the theoretical and experimental data*

In sections 4.2 and 4.3 we discussed crystals with a quite high anharmonicity constant A. It was high enough for the corresponding biphonon bands to completely split off the bands of dissociated two-particle states. In cases of low anharmonicity constants A, such that $A \lesssim T$, the levels of coupled phonon oscillations (biphonons) with an energy of approximately $2E_{1,2} - 2A$ are "submerged" in a band of two-particle dissociated states, i.e. their decay with the formation of two free quasi-particles is found to be possible. In cases when the density of two-particle states at the biphonon frequency is quite low (for instance, if the biphonon level is close to the zone edge), the biphonon lifetime can be quite large. In this case the presence of such a biphonon (more correctly called a quasi-biphonon) leads to a quite sharp density-of-states peak in the zone of two-particle states.

Such, evidently, is the situation implemented in the $^{14}NH_4Cl$ crystal. The doping of such a crystal with the isotope ^{15}N leads to an interesting evolution of the Raman spectrum (Belousov and Pogarev 1978, Belousov 1982). We shall now turn to a discussion of the results obtained, following the line of reasoning employed by Agranovich et al. (1979).

Observed in the spectrum of one-phonon states of the $^{14}NH_4Cl$ crystal at frequency 1402 cm^{-1} is the narrow line of the TO phonon (not more than 0.2 cm^{-1} wide). We can evidently assume that this frequency corresponds to the wavevector value $\boldsymbol{k} = 0$. When the crystal is doped with the isotope ^{15}N (up to 10%) the line is only negligibly broadened with a shift towards the low-frequency side. In addition, a weak continuous spectrum appears in the region from 1400 to 1450 cm^{-1}. The formation of this continuous spectrum is due to the effect of the impurity, violating the wavevector selection rules. It follows from this observation that the width of the whole phonon zone is approximately 50 cm^{-1}. No local vibrations are set up in this region. This is due to the small amount of isotopic shift $E_1 - E_2 = 6\ \text{cm}^{-1}$ compared to the zone width (Price et al. 1960). In the region of the two-phonon transition, a band from 2800 to 2900 cm^{-1} is observed in the spectrum of Raman scattering in a pure crystal. A sharp peak is observed at the low-frequency edge of this band at frequency 2804 cm^{-1}. It has been shown in various papers (Mitin et al. 1974, Gorelik et al. 1974) that this peak is associated with the excitation of a biphonon; however, since the peak at 2804 cm^{-1} in the spectrum of a pure $^{14}NH_4Cl$ crystal lies within the zone of dissociated states (at its edge), a quasi-biphonon is being observed.

Doping with the isotope ^{15}N, as shown by Belousov and Pogarev (1978), does not lead to the formation of a separate additional line in the spectrum, i.e. it does not lead to the formation of a local biphonon, but it is the cause of a narrow gap in the biphonon band. The formation of a gap in the biphonon band upon doping with impurities or, what amounts to the same thing, weak splitting of the biphonon peak, is evidently an indication of the excitation of a quasi-bound local state (local quasi-biphonon).

The theory developed above enables us to make a direct analysis of the situation which, evidently, exists in the NH_4Cl crystal, and to investigate the concentration dependence of the spectrum.

As in section 4.3, we carried out calculations of the function $I(\Sigma, E, \boldsymbol{K})$, numerical integration with respect to $\boldsymbol{K}$ in eq. (137), and a numerical solution of the system of transcendental equations (136) to determine Σ' and Σ''. As has already been pointed out, parameters Δ and T are known from experimental investigations: $2T = 50\ \text{cm}^{-1}$ and $\Delta = -6\ \text{cm}^{-1} = -0.24\ T$, whereas the anharmonicity constant A was determined from the condition of the best agreement between the experimental and theoretical data. We found by our calculations that best agreement with experimental data corresponds to $A = 0.7$ T.

The graphs in fig. 24 illustrate the results obtained in calculating the real and imaginary components of the self-energy parts $\Sigma(E)$ for the two impurity concentration values (5% and 10%) investigated by Belousov and Pogarev (1978). For the values of the parameters, A, T and Δ, shown above, the energy of the high-frequency edge of the biphonon band of the quasi-biphonon, corresponding to $\Sigma'' = 0$ and equal (in accordance to calculations carried out above) to the quantity $E^{(0)} = 2E_1 - 2A$, is within the zone of energies of dissociated states $[2(E_1 - T) < E^{(0)}]$. The energy of this edge corresponds to a clearly defined minimum of the imaginary component of the self-energy part as a function of E. It can be seen that on decreasing the impurity concentration, the low-energy maximum $\Sigma''(E)$ is narrowed and its intensity decreases. Here the maxima is shifted somewhat toward the low-frequency region. The quasi-biphonon region is also visible on the graph of the energy dependence of the real component $\Sigma'(E)$ of the self-energy part. The characteristic maximum of this dependence decreases with the impurity concentration and is shifted

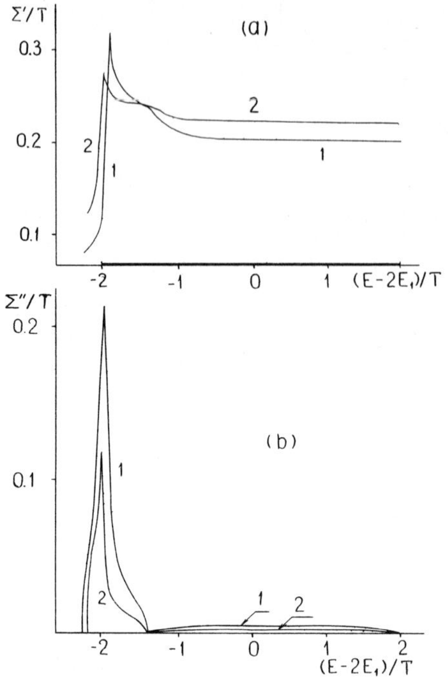

Fig. 24. The real (a) and imaginary (b) components of the self-energy parts $\Sigma(E)=\Sigma'(E)+\mathrm{i}\Sigma''(E)$ for various concentrations of impurity ^{15}N in a $^{15}N_x{}^{14}N_{1-x}H_4Cl$ crystal. The heavy line indicates the zone of two-particle states in a pure $^{14}NH_4Cl$ crystal. Concentrations: (1) $x=0.1$ and (2) $x=0.05$.

toward the low-frequency region. Also evident are the flex points on the high-frequency edge of the quasi-biphonon band.

The dependences $\rho(E)$, i.e. the energy dependences of the density of two-particle states with total wavevector $\boldsymbol{K}=0$, are shown in fig. 25. As has already been mentioned, the shape of the spectrum of Raman scattering is determined mainly by the quantity $P(E)$. Figure 25a shows the quantity $P(E)$ at an impurity concentration of 10% (continuous curve). The same figure also shows the experimental energy dependence of the intensity of Raman scattering (dashed curve) for the concentration $x=0.1$ obtained by Belousov and Pogarev (1978).

By selecting only a single parameter: $A=0.7\ T$, it proves possible, in good agreement with the experimental data, to provide explanations for three characteristic features observed in a Raman scattering spectrum. Thus, this theory yields the correct ratio of the intensities of Raman scattering in two peaks, indicates the correct distance between the peaks, and finally, correctly determines the position of the peaks with respect to the edge of the zone of dissociated states.

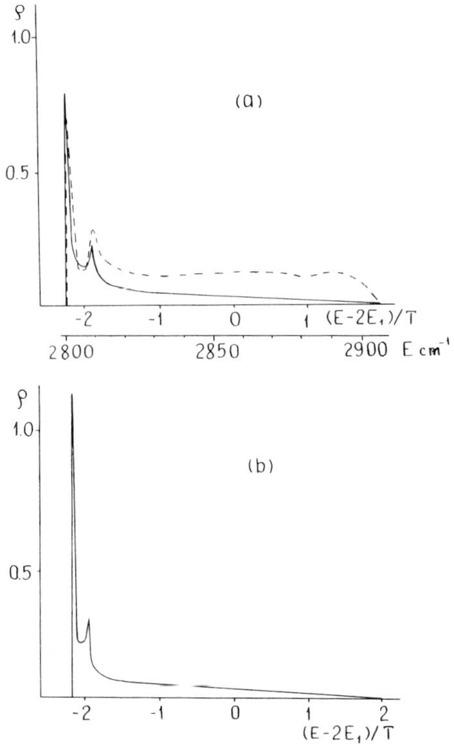

Fig. 25. The density of two-phonon vibrational states in a $^{15}N_x{}^{14}N_{1-x}H_4Cl$ crystal. The density-of-states peak corresponds to quasi-bound biphonon vibrations. Concentrations are (a) $x = 0.1$ and (b) $x = 0.05$.

It is also possible to explain the concentration dependence of Raman scattering spectra. Since Belousov and Pogarev (1978) gave only an oscillogram for the isotope concentration $x = 0.05$, fig. 25b shows only the theoretical curve. The main tendencies, however, of the concentration dependence of the density of two-photon states found by us agree well with the law observed in experimental investigations. In agreement with the experiments, a comparison of figs. 25a and 25 b shows that as the concentration is reduced, the peaks come close together and their intensity increases. Here the low-energy peak shifts in the high-frequency direction and vice versa, but so that their "center of gravity" remains practically stationary.

Hence, comparison with the experimental data enables us to assume that the above-developed theory of two-particle vibrations in disordered crystals having an isotopic impurity [with anharmonicity taken into account and in the case of intermediate values of the parameters ($A \lesssim T$)] can be applied to analyze spectra in the overtone region. The fact that CPA theory provides a rather good description of the special features in the spectra of two-particle vibrations in crystals having isotopic defects is also confirmed by a comparison with the results of a straightforward numerical solution of the system of eqs. (49).

4.5. Two-particle vibrations of disordered one-dimensional crystals: a numerical solution of the secular equation

The theory developed in sections 4.1–4.4 can be supplemented and checked by means of straightforward numerical calculations of vibration spectra by directly solving the system of equations (49) using computer techniques. It is well known that numerous similar calculations were of prime importance for the theory of one-particle states (Elliott et al. 1974). These calculations enabled the degree of validity to be assessed for a number of theoretical models, including the coherent potential approximation method. In this section we intend to discuss the results of analogous calculations, but for two-particle states with anharmonicity taken into account for a linear closed chain of a molecular crystal with an arbitrary concentration of an isotopic impurity. Specifically, we carried out straightforward calculations of two-phonon spectra of vibrations of the above-mentioned crystal with all configurations of arrangements of impurity and base molecules allowed by the given concentration. The number of molecules was taken equal to $N = 20$, the number of independent equations was 110 and use was made of the QREIG program for a BESM-6 computer.

It follows from fig. 5 that in a crystal with an isolated single impurity, three regions can be distinguished on the basis of their Δ values for a given anharmonicity constant A. The first region corresponds to low $|\Delta|$ values $(0 < |\Delta/V| < |\Delta_1/V|)$ where the zone F + L of states for a linear crystal has not yet completely split off the zone of unbound two-phonon states. The second region contains a range of $|\Delta/V|$ values, where the F + L band has already completely split off the zone of dissociated states, but does not yet intersect the zone of biphonon oscillations (i.e. $|\Delta_1/V| < |\Delta/V| < |\Delta_3/V|$). The third and final region contains a range of $|\Delta/V|$ values greater than the quantity $|\Delta_3/V|$, where the zones BP and F + L intersect. The spectrum of a disordered crystal with an arbitrary impurity concentration has a different appearance in each of these Δ ranges.

Concentration dependences of the zone boundaries for two-phonon vibrations, obtained as a result of computer calculations, are presented in figs. 26 and 27. Figure 26 shows the concentration dependence of the zone boundaries for the values of parameters A/V and Δ/V that lie within the first of the ranges mentioned above. The amount of isotopic shift in this case was taken equal to $\Delta/V = -2.66$, i.e the same value as for the linear crystal investigated

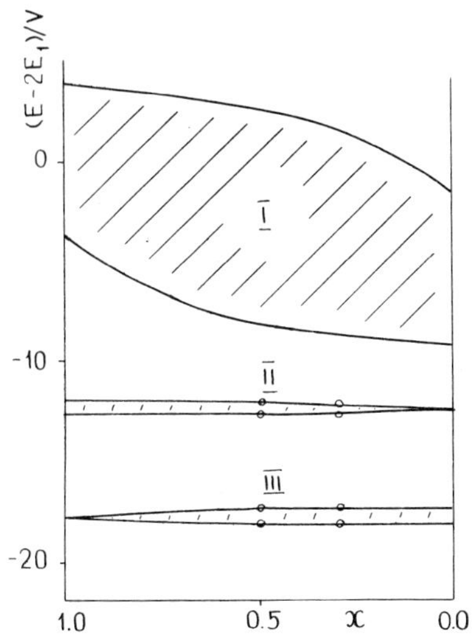

Fig. 26. Concentration dependence of the zone boundaries for two-particle states in a linear disordered crystal, plotted from the results of numerical calculations. The F + F and F + L states are not separated ($A/V = 6.0$ and $\Delta/V = -2.66$).

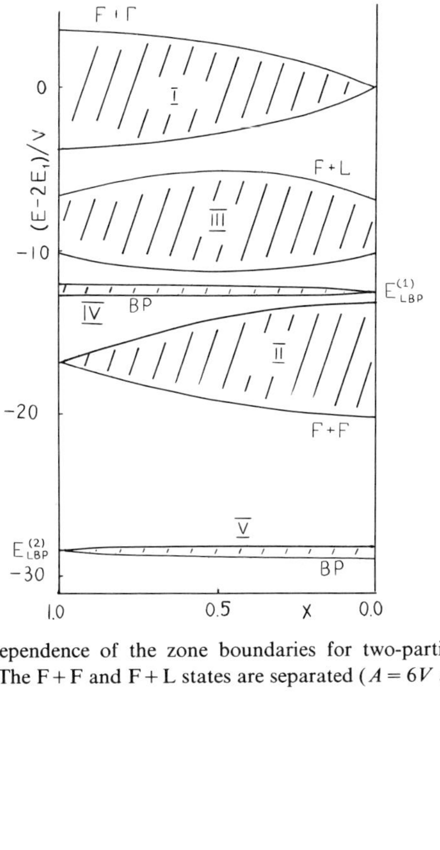

Fig. 27. Concentration dependence of the zone boundaries for two-particle states in a linear disordered crystal. The F + F and F + L states are separated ($A = 6V$ and $\Delta = -8V$).

previously by the coherent potential approximation (see figs. 19 and 20). It follows from fig. 26 that in the given case the joint zone I (unbound states F + F plus the F + L type of zone of states, see fig. 5) displays one-mode behavior, i.e., like a unified band, smoothly shifting in frequency from certain limiting values at $x \ll 1$ to limiting values as $x \to 1$. At the same time, two-mode behavior is observed for biphonons (bands II and III in fig. 26). In the presence of an impurity of low concentration $x \lesssim 1$, the splitting off of the level $E^{(2)}_{\mathrm{LBP}}$ of the local biphonon can be seen. With an increase in the impurity concentration the width of the basic band decreases (band II—zone BP in fig. 5). However, the width of the impurity band, genetically associated with level $E^{(2)}_{\mathrm{LBP}}$ of the local biphonon, increases, and when $x \to 0$, the reverse situation occurs. Here, level $E^{(1)}_{\mathrm{LBP}}$ of the local biphonon lies above the biphonon zone of the basic substance (see also fig. 5 at $\Delta > 0$).

Open circles in fig. 26 show the boundaries of the biphonon zones. They were obtained for the indicated concentrations and for the same values of parameters Δ, A and T by the coherent potential approximation method (see figs. 19 and 20). Although the applicability of the coherent potential approximation to linear crystals is restricted (see Elliott et al. 1974), the agreement between the numerical and analytical results of zone boundary calculations should evidently be regarded as satisfactory.

Shown in fig. 27 is the concentration dependence of the boundaries of all zones of two-phonon vibrations in the second of the above-mentioned ranges of values of the parameter Δ/V. Here there is an appreciable increase in the number of bands of different origins.

The zone F + F of unbound states and the zone F + L of states for these Δ values do not overlap (see the vertical dashed line at $\Delta = \Delta_4$ in fig. 5), and when x is reduced these zones behave in entirely different ways. Zone F + F in this case is drastically narrowed and disappears. At the same time, zone F + L is retained along the whole x range, although it changes its width. At the limiting values $x = 1$ and $x = 0$ its width is equal to the same width value $2T$ of the phonon zone in a perfect crystal, whereas the width of zone F + L increases in the region of intermediate x values because of the peculiar concentration broadening. With a reduction in x the zone F + F of a perfect crystal (with $x = 0$) grows, as does the zone of its free biphonons. The zone of free biphonons of a perfect crystal (with $x = 1$) narrows as x is reduced and disappears at $x = 0$, with a transition into the state of the local biphonon. The relative arrangement of the zones of a perfect crystal (with $x = 1$) can be seen from the intersections of the vertical dashed line in fig. 5 at $\Delta = \Delta_4$. In a similar way the arrangement of the zones in a crystal (with $x = 0$) is shown by the intersections of a vertical dashed line in the same figure, but at $\Delta = \Delta_5 = -\Delta_4$.

Although the evolution of two-particle spectra turns out to be even more complex in crystals having two molecules per unit lattice, the results given above may prove useful in analyzing the general pattern of the spectrum in such a disordered crystal.

4.6. *Another limiting case of an isotopically disordered crystal: the van Kranendonk model*

In this section we shall discuss the special features of a spectrum of two-particle states in the energy region of the combination tone:

$$E(\boldsymbol{k}_1, \boldsymbol{k}_2) = \varepsilon^{(1)}(\boldsymbol{k}_1) + \varepsilon^{(2)}(\boldsymbol{k}_2),$$

where (1) and (2) are indices corresponding to different zones of optical phonons. In this case the equation for the function $\Psi_{\boldsymbol{nm}}$, analogous to eq. (49), is of the form

$$\left[E - \hbar\omega_{\boldsymbol{n}}^{(1)} - \hbar\omega_{\boldsymbol{m}}^{(2)} + 2A\delta_{\boldsymbol{nm}}\right]\Psi_{\boldsymbol{nm}} = \sum_{\boldsymbol{p}}\left[V_{\boldsymbol{np}}^{(1)}\Psi_{\boldsymbol{pm}} + \Psi_{\boldsymbol{np}}V_{\boldsymbol{pm}}^{(2)}\right], \tag{151}$$

where $\hbar\omega_{\boldsymbol{n}}^{(1)}$ is the energy of a quantum of type (1) at site $\boldsymbol{n}$, $\hbar\omega_{\boldsymbol{m}}^{(2)}$ is the energy of a quantum of type (2) at site $\boldsymbol{m}$, and $V_{\boldsymbol{nm}}^{(1)}$ and $V_{\boldsymbol{nm}}^{(2)}$ are the transfer matrices. In an isotopically disordered crystal the energies $\hbar\omega_{\boldsymbol{n}}^{(1)}$ and $\hbar\omega_{\boldsymbol{m}}^{(2)}$ of the two quanta are random quantities. We shall deal, as before, with a binary solid solution, but shall assume that in a perfect crystal the width of the phonon zone (2) is much less than that of phonon zone (1). Such a limiting case (though for perfect crystals) was first investigated by van Kranendonk (1959). Naturally, in this case, there is a substantial simplification: in the system of equations (151) the transfer matrix $V_{\boldsymbol{nm}}^{(2)}$ can be put equal to zero. In this approximation a quantum of type (2) is fixed. If, for example, we assume that this quantum is localized at the site $\boldsymbol{m} = 0$, the wave function

$$\Psi_{\boldsymbol{nm}} = \Psi_{\boldsymbol{n}}\delta_{\boldsymbol{m}0}. \tag{152}$$

Consequently, the system of equations (151) can be rewritten in the form

$$\hat{L}_{\boldsymbol{nm}}\Psi_{\boldsymbol{m}} = 0, \tag{153}$$

where

$$\hat{L}_{\boldsymbol{nm}} = \left[E - \hbar\omega_{\boldsymbol{n}}^{(1)} - \hbar\omega_0^{(2)}\right]\delta_{\boldsymbol{nm}} - V_{\boldsymbol{nm}}^{(1)} + 2A\delta_{\boldsymbol{n}0}\delta_{\boldsymbol{m}0}\delta_{\boldsymbol{nm}}.$$

The corresponding Green function $G_{\boldsymbol{nm}}(E)$ satisfies the equation

$$L_{\boldsymbol{np}}G_{\boldsymbol{pm}} = \delta_{\boldsymbol{nm}}$$

or

$$\left[E - \hbar\omega_n^{(1)} - \hbar\omega_0^{(2)}\right] G_{nm} - \sum_p V_{np}^{(1)} G_{pm} = \delta_{nm}(1 - 2A\delta_{n0}\delta_{m0}G_{00}). \tag{154}$$

Consequently,

$$G_{nm}(E) = F_{nm}\left(E - \hbar\omega_0^{(2)}\right)\left[1 - 2A\delta_{n0}\delta_{m0}G_{00}\left(E - \hbar\omega_0^{(2)}\right)\right], \tag{155}$$

where $F_{nm}(E - \hbar\omega_0^{(2)})$ is the Green function of eq. (154) at $A = 0$.

Putting $n = m = 0$, we find that

$$G_{00}\left(E - \hbar\omega_0^{(2)}\right) = \frac{F_{00}\left(E - \hbar\omega_0^{(2)}\right)}{1 + 2AF_{00}\left(E - \hbar\omega_0^{(2)}\right)}. \tag{156}$$

Thus, for a given distribution of molecules in a solid solution, the Green function of an isotopically disordered crystal in the van Kranendonk model is completely determined. However, it is necessary to compare with the experimental data a Green function $G_{nm}(E)$ that has been averaged over all possible distributions of isotopes in the crystal and over all possible sites of localizations of quantum (2). We shall denote such an averaged function by $\langle\langle G_{nm}(E)\rangle\rangle$. It is clear that this function no longer depends upon the site of localization of the "heavy" quantum, so that

$$\langle\langle G_{nm}(E)\rangle\rangle = \left\langle F_{nm}(E)\frac{1 + 2AF_{00}(E)(1 - \delta_{n0}\delta_{m0})}{1 + 2AF_{00}(E)}\right\rangle \tag{157}$$

and the averaging in eq. (156) is to be carried out only over the configurations of the isotopes. If phonon (2) (i.e. the "heavy" phonon) has no isotopic shift [this being the situation dealt with by Klafter and Jortner (1980)], then the Green function $F_{nm}(E)$ in eq. (157) actually depends [see also eq. (154)] on the difference $E - \hbar\omega_0^{(2)}$.

It follows from eq. (157) that the density of states

$$\begin{aligned}
\rho(E) &= \frac{1}{\pi} \operatorname*{Im}_{\varepsilon \to +0} \left\{\frac{1}{N}\sum_n G_{nn}(E - \mathrm{i}\varepsilon)\right\} \\
&= \frac{1}{\pi}\mathrm{Im}\left\langle \frac{1}{N}\left\{\frac{F_{00}(E - \mathrm{i}\varepsilon)}{1 + 2AF_{00}(E - \mathrm{i}\varepsilon)} + \sum_{n \neq 0} F_{nn}(E - \mathrm{i}\varepsilon)\right\}\right\rangle \\
&= \frac{1}{N}\mathrm{Im}\left\{\frac{1}{N}\left\langle\frac{F_{00}(E - \mathrm{i}\varepsilon)}{1 + 2AF_{00}(E - \mathrm{i}\varepsilon)}\right\rangle + \frac{N-1}{N}\langle F_{00}(E - \mathrm{i}\varepsilon)\rangle\right\}.
\end{aligned}$$

Here we made use of the fact that function $F_{nm}(E - \mathrm{i}\varepsilon)$, determined in the harmonic approximation, should not depend upon the site of localization of the "heavy" phonon. In the CPA, the one-particle Green function $F_{00}(E - \mathrm{i}\varepsilon)$ is determined by the relation

$$\langle F_{00}(E - \mathrm{i}\varepsilon)\rangle = \frac{1}{N}\sum_{k} \frac{1}{E - \hbar\omega_0^{(2)} - \varepsilon^{(1)}(\boldsymbol{k}) - \Sigma(E)},$$

where $\Sigma(E)$ is the self-energy part, determined from the condition of self-consistency, and similar to that used by us previously. We shall refrain from dwelling on this question in more detail because the results of calculations of function $\Sigma(E)$ for one-particle states can be found, for instance, in the previously cited review by Elliott et al. (1974).

If the isotopic shift for the "heavy" phonon is also appreciable, then

$$\langle\langle F_{00}(E - \mathrm{i}\varepsilon)\rangle\rangle = \frac{1}{N}\sum_{k}\left[\frac{1-x}{E - \hbar\omega_{01}^{(2)} - \varepsilon^{(1)}(\boldsymbol{k}) - \Sigma[E - \hbar\omega_{01}^{(2)}]} + \frac{x}{E - \hbar\omega_{02}^{(2)} - \varepsilon^{(1)}(\boldsymbol{k}) - \Sigma[E - \hbar\omega_{02}^{(2)}]}\right].$$

Clearly, in the CPA:

$$\rho(E) = \frac{1}{\pi}\mathrm{Im}\left\{\frac{1}{N}\frac{\langle\langle F_{00}(E - \mathrm{i}\varepsilon)\rangle\rangle}{1 + 2A\langle\langle F_{00}(E - \mathrm{i}\varepsilon)\rangle\rangle} + \frac{(N-1)}{N}\langle\langle F_{00}(E - \mathrm{i}\varepsilon)\rangle\rangle\right\}.$$

5. *Bound three-phonon complexes (triphonons)*

5.1. *The perfect crystal*

At present no information is available to us on any experimental data whatsoever that indicate that bound three-phonon states (triphonons) have been observed. Nevertheless, interest is being shown in such states because, as has been previously pointed out, the two-photon states (biphonons) discussed above are only the very simplest of phonon complexes.

Before turning our attention to an analysis of triphonons it is necessary, first of all, to generalize the form of the Hamiltonian. Assuming, as before, that the basic factor is intramolecular anharmonicity, it is necessary to employ, instead of Hamiltonian (4), a more general Hamiltonian of the form:

$$\hat{H} = \sum_n \hbar\omega B_n^+ B_n + \sum_{n,m}{}' V_{nm} B_n^+ B_m - A \sum_n B_n^{+2} B_n^2 - \tilde{A} \sum_n B_n^{+3} B_n^3. \tag{158}$$

Here the new anharmonicity constant $\tilde{A}$ determines the intensity of the three-particle interaction, which, like the two-particle interaction with constant A, is of the contact type. Constants A and $\tilde{A}$ should obviously be selected in such a way that the energy values of an isolated molecule in a state with two and three quanta coincide with the experimental values E_2^{exp} and E_3^{exp}. This means that the value of A is determined, as before, from the relation $E_2^{\text{exp}} = 2\hbar\omega - 2A$. At the same time, the value of $\tilde{A}$ should be determined from the relation $E_3^{\text{exp}} = 3\hbar\omega - 6(A + \tilde{A})$. In general, A and $\tilde{A}$ are quantities of the same order of magnitude, so that both components of anharmonicity should be taken into account in eq. (158) in investigating the states of triphonons.

Taking the aforesaid into consideration, we shall seek the wave function of three-phonon states in the form

$$|3\rangle = \sum_{nmp} \Psi_{nmp} B_n^+ B_m^+ B_p^+ \,|0\rangle, \tag{159}$$

where Ψ_{nmp} is a function symmetrical with respect to any permutation of the indices $\boldsymbol{n}$, $\boldsymbol{m}$ and $\boldsymbol{p}$. By analogy with the way it was done for two-phonon states, we can obtain a system of equations that determine the quantity Ψ_{nmp}. It is quite clear that these quantities have the meaning of the wave function of three phonons in the coordinate representation.

The substitution of eq. (159) into the Schrödinger equation for the Hamiltonian (158) and the use of the appropriate commutation rules for the Bose operators B_n^+ and B_n lead to the following system of equations:

$$\left[E - 3\hbar\omega + 2A\left(\delta_{nm} + \delta_{np} + \delta_{mp}\right) + 6\tilde{A}\delta_{nm}\delta_{mp}\right]\Psi_{nmp}$$
$$= \sum_l \left[V_{nl}\Psi_{lmp} + V_{ml}\Psi_{nlp} + V_{pl}\Psi_{nml}\right]. \tag{160}$$

If, for the wave function, we now go over to the Fourier representation

$$\Psi_{nmp} = \sum_{k_1 k_2 k_3} \Psi_{k_1 k_2 k_3} \exp\{\mathrm{i}(\boldsymbol{k}_1\boldsymbol{n} + \boldsymbol{k}_2\boldsymbol{m} + \boldsymbol{k}_3\boldsymbol{p})\}, \tag{161}$$

then from eq. (160) we find that the quantities $\Psi_{k_1k_2k_3}$ satisfy the system of equations:

$$\Psi_{k_1k_2k_3} + G_{k_1k_2k_3}(E)$$

$$\times\Big\{(2A/N)\sum_q\left[\Psi_{k_1-q,k_2+q,k_3} + \Psi_{k_1-q,k_2,k_3+q} + \Psi_{k_1,k_2-q,k_3+q}\right]$$

$$+(6\tilde{A}/N^2)\sum_{q_1q_2}\Psi_{k_1+k_2+k_3-q_1-q_2,q_1,q_2}\Big\} = 0, \tag{162}$$

where

$$G_{k_1k_2k_3}(E) = \left[E - 3\hbar\omega - V(k_1) - V(k_2) - V(k_3)\right]^{-1}. \tag{163}$$

Of interest to us in the following are states with a specified value of the total wavevector $\boldsymbol{K} = \boldsymbol{k}_1 + \boldsymbol{k}_2 + \boldsymbol{k}_3$. For these states, as follows from eq. (162), the function $\Psi_{k_1k_2k_3}(\boldsymbol{K})$ is determined by the values of the linear combinations

$$S(\boldsymbol{k}) = N^{-1}\sum_q \Psi_{k,q,K-k-q},$$

$$\Pi = N^{-2}\sum_{q_1q_2}\Psi_{K-q_1-q_2,q_1,q_2} \equiv N^{-1}\sum_k S(\boldsymbol{k}), \tag{164}$$

so that eq. (162) can also be written in the form

$$\Psi_{k_1k_2K-k_1-k_2} + G_{k_1k_2K-k_1-k_2}(E)$$

$$\times\left\{2A\left[S(\boldsymbol{k}_1) + S(\boldsymbol{k}_2) + S(\boldsymbol{K}-\boldsymbol{k}_1-\boldsymbol{k}_2)\right] + 6\tilde{A}\Pi\right\} = 0.$$

Summing this equation over $\boldsymbol{k}_2$ and taking eq. (164) into account, we have the following integral equation for $S(\boldsymbol{k})$:

$$S(\boldsymbol{k})\left[1 + 2AN^{-1}\sum_q G_{k,q,K-k-q}(E)\right] + 4AN^{-1}\sum_q G_{k,q,K-k-q}(E)S(\boldsymbol{q})$$

$$+ 6\tilde{A}\Pi N^{-1}\sum_q G_{k,q,K-k-q}(E) = 0. \tag{165}$$

Below we shall first consider the special case when $A = 0$, but $\tilde{A} \neq 0$. In this case, after making use of the relation between the quantities $S(\boldsymbol{k})$ and Π, we find that the energy of three-phonon states is determined by the equation

$$1 + 6\tilde{A}\frac{1}{N^2}\sum_{q_1 q_2} \frac{1}{E - \varepsilon(\boldsymbol{q}_1) - \varepsilon(\boldsymbol{q}_2) - \varepsilon(\boldsymbol{K} - \boldsymbol{q}_1 - \boldsymbol{q}_2)} = 0$$

or

$$1 + 6\tilde{A}\int \frac{\rho_3(\varepsilon, \boldsymbol{K})}{E - \varepsilon}\,\mathrm{d}\varepsilon = 0, \tag{166}$$

where $\rho_3(\varepsilon, \boldsymbol{K})$ is the density of unbound three-particle states with total wavevector $\boldsymbol{K}$. This equation is analogous to eq. (20) for determining the energy of a biphonon. An essential, though natural, difference, is the appearance of the three-particle density of states in eq. (166); for three-dimensional crystals, this leads to a comparatively large critical (threshold) value of the relation $\alpha = (2\tilde{A}/T_1)_c$ (where T_1 is the phonon zone width), beyond which [i.e. at $(2\tilde{A}/T_1) > (2\tilde{A}/T_1)_c$] the formation of a triphonon becomes feasible. By definition the quantity α is equal to the ratio of the coefficient preceding the integral in eq. (166) to the whole width $3T_1$ on the zone of unbound three-particle states. The analogous ratio for the biphonon [see eq. (20)] $\beta = (A/T_1)_c < \alpha$, because the density $\rho_3(\varepsilon, \boldsymbol{K})$ of three-particle states at the boundary of the three-particle continuum ε_c as $\varepsilon \to \varepsilon_c$ in three-dimensional crystals vanishes more rapidly than the density of two-particle states at the boundary of a two-particle continuum (for one-dimensional and two-dimensional crystals $\alpha = \beta = 0$).

Let us now turn to the situation when both quantities $\tilde{A}$ and A are nonzero, but the value of A is large compared to the phonon zone width T_1. In this case the formation of a triphonon can be conveniently regarded as the result of the bonding of a biphonon and a phonon. Assuming $(T_1^2/A) \ll T_1$, a biphonon, compared to a free phonon, can be assumed to be localized in a certain lattice site. Hence, in this limiting case (high values of A and an arbitrary value of $\tilde{A}$) the formation of a triphonon is formally analogous to the formation of a phonon localized at an isotopic defect. Since the biphonon energy $E_2 \approx 2\hbar\omega - 2A + \mathrm{O}(T_1^2/A)$, and the energy of a triply excited molecule is $E_3^{\exp} = 3\hbar\omega - 6(A + \tilde{A})$, the amount of effective "isotopic" shift $\Delta = -(4A + 6\tilde{A})$. After taking into account the aforesaid, as well as eq. (42), we reach the conclusion that the equation for determining the energy E of a triphonon is of the form

$$1 + (4A + 6\tilde{A})\int \frac{\rho_1(\varepsilon)}{E - 2\hbar\omega + 2A - \varepsilon}\,\mathrm{d}\varepsilon = 0, \tag{167}$$

where $\rho_1(\varepsilon)$ is the density of one-phonon states.

Next we shall show how eq. (167) follows from eq. (165) and, at the same time, we shall take the dispersion of the triphonon into account. In order to do this we shall first discuss the energy zone $\bar{E}(\boldsymbol{K}, \boldsymbol{k})$, determined by the equation

$$F_{\boldsymbol{k}}(\bar{E}, \boldsymbol{K}) \equiv 1 + 2AN^{-1}\sum_{\boldsymbol{q}} G_{\boldsymbol{k},\boldsymbol{q},\boldsymbol{K}-\boldsymbol{k}-\boldsymbol{q}}(\bar{E}) = 0. \tag{168}$$

After comparing this equation with eq. (20), we come to the conclusion that

$$\bar{E}(\boldsymbol{K}, \boldsymbol{k}) = E_2(\boldsymbol{K}-\boldsymbol{k}) + \varepsilon(\boldsymbol{k}), \tag{169}$$

where $E_2(\boldsymbol{K}-\boldsymbol{k})$ is the energy of a biphonon with wavevector $\boldsymbol{K}-\boldsymbol{k}$. Thus the energy zone (169) corresponds to the superposition of the states of a free biphonon and a free phonon.

We shall assume in the following that the anharmonicity constants A and $\tilde{A}$ are such that the triphonon state under discussion is split off the two-particle continuum (169) under conditions in which the dispersion of states in the three-particle continuum can be ignored. It is clear that this is feasible only for triphonons with energy E, for which $|E - 3\hbar\omega| \gg 3T_1$, as shall be assumed below. In this approximation we have

$$\frac{1}{N}\sum_{\boldsymbol{q}} G_{\boldsymbol{k},\boldsymbol{q},\boldsymbol{K}-\boldsymbol{k}-\boldsymbol{q}}(E)S(\boldsymbol{q}) \approx \frac{1}{E-3\hbar\omega}\Pi,$$

$$\frac{1}{N}\sum_{\boldsymbol{q}} G_{\boldsymbol{k},\boldsymbol{q},\boldsymbol{K}-\boldsymbol{k}-\boldsymbol{q}}(E) \approx \frac{1}{E-3\hbar\omega},$$

so that eq. (165) takes the form

$$S(\boldsymbol{k})(E-3\hbar\omega)F_{\boldsymbol{k}}(E, \boldsymbol{K}) + (4A + 6\tilde{A})\Pi = 0.$$

After taking eq. (164) into consideration, we obtain from the preceding equation the dispersion equation

$$1 + \frac{(4A+6\tilde{A})}{N}\sum_{\boldsymbol{k}} \frac{1}{(E-3\hbar\omega)F_{\boldsymbol{k}}(E, \boldsymbol{K})} = 0. \tag{170}$$

Making use of the identity [see eq. (20)]

$$\frac{2A}{N}\sum_{\boldsymbol{q}} \frac{1}{E_2(\boldsymbol{K}-\boldsymbol{k}) - \varepsilon(\boldsymbol{q}) - \varepsilon(\boldsymbol{K}-\boldsymbol{k}-\boldsymbol{q})} = -1,$$

we find that

$$F_k(E, \boldsymbol{K}) \equiv -2A[E - \varepsilon(\boldsymbol{k}) - E_2(\boldsymbol{K} - \boldsymbol{k})]$$
$$\times N^{-1} \sum_{\boldsymbol{q}} [E - \varepsilon(\boldsymbol{k}) - \varepsilon(\boldsymbol{q}) - \varepsilon(\boldsymbol{K} - \boldsymbol{k} - \boldsymbol{q})]^{-1}$$
$$\times [E_2(\boldsymbol{K} - \boldsymbol{k}) - \varepsilon(\boldsymbol{q}) - \varepsilon(\boldsymbol{K} - \boldsymbol{k} - \boldsymbol{q})]^{-1}.$$

Assuming $|E - 3\hbar\omega| \gg 3T_1$ and $A \gg T_1$, we have

$$E - \varepsilon(\boldsymbol{k}) - \varepsilon(\boldsymbol{q}) - \varepsilon(\boldsymbol{K} - \boldsymbol{k} - \boldsymbol{q}) \approx E - 3\hbar\omega,$$
$$E_2(\boldsymbol{K} - \boldsymbol{k}) - \varepsilon(\boldsymbol{q}) - \varepsilon(\boldsymbol{K} - \boldsymbol{k} - \boldsymbol{q}) \approx -2A,$$

so that instead of eq. (170) we obtain the sought-for dispersion equation:

$$1 + \frac{4A + 6\tilde{A}}{N} \sum_{\boldsymbol{k}} \frac{1}{E - E_2(\boldsymbol{K} - \boldsymbol{k}) - \varepsilon(\boldsymbol{k})} = 0, \qquad (171)$$

which coincides to an accuracy of small corrections of the order of $(T_1/A)^2$ with eq. (167). However, in contrast to eq. (167), this equation enables the dispersion curve of a triphonon to be determined, i.e. the dependence $E(\boldsymbol{K})$. We point out that since quantities A and $\tilde{A}$ can, in general, have different signs, both in eq. (167) and in eq. (171), there is a possibility of compensation of the contributions due to two-particle and three-particle anharmonicity (i.e., for example, that $|4A + 6\tilde{A}| \lesssim T_1$ is feasible). In this case the triphonon energy, as follows from eq. (171), is close to the zone of disassociated states [see eq. (169)], and the radius of the triphonon can substantially increase.

For arbitrary A and $\tilde{A}$ values, the solution of the system of equations (165) can be found only if numerical methods are applied. For the case $\tilde{A} = 0$, i.e. for a crystal with zero three-particle anharmonicity [see eq. (158)], this kind of result was discussed by Mattis and Rudin (1984). Employing numerical methods they determined the dependence of the energy of bound three-particle states in a crystal on the anharmonicity constant A. This showed, among other matters, that bound three-phonon states can be formed only when constant A exceeds the critical value $A_c = 0.22\ T_1$. When $A > A_c$ a band of bound three-particle states is split away from the zone of unbound three-particle states. Then, upon increasing the values of the anharmonicity constants, in the region where $A = 0.33\ T_1$ the formation of bound three-particle states, described by Efimov (1970), becomes feasible. Finally, at $A/T_1 = 1/3$, a zone of unbound states [see eq. (169)] is split off the zone of unbound three-particle states. The above-mentioned paper by Mattis and Rudin (1984) does not, however, include an analysis of the dependence of triphonon energy on the wavevector $\boldsymbol{K}$. Since such a dependence is necessary in order to discuss the possibility of the localization of a triphonon at an isotopic defect (see below), we have one comment to make in this connection. It concerns the dependence of the widths of the biphonon and triphonon zones on the parameters A, $\tilde{A}$ and T_1 in the limiting case of strong anharmonicity ($A > T_1$ and $\tilde{A} > 0$). In this case, both the biphonon and the triphonon correspond with high accuracy to the state of a crystal in which two phonons (biphonon) or three phonons (triphonon) are localized at one site, and this excited state of the molecule propagates coherently through the crystal. Therefore, to estimate the zone width of the biphonon or that of the triphonon, it is sufficient to estimate the amplitude of the transition of such a localized state from one lattice site to the neighboring one, using perturbation theory with respect to interaction V_{nm} [see eq. (160)].

For a biphonon such an amplitude is determined by the matrix element

$$W_{nm} = \sum_{\mathrm{S}} \frac{\langle \mathrm{i} | \hat{H} | \mathrm{S} \rangle \langle \mathrm{S} | \hat{H} | \mathrm{f} \rangle}{E_{\mathrm{i}} - E_{\mathrm{S}}},$$

where i is the initial state: both phonons "sit" at site $\boldsymbol{n}$; S is the virtual state: one phonon "sits" at site $\boldsymbol{n}$ and the other at site $\boldsymbol{m}$; f is the final state: both phonons "sit" at site $\boldsymbol{m}$. It is clear that $E_{\mathrm{i}} = 2\hbar\omega - 2A$, $E_{\mathrm{S}} = 2\hbar\omega$ and $\langle \mathrm{i} | \hat{H} | \mathrm{S} \rangle = \langle \mathrm{S} | \hat{H} | \mathrm{f} \rangle = V_{nm}$, so that $W_{nm} = -(V_{nm})^2/2A$. Consequently, the biphonon zone width $T_2 \sim T_1^2/2A$, as previously indicated.

Similarly, for the triphonon

$$W_{nm} = \sum_{\mathrm{S'S''}} \frac{\langle \mathrm{i} | \hat{H} | \mathrm{S'} \rangle \langle \mathrm{S'} | \hat{H} | \mathrm{S''} \rangle \langle \mathrm{S''} | \hat{H} | \mathrm{f} \rangle}{(E_{\mathrm{i}} - E_{\mathrm{S'}})(E_{\mathrm{i}} - E_{\mathrm{S''}})},$$

in which there is one phonon in state S′ at site $\boldsymbol{m}$, two phonons in state S″, etc. An elementary consideration leads to the estimate

$$W_{nm} = \frac{T_1^3}{(4A + 6\tilde{A})^2}.$$

It follows from this estimate that for a triphonon the condition of strong anharmonicity corresponds to the inequality

$$T_1 \ll |4A + 6\tilde{A}|.$$

In the preceding discussion we mentioned triphonons formed from similar phonons. It is clear, of course, that analogous results can also be obtained for triphonons formed of phonons of different kinds. Also of interest is an analysis of the effect of Fermi resonance on triphonon structure (for example, resonance of a phonon and a triphonon, of a biphonon and a triphonon, etc.).

5.2. *Triphonon localization at an isotopic defect*

Let us now discuss a crystal with a defect: an isotopic substitution impurity with an excitation energy shift equal to Δ. An exact solution of the secular equation for a crystal with an impurity is even more difficult than for a perfect crystal. However, with strong anharmonicity where $|4A + 6\tilde{A}| \gg T_1$, when there is strong binding of three phonons and when the vibrational state is of limitingly small radius, the local triphonon energy can be determined in the same way as for the local biphonon [see section 3], from the equation

$$1 = \frac{3\Delta}{N} \sum_{\boldsymbol{K}} \frac{1}{E - E(\boldsymbol{K})}, \tag{172}$$

where $E(\boldsymbol{K})$ is the energy of a triphonon in a perfect crystal. It follows from eq. (172) that the condition for the existence of triphonons localized at impurities is

$$|\Delta| \gg \frac{T_1^3}{|4A + 6\tilde{A}|^2}, \tag{173}$$

in which insignificant numerical factors have been omitted.

As has already been pointed out in section 3, upon strong anharmonicity, local biphonons are formed if condition $|\Delta| \gg T_1^2/A$ is complied with; whereas local phonons are formed if condition $|\Delta| \gg T_1$ is complied with. Thus, for crystals with strong anharmonicity, three consecutively increasing threshold values, $\Delta_c^{(3)}$, $\Delta_c^{(2)}$ and $\Delta_c^{(1)}$, exist for the absolute value of the isotopic shift $|\Delta|$. As the quantity $|\Delta|$ surpasses the values $\Delta_c^{(3)}$, $\Delta_c^{(2)}$ and $\Delta_c^{(1)}$ it is possible for local triphonons, local biphonons and local phonons, respectively to be formed. Here, in certain crystals, the absolute value of the isotopic shift $|\Delta|$ can, in principle, be between the first and second threshold values, i.e. the following inequality can be complied with:

$$\frac{T_1^3}{|4A + 6\tilde{A}|^2} \ll |\Delta| \ll \frac{T_1^2}{2A} \ll T_1. \tag{174}$$

When condition (174) is complied with, triphonons localized at the impurities are formed because condition (173) is also complied with, but the formation of either local biphonons or local phonons is impossible, since the corresponding conditions given above are violated. It was indicated in section 3 that at strong anharmonicity of vibrations, a biphonon can be localized at a defect under conditions at which phonons are not localized. It was mentioned that such a situation had actually been experimentally observed. The relations of the parameters in eq. (174) determine the conditions for the manifestation of analogous laws for triphonons.

Appropriate analysis indicates that for more complicated complexes (four-phonon, five-phonon, etc.) the threshold for the formation of local bound multiphonon states is further lowered because the zone width of the corresponding free vibrations is successively reduced. However, a detailed investigation of this problem requires separate research.

5.3. A new type of three-phonon complexes

It must be emphasized that the existence of a finite one-phonon band width T_1 introduces a unique feature into the many-particle problem in crystals. The parameter governing the formation of bound phonon states is the ratio $\eta = A/T_1$ of the anharmonicity constant A and the one-phonon band width. In crystals, generally speaking, arbitrary values of η are possible, in particular the value $\eta \gg 1$. On the other hand, the "band width" of the kinetic energy of the free particles in vacuum is $T_1 = \infty$, so that an analysis of the bound states of quasiparticles is equivalent to that for free particles in vacuum only for low values of A, when the relation $\eta \ll 1$ is valid. In view of this, one can understand the main reason for the wide variety of multiparticle complexes in crystals as is illustrated below by the triphonon example.

Sections 5.1 and 5.2 (see, also Agranovich and Dubovsky 1986, Mattis and Rudin 1984) dealt with the ground state of a triphonon which corresponds to localization of three optical phonons at one and the same lattice site in the case of large anharmonicity. For this case, the wavevector dependence of the vibrational energy and the type of wave function were determined, and the possibility of triphonon localization at lattice defects was discussed. In particular, numerical calculations of the spectra of bound three-particle complexes in three- and two-dimensional crystals were made by means of the Monte Carlo technique (Mattis and Rudin 1984). These authors pointed out the possibility of the existence of three-particle states analogous to the "Efimov" ones (see Efimov 1970), but this problem was not discussed in detail. For a one-dimensional system, bound states of three particles were studied by Dodd (1970); by solving the Faddeev equations, the binding energy of the ground state was found for a three-particle complex at small values of the interparticle interaction constant ($\eta \ll 1$).

This section shows that, due to the interaction between molecules, triphonons of a new type can exist which have higher energies than the ones studied before, and consequently they may be considered as triphonons in excited states (Agranovich et al. 1986). At strong anharmonicity the ground state of the triphonon studied previously has a binding energy $\approx 6(A+\tilde{A})$. The excited states of the new type of triphonon correspond at $\eta \gg 1$ to the localization of two phonons of the complex at one and the same lattice site, the third phonon of the complex being bound to this pair at one of the neighbouring sites. Since, for these states, localization of two phonons at one site is characteristic (with binding energy $\approx 2A$), in the vibrational spectrum isolated terms corresponding to these states are close to a continuous band of dissociated states biphonon + free phonon (BP + P), also with energy $\approx 2A$. They are split from the edges of this band by a finite value. For a one-dimensional crystal the splitting occurs from both edges of the BP + P band, but for two- and three-dimensional crystals such states were studied only near the low-frequency edge of the BP + P band. The analytical calculation of the triphonon spectrum of this type, for the case of strong anharmonicity ($A \gg T_1$) was confirmed by numerical calculation of a triphonon spectrum at arbitrary anharmonicity. A reverse problem solution method was used: the "spectrum of eigenvalues" of the anharmonicity constant was determined at fixed binding energy of a three-phonon complex.

In order to investigate this type of triphonons let us consider eq. (165) for the wave function $S(\boldsymbol{k})$. Since we are interested in the triphonon with energy near the band BP + P for which the value $\Psi_{nnn} = 0$ we shall assume that $A \neq 0$, $\tilde{A} = 0$. In this case, by means of an identical transformation $S = \tilde{S}F^{-1/2}$ (see eq. (168)) eq. (165) may be reduced to the form

$$\tilde{S}(\boldsymbol{k}_1) = \sum_{\boldsymbol{k}_2} R_{\boldsymbol{k}_1\boldsymbol{k}_2}(A, E)\tilde{S}(\boldsymbol{k}_2),$$

where $R_{\boldsymbol{k}_1\boldsymbol{k}_2}(A, E)$ is a complicated kernel which depends nonlinearly on A and E. This integral equation was solved by Mattis and Rudin (1984) using the iteration scheme by means of the Monte Carlo technique. However, the representation (165) is more convenient for further analysis since it is linear in A. Note also that for a numerical calculation it is convenient to represent eq. (165) as follows:

$$A^{-1}S(\boldsymbol{k}_1) = \sum_{\boldsymbol{k}_2} T_{\boldsymbol{k}_1\boldsymbol{k}_2}(E)S(\boldsymbol{k}_2), \tag{175}$$

where the symmetrical matrix $T_{\boldsymbol{k}_1\boldsymbol{k}_2}(E)$ depends only on E

$$T_{\boldsymbol{k}_1\boldsymbol{k}_2}(E) = -2\delta_{\boldsymbol{k}_1\boldsymbol{k}_2}N^{-1}\sum_{\boldsymbol{k}_2'} G_{\boldsymbol{k}_1\boldsymbol{k}_2'\boldsymbol{K}-\boldsymbol{k}_1-\boldsymbol{k}_2'}(E) - 4N^{-1}G_{\boldsymbol{k}_1\boldsymbol{k}_2\boldsymbol{K}-\boldsymbol{k}_1-\boldsymbol{k}_2}(E). \tag{176}$$

It can be seen from eq. (175) that diagonalization of $T_{k_1k_2}(E)$ at fixed energy E, using standard programs, may yield the "spectrum of eigenvalues" for the value $1/A$, and from this inverse dependence one may then reconstruct the direct dependence $E(A)$. Furthermore, also for simplicity, we shall assume that $K=0$ and take into account only the interaction of the closest neighbours in the lattice and $V_{nm}<0$.

In a one-dimensional crystal the energy of one-phonon oscillations is $\varepsilon(k)=E_0-2V\cos(ka)$. For the high-frequency edge ($ka=\pi$) and the low-frequency edge ($k=0$) of the BP + P band the relationship of the anharmonicity constant $A=\tilde{A}_{1,2}(E)$ and the energy $\varepsilon\equiv 3E_0-E>0$ has [in accordance with (168)] the form

$$\tilde{A}_1(\varepsilon)=\tfrac{1}{2}(\varepsilon+2V),$$

$$\tilde{A}_2(\varepsilon)=\tfrac{1}{2}\sqrt{(\varepsilon-2V)^2-(4V)^2}\,. \tag{177}$$

We seek the solution of eq. (165) with energies close to the edges of the BP + P band at $V\ll\varepsilon$ as an expansion in V/ε up to second order:

$$S_{1,2}(\boldsymbol{k})=S_{1,2}^{(0)}(\boldsymbol{k})+S_{1,2}^{(1)}(\boldsymbol{k})+S_{1,2}^{(2)}(\boldsymbol{k}),$$

$$A_{1,2}(\varepsilon)=\tilde{A}_{1,2}^{(0)}(\varepsilon)+\tilde{A}_{1,2}^{(1)}(\varepsilon)\pm\beta_{1,2}V+A_{1,2}^{(2)}(\varepsilon), \tag{178}$$

where $S^{(n)}$, $A^{(n)}$ are the nth order terms in the expansion and we have introduced the dimensionless parameters $\beta_{1,2}$ which determine the splitting from the edges (177) linear in V. Substitution of (178) into (165) and comparison of the term of the same order in V/ε lead after some calculations to the equations for $S_{1,2}(\boldsymbol{k})$ and $A_{1,2}(\varepsilon)$. It follows from zero-order terms of eq. (165) and eq. (177) that

$$\tilde{A}_1^{(0)}=\tfrac{1}{2}\varepsilon,\qquad \sum_{\boldsymbol{k}_2}S_1^{(0)}(\boldsymbol{k}_2)=\Psi_{nnn}^{(0)}=0. \tag{179a}$$

Comparison of first-order terms in eq. (168) gives the equation

$$S_1^{(0)}(\boldsymbol{k}_1)\left[\tilde{A}_1^{(1)}+\beta_1V+2\tilde{A}_1^{(0)}(\varepsilon)(V/\varepsilon)\cos(\boldsymbol{k}_1a)\right]$$

$$=-2\tilde{A}_1^{(0)}(\varepsilon)\Big\{2(V/\varepsilon)N^{-1}\sum_{\boldsymbol{k}_2}S_1^{(0)}(\boldsymbol{k}_2)\left[\cos(\boldsymbol{k}_2a)+\cos((\boldsymbol{k}_1+\boldsymbol{k}_2)a)\right]$$

$$+N^{-1}\sum_{\boldsymbol{k}_2}S_1^{(1)}(\boldsymbol{k}_2)\Big\}. \tag{179b}$$

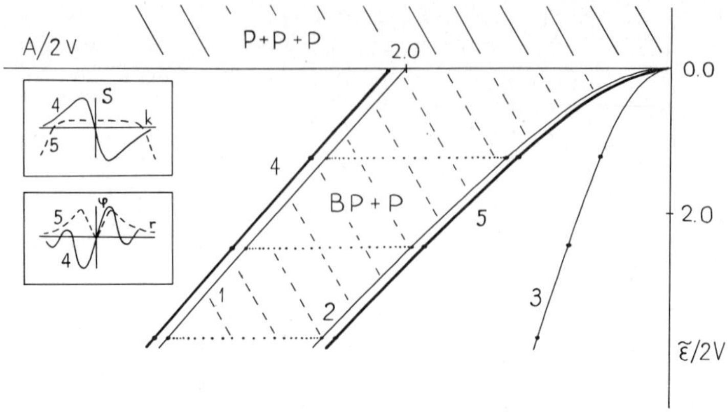

Fig. 28. Dependence of the energy of three-phonon oscillations on the anharmonicity constant in a one-dimensional crystal.

Since $A_1^{(1)} = V$ (see eq. (177)) the antisymmetric solution of eq. (179b) is

$$S_1^{(0)}(\boldsymbol{k}_1) = \frac{\sin(\boldsymbol{k}_1 a)}{1 + \beta_1 + \cos(\boldsymbol{k}_1 a)}; \qquad \sum_{\boldsymbol{k}_2} S_1^{(1)}(\boldsymbol{k}_2) = 0. \tag{179c}$$

Substituting (179c) into (179b) we find that the parameter β_1 must be determined from the appropriate equation

$$1 = 2N^{-1} \sum_{\boldsymbol{k}_2} \frac{\sin^2(\boldsymbol{k}_2 a)}{1 + \beta_1 + \cos(\boldsymbol{k}_2 a)}. \tag{179d}$$

The exact solution upon integration is $\beta_1 = 1/4$. Analogously, β_2 is determined from the equation

$$\left(\tfrac{1}{2} + \beta_2\right)^{-1} - N^{-1} \sum_{\boldsymbol{k}} \left[1 + \beta_2 - \cos(\boldsymbol{k}a)\right]^{-1} \tag{180}$$

and is equal to $\beta_2 = \beta_1 = 1/4$, but the wave function has the form

$$S_2^{(0)}(\boldsymbol{k}) = \left[\tfrac{1}{2} - \cos(\boldsymbol{k}a)\right]\left[1 + \beta_2 - \cos(\boldsymbol{k}a)\right]^{-1}, \tag{181}$$

$$\sum_{\boldsymbol{k}_2} S_2^{(0)}(\boldsymbol{k}_2) = 0.$$

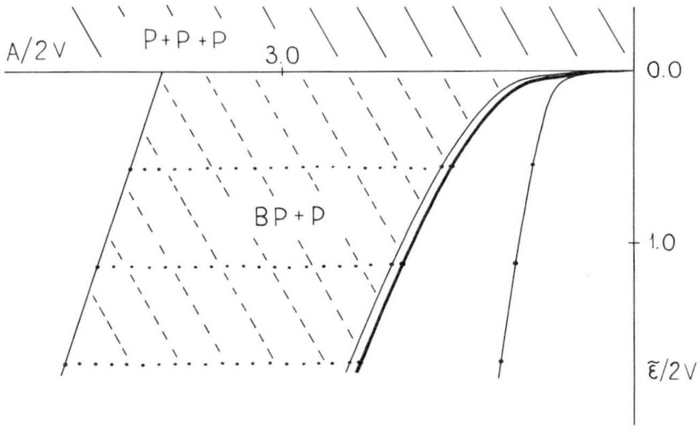

Fig. 29. Dependence of the energy of three-phonon oscillations on the anharmonicity constant in a two-dimensional crystal.

In this approximation the wave function in coordinate representation is

$$\Psi_{nmp} = \delta_{nm}\varphi_{p-n} + \delta_{np}\varphi_{m-n} + \delta_{mp}\varphi_{n-m},$$

$$\varphi_r = N^{-1}\sum_{k} S^{(0)}(k)\exp(ikr). \tag{182}$$

Hence it follows that Ψ_{nmp} differs from zero only when any two of the three indices n, m and p coincide. For the type 1 triphonon, split from the upper edge of the BP + P band, the function of relative motion φ_r is antisymmetric, i.e. $\varphi_r = -\varphi_{-r}$; for the type 2 triphonon, split from the lower edge of the

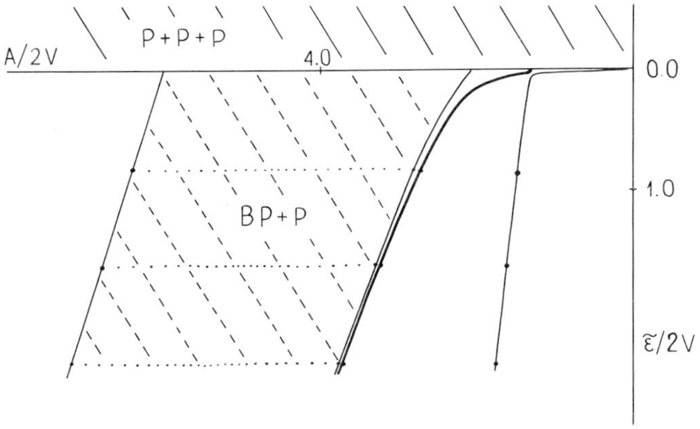

Fig. 30. Dependence of the energy of three-phonon oscillations on the anharmonicity constant in a three-dimensional crystal.

BP + P band, it is symmetric, i.e. $\varphi_r = \varphi_{-r}$. It can easily be shown by integration of (182) that in both cases φ_r equals zero at $r=0$ and has maximum values on the nearest sites. At $|r| \to \infty$, for antisymmetric states φ_r decreases to zero while oscillating, whereas for symmetric states it decreases monotonically. Subsequent calculations yield $A_1^{(2)}(\varepsilon) = -V^2/16\varepsilon$, $A_2^{(2)}(\varepsilon) = -45V^2/16\varepsilon$, i.e. second-order corrections are appreciably less significant for antisymmetric states.

The above analytical consideration at $A \gg V$ is confirmed by the results of the numerical solution of eq. (165) in the form (175) by diagonalization of $T(E)$ and the determination of the "spectrum" $A_i(E)$. In the calculation a mesh of $N_{max} = 512$ points in the first Brillouin zone was used. In fig. 28, the dependence of the binding energy $\tilde{\varepsilon} = \varepsilon - 6V$ of the triphonon on A is shown for a one-dimensional crystal; this dependence was obtained using a computer. Dots represent numerical solutions $A_i(\tilde{\varepsilon})$ (for the sake of clarity only for some $\tilde{\varepsilon}$ and $N < N_{max}$). The axis $\tilde{\varepsilon} = 0$ is the lower edge of the P + P + P band. The curves 1 and 2, i.e. the edges of the BP + P band are determined by the relationship (177), curve 3 gives the triphonon ground-state energy. It can be seen from fig. 28 that there actually exist triphonon states split from both edges of the BP + P band, namely the curves 4 and 5 with constant splitting at large A, almost parallel to 2 and 1, as predicted by the theory. While computing the "spectrum" $A_i(E)$ a computer was used to calculate simultaneously the wave functions S, φ given in the inserts of fig. 28 (solid and dashed lines represent the antisymmetric and symmetric triphonons). From the functions $S(k)$ and φ_r obtained numerically using eqs. (179), (181) and (182) we can confirm what was said above about the physical nature of the triphonon.

The theory developed above was also used to investigate the triphonon in two- and three-dimensional crystals (Agranovich et al. 1986). In that article some features of the spectrum of the new type of triphonon were discussed (see figs. 29 and 30). Investigations of the possibility of observing the above states by optical, electron and neutron spectroscopy methods are important.

References

Agranovich, V.M., 1968, Teoriya Eksitonov (Theory of Excitons) (Nauka Publishers, Moscow) (in Russian).

Agranovich, V.M., 1970, Fiz. Tverd. Tela **12**, 562 [Sov. Phys.-Solid State **12**, 430].

Agranovich, V.M., 1974, Usp. Fiz. Nauk **112**, 143 [Sov. Phys.-Usp. **17**, 103].

Agranovich, V.M., 1983, in: Biphonons and Fermi Resonance in Vibrational Spectra of Crystals, in: Spectroscopy and Excitation Dynamics of Condensed Molecular Crystals, eds V.M. Agranovich and R.M. Hochstrasser (North-Holland, Amsterdam) p. 83.

Agranovich, V.M., 1984, Surface Excitons, in: Surface Excitations, eds V.M. Agranovich and R. Loudon (North-Holland, Amsterdam) p. 513.

Agranovich, V.M., and O.A. Dubovsky, 1980, Mol. Cryst. Liq. Cryst. **57**, 175.

Agranovich, V.M., and O.A. Dubovsky, 1981, Fiz. Tverd. Tela **23**, 2197.
Agranovich, V.M., and O.A. Dubovsky, 1986, Int. Rev. Phys. Chem. **5**, 93.
Agranovich, V.M., and M.D. Galanin, 1982, Electronic Excitation Energy Transfer in Condensed Matter (North-Holland, Amsterdam).
Agranovich, V.M., and I.I. Lalov, 1971, Fiz. Tverd. Tela **13**, 1032 [Sov. Phys.-Solid State **13**, 859].
Agranovich, V.M., and I.I. Lalov, 1985, Usp. Fiz. Nauk **146**, 267 [Sov. Phys.-Usp. **28**, 484].
Agranovich, V.M., O.A. Dubovsky and K.C. Stoichev, 1979, Fiz. Tverd. Tela **21**, 3012 [Sov. Phys.-Solid State **21**, 1795].
Agranovich, V.M., O.A. Dubovsky and A.V. Orlov, 1983, Fiz. Tverd. Tela **25**, 1568.
Agranovich, V.M., O.A. Dubovsky and A.V. Orlov, 1986, Phys. Lett. A **119**, 83.
Austern, N., 1970, Direct Nuclear Reaction Theories (Wiley Interscience, New York).
Belousov, M.V., 1982, Vibrational Frenkel Excitons, in: Excitons, eds E.I. Rashba and M.D. Sturge (North-Holland, Amsterdam) p. 771.
Belousov, M.V., and D.E. Pogarev, 1978, Pis'ma v Zh. Eksp. & Teor. Fiz. **28**, 692 [JETP Lett. **28**, 646].
Belousov, M.V., B.E. Wolf, E.A. Ivanova and D.E. Pogarev, 1982, Pis'ma v Zh. Eksp. & Teor. Fiz. **35**, 457.
Brout, R., and W.M. Visscher, 1962, Phys. Rev. Lett. **9**, 54.
Butler, S.T., 1957, Nuclear Stripping Reactions (Wiley, New York).
Chang, I.F., and S.S. Mitra, 1971, Adv. Phys. **20**, 359.
Dodd, L.R., 1970, J. Math. Phys. **11**, 207.
Drchal, V., and B. Velicky, 1976, J. Phys. Chem. Solids **37**, 655.
Dubovsky, O.A., 1985, Solid State Commun. **54**, 261.
Dubovsky, O.A., and Yu.V. Konobeev, 1965, Fiz. Tverd. Tela **7**, 946.
Efimov, V., 1970, Phys. Lett. B **33**, 563.
Elliott, R.J., J.A. Krumhansl and P.L. Leath, 1974, Rev. Mod. Phys. **46**, 465.
Gorelik, V.S., G.G. Mitin and M.M. Sushchinsky, 1974, Fiz. Tverd. Tela **16**, 1562.
Gorelik, V.S., G.G. Mitin and M.M. Sushchinsky, 1975, Zh. Eksp. & Teor. Fiz. **69**, 823.
Gradshtein, I.S., and I.M. Ryzhik, 1971, Tablitsy Integralov, Summ, Ryadov i Proizvedenii (Tables of Integrals, Sums, Series and Products) (Nauka, Moscow) (in Russian).
Herzberg, G., 1945, Infrared and Raman Spectra of Polyatomic Molecules (Van Nostrand Co., New York).
Jindal, V.K., and K.N. Pathak, 1977, Phys. Rev. B **15**, 1202.
Kagan, Yu.M., and Ya.A. Ioselevsky, 1962, Zh. Eksp. & Teor. Fiz. **42**, 259.
Klafter, J., and J. Jortner, 1980, J. Chem. Phys. **47**, 25.
Lalov, I.I., 1974, Fiz. Tverd. Tela **16**, 2494.
Landau, L.D., and E.M. Lifshitz, 1959, Quantum Mechanics (Pergamon Press, New York).
Leath, P.L., 1973, J. Phys. C **6**, 1559.
Lifshitz, I.M., 1947, Zh. Eksp. & Teor. Fiz. **17**, 1017.
Lifshitz, I.M., 1956, Nuovo Cimento Suppl., **3**, 716.
Lifshitz, I.M., S.A. Gredeskul and L.A. Pastur, 1982, Vvedeniye v Teoriyu Neuporyadochenykh Sistem (Introduction to the Theory of Disordered Systems) (Nauka, Moscow) (in Russian).
Lisitsa, M.P., A.M. Yaremko and A.P. Kucherov, 1978, Fiz. Tverd. Tela **20**, 3276.
Maradudin, A.A., 1966, Theoretical and Experimental Aspects of the Effects of Point Defects and Disorder on the Vibrations of Crystals (Academic Press, New York, London).
Mattis, D., and S. Rudin, 1984, Phys. Rev. Lett. **52**, 755.
Mitin, G.G., V.S. Gorelik and M.M. Sushchinsky, 1974, Fiz. Tverd. Tela **16**, 2956.
Montroll, E.W., and R.B. Potts, 1955, Phys. Rev. **100**, 525.
Montroll, E.W., and R.B. Potts, 1956, Phys. Rev. **102**, 72.
Onodera, Y., and Y. Toyozawa, 1968, J. Phys. Soc. Jpn. **24**, 341.
Pathak, K.N., and V.K. Jindal, 1974, Lett. Nuovo Cimento **10**, 409.

Price, W.C., W.F. Sherman and G.R. Wilkinson, 1960, Proc. R. Soc. **255**, 5.
Ruvalds, J., and A. Zawadowski, 1970, Phys. Rev. B **2**, 1172.
Soven, P., 1967, Phys. Rev. **156**, 809.
Taylor, D.W., 1967, Phys. Rev. **156**, 1017.
Van Kranendonk, J., 1959, Physica **25**, 1080.
Velicky, B., S. Kirkpatrick and A. Ehrenreich, 1968, Phys. Rev. **175**, 747.
Yonezawa, F., and T. Matsubara, 1966a, Prog. Theor. Phys. Jpn. **35**, 357.
Yonezawa, F., and T. Matsubara, 1966b, Prog. Theor. Phys. Jpn. **35**, 759.

AUTHOR INDEX

Abbi, S.C., *see* Soni, R.K. 127
Abdullaev, G.B. 111
Ablyazov, N.N. 157
Abragam, A. 260
Abrahams, E. 275
Abrikosov, A.A. 284
Adams, A.R., *see* Greene, P.D. 135, 165
Agranovich, V.M. 218, 224, 225, 300, 301, 304, 310, 313, 318–320, 330, 331, 336, 337, 349, 364, 365, 374, 391, 396
Agrawal, B.K. 56, 85
Aharony, A. 210
Aharony, A., *see* Fishman, S. 210, 211
Ahlgren, D.C. 112, 279
Ahmad, C.N., *see* Greene, P.D. 135, 165
Akperov, Ya.G., *see* Tzarenkov, B.V. 4
Alben, R. 187, 207
Alben, R., *see* Thorpe, M.F. 187, 188, 207
Aleshchenko, Y.A., *see* Vodopyanov, L.K. 124
Aleshenko, U.A., *see* Allakhverdiev, K.R. 126
Alexandre, F., *see* Jusserand, B. 100
Alferov, Zh.I. 3, 4, 135, 141
Aliev, M.I., *see* Gasanly, N.M. 65, 70, 104, 106
Aliev, R.A. 112
Alieva, L.N., *see* Belenkii, G.L. 112
Alimonda, A.S., *see* Lucovksy, G. 98
Alimonda, A.S., *see* Lucovsky, G. 107
Allakhverdiev, K.R. 111, 112, 113, 126
Allakhverdiev, K.R., *see* Abdullaev, G.B. 111
Allakhverdiev, K.R., *see* Aliev, R.A. 112
Allen, P.B. 281
Allen, S.J. 194
Als-Nielsen, J., *see* Birgeneau, R.J. 204
Altshuler, B.L. 164
Amirtharaj, M.P., *see* McKnight, S.W. 98
Amirtharaj, P.M. 97, 107, 108, 125
Amirtharaj, P.M., *see* Tiong, K.K. 97
Amzallag, E., *see* Gebicki, W. 98
Amzallag, E., *see* Picquart, M. 98
Anacker, L., *see* Kopelman, R. 281
Anderson, P.W. 9, 16, 164, 274
Anderson, P.W., *see* Abrahams, E. 275
Andres, K., *see* Rosenbaum, T.F. 162
Anedda, A. 112
Angress, J.F. 88, 89
Angress, J.F., *see* Gledhill, G.A. 88
Anselm, A.I. 9
Antoniou, P.D., *see* Economou, E.N. 280
Arai, T., *see* Wakamura, K. 112
Araujo, C.B. de, *see* Montarroyos, E. 197
Areshkin, A.G. 152
Areshkin, A.G., *see* Mach, R. 4
Areshkin, A.G., *see* Maslov, A.U. 138
Areshkin, A.G., *see* Suslina, L.D. 4
Areshkin, A.G., *see* Suslina, L.G. 135, 152, 161
Aronov, A.G., *see* Altshuler, B.L. 164
Artamonov, V.V. 94
Artamonov, V.V., *see* Azhnyuk, Y.N. 128
Askenazy, S., *see* Portal, J.C. 107
Askin, M., *see* Feldman, D.W. 75, 85
Auderset, H., *see* Geick, R. 86
Austern, N. 334
Avakian, P. 281
Azhnyuk, Y.N. 128

Baars, J. 96
Babaev, S.S., *see* Allakhverdiev, K.R. 111
Bairamov, B.H. 98, 102
Bairamov, B.Th., *see* Jahne, E. 102
Bajor, G., *see* McGlinn, T.C. 125
Baker, J.M. 198

Bakhyshov, N.A. 112
Bakhyshov, N.A., *see* Allakhverdiev, K.R. 126
Baldareschi, A. 142, 143
Balkanski, M. 45, 95, 100
Balkanski, M., *see* Beserman, R. 102, 123
Balkanski, M., *see* Charfi, F. 101
Balkanski, M., *see* Hirlimann, C. 102, 105, 123
Balkanski, M., *see* Nusimovici, M.A. 95
Balkanski, M., *see* Picquart, M. 98
Balkanski, M., *see* Polian, A. 96, 97
Balkanski, M., *see* Soni, R.K. 127
Bank, H., *see* Doberer, U. 226, 237, 238
Bansal, M.L., *see* Sahni, V.C. 112
Bansil, A., *see* Prasad, R. 53
Baranovskii, S.D. 13, 141, 152
Barker, A.S. 18, 38, 42, 108, 109, 110, 180, 199
Barker, A.S., *see* Verleur, H.W. 18, 58, 59, 65, 71, 72, 73, 94, 95, 97, 104, 105, 108, 124
Barnett, S.A., *see* McGlinn, T.C. 125
Barta, C. 112
Bartzokas, A. 112
Bauhofer, W. 87, 88
Bauhofer, W., *see* Genzel, L. 82, 83
Bauhofer, W., *see* Perry, C.H. 88
Bechstedt, F., *see* Kleinert, P. 50
Becker, C.R. 181, 188–190
Becker, C.R., *see* Geis, G. 188–190, 190
Becker, C.R., *see* Richter, W. 112
Becker, E.D., *see* Farrar, T.G. 260
Bedel, E. 102, 103, 106, 125
Behera, S.N. 77
Belanger, R.M. 191, 192
Belanger, R.M., *see* Sanders, R.W. 191
Belenki, G.L., *see* Abdullaev, G.B. 111
Belenkii, G.L. 112
Belousov, M.V. 18, 285, 287, 301, 320, 321, 359, 374–377
Benchimol, J.L., *see* Soni, R.K. 127
Benderskii, V.A. 238
Benedek, G. 57, 58
Benk, H. 236
Benner, R.E., *see* Cohen, R.M. 125
Berezinskii, V.L. 275
Bergstresser, T.K., *see* Cohen, M.L. 135
Bergstresser, T.K., *see* Van Vechten, J.A. 145
Berndt, V. 98, 135, 169
Berolo, O. 145
Bertrand, D. 210
Bertrand, D., *see* Mischler, G. 128, 210
Bertrand, D., *see* Tuchendler, J. 210
Beserman, R. 94, 95, 102, 103, 105, 123, 126
Beserman, R., *see* Balkanski, M. 95
Beserman, R., *see* Hirlimann, C. 102, 105, 123
Beserman, R., *see* Newman, K.E. 125
Beserman, R., *see* Schmeltzer, D. 76, 82, 93, 94, 105
Beserman, R., *see* Teicher, M. 105
Bessolov, V.N., *see* Bairamov, B.H. 98
Besson, J.M., *see* Balkanski, M. 95
Betz, E. 203
Bevaart, L. 203
Bhatt, R.N., *see* Rosenbaum, T.F. 162
Bilz, H. 39, 42
Birgeneau, R.J. 204, 205, 207
Birgeneau, R.J., *see* Cowley, R.A. 181, 182, 187, 188, 193, 197, 206, 207
Birgeneau, R.J., *see* Wong, P. 203, 210, 211
Birman, J.L., *see* Ganguly, B.N. 50
Birman, J.L., *see* Nusimovici, M.A. 95
Biryulin, Yu.F. 101
Blackman, M. 227
Bletskan, D.I., *see* Vodopyanov, L.K. 112, 113
Blinov, A.M., *see* Vodopyanov, L.K. 93
Bloor, D. 180
Blume, M., *see* Birgeneau, R.J. 204
Bodnar, I.V. 112
Bodnar, I.V., *see* Azhnyuk, Y.N. 128
Bodnar, I.V., *see* Sirota, N.N. 106
Bokhenkov, E.I., *see* Natkaniec, I. 286
Bongiovanni, G., *see* Anedda, A. 112
Bonneville, R. 46, 48, 61, 62, 65, 100
Bootz, B. 91
Born, M. 44
Bosomworth, D.R., *see* Sennett, C.T. 98
Bottger, H. 71
Bottler, B.J. 269, 270
Boudou, A. 112
Bour, D.P. 128
Boyce, J.B., *see* Mikkelsen, J.C. 125
Brafman, O. 40, 93
Brafman, O., *see* Hayek, M. 95, 111
Brafman, O., *see* Livescu, G. 91, 124
Brafman, O., *see* Parrish, J.F. 95
Brafman, O., *see* Vardeny, Z. 91
Braunstein, R. 85
Breiland, W.G. 264, 267

Breiland, W.G., *see* Harris, C.B. 261, 263, 268
Brenner, H.C. 271
Brenner, H.C., *see* Breiland, W.G. 264
Breton, P., *see* Lemieux, M.A. 123
Breton, P., *see* Tremblay, A.M.S. 123
Brezin, E. 146, 152
Brice, J.F., *see* Vergnat-Grandjean, D. 108
Brikenstein, V.Kh., *see* Benderskii, V.A. 238
Brock, J.C., *see* Brenner, H.C. 271
Brodsky, M.H. 62, 64, 104
Brodsky, M.H., *see* Lucovsky, G. 80, 81, 101, 102
Broude, V.L. 18, 218, 223, 232, 233, 234, 237, 238, 240, 242, 243, 244, 248, 249, 251, 255, 279
Brout, R. 300
Brovchenko, I.V. 242
Brunel, L.C., *see* Bureau, J.C. 112
Buchanan, D.N.E., *see* Werthein, G.K. 198
Buchanan, M. 194, 195
Buchheiser, K., *see* Sobotta, H. 98
Buckley, M.J. 258
Budennaya, L.D., *see* Mityagin, Y.A. 95
Bureau, J.C. 112
Burlakov, V.M. 127, 128
Burlakov, V.M., *see* Nagiev, V.M. 32
Burland, D.M. 283, 285
Burnham, R.D., *see* Lucovksy, G. 98
Burnham, R.D., *see* Lucovsky, G. 107
Burns, G. 112
Burstein, E., *see* Lucovsky, G. 80, 81
Burstein, E., *see* Pinczuk, A. 49, 51
Burton, C.H., *see* Wiltshire, M.C.K. 210, 211
Buss, D.D., *see* Carter, D.L. 96
Butler, S.T. 334
Butler, W.H., *see* Nickel, B.G. 198
Butt, N.M., *see* Renker, B. 88
Buyers, W.J.L. 186, 193, 194, 196, 198, 207
Buyers, W.J.L., *see* Buchanan, M. 194, 195
Buyers, W.J.L., *see* Coombs, G.J. 181, 187, 197, 198
Buyers, W.J.L., *see* Cowley, R.A. 181–184, 190, 191
Buyers, W.J.L., *see* Svensson, E.C. 193
Byra, W.J. 75, 85

Cable, J.W., *see* Wilkinson, M.K. 204, 205
Cadien, K.C., *see* Krabach, T.N. 109
Cardona, M., *see* Kumazaki, K. 128
Cardona, M., *see* Renucci, M.A. 74, 84, 85
Cardona, M., *see* Shen, S.C. 85
Carles, R. 100, 106
Carles, R., *see* Bedel, E. 102, 103, 106, 125
Carles, R., *see* Saint-Cricq, N. 99
Carles, R., *see* Zwick, A. 110, 111
Carles, R.N. 100
Carlone, C., *see* Jandl, S. 113
Carlone, C., *see* Provencher, R. 110
Carnall, E., *see* Brafman, O. 40, 93
Carter, D.L. 96
Catalano, E., *see* Enders, B. 180, 181, 191, 192
Cerdeira, F. 113
Cerdeira, F., *see* Lemos, V. 112
Chambers, W.G., *see* Angress, J.F. 88, 89
Chang, I.F. 18, 20, 42, 45, 65, 77, 95, 301
Chang, I.F., *see* Brafman, O. 40, 93
Chang, I.F., *see* Parrish, J.F. 95
Chang, R.K. 100, 108
Chao, K.S., *see* Syme, R.W.G. 128
Çhaplik, A.V., *see* Petukhov, V.B. 8
Charfi, F. 101
Chen, M.F., *see* Brodsky, M.H. 62, 64, 104
Chen, M.F., *see* Lucovsky, G. 101, 102, 104–106
Chen, Y.S. 65, 71, 105, 244
Cheng, K.Y., *see* Lucovksy, G. 100
Cherng, M.J., *see* Cohen, R.M. 125
Chernomorets, M.P., *see* Ostapenko, N.I. 227
Chevallier, J., *see* Beserman, R. 102, 123
Chevallier, J., *see* Galtier, P. 103
Chevallier, J., *see* Hirlimann, C. 102
Chevallier, J., *see* Vodopyanov, L.K. 92
Chevallier, J., *see* Zigone, M. 92
Chicotka, R.J., *see* Lucovsky, G. 101, 102
Chigirev, A.R., *see* Osad'ko, I.S. 234
Chikotka, P.J., *see* Onton, A. 4
Chronister, E.L., *see* Dlott, D. 287
Chu, J.H., *see* Shen, S.C. 97, 124
Ciepielewski, P. 109
Clark, J.D., *see* Angress, J.F. 88, 89
Cochanour, C.R., *see* Burland, D.M. 283, 285
Cohen, E. 4, 152, 157, 162
Cohen, E., *see* Birgeneau, R.J. 204, 205, 207
Cohen, M.H. 275
Cohen, M.L. 135
Cohen, R.M. 125
Coldwell-Horsfall, R.A., *see* Xinh, N.X. 57, 58

Colson, S.D. 233, 249, 255, 278
Coombs, G.J. 181, 187, 197, 198
Coombs, G.J., *see* Lockwood, D.J. 199
Cooper, D.E., *see* Burland, D.M. 283, 285
Copland, G.M., *see* Bloor, D. 180
Cosand, A.E. 85
Cottam, M.G., *see* Lockwood, D.J. 128, 211
Cottam, M.G., *see* Mischler, G. 128, 210
Cowley, R.A. 181–184, 182, 187, 188, 190, 191, 193, 197, 206, 207
Cowley, R.A., *see* Coombs, G.J. 181, 187, 197, 198
Cowley, R.A., *see* Lockwood, D.J. 199
Cowley, R.A., *see* Svensson, E.C. 193
Cruege, F., *see* Gard, P. 126
Curti, M., *see* Razzetti, C. 127

Dacol, F.H., *see* Burns, G. 112
Damen, T.C., *see* Hurrell, J.P. 89
Damen, T.C., *see* Shah, J. 100
Daniels, P.R., *see* Katnani, A.D. 139
Davies, E.A., *see* Lucovsky, G. 39, 92
Davis, E.A., *see* Mott, N.F. 161, 280
Davydov, A.S. 217, 218, 219, 227
Day, P., *see* Haywood, S.K. 211
Day, P., *see* Wood, T.E. 208, 211
de Araujo, C.B., *see* Vianna, S. 126
de Haas, W.J., *see* Obreimov, I.B. 217
de Jongh, L.J., *see* Bevaart, L. 203
de Leeuw, S.W., *see* Thorpe, M.F. 122
Dean, P. 249
Dean, P.J. 135
Decola, R.L. 287
Della Valle, R.G., *see* Rigini, R. 287
Delpech, P., *see* Jusserand, B. 100
Demchuk, M.I., *see* Bairamov, B.H. 102
Deslandes, J. 110, 127
Dietrich, G., *see* Mitlehner, H. 194
Dietz, R.E., *see* Moch, P. 191
Dietz, R.E., *see* Parisot, G. 199
Digiovanni, A.E., *see* Shah, J. 100
Ditzenberger, J.A., *see* Barker, A.S. 109, 110, 199
Dlott, D. 287
Dlott, D.D. 259
Doberer, U. 226, 237, 238
Dobrzhanskii, G.F., *see* Barta, C. 112
Dodd, L.R. 391
Dolganov, V.K. 233
Dorner, D., *see* Natkaniec, I. 286
Dow, J.D., *see* Kobayashi, A. 122
Dow, J.D., *see* Newman, K.E. 125
Dow, J.D., *see* O'Hara, M.J. 75
Drchal, V. 314, 358, 359, 362, 364, 371
Drexel, W., *see* Beserman, R. 94
Dubovsky, O.A. 320, 345
Dubovsky, O.A., *see* Agranovich, V.M. 330, 337, 364, 365, 374, 391, 396
Dugautier, C., *see* Moch, P. 191
Dugautier, C., *see* Parisot, G. 199
Durr, U. 191, 192
Durr, U., *see* Betz, E. 203
D'yakonov, M.I. 135
Dzhavadov, B.D., *see* Gasanly, N.M. 112
Dzhavadov, B.M., *see* Vinogradov, E.A. 112
Dzyaloshinskii, I.E., *see* Abrikosov, A.A. 284
Dzyub, I.P. 187, 203

Economou, E.N. 280
Efendiev, Sh.M., *see* Bakhyshov, N.A. 112
Efendiev, Sh.M., *see* Nagiev, V.M. 32
Efimov, V. 388, 391
Efros, A.L. 13
Efros, A.L., *see* Ablyazov, N.N. 157
Efros, A.L., *see* Baranovskii, S.D. 13, 141, 152
Efros, A.L., *see* Gel'mont, B.L. 166, 168
Efros, A.L., *see* Shklovskii, B.I. 13, 156, 161, 164, 167
Efros, A.L., *see* Shlimak, I.S. 138, 165
Efros, A.L., *see* Skal, A.S. 167
Efros, Al.L., *see* Raikh, M.E. 161
Ehrenreich, A., *see* Velicky, B. 301
Ehrenreich, H. 4
El-Sabbahy, A.N., *see* Greene, P.D. 135, 165
Eliashberg, M.B., *see* Stekhanov, A.I. 89
Elliott, R.J. xv, 19, 55, 81, 87, 135, 181, 184–186, 197, 249, 301, 358, 359, 361, 366, 378, 380, 383
Elliott, R.J., *see* Buchanan, M. 194, 195
Elliott, R.J., *see* Buyers, W.J.L. 186, 193, 194, 196, 198, 207
Elliott, R.J., *see* Hayes, W. 196, 202
Elliott, R.J., *see* Tahir-Kheli, R.A. 198
Elliott, R.J., *see* Ziman, T.A.L. 206, 207
Enders, B. 180, 181, 191, 192
Esaki, L., *see* Kawamura, H. 99
Esaki, L., *see* Tsu, R. 63, 99
Evdokimov, V.M., *see* Gasanly, N.M. 65, 70, 104, 106
Evtichiev, V.G. 4

Falikov, L.M., *see* Balkanski, M. 100
Farge, Y. 39
Farr, M.K. 76
Farrar, T.G. 260
Faurie, J.P., *see* Olego, D.J. 123
Fayer, M.D. 271
Fayer, M.D., *see* Brenner, H.C. 271
Fayer, M.D., *see* Burland, D.M. 283, 285
Fayer, M.D., *see* Dlott, D.D. 259
Faymonville, R., *see* Mutzenich, G. 112
Fedorov, A.A., *see* Ovsyankin, V.V. 163, 281
Fedorov, D.L., *see* Areshkin, A.G. 152
Fedorov, D.L., *see* Maslov, A.U. 138
Fedorov, D.L., *see* Suslina, L.D. 4
Fedorov, D.L., *see* Suslina, L.G. 135, 152, 161
Feldkamp, L.A., *see* Venkataraman, G. 46
Feldman, D.W. 75, 85
Fert, A.R., *see* Bertrand, D. 210
Fert, A.R., *see* Tuchendler, J. 210
Fertel, J.H. 64, 65, 88, 89, 124
Fetter, A.L. 47
Feynman, R.P. 261
Fibich, M., *see* Beserman, R. 103
Fibich, M., *see* Herscovici, C. 76, 81
Finkenrath, H. 109
Fishman, S. 210, 211
Fishman, S., *see* Aharony, A. 210
Fleury, P.A. 179, 180, 200, 204
Flögel, P., *see* Mach, R. 4
Flögel, P., *see* Permogorov, S. 152, 162, 163
Foner, S. 197, 198
Fontana, M.P., *see* Farge, Y. 39
Fortin, E., *see* Anedda, A. 112
Fracassi, P.F., *see* Rigini, R. 287
Francis, A.H. 257–260
Franck, J. 228
Franz, G., *see* Finkenrath, H. 109
Frenkel, J. 217
Freund, G.A. 128
Friesner, R. 239
Frikee, E., *see* Bevaart, L. 203
Fujii, A. 91
Fujiwara, T. 205–207
Fujiwara, T., *see* Tahir-Kheli, R.A. 198
Fukumoto, T., *see* Nakashima, S. 92, 94

Gadgiev, A.R., *see* Gel'mont, B.L. 166, 168
Galanin, M.D., *see* Agranovich, V.M. 301
Galaska, R.R., *see* Venugopalan, S. 98
Galtier, P. 103
Gamarts, E.M. 111
Gan, T.T., *see* Lee, K.O. 242, 244
Ganguly, B.N. 50
Garbuzov, D.Z., *see* Alferov, Zh.I. 4
Garbuzov, D.Z., *see* Evtichiev, V.G. 4
Gard, P. 126
Garg, A.K. 128
Gasanly, N.M. 65, 70, 104, 106, 111, 112, 126, 128
Gasanly, N.M., *see* Bakhyshov, N.A. 112
Gasanly, N.M., *see* Musaeva, L.G. 112
Gasanly, N.M., *see* Vinogradov, E.A. 112
Gashimzade, F.M., *see* Allakhverdiev, K.R. 126
Gebicki, W. 93, 97, 98
Gebicki, W., *see* Picquart, M. 98
Geick, R. 86
Geick, R., *see* Becker, C.R. 181, 188–190
Geick, R., *see* Geis, G. 188 190, 190
Geick, R., *see* Mitlehner, H. 194
Geis, G. 188–190, 190
Gel'mont, B.L. 166, 168
Gentry, S.T. 277, 278
Genzel, L. 71, 82, 83, 110
Genzel, L., *see* Bauhofer, W. 87, 88
Genzel, L., *see* Perry, C.H. 88
George, S., *see* Colson, S.D. 278
Giehler, M., *see* Jahne, E. 68, 72, 102
Gielisse, P.J. 107
Gilat, G., *see* Beserman, R. 102, 105
Ginzburg, V.L. 224
Ginzburg, V.L., *see* Agranovich, V.M. 225
Giriat, W., *see* Peterson, D.L. 124
Gledhill, G.A. 88
Gledhill, G.A., *see* Angress, J.F. 88, 89
Glockner, E. 237
Gobeau, J. 109
Godzhaev, M.M., *see* Allakhverdiev, K.R. 112
Goede, O. 4, 152
Goede, O., *see* Suslina, L.G. 4, 156
Gogolin, A.A., *see* Rashba, E.I. 275
Golubev, L.V. 112
Golubev, L.V., *see* Allakhverdiev, K.R. 112
Golubev, L.V., *see* Plotnichenko, V.G. 95
Golubev, L.V., *see* Vodopyanov, L.K. 112, 113
Goncharov, A.F., *see* Gasanly, N.M. 111, 126
Goncharov, A.F., *see* Vinogradov, E.A. 112

Gorbatkin, S., *see* McGlinn, T.C. 125
Gorelenok, A.G., *see* Evtichiev, V.G. 4
Gorelik, V.S. 375
Gorelik, V.S., *see* Mitin, G.G. 375
Gor'kov, L.P. 275
Gor'kov, L.P., *see* Abrikosov, A.A. 284
Gorochov, O., *see* Gard, P. 126
Gorska, M. 95, 96
Gosso, J.P. 193, 194, 199
Gosso, J.P., *see* Moch, P. 191
Gouskov, A., *see* Cerdeira, F. 113
Grad, J., *see* Gorska, M. 95, 96
Gradshtein, I.S. 314, 368
Gradstein, I.S. 151
Grant, R.W., *see* Kowalczyk, S.P. 139
Gredeskul, A.S., *see* Lifshitz, I.M. 12, 13
Gredeskul, S.A., *see* Lifshitz, I.M. 12, 301
Greene, J.E., *see* Beserman, R. 126
Greene, J.E., *see* Krabach, T.N. 109
Greene, J.E., *see* McGlinn, T.C. 125
Greene, P.D. 135, 165
Gregg, J.R. 122, 123
Gribnikov, Z.S. 224
Grim, A. 56
Grosse, P., *see* Mutzenich, G. 112
Grunewald, G. 61, 65, 81
Grynberg, M. 212
Gualberto, G.M., *see* Lemos, V. 112
Guggenheim, H.J., *see* Barker, A.S. 109, 110, 199
Guggenheim, H.J., *see* Birgeneau, R.J. 204
Guggenheim, H.J., *see* Cowley, R.A. 181, 182, 187, 188, 206, 207
Guggenheim, H.J., *see* Fleury, P.A. 200–202, 204
Guggenheim, H.J., *see* Moch, P. 191
Guggenheim, H.J., *see* Parisot, G. 199
Guggenheim, H.J., *see* Werthein, G.K. 198
Guillaume, J.C., *see* Vodopyanov, L.K. 92
Gupta, H.C. 127
Guseinov, R.E., *see* Gasanly, N.M. 112

Haarer, D.P. 263
Haas, M. 75
Haken, H., *see* Benk, H. 236
Halperin, B.I. 10, 140, 143, 147
Hanson, D., *see* Leong, B. 256, 257
Hanson, D.M. 224, 237, 281
Hanson, D.M., *see* Colson, S.D. 233, 249, 255
Hanson, D.M., *see* Patel, J.S. 281, 282
Harada, H. 65, 77, 93
Harbec, J.Y. 113
Harbec, J.Y., *see* Walsh, D. 113
Harbec, Y.H., *see* Jandl, S. 113
Hardy, J.R., *see* Jaswal, S.S. 87
Hardy, J.R., *see* Montgomery, D.J. 86
Harley, R.T., *see* Buchanan, M. 194, 195
Harman, T.C., *see* Mooradian, A. 97
Harris, A.B., *see* Holcomb, W.K. 198
Harris, C.B. 261, 263, 264, 266, 268
Harris, C.B., *see* Breiland, W.G. 264, 267
Harris, C.B., *see* Brenner, H.C. 271
Harris, C.B., *see* Buckley, M.J. 258
Harris, C.B., *see* Fayer, M.D. 271
Harris, C.B., *see* Francis, A.H. 257–260
Harris, C.B., *see* Lewellyn, M.T. 272
Harrison, W. 139
Harrison, W.A. 46
Hartmann, W.M., *see* Sen, P.N. 61, 81, 123
Hashimoto, A., *see* Kamijoh, T. 125
Hassler, J., *see* Geick, R. 86
Hayek, M. 95, 111
Hayes, W. 49, 51, 179, 180, 196, 202, 205–208
Hayes, W., *see* Buchanan, M. 194, 195
Hayes, W., *see* Fleury, P.A. 200–202
Hayes, W., *see* Sennett, C.T. 98
Hayes, W., *see* Wiltshire, M.C.K. 205, 207
Haywood, S.K. 211
Heap, B.R., *see* Elliott, R.J. 197
Hellworth, R.W., *see* Feynman, R.P. 261
Henning, D., *see* Goede, O. 4, 152
Henning, D., *see* Suslina, L.G. 4, 156
Hennion, B., *see* Wagner, V. 189, 190
Herscovici, C. 76, 81
Herscovici, C., *see* Beserman, R. 103
Herzberg, G. 349
Hesp, B.H. 287
Heuret, M., *see* Gobeau, J. 109
Hibner, K., *see* Kosacki, I. 128
Hildisch, L., *see* Bairamov, B.H. 102
Hildisch, L., *see* Bairamov, B.H. 102
Hildisch, L., *see* Jahne, E. 68, 72, 102
Hirlimann, C. 102, 105, 123
Hirlimann, C., *see* Balkanski, M. 100
Hirlimann, C., *see* Beserman, R. 102, 105, 123
Hirlimann, C., *see* Zigone, M. 92
Hirlimann, Ch., *see* Charfi, F. 101
Ho, F. 287, 288
Hochstrasser, R.H., *see* Ho, F. 288

Hochstrasser, R.M. 225, 235, 236, 237
Hochstrasser, R.M., *see* Agranovich, V.M. 218
Hochstrasser, R.M., *see* Decola, R.L. 287
Hochstrasser, R.M., *see* Ho, F. 287, 288
Hoclet, M. 96
Holah, G.D., *see* Amirtharaj, P.M. 107, 108
Holcomb, W.K. 187, 198, 201, 207
Holden, T.M., *see* Coombs, G.J. 181, 187, 197, 198
Holonyak, N., *see* Nelson, R.J. 4
Hong, H.-K. 237, 242, 250, 255, 280
Hopfield, J.J. 224
Horn, P.M., *see* Wong, P. 203, 210, 211
Hoshen, J. 250
Houghton, A. 148, 171
Huang, K. 224
Huang, K., *see* Born, M. 44
Huber, D.L. 182
Hughes, A.E. 188
Hurrell, J.P. 89
Hutchings, M.T. 205
Hutchings, M.T., *see* Ikeda, H. 203

Ichimori, T., *see* Nakahara, J. 125
Ichkitidze, R.R., *see* Ipatova, I.P. 100, 101
Ignatev, B.V. 112
Ikeda, H. 203
Ilegems 65, 99
Ilin, M.A., *see* Zinger, G.M. 107
Imenkov, A.H., *see* Tzarenkov, B.V. 4
Inoshita, T. 127
Inoue, K., *see* Ishidate, T. 84, 85
Ioselevsky, Ya.A., *see* Kagan, Yu.M. 300
Ipatova, I.P. 27, 100, 101, 156
Ipatova, I.P., *see* Biryulin, Yu.F. 101
Ipatova, I.P., *see* Grim, A. 56
Ipatova, I.P., *see* Maradudin, A.A. 43, 56
Ipatova, I.P., *see* Zinger, G.M. 18, 20, 65, 73, 74, 85, 101, 127, 143, 158, 163
Ishidate, T. 84, 85
Ishikawa, T., *see* Wada, K. 206, 208
Itoh, K., *see* Nakashima, S. 92
Ivanova, E.A., *see* Belousov, M.V. 301, 320, 321
Ivchenko, E.L. 275

Jaccarino, V., *see* Sanders, R.W. 191
Jacobs, I.S. 204, 205, 207
Jahn, I.R., *see* Bauhofer, W. 87, 88
Jahn, I.R., *see* Perry, C.H. 88
Jahne, E. 68, 70, 72, 101, 102, 103, 105
Jahne, E., *see* Bairamov, B.H. 98, 102
Jahne, E., *see* Kleinert, P. 103
Jahne, E., *see* Ulrici, B. 103
Jain, K.P., *see* Balkanski, M. 100
Jain, K.P., *see* Soni, R.K. 127
Jandl, S. 110, 113
Jandl, S., *see* Deslandes, J. 110, 127
Jandl, S., *see* Harbec, J.Y. 113
Jandl, S., *see* Provencher, R. 110
Jandl, S., *see* Walsh, D. 113
Jaswal, S.S. 87
Jindal, V.K. 304
Jindal, V.K., *see* Pathak, K.N. 304
John, L., *see* Goede, O. 4, 152
Johnson, C.K., *see* Stevenson, S.H. 226
Johnstonc, I.W. 112
Johnstone, I.W., *see* Lockwood, D.J. 128, 211
Joncs, D.A., *see* Coombs, G.J. 181, 187, 197, 198
Jones, D.A., *see* Svensson, E.C. 193
Jones, G.D., *see* Johnstone, I.W. 112
Jortner, J., *see* Hoshen, J. 250
Jortner, J., *see* Klafter, J. 279, 382
Joshi, S.K., *see* Srivastava, V. 85
Jouanne, M., *see* Zigone, M. 92
Joullie, A., *see* Charfi, F. 101
Julien, Ch., *see* Gebicki, W. 98
Julien, Ch., *see* Picquart, M. 98
Jusserand, B. 63, 100, 102, 107, 122, 123, 125

Kagan, Yu.M. 300
Kaiser, W., *see* Laubereau, A. 287
Kakimoto, K. 125
Kalabukhova, V.F., *see* Ignatev, B.V. 112
Kalus, J., *see* Natkaniec, I. 286
Kamigaichi, T., *see* Watanabe, J. 128
Kamijoh, T. 125
Kaminsky, G., *see* Dean, P.J. 135
Kamitohara, W.A. 18
Kaplan, T. 61
Karapetyan, V.E. 112
Karoza, A.G., *see* Bodnar, I.V. 112
Katagiri, S., *see* Ishidate, T. 84, 85
Katnani, A.D. 139
Katoda, I., *see* Yamazaki, S. 104, 125
Katoda, T., *see* Kakimoto, K. 125
Kawamura, H. 99
Kawamura, H., *see* Tsu, R. 63, 99

Keating, D.E., *see* Chang, R.K. 100
Keezer, R.C. 86
Kekelidze, G.P., *see* Kekelidze, N.P. 106
Kekelidze, N.P. 106
Keyes, T., *see* Colson, S.D. 278
Khilko, G.I., *see* Novik, A.E. 92
Khmel'nitskii, D.E., *see* Gor'kov, L.P. 275
Khmel'nitskii, D.E., *see* Larkin, A.I. 27
Khomutova, M.D., *see* Musaeva, L.G. 112
Khotyaintseva, G.Yu., *see* Ostapenko, N.I. 227
Kim, O.K. 99
Kim, R. 96
Kim, S.M., *see* Svensson, E.C. 193
Kinch, M.A., *see* Carter, D.L. 96
Kirby, R.D., *see* Freund, G.A. 128
Kirkpatrick, S., *see* Shante, V.K.S. 206, 208, 210
Kirkpatrick, S., *see* Velicky, B. 301
Kjekshus, A., *see* Zwick, A. 110
Klafter, J. 279, 382
Klanjsek Gunde, M. 128
Klein, M.V., *see* Beserman, R. 126
Klein, M.V., *see* Krabach, T.N. 109
Klein, M.V., *see* Lai, S. 152, 157
Klein, M.V., *see* McGlinn, T.C. 125
Klein, M.V., *see* Shui Lai 4
Klein, M.V., *see* Teicher, M. 105
Kleinert, P. 50, 61, 62, 64, 65, 75, 80, 81, 103, 104, 105, 122
Kleinert, P., *see* Jahne, E. 102
Kleinhert, P. 122, 123
Kliche, G. 128
Klymko, P.W., *see* Kopelman, R. 281
Knierim, W., *see* Betz, E. 203
Kobayashi, A. 122
Kobayashi, A., *see* Newman, K.E. 125
Kochubei, S.M., *see* Broude, V.L. 248
Koda, T., *see* Murahashi, T. 90
Koda, T., *see* Tokura, Y. 253
Koehler, W.C., *see* Birgeneau, R.J. 204
Koehler, W.C., *see* Wilkinson, M.K. 204, 205
Kohler, H., *see* Richter, W. 112
Konig, W., *see* Bauhofer, W. 88
Konnikov, S.G., *see* Alferov, Zh.I. 135
Konobeev, Yu.V., *see* Dubovsky, O.A. 345
Konovets, N.K., *see* Lisitsa, M.P. 92
Konzelmann, U., *see* Burland, D.M. 283
Kopelman, R. 276, 277, 278, 279, 281
Kopelman, R., *see* Ahlgren, D.C. 112, 279
Kopelman, R., *see* Colson, S.D. 233, 249, 255
Kopelman, R., *see* Gentry, S.T. 277, 278
Kopelman, R., *see* Hong, H.-K. 237, 242, 255, 280
Kopelman, R., *see* Ochs, F.W. 233, 234
Kop'ev, P.S., *see* Alferov, Zh.I. 4
Kopilov, A.A., *see* Berndt, V. 135, 169
Kopylov, A.A., *see* Berndt, V. 98
Korshunov, V.V., *see* Benderskii, V.A. 238
Korshunov, V.V., *see* Broude, V.L. 237
Kosacki, I. 128
Kosacki, I., *see* Ciepielewski, P. 109
Kosacki, I., *see* Valakh, M.Y. 127, 128
Koster, G.F. 229
Kowalczyk, S.P. 139
Kozyrev, S.P. 124
Kozyrev, S.P., *see* Vodopyanov, L.K. 124
Krabach, T.N. 109
Krabach, T.N., *see* Beserman, R. 126
Krabach, T.N., *see* McGlinn, T.C. 125
Kramer, B., *see* McGlinn, T.C. 125
Kraut, E.A., *see* Kowalczyk, S.P. 139
Krivenko, T.A. 240, 241, 251, 252
Krivoglas, M.A. 22
Krol, A., *see* Gebicki, W. 93
Krol, A.V., *see* Karapetyan, V.E. 112
Krol, A.W. 57, 92, 112
Kronenberger, A. 217
Krozyrev, S.P. 96, 97
Krumhansl, J.A., *see* Elliott, R.J. xv, 19, 81, 135, 181, 184–186, 249, 301, 358, 359, 361, 366, 378, 380, 383
Kucherov, A.P., *see* Lisitsa, M.P. 349
Kudo, K., *see* Wakamura, K. 112
Kukharaskii, A.A., *see* Gasanly, N.M. 65, 70, 104, 106
Kukimoto, H., *see* Nakahara, J. 125
Kumazaki, K. 128
Kunc, K. 93
Kunc, K., *see* Jusserand, B. 122
Kunc, K., *see* Talwar, D.N. 91
Kushida, T., *see* Murahashi, T. 90
Kusmartsev, F.V. 27, 28, 152
Kutty, A.P.G. 71, 127

Lacina, B., *see* Chang, R.K. 108
Lacina, W.B. 58, 59, 108
Lacina, W.B., *see* Pershan, P.S. 57
Lai, S. 152, 157
Lakshmi, G., *see* Srinivasan, R. 89

Lalov, I.I. 353
Lalov, I.I., *see* Agranovich, V.M. 349
Landa, G., *see* Bedel, E. 102, 103
Landa, G., *see* Zwick, A. 110, 111
Landau, L.D. 5, 6, 10, 150, 303, 345
Langer, J.S., *see* Zittartz, J. 10, 140
Lannin, J. 85, 86, 87
Larkin, A.I. 27
Larkin, A.I., *see* Gor'kov, L.P. 275
Lastras, A., *see* McGlinn, T.C. 125
Lau, Ph., *see* Becker, C.R. 181, 188–190
Laubereau, A. 287
Laurens, J.A.J., *see* Baker, J.M. 198
Lawrence, P.E., *see* Jacobs, I.S. 204, 205, 207
Lax, M., *see* Halperin, B.I. 10, 140, 143, 147
Le Postollec, M., *see* Gebicki, W. 98
Le Youllec, R., *see* Polian, A. 96, 97
Leath, P.L. 20, 360
Leath, P.L., *see* Elliott, R.J. xv, 19, 55, 81, 135, 181, 184–186, 249, 301, 358, 359, 361, 366, 378, 380, 383
Lebesque, J.V., *see* Bevaart, L. 203
Lee, K.O. 242, 244
Legrand, S., *see* Bertrand, D. 210
Legrand, S., *see* Lockwood, D.J. 128, 211
Legrand, S., *see* Mischler, G. 128, 207, 210
Lehmann, W., *see* Mitlehner, H. 194
Leiderman, A.V., *see* Broude, V.L. 237, 249, 251
Leiderman, A.V., *see* Krivenko, T.A. 240, 241, 251, 252
Leith, R.M.A., *see* Hayek, M. 111
Lemieux, M.A. 123
Lemos, V. 112
Lengyel, G., *see* Brafman, O. 40, 93
Leong, B. 256, 257
Leotin, J., *see* Lockwood, D.J. 128, 211
Levanyuk, A.P. 27
Leveque, R., *see* Vergnat-Grandjean, D. 108
Levichev, N.V., *see* Krol, A.W. 57, 92, 112
Levy, F., *see* Riede, V. 111
Levy, F., *see* Taguchi, I. 128
Lewellyn, M.T. 272
Li, S.S. 4
Licciardello, D.C., *see* Abrahams, E. 275
Lifshitz, E.M., *see* Landau, L.D. 5, 6, 10, 150, 303, 345
Lifshitz, I.M. 10, 12, 13, 20, 229, 250, 279, 300, 301, 318
Limonov, M.F. 112, 126
Limonov, M.F., *see* Barta, C. 112
Lind, E., *see* Lucovsky, G. 39, 92
Lisita, L.M., *see* Valakh, M.Y. 93
Lisitsa, M.P. 92, 349
Litvinchuk, A.P., *see* Burlakov, V.M. 127, 128
Litvinchuk, A.P., *see* Kosacki, I. 128
Litvinchuk, A.P., *see* Valakh, M.Y. 97, 112, 127, 128
Livescu, G. 91, 124
Llinares, C., *see* Charfi, F. 101
Lockwood, D.J. 128, 179, 182, 199, 211
Lockwood, D.J., *see* Johnstone, I.W. 112
Lockwood, D.J., *see* Mischler, G. 128, 207, 210
Lockwood, D.J., *see* Syme, R.W.G. 128
Lockwood, D.J., *see* Vickers, R.E.M. 128
Lopez Castillo, J.M. 123, 124, 127
Lottici, P.P. 127
Lottici, P.P., *see* Parisini, A. 127
Lottici, P.P., *see* Razzetti, C. 127
Loudon, R., *see* Allen, S.J. 194
Loudon, R., *see* Fleury, P.A. 179, 180
Loudon, R., *see* Hayes, W. 49, 51, 179, 180
Lovesey, S.W. 183
Lucovsky, G. 39, 80, 81, 92, 98, 100, 101, 102, 104–106, 107, 112
Lucovsky, G., *see* Brodsky, M.H. 62, 64, 104
Lucovsky, G., *see* Keezer, R.C. 86
Lucovsky, G., *see* Sen, P.N. 60, 106
Lysenko, E.E. 227
Lysenko, V., *see* Permogorov, S. 163, 164
Lysenko, V.G., *see* Permogorov, S.A. 4

Macfarlane, R.M., *see* Burland, D.M. 283
Mach, R. 4
Maegle, J., *see* Mach, R. 4
Magarino, J., *see* Bertrand, D. 210
Magarino, J., *see* Tuchendler, J. 210
Makharadze, Z.D., *see* Kekelidze, N.P. 106
Maki, A.H., *see* Buckley, M.J. 258
Makovsky, J., *see* Birgeneau, R.J. 204, 205, 207
Maksimov, A.A., *see* Broude, V.L. 237
Malhotra, J., *see* Gupta, H.C. 127
Mandelbrot, B.B. 281
Mansur, L.C., *see* Chang, I.F. 45
Mansur, L.C., *see* Gielisse, P.J. 107
Maradudin, A.A. 39, 43, 56, 300

Maradudin, A.A., *see* Grim, A. 56
Maradudin, A.A., *see* Xinh, N.X. 57, 58
Marfaing, Y., *see* Hoclet, M. 96
Marfaing, Y., *see* Vodopyanov, L.K. 124
Margaritondo, G., *see* Katnani, A.D. 139
Markov, Y.F., *see* Barta, C. 112
Markov, Y.F., *see* Limonov, M.F. 112, 126
Marshall, R., *see* Gielisse, P.J. 107
Marti, C., *see* Beserman, R. 94
Martin, D.H. 180
Martin, M.S., *see* Zigone, M. 92
Martin, T.P. 52, 71
Martin, T.P., *see* Genzel, L. 71, 82, 110
Martinez, G., *see* Galtier, P. 103
Mascarenhas, S., *see* Hurrell, J.P. 89
Maschke, K., *see* Baldareschi, A. 142, 143
Maslov, A.U. 138
Maslov, A.U., *see* Ipatova, I.P. 156
Massa, N.E. 71
Massa, N.E., *see* Renker, B. 88
Mathieu, J.P., *see* Poulet, R. 49
Matossi, F. 75
Matsubara, T., *see* Yonezawa, F. 301
Mattis, D. 388, 391, 392
McGill, R.E., *see* Haas, M. 75
McGill, R.E., *see* Rosenstock, H.B. 75
McGlinn, T.C. 125
McGurn, A.R. 198, 206
McGurn, A.R., *see* Tahir-Kheli, R.A. 198
McKenzie, G., *see* Natkaniec, I. 286
McKnight, S.W. 98
McWhorter, A.L., *see* Strahm, N.D. 105
Melekhin, V.G., *see* Maslov, A.U. 138
Meletov, K.P. 234, 235
Melnik, N.N., *see* Gasanly, N.M. 111, 126, 128
Melnik, N.N., *see* Vinogradov, E.A. 112
Melnik, N.N., *see* Vodopyanov, L.K. 92
Mel'nikov, V.I., *see* Rashba, E.I. 275
Memon, A. 124
Meneses, E.A., *see* Cerdeira, F. 113
Mercier, A. 111
Meredith, G.R., *see* Hochstrasser, R.M. 225
Merrifield, R.E., *see* Avakian, P. 281
Mikkelsen, J.C. 125
Mikkelsen, J.C., *see* Lucovsky, G. 112
Miller, B.J., *see* Shah, J. 100
Mil'man, P.D. 249, 252
Miloslavskii, V.K., *see* Wo Hoang Thay 4
Minomura, S., *see* Ishidate, T. 84, 85
Minomura, S., *see* Nakahara, J. 125
Mischler, G. 128, 207, 210
Mischler, G., *see* Lockwood, D.J. 128, 211
Mishima, H., *see* Nakashima, S. 112
Misochko, Ye.Ya., *see* Benderskii, V.A. 238
Misra, A.K., *see* Agrawal, B.K. 85
Misra, T.N., *see* Talapatra, G.B. 236
Mitin, G.G. 375
Mitin, G.G., *see* Gorelik, V.S. 375
Mitlehner, H. 194
Mitra, S.S., *see* Brafman, O. 40, 93
Mitra, S.S., *see* Chang, I.F. 18, 20, 42, 45, 65, 77, 95, 301
Mitra, S.S., *see* Gielisse, P.J. 107
Mitra, S.S., *see* Massa, N.E. 71
Mitra, S.S., *see* Parrish, J.F. 95
Mitsuishi, A., *see* Nakashima, S. 92, 94
Mityagin, Y.A. 92, 95
Mityagin, Yu.A., *see* Vinogradov, E.A. 92
Moch, P. 191
Moch, P., *see* Gosso, J.P. 193, 194, 199
Moch, P., *see* Parisot, G. 199
Moiseenko, V., *see* Klanjsek Gunde, M. 128
Mokan, I.I., *see* Alferov, Zh.I. 135
Mon, J.P., *see* Gobeau, J. 109
Monberg, E.M., *see* Kopelman, R. 276, 278
Montarroyos, E. 197
Montgomery, D.J. 86
Montroll, E.W. 300
Montroll, E.W., *see* Maradudin, A.A. 43, 56
Mooradian, A. 97
Morkoc, H., *see* Teicher, M. 105
Mostoller, M., *see* Kaplan, T. 61
Motokawa, M., *see* Belanger, R.M. 191, 192
Motokawa, M., *see* Sanders, R.W. 191
Mott, N.F. 161, 274, 280
Müller, G.O., *see* Permogorov, S. 152, 162, 163
Murahashi, T. 90
Murakami, Y., *see* Osamura, K. 105
Musaeva, L.G. 112
Mutzenich, G. 112
Mykolajewycz, R., *see* Gielisse, P.J. 107
Myles, C.V. 75
Myles, C.W., *see* Gregg, J.R. 122, 123
Myles, C.W., *see* O'Hara, M.J. 75

Nadzhafov, A.I., *see* Allakhverdiev, K.R. 112
Nagiev, V.M. 32
Nahory, R.E., *see* Pinczuk, A. 107, 108

Nair, I. 89, 90
Nakada, I., *see* Tokura, Y. 253
Nakahara, J. 125
Nakashima, S. 92, 94, 112
Nani, R.K., *see* Abdullaev, G.B. 111
Nani, R.Kh., *see* Belenkii, G.L. 112
Nardelli, G.F., *see* Benedek, G. 57, 58
Narita, S., *see* Harada, H. 65, 77, 93
Narita, S., *see* Kim, R. 96
Natadze, A.L., *see* Krol, A.W. 57, 92
Natkaniec, I. 286
Nayak, P., *see* Behera, S.N. 77
Nazarewicz, W., *see* Gebicki, W. 93, 97, 98
Nazarewicz, W., *see* Gorska, M. 95, 96
Nazarewicz, W., *see* Olszewski, A. 97, 124
Nazarewicz, W., *see* Picquart, M. 98
Nelson, R.J. 4, 135
Neumann, H., *see* Riede, V. 111
Newhouse, J.S., *see* Kopelman, R. 281
Newman, K.E. 125
Newman, K.E., *see* Kobayashi, A. 122
Newman, R.C. 39
Nicholas, R.J., *see* Carles, R. 106
Nicholas, R.J., *see* Portal, J.C. 107
Nickel, B.G. 198
Nikiforova, M., *see* Permogorov, S. 152, 162, 163
Nishiguchi, N., *see* Kumazaki, K. 128
Nizametdinova, M.A., *see* Allakhverdiev, K.R. 112
None, D., *see* Belanger, R.M. 191, 192
Nonhof, C.J., *see* Bottler, B.J. 269, 270
Nordheim, L. 18
Nouvel, G. 126
Novik, A.E. 92
Nusimovici, M.A. 95

Obreimov, I.B. 217
Ochs, F.W. 233, 234
Ochs, F.W., *see* Kopelman, R. 276, 278
O'Hara, M.J. 75
Ohbayashi, K., *see* Watanabe, J. 128
Oitmaa, J., *see* Maradudin, A.A. 39
Oka, Y., *see* Murahashi, T. 90
Olego, D.J. 123
Oleinik, G.S., *see* Vinogradov, E.A. 95
Oles, B. 112, 124
Olszewski, A. 97, 124
Onari, S., *see* Wakamura, K. 112
Onodera, Y. 250, 301, 314
Onoprienko, M.I., *see* Broude, V.L. 242, 243
Onton, A. 4
O'Reilly, E.P., *see* Kobayashi, A. 122
Orel, B., *see* Klanjsek Gunde, M. 128
Orlov, A.V., *see* Agranovich, V.M. 337, 365, 396
Osad'ko, I.S. 234
Osamura, K. 105
Oseroff, A. 191
Ostapenko, N.I. 227, 237, 242
Ostapenko, N.I., *see* Brovchenko, I.V. 242
Oswald, F. 106
Ouen, R.T., *see* Li, S.S. 4
Ovander, L.N. 52
Ovsyankin, V.V. 163, 281

Painter, R.D., *see* O'Hara, M.J. 75
Paquet, D., *see* Jusserand, B. 122
Parayanthal, P. 123
Parayanthal, P., *see* Amirtharaj, P.M. 125
Parayanthal, P., *see* Tiong, K.K. 97
Parisi, G., *see* Brezin, E. 146, 152
Parisini, A. 127
Parisini, A., *see* Lottici, P.P. 127
Parisot, G. 199
Parisot, G., *see* Moch, P. 191
Parker, J.H., *see* Feldman, D.W. 75, 85
Parrish, J.F. 95
Pastur, L.A., *see* Lifshitz, I.M. 12, 13, 301
Patel, J.S. 281, 282
Pathak, K.N. 304
Pathak, K.N., *see* Jindal, V.K. 304
Patnaik, K., *see* Behera, S.N. 77
Pawley, G.S., *see* Natkaniec, I. 286
Pearsall, T., *see* Portal, J.C. 107
Pearson, G.L., *see* Chen, Y.S. 65, 71, 105, 244
Pearson, G.L., *see* Ilegems 65, 99
Pearson, G.L., *see* Lucovsky, G. 100
Pekar, S.I. 26, 224, 225, 227
Pepper, D.E., *see* Buyers, W.J.L. 186, 193, 194, 196, 198, 207
Perkowitz, S., *see* Amirtharaj, P.M. 107, 108
Perkowitz, S., *see* McKnight, S.W. 98
Permogorov, S. 152, 162, 163, 164
Permogorov, S.A. 4
Perrier, P., *see* Portal, J.C. 107
Perry, A.M., *see* Buchanan, M. 194, 195
Perry, C.H. 88, 109
Perry, C.H., *see* Bauhofer, W. 88
Perry, C.H., *see* Fertel, J.H. 64, 65, 88, 89, 124
Perry, C.H., *see* Genzel, L. 71, 82, 110

Perry, C.H., *see* Parrish, J.F. 95
Pershan, P.S. 57
Pershan, P.S., *see* Chang, R.K. 108
Pershan, P.S., *see* Lacina, W.B. 58, 59, 108
Pershan, P.S., *see* Oseroff, A. 191
Peterson, D.L. 124
Petrou, A., *see* Peterson, D.L. 124
Petrou, A., *see* Venugopalan, S. 98
Petukhov, V.B. 8
Pevnitskii, I.V., *see* Novik, A.E. 92
Philpott, M.R. 235
Pickering, C. 123, 127, 128
Picquart, M. 98
Picquart, M., *see* Gebicki, W. 98
Picquart, M., *see* Grynberg, M. 212
Pikhtin, A.N. 4, 135
Pikhtin, A.N., *see* Berndt, V. 135, 169
Pikus, G.E., *see* Ivchenko, E.L. 275
Pilz, W., *see* Jahne, E. 68, 72, 102
Pinczuk, A. 49, 51, 107, 108
Pines, A., *see* Breiland, W.G. 264, 267
Plaskett, T.S., *see* Brodsky, M.H. 62, 64, 104
Plendl, J.N., *see* Chang, I.F. 45
Plendl, J.N., *see* Gielisse, P.J. 107
Plonitchenko, V.G., *see* Vodopyanov, L.K. 92
Plotnichenko, V.G. 95
Plotnichenko, V.G., *see* Mityagin, Y.A. 95
Plumelle, P., *see* Hoclet, M. 96
Plyukhin, A.G., *see* Suslina, L.D. 4
Plyukhin, A.G., *see* Suslina, L.G. 4, 156
Plyukhin, A.G., *see* Suslina, L.G. 135, 152, 161
Pogarev, D.E., *see* Belousov, M.V. 301, 320, 321, 359, 374–377
Pokrovskii, V.L., *see* Petukhov, V.B. 8
Polian, A. 96, 97
Polissky, G.N., *see* Valakh, M.Y. 93
Pollack, F.H., *see* Amirtharaj, P.M. 97
Pollack, F.H., *see* Tiong, K.K. 97
Pollack, M.A., *see* Pinczuk, A. 107, 108
Pollak, F.H., *see* Amirtharaj, P.M. 125
Pollak, F.H., *see* Parayanthal, P. 123
Port, H. 224, 232, 233, 237
Port, H., *see* Doberer, U. 226, 237, 238
Portal, J.C. 107
Portnoy, E.L., *see* Alferov, Zh.I. 3, 141
Porto, S.P.S., *see* Hurrell, J.P. 89
Potter, R.F. 105
Potts, R.B., *see* Montroll, E.W. 300
Poulet, R. 49
Pozhidaev, Yu.E., *see* Biryulin, Yu.F. 101
Prasad, P.N., *see* Kopelman, R. 276
Prasad, P.N., *see* Ochs, F.W. 233, 234
Prasad, R. 53
Price, W.C. 321, 375
Prikhot'ko, A.F. 222, 243
Prikhot'ko, A.F., *see* Broude, V.L. 237
Pringsheim, P., *see* Kronenberger, A. 217
Prokhorov, A.S. 191
Provencher, R. 110
Provencher, R., *see* Jandl, S. 110
Psaltakis, G.C., *see* Lockwood, D.J. 128, 211
Pyrkov, V.N., *see* Burlakov, V.M. 127, 128

Quintero, M., *see* Anedda, A. 112

Rabin'kina, N.V. 233, 255, 285
Raccah, P.M., *see* Olego, D.J. 123
Raga, F., *see* Anedda, A. 112
Ragimov, A.S., *see* Gasanly, N.M. 111, 112, 126
Raikh, M.E. 161
Raikh, M.E., *see* Ablyazov, N.N. 157
Raikh, M.E., *see* D'yakonov, M.I. 135
Ralston, J.M., *see* Chang, R.K. 100
Ramakrishnan, T.W., *see* Abrahams, E. 275
Ramdas, A.K. 212
Ramdas, A.K., *see* Peterson, D.L. 124
Ramdas, A.K., *see* Venugopalan, S. 98
Rammal, R. 281
Rashba, E.I. 27, 149, 232, 235, 239, 255, 275
Rashba, E.I., *see* Broude, V.L. 18, 218, 223, 232, 233, 234, 237, 240, 244, 279
Rashba, E.I., *see* Davydov, A.S. 227
Rashba, E.I., *see* Gribnikov, Z.S. 224
Rashba, E.I., *see* Krivenko, T.A. 240, 241, 251, 252
Rashba, E.I., *see* Kusmartsev, F.V. 152
Rashba, E.I., *see* Meletov, K.P. 234, 235
Rashba, E.I., *see* Rabin'kina, N.V. 233, 255, 285
Rashevskaya, E.P., *see* Zinger, G.M. 107
Razbirin, B.S., *see* Ivchenko, E.L. 275
Razzetti, C. 127
Razzetti, C., *see* Lottici, P.P. 127
Renker, B. 88
Renucci, J.B., *see* Bedel, E. 102, 103, 106, 125
Renucci, J.B., *see* Carles, R. 100, 106

Renucci, J.B., *see* Carles, R.N. 100
Renucci, J.B., *see* Renucci, M.A. 74, 84, 85
Renucci, J.B., *see* Saint-Cricq, N. 99
Renucci, M.A. 74, 84, 85
Renucci, M.A., *see* Bedel, E. 106, 125
Renucci, M.A., *see* Carles, R. 100
Renucci, M.A., *see* Carles, R.N. 100
Renucci, M.A., *see* Nouvel, G. 126
Renucci, M.A., *see* Portal, J.C. 107
Renucci, M.A., *see* Saint-Cricq, N. 99
Renucci, M.A., *see* Zwick, A. 110, 111
Resnitskii, A.N., *see* Permogorov, S.A. 4
Rezende, S.M., *see* Montarroyos, E. 197
Rezende, S.M., *see* Sanders, R.W. 191
Rezende, S.M., *see* Vianna, S. 126
Reznitsky, A., *see* Permogorov, S. 152, 162, 163, 164
Richards, P.L. 194
Richards, P.L., *see* Allen, S.J. 194
Richards, P.L., *see* Enders, B. 180, 181, 191, 192
Richards, P.L., *see* Tennant, W.E. 197
Richter, W. 112
Riede, V. 111
Riede, V., *see* Sobotta, H. 98, 100
Rigini, R. 287
Riskin, A.I., *see* Zinger, G.M. 18
Roberts, S., *see* Jacobs, I.S. 205, 207
Robinette, S.L. 225
Robinson, G.W., *see* Colson, S.D. 233, 249, 255
Robinson, G.W., *see* Hong, H.-K. 250
Rodriguez, S., *see* Peterson, D.L. 124
Rodriguez, S., *see* Ramdas, A.K. 212
Rogachev, A.A., *see* Alferov, Zh.I. 3, 141
Rolandson, S., *see* Svensson, E.C. 193
Romano, T.C., *see* Beserman, R. 126
Rosenbaum, T.F. 162
Rosenstock, H.B. 75
Rosenstock, H.B., *see* Haas, M. 75
Rowell, N.L., *see* Syme, R.W.G. 128
Roy, A., *see* Kobayashi, A. 122
Rudashevskii, E.G., *see* Prokhorov, A.S. 191
Rudin, S., *see* Mattis, D. 388, 391, 392
Rukavishnikov, V.A., *see* Vodopyanov, L.K. 93
Rund, D., *see* Port, H. 224, 232, 233
Ruvalds, J. 305, 349
Ryskin, A.I., *see* Karapetyan, V.E. 112
Ryskin, A.I., *see* Krol, A.W. 57, 92, 112
Ryskin, A.I., *see* Novik, A.E. 92
Ryskin, A.I., *see* Zinger, G.M. 107, 127
Ryzhik, I.M., *see* Gradshtein, I.S. 314, 368
Ryzhik, I.M., *see* Gradstein, I.S. 151

Safarov, N., *see* Aliev, R.A. 112
Safarov, N.Yu., *see* Allakhverdiev, K.R. 112
Safinya, C.R., *see* Wong, P. 203, 210, 211
Sahni, V.C. 112
Sahni, V.C., *see* Venkataraman, G. 46
Saint-Cricq, N. 99
Saint-Cricq, N., *see* Carles, R. 100, 106
Sakuta, M., *see* Kamijoh, T. 125
Salaev, E.Y., *see* Abdullaev, G.B. 111
Salaev, E.Yu., *see* Belenkii, G.L. 112
Salzberg, J.B., *see* Lemos, V. 112
Sanders, R.W. 191
Sapriel, J., *see* Boudou, A. 112
Sapriel, J., *see* Jusserand, B. 63, 100, 123, 125
Sardarly, R.M., *see* Aliev, R.A. 112
Sardarly, R.M., *see* Allakhverdiev, K.R. 112, 126
Sarkisov, L.A., *see* Vodopyanov, L.K. 92
Saville, I.D., *see* Buchanan, M. 194, 195
Schäfer, L., *see* Houghton, A. 148, 171
Schaff, W.J., *see* Bour, D.P. 128
Schaffer, W.J., *see* Kowalczyk, S.P. 139
Scheibe, G. 228
Schlupp, R.L., *see* Harris, C.B. 266, 268
Schmeltzer, D. 76, 82, 93, 94, 105
Schmeltzer, D., *see* Beserman, R. 103
Schmelzer, U., *see* Natkaniec, I. 286
Schmidt, J. 258, 261, 264, 265, 267
Schmidt, J., *see* Bottler, B.J. 269, 270
Schmidt, J., *see* Van Strien, A.J. 259, 282
Schmidt, J., *see* Van 't Hof, C.A. 269
Schoenherr, H., *see* Mitlehner, H. 194
Schosser, C.L., *see* Dlott, D. 287
Schuch, H., *see* Harris, C.B. 266, 268
Schwartz, D., *see* Ehrenreich, H. 4
Schwoerer, M. 270
Sen, P.N. 60, 61, 81, 106, 123
Sennett, C.T. 98
Shah, J. 100
Shah, S.I., *see* Beserman, R. 126
Shante, V.K.S. 206, 208, 210
Shchukin, V.A. 27
Shchukin, V.A., *see* Ipatova, I.P. 27
Shchukin, V.A., *see* Kusmartsev, F.V. 27, 28

Shealy, J.R., *see* Bour, D.P. 128
Sheka, E.F. 227, 231, 233, 247, 253, 285
Sheka, E.F., *see* Broude, V.L. 218, 223, 232, 233, 234, 237, 240, 255, 279
Sheka, E.F., *see* Dolganov, V.K. 233
Sheka, E.F., *see* Krivenko, T.A. 240, 241
Sheka, E.F., *see* Meletov, K.P. 234, 235
Sheka, E.F., *see* Natkaniec, I. 286
Sheka, E.F., *see* Rabin'kina, N.V. 233, 255, 285
Shen, S.C. 85, 97, 124
Sherman, W.F., *see* Price, W.C. 321, 375
Shibuya, M., *see* Ishidate, T. 84, 85
Shirane, G., *see* Birgeneau, R.J. 204
Shirane, G., *see* Cowley, R.A. 181, 182, 187, 188, 193, 197, 206, 207
Shirane, G., *see* Wong, P. 203, 210, 211
Shklovskii, B.I. 13, 156, 161, 164, 167
Shklovskii, B.I., *see* Gel'mont, B.L. 166, 168
Shklovskii, B.I., *see* Skal, A.S. 167
Shlimak, I.S. 138, 165
Shlimak, I.S., *see* Gel'mont, B.L. 166, 168
Shmartsev, Yu.V., *see* Biryulin, Yu.F. 101
Shmartsev, Yu.V., *see* Ipatova, I.P. 100, 101
Shockley, W., *see* Chen, Y.S. 65, 71, 105, 244
Shoenfeld, D.W., *see* Li, S.S. 4
Shpak, M.T., *see* Broude, V.L. 255
Shpak, M.T., *see* Brovchenko, I.V. 242
Shpak, M.T., *see* Ostapenko, N.I. 227, 237, 242
Shpakovskaja, L.G., *see* Broude, V.L. 255
Shteinshraiber, V.Yu., *see* Belenkii, G.L. 112
Shteinshraiber, Y.A., *see* Allakhverdiev, K.R. 126
Shui Lai 4
Siapkas, D., *see* Bartzokas, A. 112
Siderenko, V.I., *see* Valakh, M.Y. 93
Sidorenko, V.I., *see* Valakh, M.Y. 127
Sievers, A.J., *see* Barker, A.S. 18, 38, 42, 180
Sikora, L.I., *see* Prikhot'ko, A.F. 243
Silbey, R., *see* Friesner, R. 239
Sinai, J.J. 121
Sinha, S.K., *see* Farr, M.K. 76
Sirota, N.N. 106
Six, H.A., *see* Lucovsky, G. 107
Sixi, H., *see* Benk, H. 236
Skal, A.S. 167
Skorobogat'ko, A.F., *see* Prikhot'ko, A.F. 243
Slade, M.L., *see* Keezer, R.C. 86
Slamovits, D., *see* Schmeltzer, D. 93, 94
Slater, J.C., *see* Koster, G.F. 229
Slempkes, S., *see* Jusserand, B. 102, 107
Slempkes, S., *see* Soni, R.K. 127
Smakula, A., *see* Gielisse, P.J. 107
Small, G.J., *see* Hochstrasser, R.M. 235, 236
Small, G.J., *see* Robinette, S.L. 225
Small, G.J., *see* Stevenson, S.H. 225, 226
Smirnova, G.F., *see* Bodnar, I.V. 112
Smirnova, G.F., *see* Sirota, N.N. 106
Smith, W., *see* Angress, J.F. 88, 89
Sobol, A.A., *see* Ignatev, B.V. 112
Sobotta, H. 98, 100
Sobotta, H., *see* Riede, V. 111
Sobyanin, A.A., *see* Levanyuk, A.P. 27
Sochilina, I.N., *see* Ipatova, I.P. 100, 101
Sokoloff, J.B., *see* Perry, C.H. 88
Soni, R.K. 127
Sood, G., *see* Gupta, H.C. 127
Sorger, F., *see* Baars, J. 96
Sourisseau, C., *see* Gard, P. 126
Soven, P. 186, 301
Spitzer, W.G., *see* Cosand, A.E. 85
Spitzer, W.G., *see* Kim, O.K. 99
Spray, A.R.L., *see* Sennett, C.T. 98
Srinivasan, R. 89
Srivastava, V. 85
Srivastava, V., *see* Weaire, D. 280
Starukhin, A.I., *see* Ivchenko, E.L. 275
Starukin, A.N., *see* Gamarts, E.M. 111
Steigmeier, E.F., *see* Geick, R. 86
Stekhanov, A.I. 89
Stevenson, R.W.H., *see* Baker, J.M. 198
Stevenson, S.H. 225, 226
Stierwalt, D.L., *see* Potter, R.F. 105
Stoichev, K.C., *see* Agranovich, V.M. 364, 374
Stolz, H., *see* Fujii, A. 91
Stolz, H.J., *see* Oles, B. 112
Strahm, N.D. 105
Strauch, D., *see* Bilz, H. 39, 42
Stringfellow, G.B., *see* Cohen, R.M. 125
Sturge, M.D., *see* Cohen, E. 4, 152, 157, 162
Subashiev, A.V., *see* Grim, A. 56
Subashiev, A.V., *see* Ipatova, I.P. 27, 156
Subashiev, A.V., *see* Zinger, G.M. 18, 20, 65, 73, 74, 85, 101, 143, 158, 163
Subashiev, V.K., *see* Gasanly, N.M. 65, 70, 104, 106
Sugakov, V.I. 224, 227

Sugakov, V.I., *see* Ostapenko, N.I. 237
Sumi, H. 239
Sushchinsky, M.M., *see* Gorelik, V.S. 375
Sushchinsky, M.M., *see* Mitin, G.G. 375
Suslina, L.D. 4
Suslina, L.G. 4, 135, 152, 156, 161
Suslina, L.G., *see* Areshkin, A.G. 152
Suslina, L.G., *see* Mach, R. 4
Suslina, L.G., *see* Maslov, A.U. 138
Svensson, E.C. 193
Svensson, E.C., *see* Coombs, G.J. 181, 187, 197, 198
Svensson, E.C., *see* Cowley, R.A. 197
Syme, R.W.G. 128
Syme, R.W.G., *see* Vickers, R.E.M. 128
Sysoev, L.A., *see* Novik, A.E. 92
Sysoev, L.A., *see* Vodopyanov, L.K. 92

Tagiev, M.M., *see* Abdullaev, G.B. 111
Tagirov, V.I., *see* Bakhyshov, N.A. 112
Tagirov, V.I., *see* Gasanly, N.M. 112, 126, 128
Taguchi, I. 128
Tagyev, M.M., *see* Abdullaev, G.B. 111
Tagyev, M.M., *see* Allakhverdiev, K.R. 111, 113
Tahir-Kheli, R.A. 187, 198
Tahir-Kheli, R.A., *see* McGurn, A.R. 198, 206
Tai, H., *see* Nakashima, S. 112
Takahashi, T., *see* Wakamura, K. 112
Talapatra, G.B. 236
Talwar, D.N. 57, 58, 91, 100, 106
Talwar, D.N., *see* Agrawal, B.K. 85
Tanner, D.B., *see* Memon, A. 124
Tarasov, G.G., *see* Burlakov, V.M. 127, 128
Tarasov, G.G., *see* Valakh, M.Y. 112
Tartakovskii, I.I., *see* Broude, V.L. 237
Taylor, D.W. 18, 39, 57, 58, 62, 63, 65, 186, 301
Taylor, D.W., *see* Elliott, R.J. 87
Taylor, D.W., *see* Kamitohara, W.A. 18
Teicher, M. 105
Teller, E., *see* Franck, J. 228
Tennant, W.E. 197
Tennant, W.E., *see* Enders, B. 180, 181, 191, 192
Terenetskaya, I.P., *see* Sheka, E.F. 253
Thomas, G., *see* Rosenbaum, T.F. 162
Thorpe, M.F. 122, 187, 188, 207
Thorpe, M.F., *see* Alben, R. 187, 207
Tinkham, M. 179
Tiong, K.K. 97
Tiong, K.K., *see* Amirtharaj, P.M. 97, 125
Tocchetti, D., *see* Wagner, V. 189, 190
Tokura, Y. 253
Tolpygo, K.B. 224
Tonegawa, T. 183
Toporov, V.V., *see* Bairamov, B.H. 102
Toporov, V.V., *see* Jahne, E. 102
Torporov, V.V., *see* Bairamov, B.H. 98
Toulouse, G., *see* Rammal, R. 281
Toyozawa, Y. 227
Toyozawa, Y., *see* Onodera, Y. 250, 301, 314
Tratas, T.G., *see* Broude, V.L. 237
Travnikov, V., *see* Permogorov, S. 152
Travnikov, V., *see* Permogorov, S.A. 4
Traylor, J.G., *see* Farr, M.K. 76
Tremblay, A.M.S. 123
Tremblay, A.-M.S., *see* Lemieux, M.A. 123
Tremblay, A.-M.S., *see* Lopez Castillo, J.M. 123, 124, 127
Triboulet, R., *see* Hoclet, M. 96
Triboulet, R., *see* Kozyrev, S.P. 124
Triboulet, R., *see* Krozyrev, S.P. 96, 97
Triboulet, R., *see* Vodopyanov, L.K. 124
Tripathi, B.B., *see* Gupta, H.C. 127
Tripathi, S., *see* Agrawal, B.K. 85
Trommsdorff, H.P., *see* Decola, R.L. 287
Trout, J., *see* Ho, F. 287, 288
Tsay, W.-S., *see* Ho, F. 287, 288
Tsu, R. 63, 99
Tsu, R., *see* Kawamura, H. 99
Tsuji, K., *see* Ishidate, T. 84, 85
Tuchendler, J. 210
Tuchendler, J., *see* Bertrand, D. 210
Twose, W.D., *see* Mott, N.F. 274
Tyu, N.S., *see* Ovander, L.N. 52
Tzarenkov, B.V. 4

Ubaidullaev, S.B., *see* Bairamov, B.H. 98, 102
Ubaidullaev, Sh.B., *see* Bairamov, B.H. 102
Udagawa, M., *see* Watanabe, J. 128
Uhle, N., *see* Bootz, B. 91
Uhle, N., *see* Finkenrath, H. 109
Ukhanov, Yu.I., *see* Ipatova, I.P. 100, 101
Ulin, V.P., *see* Alferov, Zh.I. 135
Ulrici, B. 103
Ulrici, B., *see* Jahne, E. 103, 105
Umanskii, V.E., *see* Alferov, Zh.I. 135

Umarov, B.S., *see* Vavilov, V.S. 92
Umarov, B.S., *see* Vodopyanov, L.K. 92
Ushirokawa, A., *see* Yamazaki, S. 104, 125
Uwira, B., *see* Durr, U. 191, 192

Vaida, V., *see* Colson, S.D. 278
Valakh, M.Y. 93, 97, 112, 127, 128
Valakh, M.Y., *see* Lisitsa, M.P. 92
Valakh, M.Ya., *see* Artamonov, V.V. 94
Valakh, M.Ya., *see* Kosacki, I. 128
Valkunas, L.L., *see* Brovchenko, I.V. 242
van der Waals, J.H., *see* Bottler, B.J. 269, 270
van der Waals, J.H., *see* Schmidt, J. 258, 261, 264, 265
van Dorp, W.G., *see* Schmidt, J. 264, 265
van Kooten, J.F.C., *see* Van Strien, A.J. 259, 282
Van Kranendonk, J. 381
Van Strien, A.J. 259, 282
Van Strien, A.J., *see* Bottler, B.J. 270
Van 't Hof, C.A. 269
Van Vechten, J.A. 145
Van Vechten, J.A., *see* Berolo, O. 145
Vandevyver, M., *see* Hoclet, M. 96
Vandevyver, M., *see* Talwar, D.N. 57, 58, 91, 100, 106
Vankan, J.M.J. 281
Vardeny, Z. 91
Vardeny, Z., *see* Livescu, G. 91
Vaterlaus, H.P., *see* Taguchi, I. 128
Vavilov, V.S. 92
Veeman, W.S., *see* Vankan, J.M.J. 281
Vektaris, G.B., *see* Brovchenko, I.V. 242
Velicky, B. 301
Velicky, B., *see* Drchal, V. 314, 358, 359, 362, 364, 371
Velsko, S., *see* Ho, F. 288
Venkataraman, G. 46
Venugopalan, S. 98
Verbin, S., *see* Permogorov, S. 152, 162, 163, 164
Verbin, S.Yu., *see* Permogorov, S.A. 4
Vergnat, P., *see* Vergnat-Grandjean, D. 108
Vergnat-Grandjean, D. 108
Verleur, H.W. 18, 58, 59, 65, 71, 72, 73, 94, 95, 97, 104, 105, 108, 124
Verleur, H.W., *see* Barker, A.S. 108
Vernon, F.L., *see* Feynman, R.P. 261
Vetelino, J.F., *see* Massa, N.E. 71
Vianna, S. 126
Vickers, R.E.M. 128
Vidmont, N.A., *see* Benderskii, V.A. 238
Vinogradov, E.A. 92, 93, 95, 112
Vinogradov, E.A., *see* Gasanly, N.M. 112, 126
Vinogradov, E.A., *see* Mityagin, Y.A. 92
Vinogradov, E.A., *see* Vodopyanov, L.K. 92, 93
Vinogradov, V.S. 58
Vinogradov, V.S., *see* Vavilov, V.S. 92
Vishnevskii, V.N., *see* Bairamov, B.H. 102
Visscher, W.M., *see* Brout, R. 300
Vitrikhovskii, N.I., *see* Artamonov, V.V. 94
Vitrikhovskii, N.I., *see* Burlakov, V.M. 127, 128
Vitrikhovskii, N.I., *see* Valakh, M.Y. 127
Vlasenko, A.I., *see* Broude, V.L. 234
Vodopyanov, L.K. 92, 93, 112, 113, 124
Vodopyanov, L.K., *see* Allakhverdiev, K.R. 112
Vodopyanov, L.K., *see* Golubev, L.V. 112
Vodopyanov, L.K., *see* Kozyrev, S.P. 124
Vodopyanov, L.K., *see* Krozyrev, S.P. 96, 97
Vodopyanov, L.K., *see* Mityagin, Y.A. 92, 95
Vodopyanov, L.K., *see* Plotnichenko, V.G. 95
Vodopyanov, L.K., *see* Vavilov, V.S. 92
Vodopyanov, L.K., *see* Vinogradov, E.A. 93, 95
Vodopyanov, L.P., *see* Allakhverdiev, K.R. 126
Vogel, D., *see* Port, H. 237
Voight, G., *see* Mach, R. 4
Voitchovsky, J.P., *see* Mercier, A. 111
von der Osten, W., *see* Bootz, B. 91
von der Osten, W., *see* Fujii, A. 91
Von Schnering, H.G., *see* Oles, B. 112
Von Schnering, H.G., *see* Oles, B. 124

Wada, K. 206, 208
Wada, N., *see* Krabach, T.N. 109
Wagner, V. 189, 190
Wagner, V., *see* Becker, C.R. 181, 188–190
Wagner, V., *see* Geis, G. 188, 190
Wagner, V., *see* Perry, C.H. 88
Wakamura, K. 112
Walecka, J.D., *see* Fetter, A.L. 47
Walker, C.T., *see* Nair, I. 89, 90
Walker, L.R., *see* Birgeneau, R.J. 204

Walker, P.J., *see* Hayes, W. 205–208
Wallis, R.F., *see* Wanser, K.H. 75
Walsh, D. 113
Wang, X.-J. 125
Wanser, K.H. 75
Ward, A.T., *see* Lucovsky, G. 101, 102
Watanabe, J. 128
Watanabe, N., *see* Kamijoh, T. 125
Watson, P.C., *see* Zitter, R.N. 86
Weaire, D. 280
Weber, R. 191, 192
Weber, R., *see* Mitlehner, H. 194
Wehner, R.K., *see* Bilz, H. 39, 42
Weiss, G.H., *see* Maradudin, A.A. 43, 56
Weisskopf, V. 227, 284
Werthein, G.K. 198
Wheller, S.A., *see* Greene, P.D. 135, 165
Whiteman, J.D., *see* Hochstrasser, R.M. 237
Wicks, G.W., *see* Bour, D.P. 128
Wiersma, D.A., *see* Hesp, B.H. 287
Wigner, E., *see* Weisskopf, V. 227, 284
Wildmont, N.A., *see* Broude, V.L. 237
Wilkinson, G.R., *see* Price, W.C. 321, 375
Wilkinson, M.K. 204, 205
Wiltshire, M.C.K. 197, 198, 205, 207, 209, 210, 211
Wiltshire, M.C.K., *see* Hayes, W. 205–208
Wo Hoang Thay 4
Wojdowski, W., *see* Olszewski, A. 97, 124
Wolf, B.E., *see* Belousov, M.V. 301, 320, 321
Wolf, H.C., *see* Glockner, E. 237
Wolf, H.C., *see* Haarer, D.P. 263
Wolf, H.C., *see* Port, H. 224, 232, 233, 237
Wolf, H.C., *see* Schwoerer, M. 270
Wollan, E.C., *see* Wilkinson, M.K. 204, 205
Wolland, E.O., *see* Wilkinson, M.K. 205
Wolley, J.C., *see* Berolo, O. 145
Wong, P. 203, 210, 211
Wood, T.E. 208, 211
Wood, T.E., *see* Haywood, S.K. 211
Worlock, J.M., *see* Pinczuk, A. 107, 108
Wu, S.Y., *see* Sinai, J.J. 121

Xinh, N.X. 57, 58

Yakovlev, Y.P., *see* Bairamov, B.H. 98
Yakovlev, Yu.G., *see* Tzarenkov, B.V. 4
Yamazaki, S. 104, 125
Yanchev, I.V., *see* Shlimak, I.S. 138, 165
Yaremko, A.M., *see* Lisitsa, M.P. 349
Yavadov, B.M., *see* Bakhyshov, N.A. 112
Yavich, B.S., *see* Alferov, Zh.I. 135
Ye, H.J., *see* Shen, S.C. 124
Yelon, W.B., *see* Birgeneau, R.J. 204, 205, 207
Yonezawa, F. 301
Young, E.F., *see* Perry, C.H. 109
Yushin, A.A., *see* Gasanly, N.M. 128

Zadokhin, B.S., *see* Gamarts, E.M. 111
Zanotti, L., *see* Razzetti, C. 127
Zawadowski, A., *see* Ruvalds, J. 305, 349
Zetterstrom, R.B., *see* Dean, P.J. 135
Zewail, A.H., *see* Lewellyn, M.T. 272
Zhang, X.-Y., *see* Wang, X.-J. 125
Zhao, Fe-Xio, *see* Katnani, A.D. 139
Zheng, Z.-B., *see* Sinai, J.J. 121
Zhilyaev, Yu.V., *see* Alferov, Zh.I. 4
Zhingarev, M.Z., *see* Alferov, Zh.I. 135
Zigone, M. 92
Zigone, M., *see* Beserman, R. 94
Zigone, M., *see* Galtier, P. 103
Zigone, M., *see* Talwar, D.N. 57, 58, 91, 100, 106
Ziman, T.A.L. 206, 207
Zinger, G.M. 18, 20, 65, 73, 74, 85, 101, 107, 127, 143, 158, 163
Zinger, G.M., *see* Barta, C. 112
Zinger, G.M., *see* Biryulin, Yu.F. 101
Zittartz, J. 10, 140
Zitter, R.N. 86
Zouaghi, M., *see* Charfi, F. 101
Zwick, A. 110, 111
Zwick, A., *see* Bedel, E. 106, 125
Zwick, A., *see* Carles, R. 100
Zwick, A., *see* Carles, R.N. 100
Zwick, A., *see* Lockwood, D.J. 128, 211
Zwick, A., *see* Mischler, G. 207
Zwick, A., *see* Nouvel, G. 126

SUBJECT INDEX

absorption linewidth 157
activation energy 136, 166–168
– of hopping conductivity 136, 166–168
activation transport 274
additive refraction 301
adiabatic approximation 97
Anderson localization 9, 13, 17, 163, 253, 300
Anderson model 15–17
Anderson transition 274, 279, 281
anharmonic coupling 123
anharmonic interactions xiv
anharmonicity
– defect 321
– arbitrary 392
– intermolecular 303, 304, 322, 349, 350, 384
– intramolecular 303
– strong 312, 318–320, 324, 330, 352, 358, 361, 389–392
– weak 335
antiferromagnetic resonance 179
antiferromagnetic transition metal fluorides 190
antiferromagnetic transition temperature 181
asymmetric line shape 93, 123, 125
average amplitude approximation (AAA) 243, 246–250
average *T*-matrix approach 18
average *T*-matrix approximation (ATA) 60
average *T*-matrix approximation (ATA) 1D 106

band discontinuities 138, 139
band-edge-renormalization 145, 149
band tailing 4, 13
band tails xiii
band-to-band transitions (BTBT)
– reconstruction of density of states 255
– reconstruction of energy bands 258
– spin BTBT (SBTBT) 257, 258, 282, 283
– transient spectra 256
– vibronic BTBT (VBTBT) 254
Bethe lattice 62, 104, 123
Bethe lattice approximation 80, 85
biaxial mixed crystals 26
binary solid solution 381
binding energy 11, 13
binomial distribution 29, 30
biphonon
– elastic scattering 340
– in imperfect crystal 300, 318
– in one-dimensional (linear) crystal 313
– in perfect crystal 302
– – dispersion relation 305, 309, 313
– in three-dimensional crystal 313
– lifetime 374
– local 300–302, 309, 312, 320–323, 340, 344, 349, 390
– localization 321
– quasi-local 302, 375
– scattering and decay processes 340, 345, 347
– scattering by an isotopic defect 345
– zone width 309, 312, 319, 321, 389
bound state
– effective constant 356
– of the phonon 309
bound three-phonon complex, *see* triphonon 383, 388
Brillouin zone, reduction 87

cadmium chloride structure 204
central cell potential 167–169
chain-like structure 110

characteristic energy 8, 12
CI model 85
cluster, Bethe lattice 75
cluster, embedded 75, 122
cluster calculations 46, 75
cluster model 18, 29, 183
cluster structure 251, 252
clustering 124, 125
clustering effects 187
cobalt impurity mode 193
coherence 260, 275
– controlled by inter-impurity transitions 271
– optical detection of relaxation times 263
– phase coherence time 269, 271, 275
– phase memory time 275
– probe pulse methods 264
– rotary echo 265, 266
– spin-locking method 267, 269
– temperature dependence 269
– transient nutation 264, 265, 271
– two-pulse echo (Hahn) 266, 267, 269, 282
coherent anti-Stokes Raman scattering (CARS) 287
coherent potential approximation (CPA) xiii, 18, 60, 65, 71, 81, 85, 100, 121, 181, 206, 249, 301, 302, 359–361, 364, 367, 368, 370, 377, 378, 380, 383
– 1D Homorphic 61
– model 80
composition
– disorder 7, 17
– fluctuations of 10, 18, 23, 28
composition fluctuations 136–139, 141, 142, 145, 152, 155–157, 164–167, 171
– nonlocal interaction with 137, 171
concentrated mixtures 190
configuration averages 54
continuum 123
continuum anharmonic coupling 102
correlation function 50
Coulomb field 46, 49
Coulomb forces 42, 47
critical scattering 197
cross section of the process of scattering a biphonon by an impurity 345
crystal-field effects 198

damping and strength 45
DATA 50
Davydov splitting 220, 222, 228, 237, 249, 250, 252, 286
– in mixed crystals 246
– in vibrational spectra 286
defect activated first-order scattering 89
defects xiii
density of states 4, 5, 7, 9, 24, 26, 136, 139–142, 145, 146, 149, 151, 152, 154, 155, 159, 161
– one-particle 314
– prefactor 140, 141, 149
– tail 156
– two-particle 325, 377
– two-phonon 377
– two-phonon vibrational 377
dephasing 260, 271, 275, 287, 288
diagonal disorder 305
dielectric constant 44–46, 48, 62, 67
diffusion edge, *see* mobility edge 274
diffusion region 7, 9, 13, 22, 32
dilute mixtures 200
dilute systems 194
dimensionless variables 13, 24
dipole-active oscillation 348
directional dependence 26
directional dispersion 26–30, 32
disorder activation 41
disordered sublattice (DSL) 70
disordered systems
– excitons 217
dispersion equation for biphonons 309
dispersion relation
– for biphonon in crystal with isotope impurity 330
– for three-phonon 387, 388
distribution function 20, 26, 30–32
diverging phonon waves 342
donor-like case 156, 159, 161
dynamical localization 275

effective anisotropy field 189
effective charge tensor 46, 49
electric field 65
electron mobility 164, 165
electron state
– delocalized (extended) 13, 15–17
– localized 11, 13
energy band reconstruction 237, 254
energy region of the combination tone 381
energy spectrum
– in amalgamation limit 250
– in persistent limit 250
exciton
– absorption bands

– – sharply polarized 222
– – weakly polarized 222
– absorption linewidth 136, 154, 155, 157, 161
– bound by an impurity 135, 154, 156, 157, 159, 160
– charge transfer 217
– dimer, *see* mini-exciton 238, 268, 269
– energy bands 222, 238
– fluorescence spectra 234, 237, 268, 276, 277
– four- or five-dimensional behaviour 290
– Frenkel 217
– in disordered systems 217
– in strong magnetic field 161
– local 229, 232, 234
– localized by composition fluctuation 152
– molecular 217
– monomer 229, 238
– multiplet 222, 228, 231
– photoluminescence 162
– reflection spectrum 135, 152
– singlet 233, 242, 248, 250, 251, 279, 285, 287
– transport critical behaviour 276, 278, 285
– triplet 224, 226, 232, 238, 257, 258, 263, 269, 279, 282, 287
– Wannier–Mott 217
excitons xi, xiv
extended states 17
extended wave function 14

Fe–Co exchange 211
Fermi resonance (FR) 348, 353, 354, 390
ferromagnetic interactions 206
ferromagnetic nearest-neighbour exchange 192
Feynman–Vernon–Hellworth method 261
fluctuation
– Fermi 12, 29–32
– Gaussian 12, 28–30
– of composition 10, 21
– of crystal potential 3
– of mixed crystal composition 21
– Poisson 12
fluctuations xiii, xiv, 137–142, 152–158, 164, 165, 167–169
– large-scale 168, 170
– small-scale 145, 147, 149
force constant matrix 42, 45
force constants, disorder 62
force constants; concentration dependence of 67
Fourier transform spectroscopy 191
fractional concentration 3, 4, 18
fundamental absorption 302
fundamental absorption edge 152

gap mode 38, 55, 76
Gaussian statistics 140, 141
generalised Lyddane–Sachs–Teller 67
generating function 127
Green function 43, 49, 50, 102, 183, 187
Green function, displacement/displacement 48

Hall mobility 165
hopping conduction 136, 166, 167, 169
– activation energy 136, 166
hot-band spectroscopy 255

impurities in the rock salt structure 58
impurity
– chemical 235, 237
– excitons 229
– isotope 235, 237
– mode 39
impurity–impurity exchange interactions 192
incoherent transport 273
increasing dimensionality 27
inelastic neutron scattering 88
inelastic neutron scattering experiments 193, 204
infinite cluster 17
infinitely connected cluster 197
infrared absorption 41, 43
infrared absorption coefficient 184, 192, 194, 205, 210
infrared fluorescence 203
infrared spectroscopy 179
inter-cell interactions 73
interaction
– intermediate 355
– intermolecular 303, 304
– strong 355
– three-particle 302, 383, 384
– two-particle 384
– weak 354
intra-atomic shell/core forces 61
intra-cell interactions 73
intra-impurity absorption 135, 169

irreducible representations 56
Ising cluster model 182
isodisplacement 121
isodisplacement model 59
isoelectronic substitution 3, 6
isoshift model 301
isotope impurity 19, 20
isotopic impurity 318
– substitutional 318
isotopic shift 331, 333, 336, 356, 367, 375, 378, 382
isotopically mixed crystals 301, 358, 359, 381, 382

Lamb shift 146
layer perovskite structure 202
linear chain 81
LO–TO splitting 122
local bound multiphonon states 391
local environment effect 61
local excitons
– aggregate centers 236
– chemical impurity 235
– conductivity tensor 230
– defect centers 237
– dependence on dimensionality 229
– dimers (mini-excitons) 236, 238, 268, 269
– energy levels 229
– monomers 229
– on isotopic impurity centres 229
– oscillator strength 230
– phonon satellites in fluorescence spectrum 234
– polarization ratio 230
– repelling 229
– trimers 236
– triplet 260
– vibronic spectra 240, 241
– wave function 230
local field 43, 46, 47, 62, 65
local mode 38, 39, 55, 76
local mode, Si 85
local phonons 390
local relaxation 45
local triphonon 390
local vibration 20, 24, 361
– frequency 20
– optical 318
localization xiii, 273, 279, 280
– length 274
– of biphonon 321
– of quasi-particle 303
– of states 82
– of triphonon 390
– threshold 9, 13, 15, 136, 161, 162, 164
localized modes xiii, 183
locator 65
long-range disorder 136
long wavelength AFMR 205
longitudinal and transverse responses 62
longitudinal optic frequencies 43
longitudinal/transverse splitting (LT) 44, 65
low-concentration theory 87

macroscopic field 51
macroscopic fluctuations 13, 22
magnetic dipole–dipole interactions 183
magnetic impurity modes 180
magnetic insulators 179
magnon band 192
magnons xi, xiv, 179
manganese impurities 191
manganese impurity mode 193
materials with the sodium chloride, rutile, perovskite and cadmium chloride structures 182
mean-field model 203
$MgCl_2$ 205
mixed binary crystals 358
mixed crystal composition 3, 5
mixed crystals xiii, xiv, 181
– classification of spectra 250
– exciton spectra 242
– – one-mode regime 246
– – two-mode regime 246
– vibronic spectra of 252
mixed cubic perovskite 199
mixed-mode behaviour 40, 68, 70, 76, 77
mixing between acoustic and optic branches 58
mixtures of magnetic ions 199, 210
mobility
– edge, *see* diffusion edge 274
– of band electron 135, 165
– of hole 165
– threshold 163
mode behaviour of zone boundary (ZB) 82
model Hamiltonian 302
molecular excitons 217
– absorption spectrum 226
– conductivity tensor 221
– density of states 221

– effect of electric field 224
– energy bands 220
– energy relaxation time 276
– energy spectrum 220, 222
– energy spectrum of relaxation time 287
– identification 228
– in a rigid lattice 219
– in doped crystals 228
– in mixed crystals 242
– in polymers 228
– local 229
– reconstruction of density of states 233
– reconstruction of energy bands 237
– scattering 285
– shape of absorption band 284
– triplet 223
– vibrational 286, 287
– vibronic interaction 239
Monte Carlo technique 391–393
Mott criterion 280
multicritical point 197
multimode 45
multimode behaviour 128
– dimers in 269
– exciton band 237, 238
– local excitons 231
– phase memory of dimers 269
– polaritons 225
– vibronic spectra 240, 241

Napier factorization method (MNF) 249
narrowing of the band gap 145
Néel temperature 204
nondiagonal disorder 305
nonhomogeneous broadening 145
normal modes xii

off-centre impurities 91
off-diagonal elements 61
one-mode behaviour 38, 40, 68, 76, 77, 369
one-site approximation 362
1D ATA 106
1D chains 122
1D embedded cluster 127
1D Homorphic CPA 61
optimum fluctuation 10, 12, 22, 29, 31, 140, 146–148, 157
– method 136, 140, 146, 158
oscillator frequency 45
oscillator strength 44, 67
overlap integral 14
overlap of defects 51
overtone frequency region 303

packing faults 138
pair correlation function 8, 20
partly two-mode behaviour 76
percolation 167, 278
percolation limit 201
perturbation matrices 54
phase-matching condition 289
phonon damping 24
phonon displacements 50
phonon Green function 52
phonon localization 27
phonons xi
– external (lattice) (LP) 219
– internal (IP) 239, 286
photoluminescence
– of exciton 162
– polarization of 163
planar stacking faults 27
plasmons xii
polar crystals 43
polarisability tensor 50, 51, 88, 98
polarisation 44, 46
polaritons 91, 224–226
prefactor 140, 141, 147, 149
probability density 10, 11, 23
process
– "pickup" 346, 348
– stripping 340, 346, 348
pseudoatom approximation 71
pseudoatom shell-model 89

quasi-biphonons 374, 375
quasi 1D materials 123
quasi-spin dynamics 261, 271
quasistatic approximation 50
quaternary systems 60, 75, 122

Raman effect xiv
Raman scattering 41, 43, 51, 57, 71, 179
Raman scattering intensity 184, 199, 211
Raman spectra 302
random element isodisplacement (REI) model 40, 65, 71, 82, 87
Rashba effect 228, 232, 237
Rb_2CoF_4 202
recursion method 76, 81, 121, 126
reflectivity 45
relaxation times 263

renormalisation group 82
replica 105
replica trick 76
resonancc
– frequencies 190
– of biphonon and triphonon 390
– of phonon and triphonon 390
resonant mode 40
resonant mode, Ge 85
resonant Raman effect 100
resonant Raman scattering 50, 92, 97, 106, 124, 125

saturated paramagnetic state 209
scattering amplitude 5, 6, 20
scattering length 6, 21
selection rule xii
shallow local states 331
shell model 46, 61
shift
– band-edge 142, 143
– of bands 156
– of energy 136, 142, 147
– of impurity level 142, 145
short-range correlation 3
short-range order 72, 83
short-range potential 5, 16
simulation method 188
simulation techniques 209
single impurity approximation 85
single-ion anisotropy 188
single-photon processes xiv
single-site approximations 186
single-site CPA 198
spacial correlation 125
spacial correlation function 123
spectral function 81
spectral response function 188
spin-echo 282
spin–spin interaction 196
spin waves xi
state
– delocalized 17
– extended, *see* nonlocalized 274, 276
– localized 274, 276
– one-particle 332, 336, 337, 354, 356, 357, 361
– three-phonon 384, 386–389
– two-particle 301, 309, 317, 326, 331, 343, 353, 354, 356, 374, 379, 381
– two-phonon 378, 384
substitutional atoms 135–137, 145
susceptibility 47, 71
symmetry selection rules 41
Szigeti effective charge 43, 66

T-matrix 301, 360, 361
t-matrix 54, 55, 100
tailing
– of the density of states 17
– of the frequency distribution function 24
three-phonon vibrations 320
transition metal oxides 188
transmission coefficient 45
transverse Coulomb force 66
transverse effective charge 44, 47
transverse exchange 207
transverse optic modes 43
triphonon 383
– zone width 389
triplet excitons
– energy relaxation time 282
– energy spectrum 257
– local 260
– optical detection of EPR 258
two-dimensional anisotropic ferromagnet 207
two-dimensional ferromagnet 206
two-level systems
– quasi-spin description 260, 261, 263
two-magnon spectrum 194, 196, 201
two-mode behaviour 38–40, 68, 76, 77, 369, 380
two-mode systems 71
two-phonon contributions 41
two-photon processes xiv

uncorrelated distribution 138
uncorrelated potential 146
uniaxial mixed crystals 26, 27, 31
universal parameters 19, 24

vacancy defect 331, 338
Van Kranendonk model 381, 382
vibronic spectra
– dynamical theory 239, 252
– of monomer centres 238
virtual crystal 60, 70, 135, 169, 170
– approximation 18, 135, 142, 185

wave vector xii
"white noise" 142
– potential 136
wurtzite/zincblende 95

zincblende structure 56, 57
zone-edge magnons 200

MATERIALS INDEX

AgBr 91
AgCl 91
Ag(Cl/Br) 77, 83, 91
AlAs 106
(Al/Ga)As 3, 62, 79, 99, 100, 122, 123, 125
(Al/Ga)P 79, 98
(Al/Ga)(P/As) 106, 107, 122
(Al/Ga)Sb 79, 83, 100
(Al/In)P 128
anthracene, *see* $C_{14}H_{10}$
(Au/Ag) 3

$Ba(H/F)_2$ 108
$Ba_2(Na/K)Nb_5O_{15}$ 112
$(Ba/Pb)TiO_3$ 112
benzene, *see* C_6H_6
Bi/Sb 86
$Bi_2(Se/Te)_3$ 112

CaF_2 107
$Ca(H/F)_2$ 108
$Ca(Mo/W)O_4$ 112
$(Ca/Pb)F_2$ 109
$(Ca/Sr)F_2$ 58, 59, 72, 108, 109
Cd(Br/I) 128
Cd(Cl/Br) 128
$C_6D_2(Cl/CD_3)_4$ 265, 267, 272
$Cd(Cr/In)_2S_4$ 128
$Cd(Ga/In)_2S_4$ 127
$CdGa_2(S/Se)_4$ 127
(Cd/Hg)Se 128
(Cd/Hg)Te 78, 95, 96, 124, 125
(Cd/Mn)Te 212
$(Cd/Pb)F_2$ 128
CdS 69, 94, 95
CdSe 94, 95
Cd(Se/Te) 78, 95, 96
Cd(S/Se) 45, 71, 72, 78, 83, 94, 152, 162
Cd(S/Se/Te) 127
Cd(S/Te) 78, 95
CdTe 93, 95, 97
C_6H_6 231, 234
$C_{10}H_8$ 222–224, 226, 231, 233–235, 238, 240, 241, 253, 263, 269, 270, 281, 285
$C_{14}H_{10}$ 222, 224, 232, 237, 263
$C_{22}H_{14}$ 242, 244
$C_{10}H_6Br_2$ 237, 256, 281
$[C_6H_4(CH_3)]_2CO$ 282
$C_6H_2Cl_4$ 258, 259, 282, 283
$C_6(H/D)_6$ 242, 243, 248, 278, 287, 288
$C_{10}(H/D)_8$ 249–251, 276–278, 287
$C_{10}(H/D)_6Br_2$ 233, 242, 248, 281
C_9H_7N 265, 267
$(C_8H_4N_2)Cl_2$ 267
$(Co/Cd)Cl_2$ 208
$CoCl_2$ 204, 205, 210, 211
CoF_2 191, 193
$(Co/Mg)Cl_2$ 207, 208
$(Co/Mn)F_2$ 191–193
(Co/Ni)O 107, 188, 189
CoO 188
$Cs(Mg/Co)Cl_3$ 112
(Cu/Ag)I 124
$Cu(Al/Ga)Se_2$ 128
Cu(Br/I) 77, 90
CuCl 91
Cu(Cl/Br) 77, 90, 91
Cu(Cl/I) 77, 90, 91
$Cu(Ga/In)S_2$ 112
$CuGa(S/Se)_2$ 112

deuterobenzenes, *see* $C_6(H/D)_6$
deuteronaphtalenes, *see* $C_{10}(H/D)_8$
dibromonaphtalene, *see* $C_{10}H_6Br_2$
dichloroquinoxaline, *see* $(C_8H_4N_2)Cl_2$
dimethylbenzophenone, *see* $[C_6H_4(CH_3)]_2$-CO
durene, *see* $C_6D_2(Cl/CD_3)_4$

$Eu(P/As)_3$ 112
$(Eu/Sr)As_3$ 112

$FeBr_2$ 210
$(Fe/Cd)Cl_2$ 207
$FeCl_2$ 204, 205, 207, 209–211
FeCl/Co 210
$(Fe/Co)Cl_2$ 128, 203, 210, 211
FeF_2 179, 180, 191, 192
$(Fe/Mg)Cl_2$ 205–207, 209–211
$(Fe/Mn)Cl_2$ 128, 210, 211
$(Fe/Mn)F_2$ 126, 192, 199
FeO 188
$(Fe/Zn)F_2$ 126, 197, 198

(Ga/Al)As 122
(Ga/Al)(As/P) 60
GaAs 77, 99, 105–107, 122, 125
$(GaAs)/(Ga_2Se_3)$ 112
(GaAs)/Ge 135
Ga(As/P) 152
Ga(As/Sb) 65, 77, 79, 105, 106, 123, 125
(Ga/In)As 76, 77, 79, 104, 105, 121, 125, 139, 152
(Ga/In)(As/Sb) 123, 128
(Ga/In)P 65, 69, 70, 72, 76, 77, 79, 82, 83, 88, 101–103, 105, 121–123, 125, 169
Ga(In/P) 68
(Ga/In)(P/As) 102, 106–108, 127
(Ga/In)Sb 62, 64, 65, 75, 77, 79, 104, 106
(Ga/In)Se 112
GaP 69, 98, 105, 106
(Ga/P)As 103
Ga(P/As) 71–74, 76, 79, 83, 105, 106, 123, 125
GaSb 76, 105, 109
$GaSb/Ge_2$ 122, 126
Ga(Se/Te) 113
Ga(S/Se) 111, 112, 128
$Ga_2(S/Se)_3$ 112
Ge(GaSb) 109
Ge/Si 3, 18, 165
(Ge/Sn)S 112
(Ge/Sn)Se 112
Ge(S/Se) 112

(H/D)Cl 112
$Hf(S/Se)_2$ 128
$Hf(S/Se)_3$ 110, 111
$Hg(Br/Cl)_2$ 3, 18
$Hg(Br/I)_2$ 112
$Hg(Cl/Br)_2$ 112, 126
HgTe 97

InAs 104, 106, 107, 125
In(As/Sb) 77, 79, 106
(In/Ga)(As/P) 165
InP 69, 106, 107
In(P/As) 79, 106, 122
InSb 105
isocyanine 228

K(Br/I) 83, 90
KCl 89
K(Cl/Br) 65, 77, 88–90
K_2CoF_4 202, 203
$K_2(Co/Fe)F_4$ 203
$K_2(Co/Mn)F_4$ 203
K_2FeF_4 203
$KMgF_3$ 110
$K(Mg/Ni)F_3$ 109, 110
$K(Mn/Co)F_3$ 199
$KMnF_3$ 199
$K_2(Mn/Fe)F_4$ 203
$K(Mn/Ni)F_3$ 199
$KNiF_3$ 200–202
K_2NiF_4 202
$K(Ni/Mg)F_3$ 200, 201
(K/Rb)Br 77, 88, 89, 123
(K/Rb)Cl 77, 88, 90
(K/Rb)(Cl/Br) 89
(K/Rb)I 63–65, 76, 77, 88, 89, 109, 124

Li(H/D) 77, 86
$(Li^6/Li^7)F$ 86

(Mg/Cd)Te 78, 92
$MgCl_2$ 205, 206, 209
$(Mg/Fe)Cl_2$ 206
MgTe 92
(Mg/Zn)S 78, 91
(Mg/Zn)Te 78, 92
$MnBr_2$ 210
(Mn/Cd)Te 78, 97, 124
$MnCl_2$ 205, 210
$(Mn/Co)F_2$ 186, 193, 194
(Mn/Co)O 188
Mn/F_2 196
MnF_2 179–181, 191–194, 197
$(Mn/Fe)F_2$ 199
(Mn/Hg)Te 78, 98, 125
MnO 188

MnTe 97
(Mn/Zn)F_2 181, 186, 187, 194, 195, 197, 198, 206
(Mn/Zn)Te 78, 124
M(S/Se)$_3$ 123

Na(Cl/Br) 75
napthalene, *see* $C_{10}H_8$
NH_4Br 69, 320, 321
NH_4Cl 69, 374, 375
NH_4(Cl/Br) 68, 69, 77, 83, 87
(NH_4/K)Cl 61
(NH_4/K)$_2CuCl_4 \cdot 2H_2O$ 112
(Ni/Co)O 188–190
NiO 188
Ni(S/Se)$_2$ 112
(^{14}N/^{15}N)H_4Br 320, 321
(^{14}N/^{15}N)H_4Cl 375–377

Pb_5(Ge/SiO_4)(VO_4)$_2$ 128
Pb(Mo/W)O_4 26, 32
Pb(Se/Te) 109
PbTe 109
pentacene, *see* $C_{22}H_{14}$
Pt(S/Se)$_2$ 128

quinoline, *see* C_9H_7N

RbBr 89
(Rb/Cl)Br 88
Rb(Cl/Br) 77, 88, 89, 123
Rb_2CoF_4 202, 203
Rb_2(Co/Mg)F_4 187, 188
Rb(Mn/Co)F_3 199
Rb_2(Mn/Mg)F_4 187, 188, 207
Rb_2(Mn/Ni)F_4 204
Rb_2(Ni/Mn)F_4 188

Sb/Bi 87
(Sb/Bi)Te_3 112
SbS(Br/I) 112
Se/Te 86
Si/Ge 41, 49, 74, 83, 84
(Si/Ge) 166
Sn(S/Se) 112
Sn(S/Se)$_2$ 128
(Sr/Ba)F_2 72, 108
(Sr/Cd)F_2 109
(Sr/Eu)S 112
Sr(H/F)$_2$ 108
(Sr/Pb)TiO_3 112
$S_2Yb_2F_8$ 163

tetrachlorobenzene, *see* $C_6H_2Cl_4$
TiGa(S/Se)$_2$ 112
(Ti/Hf)Se_2 128
Ti(Se/Te)$_2$ 128
Ti(S/Se)$_2$ 128
(Ti/Zr)Se_2 128
Tl(Ga/In)S_2 112
Tl(Ga/In)Te_2 126
TlIn(Se/Te)$_2$ 126
Tl(InS_2)/($GaSe_2$) 112
Tl(In/Tl)Se_2 126
Tl(S/Se) 112, 126

(Zn/Cd)Cr_2Se_4 112
(Zn/Cd)Ga_2S_4 127
(Zn/Cd)S 38, 39, 68, 69, 75, 78, 92, 152
(Zn/Cd)Se 78, 93, 152
(Zn/Cd)Te 78, 93, 123, 152
ZnF_2 181
$ZnGa_2$(S/Se)$_4$ 127
(Zn/Hg)Cr_2Se_4 112
(Zn/Hg)Te 78, 128
(ZnHg)Te 93
(Zn/Mn)Te 97
ZnS 69, 93
ZnSe 69, 93
(ZnSe)/(GaAs) 112
Zn(Se/Te) 76, 78, 93, 94, 128
Zn(S/Se) 26, 39, 40, 45, 69, 70, 74, 76, 78, 83, 94
Zn(S/Se)$_3$ 110
Zn(S/Se/Te) 127
ZnTe 97
(ZnTe)/(CdSe) 112
(ZrHf)O_2 112
(Zr/Hf)S_3 126
Zr(S/Se)$_3$ 123, 126
(Zr/Ti)S_3 126

CUMULATIVE INDEX, VOLUMES 1–23

Monographs

Agranovich, V.M., and M.D. Galanin, Electronic excitation energy transfer in condensed matter **3**

Brandt, N.B., S.M. Chudinov and Ya.G. Ponomarev, Semimetals – 1. Graphite and its compounds **20.1**

Chudinov, S.M., see Brandt, N.B. **20.1**

Galanin, M.D., see Agranovich, V.M. **3**
Gantmakher, V.F., and Y.B. Levinson, Carrier scattering in metals and semiconductors **19**
Gurevich, V.L., Transport in phonon systems **18**

Levinson, Y.B., see Gantmakher, V.F. **19**

Mashkova, E.S., and V.A. Molchanov, Medium-energy ion scattering by solid surfaces **11**
Molchanov, V.A., see Mashkova, E.S. **11**

Ponomarev, Ya.G., see Brandt, N.B. **20.1**

Wagner, M., Unitary transformations in solid state physics **15**

Contributed Volumes

Abelés, F., and T. Lopez-Rios, Surface polaritons at metal surfaces and interfaces **1**, 239
Agranovich, V.M., Effects of the transition layer and spatial dispersion in the spectra of surface polaritons **1**, 187
Agranovich, V.M., V.E. Kravtsov and T.A. Leskova, Scattering of surface polaritons by order parameter fluctuations near phase transition points **1**, 511
Agranovich, V.M., Biphonons and Fermi resonance in vibrational spectra of crystals **4**, 83
Agranovich, V.M., Surface excitons **9**, 513

Agranovich, V.M., and V.V. Kirsanov, Production of radiation defects by collision cascades in metals **13**, 117
Agranovich, V.M., and O.A. Dubovsky, Phonon multimode spectra: biphonons and triphonons in crystals with defects **23**, 297
Alexandrova, I.P., Nuclear quadrupole resonance of the incommensurate phase in Rb_2ZnCl_4 and Rb_2ZnBr_4 **14.1**, 277
Altshuler, B.L., and A.G. Aronov, Electron–electron interaction in disordered conductors **10**, 1
Al'tshuler, S.A., A.Kh. Khasanov and B.I. Kochelaev, Studies of spin systems by means of light scattering in paramagnetic crystals **21**, 607
Aoyagi, K., see Tanabe, Y. **2**, 603
Aronov, A.G., and A.S. Ioselevich, Exciton electrooptics **2**, 267
Aronov, A.G., see Altshuler, B.L. **10**, 1
Aronov, A.G., Yu.M. Gal'perin, V.L. Gurevich and V.I. Kozub, Nonequilibrium properties of superconductors (transport equation approach) **12**, 325
Aslamazov, L.G., and A.F. Volkov, Nonequilibrium phenomena in superconducting weak links **12**, 65
Axe, J.D., M. Iizumi and G. Shirane, Phase transformations in K_2SeO_4 and structurally related insulators **14.2**, 1

Bagaev, V.S., T.I. Galkina and N.N. Sibeldin, Interaction of EHD with deformation field, ultrasound and non-equilibrium phonons **6**, 267
Bak, P., see Pokrovsky, V.L. **17**, 71
Bar'yakhtar, V.G., and E.A. Turov, Magnetoelastic excitations **22.2**, 333
Basiev, T.T., V.A. Malyshev and A.K. Przhevuskii, Spectral migration of excitations in rare-earth activated glasses **21**, 275
Basun, S.A., see Kaplyanskii, A.A. **16**, 373
Beasley, M.R., see De Lozanne, A.L. **12**, 111
Belousov, M.V., Vibrational Frenkel excitons **2**, 771
Belyakov, V.A., and E.I. Kats, Light scattering and phase transitions in liquid crystals **5**, 227
Benoit à la Guillaume, C., see Planel, R. **8**, 353
Bereznyak, P.A., see Slyozov, V.V. **13**, 575
Berkovits, V.L., see Paget, D. **8**, 381
Bernard, L., see Currat, R. **14.2**, 161
Bersuker, I.B., and V.Z. Polinger, Theoretical background of the Jahn–Teller effect **7**, 21
Beyer, J., C.J. Pethick, J. Rammer and H. Smith, Pair breaking in nonequilibrium superconductors **12**, 129
Bhatt, R.N., see Milligan, R.F. **10**, 231
Bill, H., Observation of the Jahn–Teller effect with electron paramagnetic resonance **7**, 709
Birman, J.L., Electrodynamic and non-local optical effects mediated by exciton polaritons **2**, 27
Bishop, A.R., see Steiner, M. **17**, 783
Blinc, R., P. Prelovšek, V. Rutar, J. Seliger and S. Žumer, Experimental observations of incommensurate phases **14.1**, 143
Borovik-Romanov, A.S., and N.M. Kreines, Light scattering from spin waves **22.1**, 81
Borovik-Romanov, A.S., and S.K. Sinha, Introduction **22**, xi
Borstel, G., see Wöhlecke, M. **8**, 423
Bron, W.E., Phonon generation, transport and detection through electronic states in solids **16**, 227

Bullough, R., and M.H. Wood, Theory of microstructural evolution **13**, 189
Burns, M.J., W.K. Liu and A.H. Zewail, Nonlinear laser spectroscopy and dephasing of molecules: An experimental and theoretical overview **4**, 301
Burstein, E., see Chen, Y.J. **1**, 587
Büttner, H., see Mertens, F.G. **17**, 723

Cailleau, H., Incommensurate phases in an aromatic molecular crystal: biphenyl **14.2**, 71
Celotta, R.J., see Pierce, D.T. **8**, 259
Challis, L.J., and A.M. de Goër, Phonon spectroscopy of Jahn–Teller ions **7**, 533
Challis, L.J., Crossing effects in phonon scattering **16**, 145
Chang, Jhy-Jiun, Kinetic equations in superconducting thin films **12**, 453
Chen, Y.J., and E. Burstein, Three wave nonlinear interactions involving surface polaritons; Raman scattering, light diffraction and parametric mixing **1**, 587
Chubukov, A.V., see Kaganov, M.I. **22.1**, 1
Clarke, J., Experiments on charge imbalance in superconductors **12**, 1
Clinard Jr, F.W., and L.W. Hobbs, Radiation effects in non-metals **13**, 387
Cone, R.L., and R.S. Meltzer, Ion–ion interactions and exciton effects in rare earth insulators **21**, 481
Cottam, M.G., and A.A. Maradudin, Surface linear response functions **9**, 1
Cummins, H.Z., Brillouin scattering studies of phase transitions in crystals **5**, 359
Currat, R., see Durand, D. **14.2**, 101
Currat, R., see Dénoyer, F. **14.2**, 129
Currat, R., L. Bernard and P. Delamoye, Incommensurate phase in β-$ThBr_4$ **14.2**, 161

Davydov, A.S., Solitons in biology **17**, 1
de Goër, A.M., see Challis, L.J. **7**, 533
De Lozanne, A.L., and M.R. Beasley, Time-dependent superconductivity in SNS bridges: an example of TDGL theory **12**, 111
de Wolff, P.M., and F. Tuinstra, The incommensurate phase of Na_2CO_3 **14.2**, 253
Delamoye, P., see Currat, R. **14.2**, 161
deMartini, F., see Shen, Y.R. **1**, 629
Dénoyer, F., see Durand, D. **14.2**, 101
Dénoyer, F., and R. Currat, Modulated phases in thiourea **14.2**, 129
Dmitriev, V.M., V.N. Gubankov and F.Ya. Nad', Experimental study of enhanced superconductivity **12**, 163
Dolgopolov, V.T., and I.L. Landau, Resistive states in type-I superconductors **12**, 641
Dolino, G., The incommensurate phase of quartz **14.2**, 205
Dubovsky, O.A., see Agranovich, V.M. **23**, 297
Durand, D., F. Dénoyer, R. Currat and M. Lambert, Incommensurate phase in $NaNO_2$ **14.2**, 101
Dyakonov, M.I., and V.I. Perel', Theory of optical spin orientation of electrons and nuclei in semiconductors **8**, 11

Echegut, P., see Gervais, F. **14.1**, 337
Efros, A.L., and B.I. Shklovskii, Coulomb interaction in disordered systems with localized electronic states **10**, 483
Efros, A.L., and M.E. Raikh, Effect of composition disorder on the electronic properties of semiconducting mixed crystals **23**, 133
Ehrhart, P., K.H. Robrock and H.R. Schober, Basic defects in metals **13**, 3

Elesin, V.F., see Galitskii, V.M. **12**, 377
Eliashberg, G.M., and B.I. Ivlev, Theory of superconductivity stimulation **12**, 211
Elliott, R.J., Introduction **23**, xi
Eremenko, V.V., Yu.G. Litvinenko and E.V. Matyushkin, Optical magnetic excitations **22.1**, 175
Eremenko, V.V., and V.M. Naumenko, Magnetic impuritons in antiferromagnetic dielectric crystals **22.2**, 259
Errandonéa, G., see Tolédano, J.C. **14.2**, 233

Fayer, M.D., Exciton coherence **4**, 185
Feigel'man, M.V., see Pokrovsky, V.L. **22.2**, 67
Fischer, B., see Lagois, J. **1**, 69
Fleisher, V.G., and I.A. Merkulov, Optical orientation of the coupled electron–nuclear spin system of a semiconductor **8**, 173
Fleury, P.A., and K.B. Lyons, Central peaks near structural phase transitions **5**, 449
Flores, F., and F. Garcia-Moliner, Electronic surface excitations **9**, 441
Frait, Z., and D. Fraitová, Spin-wave resonance in metals **22.2**, 1
Fraitová, D., see Frait, Z. **22.2**, 1
Fukuyama, H., Interaction effects in the weakly localized regime of two- and three-dimensional disordered systems **10**, 155
Fulde, P., and M. Loewenhaupt, 4f Moments and their interaction with conduction electrons **22.1**, 367
Furrer, R., Electron paramagnetic resonance in the excited states of rare earth ions in crystals **21**, 641

Galaǐko, V.P., and N.B. Kopnin, Theory of the resistive state in narrow superconducting channels **12**, 543
Galitskii, V.M., V.F. Elesin and Yu.V. Kopaev, The kinetic theory of superconductors with excess quasiparticles **12**, 377
Galkina, T.I., see Bagaev, V.S. **6**, 267
Gal'perin, Yu.M., see Aronov, A.G. **12**, 325
Garcia-Moliner, F., see Flores, F. **9**, 441
Gershenzon, E.M., A.P. Mel'nikov and R.I. Rabinovich, H^--like impurity centers, molecular complexes and electron delocalization in semiconductors **10**, 483
Gervais, F., and P. Echegut, Infrared studies of incommensurate systems **14.1**, 337
Ginzburg, V.L., A.A. Sobyanin and A.P. Levanyuk, General theory of light scattering near phase transitions in ideal crystals **5**, 3
Gladkii, V.V., Macroscopic electric quadrupole moment in the incommensurate phase in ferroelectrics **14.1**, 309
Glinchuck, M.D., Paraelectric resonance of off-center ions **7**, 819
Goldburg, W.I., Light scattering investigations of the critical region in fluids **5**, 531
Goldman, A.M., see Kadin, A.M. **12**, 253
Golovko, V.A., and A.P. Levanyuk, Light scattering from incommensurate phases **5**, 169
Gor'kov, L.P., Disorder and interactions in the system of quasi one-dimensional electrons **10**, 619
Grun, J.B., B. Hönerlage and R. Lévy, Biexcitons in CuCl and related systems **2**, 459
Gubankov, V.N., see Dmitriev, V.M. **12**, 163
Gurevich, V.L., see Aronov, A.G. **12**, 325
Guseva, M.I., and Yu.V. Martynenko, Blistering **13**, 621

Haarer, D., and M.R. Philpott, Excitons and polarons in organic weak charge transfer crystals **4**, 27
Hamano, K., Thermal hysteresis phenomena in incommensurate systems **14.1**, 365
Hanson, D.M., J.S. Patel, I.C. Winkler and A. Morrobel-Sosa, Effects of electric fields on the spectroscopic properties of molecular solids **4**, 621
Harley, R.T., Spectroscopic studies of Jahn–Teller phase transitions in rare-earth crystals **21**, 557
Hayes, W., and M.C.K. Wiltshire, Infrared and Raman studies of disordered magnetic insulators **23**, 177
Hermann, C., and C. Weisbuch, Optical detection of conduction electron spin resonance in semiconductors and its application to ***k***, ***p*** perturbation theory **8**, 463
Hesselink, W.H., and D.A. Wiersma, Theory and experimental aspects of photon echoes in molecular solids **4**, 249
Hizhnyakov, V.V., and N.N. Kristoffel, Jahn–Teller mercury-like impurities in ionic crystals **7**, 383
Hobbs, L.W., see Clinard Jr, F.W. **13**, 387
Hönerlage, B., see Grun, J.B. **2**, 459
Horovitz, B., Solitons in charge and spin density wave systems **17**, 691
Huber, D.L., Energy transfer in crystals **21**, 251

Iizumi, M., see Axe, J.D. **14.2**, 1
Ioselevich, A.S., see Aronov, A.G. **2**, 267
Ipatova, I.P., Universal parameters in mixed crystals **23**, 1
Ishibashi, Y., Phenomenology of incommensurate phases in the A_2BX_4 family **14.2**, 49
Ivchenko, E.L., Spatial dispersion effects in the exciton resonance region **2**, 141
Ivchenko, E.L., see Pikus, G.E. **2**, 205
Ivlev, B.I., see Eliashberg, G.M. **12**, 211

Janssen, T., Microscopic theories of incommensurate crystal phases **14.1**, 67
Jeffries, C.D., see Wolfe, J.P. **6**, 431
Judd, B.R., Group theoretical approaches **7**, 87

Kadin, A.M., and A.M. Goldman, Dynamical effects in nonequilibrium superconductors: some experimental perspectives **12**, 253
Kaganov, M.I., and A.V. Chubukov, Spin waves in magnetic dielectrics. Current status of the theory **22.1**, 1
Kalia, R.K., see Vashishta, P. **6**, 1
Kamimura, H., Electron–electron interactions in the Anderson-localised regime near the metal–insulator transition **10**, 555
Kaplyanskii, A.A., and S.A. Basun, Multiple resonant scattering of the 29 cm^{-1} acoustic phonons in optically excited ruby **16**, 373
Kaplyanskii, A.A., and A.I. Ryskin, P.P. Feofilov and the spectroscopy of activated crystals **21**, 1
Kats, E.I., see Belyakov, V.A. **5**, 227
Keldysh, L.V., and N.N. Sibeldin, Phonon wind in highly excited semiconductors **16**, 455
Khasanov, A.Kh., see Al'tshuler, S.A. **21**, 607
Kholodar, G.A., see Vinetskii, V.L. **13**, 283
Kind, R., and P. Muralt, Unique incommensurate–commensurate phase transitions in a layer-structure perovskite **14.2**, 301

Kirsanov, V.V., see Agranovich, V.M. **13**, 117
Kivelson, S., Soliton model of polyacetylene **17**, 301
Klein, M.V., Light scattering studies of incommensurate transitions **5**, 503
Kochelaev, B.I., see Al'tshuler, S.A. **21**, 607
Kolb, D.M., The study of solid–liquid interfaces by surface plasmon polariton excitation **1**, 299
Kopaev, Yu.V., see Galitskii, V.M. **12**, 377
Kopelman, R., Energy transport in mixed molecular crystals **4**, 139
Kopnin, N.B., see Galaǐko, V.P. **12**, 543
Korenblit, I.Ya., and E.F. Shender, Theory of magnetic excitations in disordered systems **22.2**, 109
Kosevich, A.M., Dynamical and topological solitons in ferromagnets and antiferromagnets **17**, 555
Koteles, E.S., Investigation of exciton-polariton dispersion using laser techniques **2**, 83
Kozub, V.I., see Aronov, A.G. **12**, 325
Kravtsov, V.E., see Agranovich, V.M. **1**, 511
Kreines, N.M., see Borovik-Romanov, A.S. **22.1**, 81
Kristoffel, N.N., see Hizhnyakov, V.V. **7**, 383
Kulakovskii, V.D., and V.B. Timofeev, Thermodynamics of electron–hole liquid in semiconductors **6**, 95
Kurkin, M.I., and E.A. Turov, Nuclear spin excitations **22.2**, 381
Kuznetsov, E.A., see Zakharov, V.E. **17**, 503

Lagois, J., and B. Fischer, Surface exciton polaritons from an experimental viewpoint **1**, 69
Lambert, M., see Durand, D. **14.2**, 101
Landau, I.L., see Dolgopolov, V.T. **12**, 641
Larkin, A.I., and Yu.N. Ovchinnikov, Vortex motion in superconductors **12**, 493
Lemaistre, J., see Zewail, A.H. **2**, 665
Leskova, T.A., see Agranovich, V.M. **1**, 511
Levanyuk, A.P., see Ginzburg, V.L. **5**, 3
Levanyuk, A.P., A.S. Sigov and A.A. Sobyanin, Light scattering anomalies due to defects **5**, 129
Levanyuk, A.P., see Golovko, V.A. **5**, 169
Levanyuk, A.P., General ideas about incommensurate phases **14.1**, 1
Levinson, Y.B., Phonon propagation with frequency down-conversion **16**, 91
Lévy, R., see Grun, J.B. **2**, 459
Lindgård, P.-A., Theory of spin excitations in the rare earth systems **22.1**, 287
Litster, J.D., Scattering spectroscopy of liquid crystals **5**, 583
Litvinenko, Yu.G., see Eremenko, V.V. **22.1**, 175
Liu, W.K., see Burns, M.J. **4**, 301
Loewenhaupt, M., see Fulde, P. **22.1**, 367
Lopez-Rios, T., see Abelés, F. **1**, 239
Loudon, R., see Ushioda, S. **1**, 535
Loudon, R., Ripples on liquid interfaces **9**, 589
Luhman, T., see Snead Jr, C.L. **13**, 345
Lushchik, Ch.B., Free and self-trapped excitons in alkali halides: spectra and dynamics **2**, 505
Lushchik, Ch.B., Creation of Frenkel defect pairs by excitons in alkali halides **13**, 473

L'vov, V.S., and L.A. Prozorova, Spin waves above the threshold of parametric excitations **22.1**, 233
L'vov, V.S., Solitons and nonlinear phenomena in parametrically excited spin waves **17**, 241
Lynn, J.W., and J.J. Rhyne, Spin dynamics of amorphous magnets **22.2**, 177
Lyons, K.B., see Fleury, P.A. **5**, 449
Lyuksyutov, I.F., A.G. Naumovets and Yu.S. Vedula, Solitons and surface diffusion **17**, 605

Macfarlane, R.M., and R.M. Shelby, Coherent transient and holeburning spectroscopy of rare earth ions in solids **21**, 51
Maki, K., Solitons in superfluid ^{3}He **17**, 435
Malkin, B.Z., Crystal field and electron–phonon interaction in rare-earth ionic paramagnets **21**, 13
Malyshev, V.A., see Basiev, T.T. **21**, 275
Maradudin, A.A., Interaction of surface polaritons and plasmons with surface roughness **1**, 405
Maradudin, A.A., see Cottam, M.G. **9**, 1
Maris, H.J., Phonon focusing **16**, 51
Markiewicz, R.S., and T. Timusk, Interaction of electromagnetic radiation with electron–hole droplets **6**, 543
Martynenko, Yu.V., see Guseva, M.I. **13**, 621
Matyushkin, E.V., see Eremenko, V.V. **22.1**, 175
Meier, F., and D. Pescia, Spin-polarized photoemission by optical orientation **8**, 295
Mel'nikov, A.P., see Gershenzon, E.M. **10**, 483
Meltzer, R.S., see Cone, R.L. **21**, 481
Merkulov, I.A., see Fleisher, V.G. **8**, 173
Mertens, F.G., and H. Büttner, Solitons on the Toda lattice: thermodynamical and quantum-mechanical aspects **17**, 723
Mikhailov, A.V., Integrable magnetic models **17**, 623
Milligan, R.F., T.F. Rosenbaum, R.N. Bhatt and G.A. Thomas, A review of the metal–insulator transition in doped semiconductors **10**, 231
Mills, D.L., Surface spin waves on magnetic crystals **9**, 379
Mirlin, D.N., Surface phonon polaritons in dielectrics and semiconductors **1**, 3
Mirlin, D.N., Optical alignment of electron momenta in GaAs-type semiconductors **8**, 133
Mook, H.A., Neutron scattering studies of magnetic excitations in itinerant magnets **22.1**, 425
Morrobel-Sosa, A., see Hanson, D.M. **4**, 621
Moskalova, M.A., see Zhizhin, G.N. **1**, 93
Muralt, P., see Kind, R. **14.2**, 301

Nad', F.Ya., see Dmitriev, V.M. **12**, 163
Natadze, A.L., A.I. Ryskin and B.G. Vekhter, Jahn–Teller effects in optical spectra of II–VI and III–V impurity crystals **7**, 347
Naumenko, V.M., see Eremenko, V.V. **22.2**, 259
Naumovets, A.G., see Lyuksyutov, I.F. **17**, 605
Nelson, R.J., Excitons in semiconductor alloys **2**, 319
Nizzoli, F., see Stegeman, G.I. **9**, 195

Ortuño, M., see Pollak, M. **10**, 287
Osad'ko, I.S., Theory of light absorption and emission by organic impurity centers **4**, 437
Ovchinnikov, Yu.N., see Larkin, A.I. **12**, 493

Ovsyankin, V.V., Spectroscopy of collective states and cooperative transitions in disordered rare-earth activated solids **21**, 343

Paget, D., and V.L. Berkovits, Optical investigation of hyperfine coupling between electronic and nuclear spins **8**, 381
Patel, J.S., see Hanson, D.M. **4**, 621
Pearlstein, R.M., Excitons in photosynthetic and other biological systems **2**, 735
Pedersen, N.F., Solitons in Josephson transmission lines **17**, 469
Perel', V.I., and B.P. Zakharchenya, Major physical phenomena in the optical orientation and alignment in semiconductors **8**, 1
Perel', V.I., see Dyakonov, M.I. **8**, 11
Perlin, Yu.E., and M. Wagner, Introduction **7**, 1
Perlin, Yu.E., and B.S. Tsukerblat, Optical bands and polarization dichroism of Jahn–Teller centers **7**, 251
Permogorov, S., Optical emission due to exciton scattering by LO phonons in semiconductors **2**, 177
Personov, R.I., Site selection spectroscopy of complex molecules in solutions and its applications **4**, 555
Pescia, D., see Meier, F. **8**, 295
Pethick, C.J., see Beyer, J. **12**, 129
Philpott, M.R., see Haarer, D. **4**, 27
Pick, R.M., see Poulet, H. **14.1**, 315
Pierce, D.T., and R.J. Celotta, Applications of polarized electron sources utilizing optical orientation in solids **8**, 259
Pikin, S.A., Incommensurate structures in liquid crystals **14.2**, 319
Pikus, G.E., and E.L. Ivchenko, Optical orientation and polarized luminescence of excitons in semiconductors **2**, 205
Pikus, G.E., and A.N. Titkov, Spin relaxation under optical orientation in semiconductors **8**, 73
Planel, R., and C. Benoit à la Guillaume, Optical orientation of excitons **8**, 353
Pokrovskii, Ya.E., Transport phenomena in electron–hole liquid **6**, 509
Pokrovsky, V.L., A.L. Talapov and P. Bak, Thermodynamics of two-dimensional soliton systems **17**, 71
Pokrovsky, V.L., M.V. Feigel'man and A.M. Tsvelick, Excitations in low-dimensional magnetic systems **22.2**, 67
Polinger, V.Z., see Bersuker, I.B. **7**, 21
Pollak, M., and M. Ortuño, The effect of Coulomb interactions on electronic states and transport in disordered insulators **10**, 287
Pooler, D.R., Numerical diagonalization techniques in the Jahn–Teller effect **7**, 199
Poulet, H., and R.M. Pick, Light scattering in incommensurate systems **14.1**, 315
Prelovšek, P., see Blinc, R. **14.1**, 143
Prozorova, L.A., see L'vov, V.S. **22.1**, 233
Przhevuskii, A.K., see Basiev, T.T. **21**, 275

Rabinovich, R.I., see Gershenzon, E.M. **10**, 483
Raether, H., Surface plasmons and roughness **1**, 331
Raikh, M.E., see Efros, A.L. **23**, 133
Rammer, J., see Beyer, J. **12**, 129
Rashba, E.I., Self-trapping of excitons **2**, 543

Rashba, E.I., Spectroscopy of excitons in disordered molecular crystals **23**, 215
Reik, H.G., Non-adiabatic systems: analytical approach and exact results **7**, 117
Renk, K.F., Detection of high-frequency phonons by phonon-induced fluorescence **16**, 277
Renk, K.F., Optical generation and detection of 29 cm^{-1} phonons in ruby **16**, 317
Reznichenko, E.A., see Zelensky, V.F. **13**, 527
Rhyne, J.J., see Lynn, J.W. **22.2**, 177
Robrock, K.H., see Ehrhart, P. **13**, 3
Rosenbaum, T.F., see Milligan, R.F. **10**, 231
Rubenchik, A.M., see Zakharov, V.E. **17**, 503
Rutar, V., see Blinc, R. **14.1**, 143
Ryskin, A.I., see Natadze, A.L. **7**, 347
Ryskin, A.I., see Kaplyanskii, A.A. **21**, 1

Sannikov, D.G., Phenomenological theory of the incommensurate–commensurate phase transition **14.1**, 43
Schneck, J., see Tolédano, J.C. **14.2**, 233
Schneider, T., and E. Stoll, Spin dynamics of Heisenberg chains **17**, 129
Schneider, T., Classical statistical mechanics of lattice dynamic model systems **17**, 389
Schober, H.R., see Ehrhart, P. **13**, 3
Schön, G., Collective modes in superconductors **12**, 589
Scott, A.C., Experimental observation of a Davydov-like soliton **17**, 53
Scott, J.F., Raman spectroscopy of structural phase transitions **5**, 291
Scott, J.F., Statics and dynamics of incommensurate $BaMnF_4$ **14.2**, 283
Seliger, J., see Blinc, R. **14.1**, 143
Shapiro, S.M., Magnetic excitations in spin glasses **22.2**, 219
Shelby, R.M., see Macfarlane, R.M. **21**, 51
Sheleg, A.U., and V.V. Zaretskii, List of incommensurate crystals **14.2**, 367
Shen, Y.R., and F. deMartini, Nonlinear wave interaction involving surface polaritons **1**, 629
Shender, E.F., see Korenblit, I.Ya. **22.2**, 109
Shirane, G., see Axe, J.D. **14.2**, 1
Shklovskii, B.I., see Efros, A.L. **10**, 483
Shomina, E.V., see Zhizhin, G.N. **1**, 93
Shustin, O.A., see Yakovlev, I.A. **5**, 605
Sibeldin, N.N., see Bagaev, V.S. **6**, 267
Sibeldin, N.N., see Keldysh, L.V. **16**, 455
Sigmund, E., Phonon scattering at Jahn–Teller defects in semiconductors **7**, 495
Sigov, A.S., see Levanyuk, A.P. **5**, 129
Silbey, R., Theories of energy transport **4**, 1
Silin, A.P., Electron–hole liquid in a magnetic field **6**, 619
Singwi, K.S., see Vashishta, P. **6**, 1
Sinha, S.K., see Borovik-Romanov, A.S. **22**, xi
Sipe, J.E., and G.I. Stegeman, Nonlinear optical response of metal surfaces **1**, 661
Slyozov, V.V., and P.A. Bereznyak, Irradiation creep in metals **13**, 575
Small, G.J., Persistent nonphotochemical hole burning and the dephasing of impurity electronic transitions in organic glasses **4**, 515
Smith, D.D., see Zewail, A.H. **2**, 665
Smith, H., see Beyer, J. **12**, 129
Snead Jr, C.L., and T. Luhman, Radiation damage and stress effects in superconductors: Materials for high-field applications **13**, 345

Sobyanin, A.A., see Ginzburg, V.L. **5**, 3
Sobyanin, A.A., see Levanyuk, A.P. **5**, 129
Stegeman, G.I., see Sipe, J.E. **1**, 661
Stegeman, G.I., and F. Nizzoli, Surface vibrations **9**, 195
Steiner, M., and A.R. Bishop, Nonlinear effects in low-dimensional magnets **17**, 783
Stoll, E., see Schneider, T. **17**, 129
Sturge, M.D., Introduction **2**, 1
Sugakov, V.I., Excitons in strained molecular crystals **2**, 709

Talapov, A.L., see Pokrovsky, V.L. **17**, 71
Tanabe, Y., and K. Aoyagi, Excitons in magnetic insulators **2**, 603
Taylor, D.W., Phonon response theory and the infrared and Raman experiments **23**, 35
Thewalt, M.L.W., Bound multiexciton complexes **2**, 393
Thomas, G.A., see Milligan, R.F. **10**, 231
Timofeev, V.B., Free many particle electron–hole complexes in an indirect gap semiconductor **2**, 349
Timofeev, V.B., see Kulakovskii, V.D. **6**, 95
Timusk, T., see Markiewicz, R.S. **6**, 543
Titkov, A.N., see Pikus, G.E. **8**, 73
Tolédano, J.C., J. Schneck and G. Errandonéa, Incommensurate phase of barium sodium niobate **14.2**, 233
Tsukerblat, B.S., see Perlin, Yu.E. **7**, 251
Tsvelick, A.M., see Pokrovsky, V.L. **22.2**, 67
Tuinstra, F., see de Wolff, P.M. **14.2**, 253
Turov, E.A., see Bar'yakhtar, V.G. **22.2**, 333
Turov, E.A., see Kurkin, M.I. **22.2**, 381

Ulrici, W., Manifestations of the Jahn–Teller effect in the optical spectra of transition metal impurities in crystals **7**, 439
Ushioda, S., and R. Loudon, Raman scattering by surface polaritons **1**, 535

Vashishta, P., R.K. Kalia and K.S. Singwi, Electron–hole liquid: theory **6**, 1
Vedula, Yu.S., see Lyuksyutov, I.F. **17**, 605
Vekhter, B.G., see Natadze, A.L. **7**, 347
Vinetskii, V.L., and G.A. Kholodar, Quasichemical reactions involving point defects in irradiated semiconductors **13**, 283
Vinogradov, E.A., G.N. Zhizhin and V.I. Yudson, Thermally stimulated emission of surface polaritons **1**, 145
Volkov, A.F., see Aslamazov, L.G. **12**, 65

Wagner, M., see Perlin, Yu.E. **7**, 1
Wagner, M., Unitary transformation methods in vibronic problems **7**, 155
Weis, O., Phonon radiation across solid/solid interfaces within the acoustic mismatch model **16**, 1
Weisbuch, C., see Hermann, C. **8**, 463
Westervelt, R.M., Kinetics of electron–hole drop formation and decay **6**, 187
Wiedersich, H., Phase stability and solute segregation during irradiation **13**, 225
Wiersma, D.A., see Hesselink, W.H. **4**, 249
Wiltshire, M.C.K., see Hayes, W. **23**, 177

Winkler, I.C., see Hanson, D.M. **4**, 621
Wöhlecke, M., and G. Borstel, Spin-polarized photoelectrons and crystal symmetry **8**, 423
Wolfe, J.P., and C.D. Jeffries, Strain-confined excitons and electron–hole liquid **6**, 431
Wood, M.H., see Bullough, R. **13**, 189

Yakovlev, I.A., and O.A. Shustin, Light scattering in quartz and ammonium chloride and its peculiarities in the vicinity of phase transition of crystals. A retrospective view and recent results **5**, 605
Yakovlev, V.A., see Zhizhin, G.N. **1**, 93
Yakovlev, V.A., see Zhizhin, G.N. **1**, 275
Yen, W.M., Experimental studies of energy transfer in rare earth ions in crystals **21**, 185
Yudson, V.I., see Vinogradov, E.A. **1**, 145

Zakharchenya, B.P., see Perel', V.I. **8**, 1
Zakharov, V.E., E.A. Kuznetsov and A.M. Rubenchik, Soliton stability **17**, 503
Zapasskii, V.S., Optical detection of spin-system magnetization in rare-earth-activated crystals and glasses **21**, 673
Zaretskii, V.V., see Sheleg, A.U. **14.2**, 367
Zelensky, V.F., and E.A. Reznichenko, Irradiation growth of metals and alloys **13**, 527
Zewail, A.H., D.D. Smith and J. Lemaistre, Dynamics of molecular excitons: disorder, coherence and dephasing **2**, 665
Zewail, A.H., see Burns, M.J. **4**, 301
Zhizhin, G.N., M.A. Moskalova, E.V. Shomina and V.A. Yakovlev, Surface electromagnetic wave propagation on metal surfaces **1**, 93
Zhizhin, G.N., see Vinogradov, E.A. **1**, 145
Zhizhin, G.N., and V.A. Yakovlev, Resonance of transition layer excitations with surface polaritons **1**, 275
Žumer, S., see Blinc, R. **14.1**, 143